Foundations of Probability

ALFRED RÉNYI

DOVER PUBLICATIONS, INC.
Mineola, New York

Copyright

Bibliographical Note

This Dover edition, first published in 2007, is an unabridged republication of the work originally published in 1970 by Holden-Day, Inc., San Francisco, California.

Library of Congress Cataloging-in-Publication Data

Rényi, Alfréd.
Foundations of probability / Alfred Renyi.—Dover ed.
p. cm.
Originally published: San Francisco : Holden-Day, [1970]
Includes index.
ISBN-13: 978-0-486-46261-5
ISBN-10: 0-486-46261-7
1. Probabilities. I. Title.

QA273.R417 2007
519.2—dc22

2007025600

Manufactured in the United States of America
Dover Publications, Inc., 31 East 2nd Street, Mineola, N.Y. 11501

This book is dedicated to the memory of Catherine Rényi. During the 25 years which we spent together she was for me the main source of inspiration in all my work.

While this book was in press, we learned with great sadness of the death of Alfred Rényi, in his prime, at the age of 49. Thus, Foundations of Probability *constitutes the last major project he was able to complete. It was a project that was particularly close to his heart since it enabled him to present in a systematic fashion some of his own major ideas, and he worked on it with great enthusiasm. Professor Rényi was still able to read proof and make up most of the index. The publishers and editor are grateful to Dr. P. Bartfai and Dr. P. Révész for some final checking and for completing the index.*

FEBRUARY 1970

ERICH L. LEHMANN, EDITOR
Holden-Day Series in Probability and Statistics

PREFACE

The story of this book is as follows. I have been giving introductory courses of probability theory at the University of Budapest for students of mathematics each year since 1948. Out of these courses grew my first textbook of probability theory, published in German* and French† (and of course in Hungarian).

When this book was about to appear, it was suggested to me by Holden-Day, Inc. that I translate the book into English. I agreed, but instead of translating the book, I decided to revise it in many respects. One of my reasons for this was that I felt as Pascal, who said that whenever he wrote something, he usually found out with what he should have begun, only after having finished his work. My other reason was that I thought that, as there exists so many excellent introductory textbooks of probability theory in English, it would be worth while to publish a new textbook only if it was essentially different in its method of presentation and as regards its choice of material from other available books. Thus I planned to expand those parts of the book which deal with topics not discussed in detail in other textbooks, or in which the approach is different from the usual, and to shorten everything else. In spite of these plans, originally I thought only about a thoroughly revised English edition of my German book.

Owing to other obligations the carrying out of this plan was delayed until 1966, when I spent the summer term at Stanford University and gave a course in probability theory for graduate students. The lecture notes for this course were the first draft of the first three chapters of the present book. When these lecture notes were finished, I realized that what I had done was to start to write a completely new book: the overlap with my old book was negligible. Thus I started to work on the manuscript with the aim of writing a new book, which would have practically nothing to do with my first book.‡

Chapters 4 and 5 were brought into a final form during my stay in 1968 at

* *Wahrscheinlichkeitsrechnung mit einem Anhang über Informationstheorie*, VEB Deutscher Verlag der Wissenschaften, Berlin, 1st ed. 1962, 2nd ed. 1966.

† *Calcul des Probabilités avec un Appendice sur la Théorie de l'information*, Dunod, Paris 1966.

‡ My German book will be published next year in English too (by the North Holland Publishing Company), and the reader will have the opportunity to compare the two books and will see that these two books complement each other in many respects, without any essential overlap.

Cambridge University as an Overseas Fellow of Churchill College, and I managed to finish the manuscript during the summer of 1968, after returning from Cambridge to Hungary.

The new book which resulted reflects the changes and innovations which I have introduced into my courses of probability theory during recent years.

The features of the present book, which make it different from other textbooks of probability theory (including my own book mentioned above) are as follows:

(a) The treatment of probability theory is based on the notion of a conditional probability space: ordinary probability spaces in the sense of Kolmogoroff are introduced as special conditional probability spaces, generated by bounded measures.

(b) The notion of independence is introduced in two steps, the first being the introduction of the notion of qualitative independence following E. Marczewski. This approach leads also to a treatment of product spaces different from the usual, which has certain advantages.

(c) While in most other books the probability spaces in which one works are usually not effectively constructed (only their existence is established by the fundamental theorem of Kolmogoroff), in the present book, much care is given to the effective construction of these spaces: for instance, it is pointed out to what purposes a denumerable probability space is sufficient and for what purposes a nondenumerable space is necessary. The author feels that the emphasis on the effective construction of the probability spaces in which one works is of particular importance for the application of methods and results of probability theory in other branches of mathematics.

(d) Much emphasis is laid on combinatorial questions connected with the basic notions of probability theory.

(e) Many classical theorems have been presented from a novel point of view, or, in a form different from the usual and/or with proofs which are rather different from the familiar ones. For instance, the Vitali–Hahn–Saks theorem is proved by reducing it to an elementary theorem of Steinhaus about linear summation methods; Bernoulli's law of large numbers is proved by Bernoulli's original method: this leads immediately to a proof of the Moivre–Laplace theorem, in a generalized form including also a theorem about "big deviations"; new proofs are given also for Marczewski's theorem (see Problem P.3.3) and Banach's theorem, on qualitatively independent events, respectively σ-algebras.

(f) The σ-algebra of sets on which a probability measure is defined is interpreted as the set of observable events; while this seems to be only a change in terminology, there is more behind it. It is my experience that this terminology helps considerably the understanding of the foundations of probability theory.

(g) The notion of information is treated as a basic notion of probability theory; as an intermediate step in defining this concept, the notion of qualitative information is introduced.

(h) Instead of trying to include as much material as the size of the book admits, the book aims at dealing only with the most basic questions, those which in fact belong to the foundations of probability theory; however, it tries to go deeply into these questions.

(i) The basic theorems of probability theory are dealt with, but it was not the intention of the author to present these theorems in their most general form; on the contrary, we present these theorems usually in a generality which is just enough to give a clear idea of the importance of the theorem in question. In this way it was possible to include at least a special case of certain important theorems (e.g., theorems of Doob, Kakutani and McMillan) which could not be discussed in full generality in such an introductory book.

A short summary of the content of the book may be appropriate here:

Chapter 1 deals with the mathematical notion of an *experiment*, i.e., a basic set Ω interpreted as the set of outcomes of an experiment and a σ-algebra $\mathscr{A}$ of subsets of Ω, interpreted as the set of observable events concerning the experiment. The introduction of the notion of probability is postponed until Chapter 2, but many notions which are usually introduced only after the notion of probability (e.g., the notion of a random variable) are introduced at this early stage.

Chapter 2 introduces and discusses the notion of a conditional probability space; as emphasized above, we arrive at the notion of an ordinary probability space in the sense of Kolmogoroff by specialization, i.e., by considering conditional probability spaces generated by bounded measures. It is pointed out in Chapter 2 how every result on ordinary probability spaces can be translated into a corresponding result on conditional probability spaces. In view of this, from Chapter 3 onwards mainly ordinary probability spaces are considered.

Chapter 3 deals with the notion of independence—being certainly the most important notion of probability theory; as a matter of fact, from a purely mathematical point of view, probability theory has often been characterized as "measure theory plus the notion of independence." Independence is introduced starting with the notion of qualitative independence, which is a necessary, and in some sense also, a sufficient condition for the existence of a probability measure with respect to which ordinary (stochastic) independence is present. Our point of view is that qualitative independence is the basic notion, because by replacing the probability measure by another equivalent one (under which transformation most theorems of probability theory remain valid), only qualitative independence is preserved, however independent random variables usually become weakly dependent.

Chapter 4 deals with the laws of chance for independent random variables: the laws of large numbers, limit distribution theorems, and results describing the "fine structure" of random fluctuations (the law of the iterated logarithm, the arc sine law, etc.). All these results are—as pointed out before—presented without aiming at maximal generality; the level of generality is chosen so as to give an idea of the general scope and importance of the law without too much technical complication.

The main topic of Chapter 5 is dependence. Of course, dependence is understood here not as the negation of independence, but as the study of those special types of dependence under which laws of chance can be established: Markov-type dependence, the martingale-property, stability, mixing, exchangeability etc. All these types of dependence are such that if an infinite sequence of random variables exhibits dependence of one of these types, then in the limit these variables have properties similar to those possessed by independent random variables, respectively their partial sums or other functions. Here again we do not aim at full generality: for instance, we go into martingale theory only as far as needed to get clear understanding of this important concept and its applications.

Let us add a few more words about the choice of topics. All over the book we have restricted ourselves to denumerable families of random variables. Thus, the general theory of stochastic processes (with a continuous parameter) lies outside the scope of the book. Nevertheless the Poisson process and the Wiener process are discussed; we have chosen for both such a method of presentation which builds up these processes by starting from a sequence of independent identically distributed random variables.

The selection of topics discussed in the book was, of course, to some extent influenced by the personal interest of the author; especially many topics concerning which the author has himself done some work have been included either in the main text or among the exercises and problems. The book also contains some yet unpublished results of the author (e.g., Problems P.1.7(b), P.1.8(a), (b), Exercise E.2.9, Example 3.8.1, Problem P.3.4), further, unpublished joint results with P. Erdös (see Theorem 4.8.7 and Problem P.1.8(c), and with N. Friedmann (see Problem P.5.9).

A preface to a textbook should give an answer to the following questions: For whom is the book written? What profit can the reader obtain from studying the book? We will now try to answer these questions.

The book is written mainly for graduate students of mathematics and mathematicians working in other fields who are not yet familiar with probability theory, *who want to start studying probability theory as a field of mathematical research.* Accordingly, no preliminary knowledge of probability theory is assumed, however a considerable amount of mathematical maturity and a good knowledge of other chapters of mathematics, especially real analysis, is supposed. (Two short appendices on measure theory and functional analysis are added, which recapitulate those notions and theorems the knowledge of which is taken for granted throughout the book.)

The treatment is meant to be rigorous from the beginning to the end—as it should be in view of the readers envisaged; however, certain details which can be filled in with not too much effort are often left to the reader to work out for himself. In general, throughout the book the reader is encouraged to activity: this is the reason why a considerable part of the material is given in the form of exercises and problems. There are ten exercises and ten problems at the end of each chapter, the content of these should be considered as an organic part of the book, as many important notions and results are presented in them. This material is given in the form of Exercises and Problems (with hints where necessary) only in order to force the reader to active (instead of purely receptive) study of the topics in question.

A serious effort has been made to make clear that probability theory was created to furnish mathematical models of these situations in the real world which can be most suitably described as situations in which chance plays an important role. Nevertheless, the emphasis is not on the applications of probability theory in science and everyday life, but on the mathematical theory itself. Thus the book may be of interest to a reader who is interested in probability theory for the sake of its applications only if he wants to get the basic knowledge for a *creative* application of the theory and not only for applying certain ready clichés; for this purpose much less thorough knowledge of the foundations is sufficient. However, the author hopes that the book will be welcomed by mathematicians who want to get acquainted with probability theory for the sake of its application in other branches of mathematics; as is well known, the importance of such applications is rapidly increasing.

I would like to use this occasion to express my thanks to all who helped to write and publish this book. My thanks are due to Professor Paul Erdös and Dr. Nathan Friedman for their permission to include unpublished joint results, mentioned above. My thanks are due to my colleagues and friends Dr. P. Bártfai,Dr. Katalin Boghár, Dr. J. Galambos and Dr. P. Révész, who read parts of the manuscript and proofsheets and made many valuable suggestions for improvements. My most since thanks are due to Mrs. Várnai and Mrs. Palotai, secretaries of the Mathematical Institute; Mrs. Várnai did excellent work in typing the manuscript and helped to prepare the Index, while Mrs. Palotai helped by copying the formulas. Finally I want to thank Holden-Day, Inc., and all members of their staff who were involved—especially Miss Phyllis Hirsch—for their helpful and understand cooperation in producing this book.

Alfred Rényi

Budapest, November 1969

CONTENTS

1. EXPERIMENTS

2. PROBABILITY

3. INDEPENDENCE

4. THE LAWS OF CHANCE

5. DEPENDENCE

Foundations of Probability

CHAPTER 1

EXPERIMENTS

1.1 THE DEFINITION OF AN EXPERIMENT

Probability theory deals with mathematical models of situations depending on chance. We shall call such a situation an *experiment*. Thus we use the word experiment as a technical term, with a meaning different from the everyday use of the word. For instance, in our sense, observing how long one has to wait for the departure of an airplane is an experiment. We call everything an experiment which may have different *outcomes* (i.e., mutually exclusive results), and which of them occurs depends on chance. Thus, to every experiment there corresponds a nonempty set Ω, the set of possible outcomes of the experiment, which set will also be called *the basic space*, and its elements, denoted by ω, will be called outcomes of the experiment.* For instance, let us consider the life-span of a tire measured in miles. Clearly the life-spans of different tires of the same type are different: they depend on road conditions, on the driving habits of the driver and on individual properties of the tire, i.e., on chance. Thus if the experiment consists in observing the life-span of a tire of a given type, the outcomes of the experiment are nonnegative numbers. Certainly no tire lasts for more than 100,000 miles; thus we could define Ω as the interval $(0, 10^5)$. However there is no disadvantage in choosing for Ω the interval $(0, +\infty)$, as in general it does not matter if the set Ω includes impossible cases too. It is important, however, that all possible outcomes of the experiment should be included in Ω. Now clearly the exact value of the life-span of a tire is of no practical importance (and can be determined anyway only with a certain degree of exactness, say, to one decimal digit); what is of interest is the event that the life-span exceeds a certain number, e.g., 20,000 miles. In general an *event* is a statement, the validity of which depends exclusively on the outcome of the experiment. In other words, an event occurs or not, according to the (random) outcome of the experiment. Thus, to every event concerning an experiment there corresponds a set which is a subset of the basic space, namely the set of those outcomes of the experiment which imply the occurrence of the event. *In what follows we shall identify an event with the set of*

* The outcomes of an experiment are also often called "*elementary events.*"

outcomes of the experiment implying the event. For instance, we identify the event that a tire lasts for more than 20,000 miles with the set of all numbers exceeding 20,000.

Thus, in probability theory we speak about events, but mean by this sets. Somebody may ask: If we mean sets, why do we speak about events? While this is certainly somewhat strange for the beginner, one can easily get used to this, and anybody after becoming familiar with this usage will soon realize that this makes the meaning of statements of probability theory intuitively clearer and thus facilitates understanding, and helps the development of that special way of thinking which is characteristic for probability theory. At the end of this section we shall give a short "dictionary" containing the translation of expressions from the language of probability theory into the language of set theory.*

As mentioned above, to every event connected with an experiment there corresponds a subset of the basic space Ω. However, not every subset of Ω corresponds in this way to an event which can be really observed and which is of interest. We consider the specification of the class of *observable events* as a part of the definition of an experiment. This is natural as, for instance, the degree of precision with which some quantity characterizing the outcome of an experiment can be measured is an essential feature of the experiment.

The following example may elucidate the question: Let an experiment consist in throwing two dice. Then the set Ω of outcomes of the experiment consists of the thirty-six ordered pairs (a, b) where $1 \leqq a \leqq 6$, $1 \leqq b \leqq 6$. However, if the dice are exactly alike, we cannot distinguish between throwing 6 with the first and 4 with the second of the dice and throwing 4 with the first and 6 with the second of the dice. Thus, in this case only those subsets of Ω correspond to observable events which contain together with the pair (a, b) $(a \neq b)$, the pair (b, a) also. Among the 2^{36} subsets of Ω there are clearly only 2^{21} such subsets. In this particular simple case we could also proceed in a different way: Define the set Ω of outcomes of the experiment as the set of all *unordered* pairs which can be formed from the numbers 1, 2, 3, 4, 5, 6. In this case the set Ω contains only twenty-one elements and all its subsets correspond to observable events. However, in general it is more convenient to suppose that while to every observable event there corresponds uniquely a subset of the set of possible outcomes of the experiment, not every subset of Ω corresponds in this way to an observable event.

If A is an observable event concerning an experiment, the event that A does not take place is clearly also observable. We denote this event (i.e., the event *opposite* to the event A) by $\overline{A}$. Evidently for every outcome of the experiment either the event A or the event $\overline{A}$ occurs, but not both. Thus the set corresponding to the event $\overline{A}$ is the *complementary set* of A with respect

* This dictionary is continued in Section 2.12.

to the full set of outcomes Ω. If $A_1, A_2, \ldots$ is a finite or denumerably infinite sequence of observable events concerning an experiment, we may suppose that the event, that at least one of the events A_n $(n = 1, 2, \ldots)$ takes place, is also observable. We denote this event by $A_1 + A_2 + \cdots$ or by $\sum_n A_n$. The event $\sum_n A_n$ occurs if the outcome ω of the experiment belongs to at least one of the sets A_n, i.e., $\sum_n A_n$ is the *union* of the sets A_n $(n = 1, 2, \ldots)$.

We denote the family of observable events concerning an experiment (with the set of outcomes Ω) by $\mathscr{A}$. We suppose that Ω itself also belongs to $\mathscr{A}$, i.e., we interpret Ω itself also as an event (and not only as the set of outcomes of the experiment). The event Ω (i.e., the basic space Ω considered as a member of the family $\mathscr{A}$) does not specify the outcome of the experiment at all. This event occurs whatever the outcome of the experiment may be, i.e., it certainly occurs. We therefore call it *the certain event*. There are, in general, no further restrictions on the family $\mathscr{A}$; thus we suppose only that $\mathscr{A}$ is a family of subsets of Ω such that

(1) If $A \in \mathscr{A}$, then $\overline{A} \in \mathscr{A}$.
(2) If $A_n \in \mathscr{A}$ $(n = 1, 2, \ldots)$, then $\sum_n A_n \in \mathscr{A}$.
(3) $\Omega \in \mathscr{A}$.

DEFINITION 1.1.1. *A family of subsets of a set Ω having the properties* (1), (2) *and* (3) *is called a σ-algebra* (*of events*). *If* (1) *and* (3) *are supposed, but instead of* (2) *only the weaker supposition* (2′) *is made, then $\mathscr{A}$ is called an algebra* (*of events*).

(2′) If $A \in \mathscr{A}$ and $B \in \mathscr{A}$, then $A + B \in \mathscr{A}$.

Thus we arrive at the following definition of the mathematical model of an experiment.*

DEFINITION 1.1.2. *An experiment $\mathscr{E}$ is a nonempty set Ω of elements ω called outcomes of the experiment, and a σ-algebra $\mathscr{A}$ of subsets of Ω called observable events. We put for the sake of brevity $\mathscr{E} = (\Omega, \mathscr{A})$.*

The set $\overline{\Omega}$ is clearly the empty set. We shall denote this set by $\varnothing$. Clearly, $\varnothing$ as an event is such an event which never occurs. We shall call it *the impossible event*.

It is possible that $\mathscr{A}$ contains only the sets Ω and $\varnothing$; in this case we call $\mathscr{E} = (\Omega, \mathscr{A})$ *a trivial experiment*; otherwise we call it *nontrivial*. Evidently if Ω consists of a single element only, then $\mathscr{E} = (\Omega, \mathscr{A})$ is necessarily a trivial experiment.

Let us denote by AB the *intersection* of the sets A and B. We shall also use the notation $\prod_n A_n$ for the intersection of the sets $A_1, A_2, \ldots, A_n, \ldots$.

* A system $\mathscr{E} = (\Omega, \mathscr{A})$ which we call an experiment is called a *measurable space* in measure theory. See Halmos [5].

Clearly we have

$$AB = \overline{\bar{A} + \bar{B}} \tag{1.1.1}$$

Thus if $\mathscr{A}$ is an algebra, $A \in \mathscr{A}$ and $B \in \mathscr{A}$, then $AB \in \mathscr{A}$. More generally

$$\prod_{n=1}^{N} A_n = \overline{\sum_{n=1}^{N} \bar{A}_n} \tag{1.1.2}$$

Evidently AB as an event means that *both the events A and B occur* and $\prod_n A_n$ means that all the events A_n occur. It follows from (1.1.2) that if $\mathscr{A}$ is an algebra and $A_n \in \mathscr{A}$ $(n = 1, 2, \ldots, N)$, then $\prod_{n=1}^{N} A_n \in \mathscr{A}$. Equation (1.1.2) holds also for $N = +\infty$, i.e., we have

$$\prod_{n=1}^{+\infty} A_n = \overline{\sum_{n=1}^{+\infty} \bar{A}_n} \tag{1.1.2'}$$

It can be seen from (1.1.2′) that if $A_n \in \mathscr{A}$ for $n = 1, 2, \ldots$ where $\mathscr{A}$ is a σ-algebra of sets, then $\prod_n A_n \in \mathscr{A}$.

We shall often use the following notations:

$$\limsup_{n \to +\infty} A_n = \prod_{n=1}^{+\infty} \left(\sum_{k=n}^{+\infty} A_k \right)$$

and

$$\liminf_{n \to +\infty} A_n = \sum_{n=1}^{+\infty} \prod_{k=n}^{+\infty} A_k$$

Clearly the event $\limsup_{n \to +\infty} A_n$ means that an infinity of the events A_n occur, while $\liminf_{n \to +\infty} A_n$ means that all but a finite number of the events A_n occur simultaneously. Evidently if $A_n \in \mathscr{A}$ for $n = 1, 2, \ldots$, we have $\limsup_{n \to +\infty} A_n \in \mathscr{A}$ and $\liminf_{n \to +\infty} A_n \in \mathscr{A}$. If for a certain sequence A_n of events $\limsup_{n \to +\infty} A_n = \liminf_{n \to +\infty} A_n$, we denote their common value by $\lim_{n \to +\infty} A_n$. Thus, if $A_n \in \mathscr{A}$ (n $= 1, 2, \ldots$) and $\lim_{n \to +\infty} A_n$ exists, then it belongs to $\mathscr{A}$.

Note that the statement $AB = \varnothing$, i.e., that the sets A and B are disjoint, can be expressed in the language of events by saying that *the events A and B exclude each other*; further it can be said that $A \subseteq B$ (the set A is a subset of the set B) means that *the occurrence of the event A implies that of the event B.* Note that $A \subseteq B$ is equivalent to $AB = A$. Thus we can set up the dictionary below.

Clearly the set of all subsets of a set Ω (called the *power set* of Ω and denoted by $\mathscr{P}(\Omega)$) is a σ-algebra. Thus if $\mathscr{B}$ is any family of subsets of the set Ω, there exist σ-algebras $\mathscr{A}$ of subsets of Ω such that $\mathscr{B} \subseteq \mathscr{A}$; the intersection of all such σ-algebras is also a σ-algebra which contains $\mathscr{B}$. We shall call this

Language of probability theory	Language of set theory	Notation
experiment	measurable space	$\mathscr{E} = (\Omega, \mathscr{A})$
set of possible outcomes of an experiment	basic space	Ω
outcome of the experiment	point (element) of the space	$\omega \in \Omega$
observable event	measurable set	$A \in \mathscr{A}$
certain event	the full set	Ω
impossible event	empty set	$\varnothing$
σ-algebra of observable events	σ-algebra of measurable subsets	$\mathscr{A}$
occurrence of at least one of the events A_n $(n = 1, 2, \ldots)$	union of the sets A_n $(n = 1, 2, \ldots)$	$A_1 + A_2 + \cdots + A_n + \cdots$ or $\Sigma_n A_n$
joint occurrence of all the events A_n $(n = 1, 2, \ldots)$	intersection of the sets A_n $(n = 1, 2, \ldots)$	$A_1 A_2 \ldots A_n \ldots$ or $\Pi_n A_n$
opposite event of the event A	complementary set of the set A (with respect to Ω)	$\bar{A}$
the event A implies the event B	A is a subset of B	$A \subseteq B$
the events A and B exclude each other	the sets A and B are disjoint	$AB = \varnothing$

σ-algebra the *least* σ-algebra containing $\mathscr{B}$ (or the σ-algebra *generated* by $\mathscr{B}$) and denote it by $\sigma(\mathscr{B})$. If $\mathscr{F}_r$ is the family of all r-dimensional intervals $a_k \leqq x_k < b_k$ $(k = 1, 2, \ldots, r)$ of the r-dimensional Euclidean space R^r, then $\sigma(\mathscr{F}_r)$ is called the family of *Borel subsets* of R^r $(r = 1, 2, \ldots)$ (see Halmos[5]). If $\mathscr{A}$ and $\mathscr{A}'$ are σ-algebras of subsets of the same set Ω and $\mathscr{A} \subseteq \mathscr{A}'$, we shall say that $\mathscr{A}$ is a *sub-σ-algebra* of $\mathscr{A}'$.

If $\mathscr{A}$ is a σ-algebra of subsets of a set Ω and there exists a denumerable family $\mathscr{B} \subseteq \mathscr{A}$ of sets such that $\mathscr{A} = \sigma(\mathscr{B})$, then $\mathscr{A}$ is called *separable*. For instance, the family of Borel subsets of R^r is separable, because it can be generated by the family of r-dimensional intervals $a_k \leqq x_k < b_k$ $(k = 1, 2, \ldots, r)$ such that the a_k and b_k are rational numbers, and this family is denumerable.

We shall now introduce some further definitions. Let us introduce the postulate (which is a weaker form of (1)).

(1′) If $A \in \mathscr{A}$ and $B \in \mathscr{A}$, then $B\bar{A} \in \mathscr{A}$.

A nonempty family $\mathscr{A}$ of subsets of a set Ω which satisfies (1′) and (2′) is called a *ring*. A nonempty family $\mathscr{A}$ of subsets of a set Ω satisfying (1′) and (2) is called a *σ-ring*. Clearly every algebra is a ring, and every σ-algebra is a σ-ring, but not conversely. For instance, the set of finite subsets of an infinite set is a ring but not an algebra. The family of all denumerable subsets of a nondenumerable set is a σ-ring but not a σ-algebra. A ring (σ-ring) $\mathscr{A}$ of subsets of a set Ω is an algebra (σ-algebra) if and only if $\Omega \in \mathscr{A}$. As a matter of fact, if $\Omega \in \mathscr{A}$, (1′) holds, and $A \in \mathscr{A}$, then taking $B = \Omega$ it follows that $\bar{A} \in \mathscr{A}$, i.e., that (1) also holds.

A nonempty class $\mathscr{A}$ of subsets of a set Ω is called a *semiring* if it has the following properties:

(1″) If $A \in \mathscr{A}$ and $B \in \mathscr{A}$, then there exists a finite sequence of disjoint sets $C_k \in \mathscr{A}$ $(k = 1, 2, \ldots, N)$ such that $B\bar{A} = C_1 + C_2 + \cdots + C_N$.

(2″) If $A \in \mathscr{A}$ and $B \in \mathscr{A}$, then $AB \in \mathscr{A}$.

Choosing in (1″) $B = A$ we see that every semiring contains $\varnothing$. If $\mathscr{A}$ consists only of $\varnothing$, $\mathscr{A}$ is a semiring. Every ring is a semiring but not conversely. For instance, the family J_1 of all intervals $[a, b)$ on the real line R, which are closed to the left and open to the right, is a semiring, (1″) holding with $N \leqq 2$, but not a ring. Similarly, the family J_r of all r-dimensional intervals $a_k \leqq x_k < b_k$ $(k = 1, 2, \ldots, r)$ of the r-dimensional Euclidean space R^r is a semiring, (1″) holding with $N \leqq 2r$, but not a ring.

If $\Omega = \sum_k \Omega_k$ where the sets $\Omega_1, \Omega_2, \ldots, \Omega_k, \ldots$ are disjoint, then the sets Ω_k $(k = 1, 2, \ldots)$ form a semiring but not a ring. If $\mathscr{B}$ is a semiring and $\mathscr{A} = r(\mathscr{B})$ denotes the family of all sets which can be represented as the union of a finite number of sets belonging to $\mathscr{B}$, then $\mathscr{A}$ is a ring. For instance, if $\mathscr{B}$ is the semiring of all intervals $[a, b)$ of the real line R, then $r(\mathscr{B})$ is the ring consisting of all subsets of R which are the union of a finite number of disjoint intervals $[a, b)$.

1.2 ALGEBRAS OF EVENTS AS BOOLEAN ALGEBRAS

From an algebraic point of view an algebra of events is a *Boolean algebra*. (See, e.g., Halmos [6].) A Boolean algebra can be defined algebraically in different ways. It can be defined as a special case of a *lattice*, namely a distributive and complemented lattice (see Birkhoff [1]); it can be defined as an algebraic structure in which there are two binary operations: $A + B$ and AB, further, there is a unary operation $\bar{A}$, and finally, there are two different special elements denoted by Ω and $\varnothing$ for which the following rules are valid:

$$A + B = B + A \qquad AB = BA \tag{1.2.1}$$

$$A + (B + C) = (A + B) + C \qquad A(BC) = (AB)C \tag{1.2.2}$$

$$A + A = A \qquad AA = A \tag{1.2.3}$$

$$A(B + C) = AB + AC \qquad A + BC = (A + B) \cdot (A + C) \tag{1.2.4}$$

$$A + \varnothing = A \qquad A\Omega = A \tag{1.2.5}$$

$$A + \Omega = \Omega \qquad A\varnothing = \varnothing \tag{1.2.6}$$

$$A + \bar{A} = \Omega \qquad A\bar{A} = \varnothing \tag{1.2.7}$$

$$\bar{\Omega} = \varnothing \qquad \bar{\varnothing} = \Omega \tag{1.2.8}$$

$$\overline{A + B} = \bar{A} \cdot \bar{B} \qquad \overline{AB} = \bar{A} + \bar{B} \tag{1.2.9}$$

$$\bar{\bar{A}} = A \tag{1.2.10}$$

For algebras of sets the above rules can be directly verified from the definition of the operations $A + B$, AB and $\bar{A}$. The rules (1.2.1)–(1.2.4) can be stated in words as follows: Both the addition and the multiplication are commutative and associative; every element is idempotent under both operations; there are two distributive laws concerning the combination of the two operations. The rules (1.2.9) are called De Morgan's laws.

Note that in the above list of rules in every row (except in (1.2.10)) there are two rules which are in the following relation: if we replace addition by multiplication everywhere, and conversely, and replace Ω by $\varnothing$ and $\varnothing$ by Ω everywhere, then we obtain the first rule from the second, and conversely. This duality in the axioms implies the same duality in all identities deduced from the axioms. For instance, the formulas (identities) which follow easily from the axioms

$$A + AB = A \qquad A(A + B) = A \tag{1.2.11}$$

are duals of each other.

It may happen that a formula is identical with its dual; such a formula is called self-dual. For instance, (1.2.10) and also the formula (which is a consequence of (1.2.11))

$$(A + B)(A + C)(B + C) = AB + AC + BC \tag{1.2.12}$$

are self-dual.

The above set of axioms is highly redundant: for instance, (as shown by E. V. Huntington) the axioms (1.2.1), (1.2.4), (1.2.5) and (1.2.7) are alone sufficient* to deduce all the others and thus to define a Boolean algebra (for the proof see Goodstein [4]). Several other sufficient sets of axioms are known (see, e.g., Halmos [6], Birkhoff [1], Goodstein [4], Sikorski [15]). It is possible to define an algebra of events as an abstract Boolean algebra without representing events by sets, and one can build up probability theory in this way too (see Gilvenko [3], Kappos [7]). However, in the present book we do not make use of this possibility and (following Kolmogoroff [8]) define algebras of events as algebras of sets, that is, we represent every event by the set of those outcomes of the experiment which imply the occurrence of the

* Equation (1.2.7) is to be understood in the sense that to every A there exists in $\mathscr{A}$ an $\bar{A}$ such that (1.2.7) holds.

event. This seems to be at first glance a restriction of generality, but it is in reality no essential restriction. As a matter of fact, according to a famous theorem of M. Stone (see Kappos [7], Frink [2] and Rényi [13]) every Boolean algebra is isomorphic* with an algebra of sets, that is, every Boolean algebra can be realized by sets (see Halmos [6]).

In Section 1.1 we have defined an experiment $\mathscr{E}$ as a pair $(\Omega, \mathscr{A})$ where $\mathscr{A}$ is a σ-algebra of subsets of Ω. The notion of a σ-algebra of sets also has an abstract generalization, namely it is a particular case of a Boolean σ-algebra. In a Boolean algebra a partial order is given by the relation: $A \subseteq B$ if $AB = A$. A Boolean algebra is called a Boolean σ-algebra if every denumerable set of elements has a least supremum and a greatest infimum. The question of representation of Boolean σ-algebras by σ-algebras of sets is somewhat more involved. It is not true (see, e.g., Halmos [6]) that every Boolean σ-algebra is isomorphic to a σ-algebra of sets. However, according to Loomis' theorem (see Halmos [5]) every Boolean σ-algebra is isomorphic to some σ-algebra of sets modulo a σ-ideal. An ideal J of a Boolean algebra $\mathscr{A}$ is a subset of $\mathscr{A}$ such that (a) $\varnothing \in J$, (b) if $A \in J$ and $B \in J$, then $A + B \in J$, (c) if $A \in J$ and $B \in \mathscr{A}$, then $AB \in J$. Clearly $\mathscr{A}$ itself is always an ideal (it is called an improper ideal and all other ideals are proper). Further, the set consisting of $\varnothing$ alone is also an ideal; it is called the trivial ideal; all others are nontrivial. An ideal is called a σ-ideal if it contains the sum of a denumerable set of its elements. To take a σ-algebra $\mathscr{A}$ modulo a σ-ideal J means to identify elements A and B of $\mathscr{A}$ if $A\bar{B} \in J$ and $\bar{A}B \in J$. As every algebra of sets can be extended to a σ-algebra (see Section 1.1), in spite of the mentioned negative result it is not an essential restriction from the point of view of probability theory to deal only with σ-algebras of sets.

In Boolean algebras the *difference* $A - B$ of two elements is defined by

$$A - B = A\bar{B} \tag{1.2.13}$$

By this notation,

$$\bar{A} = \Omega - A \tag{1.2.14}$$

A further operation in Boolean algebras which is often used in probability theory is the *symmetric difference*, denoted by $A \circ B$ and defined by

$$A \circ B = (A - B) + (B - A) = A\bar{B} + \bar{A}B \tag{1.2.15}$$

The meaning of $A \circ B$ as an event is that one and only one of the events A and B occurs. The operation $A \circ B$ is clearly commutative and associative.

* Two Boolean algebras $\mathscr{A}$ and $\mathscr{A}'$ are called isomorphic if there exists a one-to-one mapping of their elements which preserves all relations, i.e., is such that if the element of $\mathscr{A}'$ into which $A \in \mathscr{A}$ is mapped is denoted by A', one has $(A + B)' = A' + B'$, $(AB)' = A' \cdot B'$ and $(\bar{A})' = \bar{A}'$.

The notation $A_1 \circ A_2 \circ \cdots \circ A_n$ means that the number of those of the events A_k which occur is odd. We have further,

$$A(B \circ C) = AB \circ AC$$

and the equation $A \circ X = B$ has the (unique) solution $X = A \circ B$.

A Boolean algebra can be defined also by taking the operations $A \circ B$ and AB as fundamental: A Boolean algebra is a ring* with respect to the operations $A \circ B$ and AB, in which every element is idempotent, i.e., $AA = A$, and which has a unit element (namely Ω). Clearly $\varnothing$ is the zero element of this ring. Note that if instead of the addition, the symmetric difference is chosen as a basic operation, we can express $A + B$ by the formula

$$A + B = A \circ B \circ AB \tag{1.2.16}$$

Note that the dual operation of the symmetric difference is its opposite; as a matter of fact,

$$(A + \bar{B})(\bar{A} + B) = AB + \bar{A}\bar{B} = \overline{A \circ B} \tag{1.2.17}$$

1.3 OPERATIONS WITH EXPERIMENTS

If $\mathscr{E} = (\Omega, \mathscr{A})$ is an experiment and $B \in \mathscr{A}$ is an event different from the impossible event $\varnothing$, we may modify the experiment as follows: we perform the experiment, but we take its result into account only if the event B occurred; otherwise the experiment is considered as unsuccessful and is repeated. We call this modified experiment *the experiment $\mathscr{E}$ conditioned by the event B*, and denote it by $\mathscr{E}/B$. Clearly $\mathscr{E}/B = (B, B\mathscr{A})$ where $B\mathscr{A}$ denotes the family of those $A \in \mathscr{A}$ for which $A \subseteq B$. If $\mathscr{A}$ is a σ-algebra, then $B\mathscr{A}$ is also a σ-algebra. If $\mathscr{A}$ contains only the sets $\varnothing$, B, $\bar{B}$, Ω, then $\mathscr{E}/B$ is a trivial experiment.

Let $\mathscr{E}_n = (\Omega_n, \mathscr{A}_n)$ $(n = 1, 2, \ldots)$ be a finite or denumerably infinite sequence of experiments such that the sets Ω_n and Ω_m are disjoint if $n \neq m$.

* The expression "ring" is used here in the sense of algebra. A set $\mathscr{R}$ in which two operations, called addition and multiplication, are defined, is called in algebra a *commutative ring* if both operations are commutative and associative, the equation $a + x = b$ always has a solution in x, and the distributive law $a(b + c) = ab + ac$ is valid. It is called a *commutative ring with unity* if it contains an element e such that $ea = a$ for all $a \in \mathscr{R}$. In each ring there is a unique zero-element, 0, such that $a + 0 = a$; 0 is the unit element of $\mathscr{R}$ considered as a *commutative group* with respect to addition. However, the two notions of a ring are in conformity: a ring of sets is a commutative ring in the sense of algebra with respect to the operations $A + B$ and AB; this follows from the formula

$$AB = A \cdot (\overline{A\bar{B}})$$

which shows that if $\mathscr{A}$ is a ring of sets, $A \in \mathscr{A}$, and $B \in \mathscr{A}$, then $AB \in \mathscr{A}$.

We define the experiment $\mathscr{E} = (\Omega, \mathscr{A}) = \sum_n \mathscr{E}_n$ by putting $\Omega = \sum_n \Omega_n$ and defining the σ-algebra $\mathscr{A}$ of subsets of Ω as follows: A subset A of Ω belongs to $\mathscr{A}$ if and only if $A\Omega_n \in \mathscr{A}_n$ for $n = 1, 2, \ldots$. We call $\mathscr{E} = \sum_n \mathscr{E}_n$ the *sum* (or *mixture*) of the experiments $\mathscr{E}_n$. The meaning of $\sum_n \mathscr{E}_n$ as an experiment is that we select each time (at random) one of the experiments $\mathscr{E}_n$ and observe its outcome. Thus the performing of a mixture of the experiments $\mathscr{E}_n$ can be imagined as a compound experiment consisting of two steps: first we perform an auxiliary experiment whose outcome is an integer n; as the second step we perform the experiment $\mathscr{E}_n$. The addition of experiments is clearly commutative and associative.

Clearly $\mathscr{E}_n = \sum_k \mathscr{E}_k/\Omega_n$, i.e., $\mathscr{E}_n$ is the experiment $\mathscr{E} = \sum_k \mathscr{E}_k$ conditioned by Ω_n. (Remember that we supposed that the sets Ω_n are disjoint!)

If $\mathscr{E} = (\Omega, \mathscr{A})$ is an experiment and $\Omega = \sum_n \Omega_n$ (the sum may be finite or denumerably infinite) where $\Omega_n \in \mathscr{A}$, $\Omega_n \neq \varnothing$ $(n = 1, 2, \ldots)$ and $\Omega_j \Omega_k = \varnothing$ if $j \neq k$, we say that the sets Ω_n form a *partition** of the set of outcomes Ω. The labeling of the sets Ω_n in a partition is considered irrelevant, i.e., two partitions of Ω into the same sets, only in different order, are considered as identical.

If $\mathscr{E} = (\Omega, \mathscr{A})$ and $\Omega = \sum_n \Omega_n$ is a partition of Ω, then $\mathscr{E} = \sum_n \mathscr{E}/\Omega_n$ and we call such a decomposition a *partition* of the experiment $\mathscr{E}$.

We call a finite or denumerably infinite sequence $A_n \neq \varnothing$ $(n = 1, 2, \ldots)$ of events *a partition of the event* A if $\sum_n A_n = A$ and $A_j A_k = \varnothing$ for $j \neq k$. Clearly if $\Omega = \sum_n \Omega_n$ is a partition of the basic space Ω, then for every $A \in \mathscr{A}$ those among the events $A\Omega_n$ $(n = 1, 2, \ldots)$ which are not equal to $\varnothing$ form a partition of the event A.

If $\mathscr{P} = (A_1, \ldots, A_n, \ldots)$ and $\mathscr{Q} = (B_1, \ldots, B_m, \ldots)$ are two partitions of the space Ω, we define their *product* $\mathscr{P} . \mathscr{Q}$ as the partition of Ω into those of the sets $A_n B_m$ $(n = 1, 2, \ldots; m = 1, 2, \ldots)$ which are not empty. The multiplication of partitions is a commutative and associative operation.

If $\mathscr{E} = (\Omega, \mathscr{A})$ and $\mathscr{E}' = (\Omega, \mathscr{A}')$ are two experiments having the same set of outcomes, and such that $\mathscr{A} \subseteq \mathscr{A}'$, we shall say that the experiment $\mathscr{E}'$ is a *refinement* of the experiment $\mathscr{E}$, and denote this relation by $\mathscr{E} \ll \mathscr{E}'$. Thus the refinement of an experiment is another experiment which has the same set of outcomes but a richer set of observable events. For instance, if we perform the same experiment but use a finer measuring instrument, or use the same instrument but note the result with greater accuracy, we get a refinement of the experiment.

If $\mathscr{E}_1 = (\Omega_1, \mathscr{A}_1)$ and $\mathscr{E}_2 = (\Omega_2, \mathscr{A}_2)$ are two arbitrary experiments, we may consider the carrying out of both experiments as a single experiment. We shall denote this compound experiment by $\mathscr{E}_1 \times \mathscr{E}_2$. Thus $\mathscr{E}_1 \times \mathscr{E}_2 = (\Omega_1 \times \Omega_2, \mathscr{A})$ where $\Omega_1 \times \Omega_2$ is the set of all (ordered) pairs (ω_1, ω_2) with

* A set of events forming a partition of Ω is sometimes called a *complete set of events*.

$\omega_1 \in \Omega_1$ and $\omega_2 \in \Omega_2$, and $\mathscr{A}$ is the least σ-algebra of subsets of $\Omega_1 \times \Omega_2$ containing all sets of the form $A_1 \times A_2$, where $A_1 \in \mathscr{A}_1$, $A_2 \in \mathscr{A}_2$, and $A_1 \times A_2$ denotes the set of those pairs (ω_1, ω_2) for which $\omega_1 \in A_1$ and $\omega_2 \in A_2$. We shall call $\mathscr{E}_1 \times \mathscr{E}_2$ the *product* (or *Cartesian product*) of the experiments $\mathscr{E}_1$ and $\mathscr{E}_2$. The product of any number of experiments is defined similarly: If $\mathscr{E}_n = (\Omega_n, \mathscr{A}_n)$ $(n = 1, 2, \ldots)$ is any finite or infinite sequence of experiments, we put $\prod_n \mathscr{E}_n = (\prod_n \Omega_n, \mathscr{A})$ where $\prod_n \Omega_n$ denotes the set of all sequences $(\omega_1, \ldots, \omega_n, \ldots)$ where $\omega_n \in \Omega_n$ $(n = 1, 2, \ldots)$, and $\mathscr{A}$ denotes the least σ-algebra of subsets of $\prod_n \Omega_n$ containing all sets of the form $\prod_n A_n$ where $A_n \in \mathscr{A}_n$ $(n = 1, 2, \ldots)$ and $A_n = \Omega_n$ for all but a finite number of values of n. (The set $\prod_n A_n$ consists of all sequences $(\omega_1, \ldots, \omega_n, \ldots)$ such that $\omega_n \in A_n$ for $n = 1, 2, \ldots$.)

If $\mathscr{E} = \mathscr{E}_1 \times \mathscr{E}_2$ where $\mathscr{E}_1 = (\Omega_1, \mathscr{A}_1)$ and $\mathscr{E}_2 = (\Omega_2, \mathscr{A}_2)$, then each event $A_1 \in \mathscr{A}_1$ concerning the experiment $\mathscr{E}_1$ can also be considered as an event concerning the combined experiment $\mathscr{E}$, namely we may identify $A_1 \in \mathscr{A}_1$ with the event $A_1 \times \Omega_2$ because $A_1 \times \Omega_2$ means that A_1 occurs at the first experiment and the outcome of the experiment $\mathscr{E}_2$ is not specified at all. Let us denote by $\mathscr{A}_1 \times \Omega_2$ the family of all sets of the form $A_1 \times \Omega_2$ with $A_1 \in \mathscr{A}_1$. Then the experiment $\mathscr{E}_1' = (\Omega_1 \times \Omega_2, \ \mathscr{A}_1 \times \Omega_2)$ means that while we perform both experiments $\mathscr{E}_1$ and $\mathscr{E}_2$, we do not care about the outcome of $\mathscr{E}_2$. The σ-algebras $\mathscr{A}_1$ and $\mathscr{A}_1 \times \Omega_2$ are isomorphic. Further, $\mathscr{E} = \mathscr{E}_1 \times \mathscr{E}_2$ is a refinement of $\mathscr{E}_1'$.

It is easy to see that if $A_1 \in \mathscr{A}_1$ and $A_1 \neq \varnothing$, further $A_2 \in \mathscr{A}_2$ and $A_2 \neq \varnothing$, then $\mathscr{E}_1 \times \mathscr{E}_2 \mid A_1 \times A_2 = (\mathscr{E}_1 \mid A_1) \times (\mathscr{E}_2 \mid A_2)$. We have further, $\mathscr{E}_0 \times (\sum_n \mathscr{E}_n) = \sum_n (\mathscr{E}_0 \times \mathscr{E}_n)$.

Evidently if $\mathscr{E}_2 = \mathscr{E}_1$, we may interpret the compound experiment $\mathscr{E}_1 \times \mathscr{E}_2$ as an experiment consisting in carrying out the experiment $\mathscr{E}_1$ twice. Similarly if $\mathscr{E}_n = \mathscr{E}_1$ $(n = 2, 3, \ldots)$ we may interpret the experiment $\prod_n \mathscr{E}_n$ as a sequence of *repetitions* of the experiment $\mathscr{E}_1$.

Let $\mathscr{E}_1 = (\Omega, \mathscr{A}_1)$ and $\mathscr{E}_2 = (\Omega, \mathscr{A}_2)$ be two experiments with the same set of outcomes. Let $\mathscr{E} = (\Omega, \mathscr{A})$ denote *the least common refinement* of the experiments $\mathscr{E}_1$ and $\mathscr{E}_2$. In other words, let $\mathscr{A}$ be the least σ-algebra of subsets of Ω containing both $\mathscr{A}_1$ and $\mathscr{A}_2$. The following result is valid:

THEOREM 1.3.1. *Let $\mathscr{A}_1$ and $\mathscr{A}_2$ be two σ-algebras of subsets of the same set Ω. Let $\mathscr{A}$ denote the least σ-algebra containing both $\mathscr{A}_1$ and $\mathscr{A}_2$. Let $\mathscr{A}_{12}$ denote the family of subsets C of Ω which can be represented in the form*

$$C = \sum_{k=1}^{n} A_k B_k \tag{1.3.1}$$

where $A_k \in \mathscr{A}_1$, $B_k \in \mathscr{A}_2$ $(k = 1, 2, \ldots n)$, further, the sets A_k $(k = 1, 2, \ldots, n)$ form a partition of Ω. Then $\mathscr{A}_{12}$ is an algebra of sets, and $\mathscr{A}$ is the least σ-algebra containing $\mathscr{A}_{12}$.

Proof. Suppose that $C \in \mathscr{A}_{12}$ has the representation (1.3.1). Then we have

$$\bar{C} = \sum_{k=1}^{n} A_k \bar{B}_k \tag{1.3.2}$$

Thus $\bar{C} \in \mathscr{A}_{12}$. Now let C' be another set belonging to $\mathscr{A}_{12}$, having the representation

$$C' = \sum_{j=1}^{m} A_j' B_j' \tag{1.3.3}$$

where $A_j' \in \mathscr{A}_1$, $B_j' \in \mathscr{A}_2$ $(j = 1, 2, \ldots, m)$, and the sets A_j' $(j = 1, 2, \ldots, m)$ form a partition of Ω. Then the sets $A_k A_j'$ $(1 \leqq k \leqq n, 1 \leqq j \leqq m)$ all belong to $\mathscr{A}_1$, and those of these sets which are not empty form a partition of Ω. As further,

$$C + C' = \sum_{k=1}^{n} \sum_{j=1}^{m} A_k A_j'(B_k + B_j') \tag{1.3.4}$$

and

$$CC' = \sum_{k=1}^{n} \sum_{j=1}^{m} A_k A_j'(B_k B_j') \tag{1.3.5}$$

it follows that $C + C'$ and CC' also belong to $\mathscr{A}_{12}$, i.e., $\mathscr{A}_{12}$ is an algebra. As every algebra containing both $\mathscr{A}_1$ and $\mathscr{A}_2$ has to contain all sets of the form (1.3.1), all statements of Theorem 1.3.1 are proved. ■*

1.4 CANONICAL REPRESENTATION OF POLYNOMIALS OF EVENTS

Let $\mathscr{A}$ be an algebra of events and let $p(A_1, \ldots, A_n)$ be a *polynomial* of the n variable events $A_1, A_2, \ldots, A_n$; by this we mean that the value of the function $p(A_1, \ldots, A_n)$ is itself an event which is defined by a formula containing only the variables $A_1, A_2, \ldots, A_n$ (any number of times) and built up by the basic operations $A + B$, AB and $\bar{A}$. For instance,

$$(\bar{A}_1 + A_2 A_3)(\bar{A}_2 + A_1 A_3)(\bar{A}_3 + A_1 A_2)$$

is a polynomial of the three variables A_1, A_2, A_3. (Note that this polynomial can be expressed also by the simpler formula $A_1 A_2 A_3 + \bar{A}_1 \bar{A}_2 \bar{A}_3$.)

Let us introduce the following:

DEFINITION 1.4.1. *A basic function of the variable events $A_1, \ldots, A_n$ is a product of n events of which the kth is either A_k or $\bar{A}_k$ for $k = 1, 2, \ldots, n$.*

* The end of a proof will be denoted by the sign ■.

Thus there are exactly 2^n basic functions of the n variables $A_1, A_2, \ldots, A_n$. For instance, all basic functions of the variables A_1, A_2, A_3 are the following:

$$\begin{array}{cccc} A_1A_2A_3 & A_1A_2\bar{A}_3 & A_1\bar{A}_2A_3 & A_1\bar{A}_2\bar{A}_3 \\ \bar{A}_1A_2A_3 & \bar{A}_1A_2\bar{A}_3 & \bar{A}_1\bar{A}_2A_3 & \bar{A}_1\bar{A}_2\bar{A}_3 \end{array}$$

We shall denote the 2^n basic functions of the variable events $A_1, A_2, \ldots, A_n$ by $B_{n,1}, B_{n,2}, \ldots, B_{n,2^n}$. To fix the order of the $B_{n,k}$'s we define these functions by recurrence as follows: We put $B_{1,1} = A_1$, $B_{1,2} = \bar{A}_1$; if the $B_{n,k}(k = 1, 2, \ldots, 2^n)$ are already defined for some value of n, we define the $B_{n+1,k}$ as follows:

$$B_{n+1,k} = \begin{cases} B_{n,k} \cdot A_{n+1} & \text{for } 1 \leqq k \leqq 2^n \\ B_{n,k-2^n} \cdot \bar{A}_{n+1} & \text{for } 2^n + 1 \leqq k \leqq 2^{n+1} \end{cases} \tag{1.4.1}$$

Expressed otherwise, if the binary expansion of the number $k - 1$ is

$$k - 1 = \sum_{i=1}^{n} \varepsilon_i(k) 2^{i-1} \quad (\varepsilon_i(k) = 1 \text{ or } 0,\ 1 \leqq k \leqq 2^n)$$

we put

$$B_{n,k} = \prod_{i=1}^{n} A_i^{1-2\varepsilon_i(k)} \tag{1.4.1'}$$

where A_i^1 means A_i and A_i^{-1} means $\bar{A}_i$. For instance, $11 = 1 \cdot 1 + 1 \cdot 2 + 0 \cdot 4 + 1 \cdot 8$, therefore

$$B_{4,12} = \bar{A}_1\bar{A}_2A_3\bar{A}_4$$

Now we prove the following:

THEOREM 1.4.1. *Every polynomial of n variable events can be represented uniquely as the sum of certain basic functions $B_{n,k}$ of these events. Thus there are exactly 2^{2^n} different polynomials of the events $A_1, A_2, \ldots, A_n$.*

Proof. The representation of a polynomial $p(A_1, A_2, \ldots, A_n)$ as the sum of certain basic functions is called its *canonical representation.* First we prove that if a polynomial has a canonical representation, this is unique up to the order of the terms* which of course is irrelevant. This follows from the fact that clearly

$$B_{n,k} \cdot B_{n,l} = \varnothing \qquad \text{if } k \neq l \tag{1.4.2}$$

The validity of (1.4.2) follows from the remark that if $k \neq l$, then the basic functions $B_{n,k}$ and $B_{n,l}$ differ at least in one of their factors: if the first factor

* In view of $A + A = A$ we may suppose that no term occurs more than once.

in which they differ is the jth, then the product $B_{n,k} \cdot B_{n,l}$ after appropriate rearrangement contains the factor $A_j \bar{A}_j$, and thus is equal to $\varnothing$. Now if we had

$$\sum_{k \in E_1} B_{n,k} = \sum_{j \in E_2} B_{n,j} \tag{1.4.3}$$

where E_1 and E_2 are two different subsets of the set $S_{2^n} = \{1, 2, \ldots, 2^n\}$, then there would be at least one term $B_{n,l}$ which occurs on one side of the equality (1.4.3) but not on the other side. Multipyling both sides of (1.4.3) by this $B_{n,l}$ we obtain from (1.4.2) that $B_{n,l} = \varnothing$ and this contradiction proves that (1.4.3) is impossible, i.e., that the canonical representation is unique if it exists. Now such a representation certainly exists for the (constant) polynomial $p(A_1, \ldots, A_n) = \Omega = A_1 + \bar{A}_1$, namely we have

$$\Omega = \sum_{k=1}^{2^n} B_{n,k} \tag{1.4.4}$$

Equation (1.4.4) follows by carrying out the multiplication on the right-hand side of the identity

$$\Omega = \prod_{j=1}^{n} (A_j + \bar{A}_j) \tag{1.4.5}$$

It follows that if A has a canonical representation, so has $\bar{A}$, because if

$$A = \sum_{k \in E} B_{n,k} \tag{1.4.6}$$

where E is a subset of the set $S_{2^n} = \{1, 2, \ldots, 2^n\}$, then

$$\bar{A} = \sum_{k \in \bar{E}} B_{n,k} \tag{1.4.7}$$

where $\bar{E}$ is the complementary set of E (with respect to the set S_{2^n}). Now we show that the polynomial $p(A_1, \ldots, A_n) = A_i$ has a canonical representation ($i = 1, 2, \ldots, n$). As a matter of fact, we obtain this representation by multiplying both sides of (1.4.4) by A_i. As $A_i B_{n,k} = B_{n,k}$ if the ith factor of $B_{n,k}$ is A_i and $A_i B_{n,k} = \varnothing$ if the ith factor of $B_{n,k}$ is $\bar{A}_i$, we obtain for A_i the canonical representation,

$$A_i = \sum_{k \in E_i} B_{n,k} \tag{1.4.8}$$

where the set E_i contains 2^{n-1} terms, namely the indices of those basic functions the ith factor of which is A_i (and not $\bar{A}_i$).

It is easy to see that $k \in E_i$ if and only if the ith digit (from the right) of the number $k - 1$ represented in the binary system is 0; that is, if $k - 1 = \sum_{j=1}^{n} \varepsilon_j(k) 2^{j-1}$ where $\varepsilon_j(k)$ is 0 or 1, we have $k \in E_i$ if and only if $\varepsilon_i(k) = 0$.

Now let us observe that the sum and the product of two functions which have a canonical representation, has also a canonical representation. As a matter of fact, if E_1 and E_2 are two subsets of the set S_{2^n}, further

$$C_1 = \sum_{k \in E_1} B_{n,k} \tag{1.4.9}$$

and

$$C_2 = \sum_{k \in E_2} B_{n,k} \tag{1.4.10}$$

then

$$C_1 + C_2 = \sum_{k \in E_1 + E_2} B_{n,k} \tag{1.4.11}$$

and

$$C_1 C_2 = \sum_{k \in E_1 E_2} B_{n,k} \tag{1.4.12}$$

As by definition any polynomial of the events $A_1, \ldots, A_n$ can be obtained by applying the elementary operations (addition, multiplication, complementation) starting from the events A_i, it follows that every polynomial $p(A_1, \ldots, A_n)$ of $A_1, \ldots, A_n$ has a canonical representation,

$$p(A_1, \ldots, A_n) = \sum_{k \in E} B_{n,k} \tag{1.4.13}$$

where E is some subset of the set S_{2^n}, and as we have seen such a representation is unique. (Note that the canonical representation of the polynomial identically equal to $\emptyset$ is obtained by taking the empty set for E in (1.4.13).) As the set S_{2^n} has 2^{2^n} subsets, there are 2^{2^n} different polynomials of n events. Thus Theorem 1.4.1 is proved. ■

Let $\mathscr{A}'$ be the family of all events which can be expressed as polynomials of the events* $A_1, \ldots, A_n$ and thus (by Theorem 1.4.1) which can be expressed in the form (1.4.13). Clearly $\mathscr{A}'$ is an algebra of events; in fact, it is a subalgebra of $\mathscr{A}$, namely the least subalgebra of $\mathscr{A}$ which contains the events $A_1, \ldots, A_n$, i.e., the subalgebra generated by the events $A_1, \ldots, A_n$.

An event $A \in \mathscr{A}$ such that $A \neq \emptyset$ is called an *atom* of the experiment $\mathscr{E} = (\Omega, \mathscr{A})$ (or of the algebra of events $\mathscr{A}$) if from $B \in \mathscr{A}$, $B \subseteq A$ it follows that either $B = \emptyset$ or $B = A$. In the language of sets, A is an atom of $\mathscr{E} = (\Omega, \mathscr{A})$ if no proper nonempty subset of A belongs to $\mathscr{A}$. Clearly a set A which contains only a single element is, if it belongs to $\mathscr{A}$, always an atom. If $\mathscr{A}$ is the set $\mathscr{P}(\Omega)$ of all subsets of a set Ω, then the sets with one element are the only atoms of $\mathscr{A}$. It is easy to see that the events $B_{n,k}$ which are different from $\emptyset$ are the only atoms of the algebra of events $\mathscr{A}'$ generated by the sets $A_1, \ldots, A_n$.

* Now the events $A_1, \ldots, A_n$ are considered as fixed, not as variables.

Note that in the example of Section 1.1 concerning a throwing of two indistinguishable dice, the atoms of $\mathscr{A}$ are the sets consisting of the single elements (k, k) $(k = 1, 2, 3, 4, 5, 6)$ and the sets consisting of the pairs of elements (a, b) and (b, a) where $1 \leqq a \leqq 6$, $1 \leqq b \leqq 6$, $a < b$.

1.5 QUALITATIVE INDEPENDENCE

Let $A_1, \ldots, A_n$ be any events concerning the experiment $\mathscr{E} = (\Omega, \mathscr{A})$. We define the events $B_{n,k}$ $(1 \leqq k \leqq 2^n)$ in the same way as in Section 1.4; note, however, that now the A_k are not variables but fixed events.

In this case it may happen that none of the events $B_{n,k}$ $(1 \leqq k \leqq 2^n)$ is impossible (i.e., no one of these sets is empty). In this case we call the events $A_1, \ldots, A_n$ *qualitatively independent* (or set-theoretically independent; see Marczewski [9]). In other words, the subsets $A_1, \ldots, A_n$ of the set Ω are qualitatively independent if and only if they divide the set Ω into 2^n nonempty parts.

It is easy to see that if the events $A_1, \ldots, A_n$ are qualitatively independent, then omitting some of them the remaining events $A_{k_1}, A_{k_2}, \ldots, A_{k_r}$ $(1 \leqq k_1 < k_2 \cdots < k_r \leqq n)$ will still be qualitatively independent. If the events $A_1, A_2, \ldots, A_n$ are qualitatively independent, so are the events $\bar{A}_1, A_2, \ldots, A_n$ and thus also the events $\bar{A}_1, \bar{A}_2, A_3, \ldots, A_n$, etc., and finally the events $\bar{A}_1, \bar{A}_2, \ldots, \bar{A}_n$.

An infinite sequence $A_1, A_2, \ldots, A_n, \ldots$ of events is called qualitatively independent if the events $A_1, A_2, \ldots, A_n$ are qualitatively independent for $n = 2, 3, \ldots$.

Let $\mathscr{P}_k = (A_{k,1}, A_{k,2}, \ldots, A_{k,l_k})$, where $(2 \leqq l_k \leqq +\infty)$ and $(k = 1, 2, \ldots, n)$, be a partition of the space of outcomes Ω. We call the partitions $\mathscr{P}_k$ $(k = 1, 2, \ldots, n)$ qualitatively independent if none of the sets $A_{1,i_1} \cdot A_{2,i_2} \ldots A_{n,i_n}$ is empty. These sets are elements of the product of the partitions $\mathscr{P}_k$ $(k = 1, 2, \ldots, n)$. Thus if the partitions $\mathscr{P}_1, \ldots, \mathscr{P}_n$ are qualitatively independent and each l_k is finite, they split the space of events Ω into $\prod_{k=1}^n l_k$ disjoint nonempty parts. Clearly to say that the sets $A_1, \ldots, A_n$ are qualitatively independent is the same as to say that the partitions $(A_1, \bar{A}_1)$, $(A_2, \bar{A}_2), \ldots, (A_n, \bar{A}_n)$ are qualitatively independent. Infinitely many partitions are called qualitatively independent if any finite number of them are qualitatively independent.

Let $\mathscr{E}_n = (\Omega_n, \mathscr{A}_n)$ $(n = 1, 2, \ldots)$ be a sequence of experiments and put $\mathscr{E} = \prod_n \mathscr{E}_n$. Let B_n be defined as follows:

$$B_n = \Omega_1 \times \Omega_2 \times \cdots \times \Omega_{n-1} \times A_n \times \Omega_{n+1} \times \cdots$$

where $A_n \in \mathscr{A}_n$, $A_n \neq \Omega_n$ and $A_n \neq \emptyset$. Then the events B_n $(n = 1, 2, \ldots)$ are clearly qualitatively independent.

1.6 ON THE STRUCTURE OF ALGEBRAS OF EVENTS OF A FINITE OR DENUMERABLE BASIC SPACE

Let $\mathscr{E}$ be an experiment with a finite number of observable events, i.e., $\mathscr{E} = (\Omega, \mathscr{A})$ where $\mathscr{A}$ is a finite set.* Let ω be any element of Ω and let us define the event B_ω as follows:

$$B_\omega = \prod_{\omega \in A \in \mathscr{A}} A \tag{1.6.1}$$

Thus B_ω is the intersection of all sets $A \in \mathscr{A}$ which contain the element ω. As $\mathscr{A}$ is by supposition finite, so is the product (1.6.1) and thus $B_\omega \in \mathscr{A}$. It is easy to see that B_ω is an atom of $\mathscr{A}$, because if there would exist an event $B^* \in \mathscr{A}$ such that $B^* \subseteq B_\omega$, $B^* \neq \varnothing$ and $B^* \neq B_\omega$, we would have $B_\omega = B^* + B^{**}$ where $B^{**} = B_\omega \cdot \overline{B^*}$ and we would have $B^{**} \in \mathscr{A}$, $B^{**} \subseteq B_\omega$, further, $B^{**} \neq \varnothing$, $B^{**} \neq B_\omega$ and $B^*B^{**} = \varnothing$. Now as $\omega \in B_\omega$, either $\omega \in B^*$ or $\omega \in B^{**}$; in both cases we get a contradiction because, for instance, in case $\omega \in B^*$, B^* would be one of the factors at the right-hand side of (1.6.1) and thus we would have $B_\omega \subseteq B^*$ which implies $B^* = B_\omega$. Thus, every element ω of Ω is contained in an atom of $\mathscr{A}$. It is easy to see that any two different atoms of $\mathscr{A}$ are events which exclude each other (i.e., are disjoint sets) because if B_1 and B_2 are atoms, then $B_1B_2 \subseteq B_1$ and thus either $B_1B_2 = \varnothing$ or $B_1B_2 = B_1$. The second case, however, is impossible because it would mean that $B_1 \subseteq B_2$, i.e., that B_2 is not an atom. Thus if ω_1 and ω_2 are elements of Ω, the sets B_{ω_1} and B_{ω_2} are either identical or disjoint. Thus the atoms of $\mathscr{A}$ furnish a partition of Ω. Now let A be any event and let B represent an atom in $\mathscr{A}$. Then $BA \subseteq B$ and thus either $BA = \varnothing$ or $BA = B$, i.e., $B \subseteq A$. Now let A be any event in $\mathscr{A}$ and put

$$A^* = \sum_{\omega \in A} B_\omega \tag{1.6.2}$$

Clearly $A \subseteq A^*$, because if $\omega \in A$, then $\omega \in B_\omega \subseteq A^*$. On the other hand, if $\omega_1 \in A^*$, then there is an ω_2 such that $\omega_2 \in A$ and $\omega_1 \in B_{\omega_2}$. As in this case the atom B_{ω_2} and the event are not disjoint, it follows that $B_{\omega_2} \subseteq A$ and thus $\omega_1 \in A$. This implies $A^* \subseteq A$. Thus $A^* = A$, i.e., we have for any $A \in \mathscr{A}$

$$A = \sum_{\omega \in A} B_\omega \tag{1.6.3}$$

In other words, every $A \in \mathscr{A}$ is identical with the union of certain atoms of $\mathscr{A}$. Conversely, as $\mathscr{A}$ is an algebra of events, every union of atoms is an event. Thus $\mathscr{A}$ is identical with the set of all unions of atoms of $\mathscr{A}$. If the number

* Clearly if Ω is a finite set, $\mathscr{A}$ is necessarily finite. However, $\mathscr{A}$ may be finite even if Ω contains infinitely many elements.

of different atoms of $\mathscr{A}$ is n, then clearly $\mathscr{A}$ contains 2^n events. Thus we have proved the following:

THEOREM 1.6.1. *Let $\mathscr{E} = (\Omega, \mathscr{A})$ be an experiment such that $\mathscr{A}$ is a finite set. Then there exists a partition of Ω into n atoms such that $\mathscr{A}$ is identical with the 2^n sets obtained by forming all possible unions of some of these atoms.*

Let Ω' denote the set of all atoms of $\mathscr{E}$. Let $\mathscr{A}'$ denote the set of all subsets of the set Ω'. Let $A \in \mathscr{A}$ be any event, then as we have shown, A is identical with the union of certain atoms: Let $A' \in \mathscr{A}'$ denote the set of these atoms. Clearly the correspondence $A \leftrightarrow A'$ is an isomorphism. Thus we obtain from Theorem 1.6.1 the following:

COROLLARY TO THEOREM 1.6.1. *If $\mathscr{A}$ is a finite algebra of subsets of a set Ω, then $\mathscr{A}$ is isomorphic with the algebra $\mathscr{P}(\Omega')$ of all subsets of some finite set Ω'.*

If Ω is finite, the family $\mathscr{A}$ is necessarily finite too. Thus, incidentally we have also proved that if $(\Omega, \mathscr{A})$ is an experiment such that Ω is finite, then there exists a partition of Ω into n atoms such that $\mathscr{A}$ is identical with the set of all 2^n unions of these atoms. As mentioned in Section 1.2, every finite Boolean algebra is isomorphic with a finite algebra of sets; thus it follows that every finite Boolean algebra is isomorphic to the algebra of all subsets of a finite set. This can be proved, of course, directly (see, e.g., Rényi [13]).

The situation is essentially the same for experiments $\mathscr{E} = (\Omega, \mathscr{A})$ such that Ω is denumerably infinite. We shall show that the following theorem holds (see Hanisch, Hirsch and Rényi [16]):

THEOREM 1.6.2. *If $\mathscr{E} = (\Omega, \mathscr{A})$ is an experiment such that Ω is denumerably infinite, then $\mathscr{A}$ is isomorphic with the algebra of all subsets of some finite or denumerably infinite set.*

Proof. The proof is very similar to that of Theorem 1.6.1, only the proof of the existence of atoms is somewhat more delicate. Let ω be any element of Ω. We cannot now define B_ω directly by (1.6.1), because the set of all sets $A \in \mathscr{A}$ such that $\omega \in A$ may be nondenumerable and thus if B_ω was defined by (1.6.1), it would not be guaranteed that $B_\omega \in \mathscr{A}$. Therefore we proceed as follows: Let ω' be any element of Ω different from ω; then two cases are possible; either there exists a set $A \in \mathscr{A}$ such that $\omega \in A$ and $\omega' \notin A$ (i.e., there is an observable event A which *separates* ω and ω') or there does not exist such a set. Let C_ω denote the set of all $\omega' \in \Omega$ of the first kind (i.e., which can be separated from ω by a set $A \in \mathscr{A}$) and let us choose for every $\omega' \in C_\omega$ a set $A(\omega, \omega') \in \mathscr{A}$ such that $\omega \in A(\omega, \omega')$ and $\omega' \notin A(\omega, \omega')$. Now we put

$$B(\omega) = \prod_{\omega' \in C_\omega} A(\omega, \omega') \tag{1.6.4}$$

Clearly $B(\omega) \in \mathscr{A}$ because $B(\omega)$ is the product of denumerably many observable events, further, for every $\omega' \in C_\omega$ we have by definition $\omega \in A(\omega, \omega')$; thus it follows that $\omega \in B(\omega)$; similarly we get $B(\omega)C_\omega = \emptyset$. We shall show now that $B(\omega)$ is an atom of $\mathscr{A}$. As a matter of fact, if one could split $B(\omega)$ into the sum of two proper nonempty subsets B_1 and B_2 such that $B_1 \in \mathscr{A}$, $B_2 \in \mathscr{A}$, $B_1B_2 = \emptyset$, suppose that $\omega \in B_1$ and let ω^* be any element of B_2. Then clearly ω can be separated from ω^* (namely by the event B_1) and thus $\omega^* \in C_\omega$ and therefore $\omega^* \notin B(\omega)$, which is a contradiction. Thus it follows that $B(\omega)$ is an atom. As any two atoms are either identical or disjoint, this implies that $B(\omega)$ is uniquely determined, in spite of the fact that each set $A(\omega, \omega')$ was chosen arbitrarily among the sets belonging to $\mathscr{A}$ which contain ω but not ω'. As $\omega \in B(\omega)$ and $\omega \in \Omega$ was arbitrary, it follows that every element of Ω is contained in an atom; as any two atoms are disjoint, the set of atoms of Ω is clearly denumerable. The remaining part of the proof is the same as in the case when $\mathscr{A}$ is finite. ■

Let us call an experiment $\mathscr{E} = (\Omega, \mathscr{A})$ such that Ω is the union of the denumerable set of atoms of $\mathscr{A}$, *purely atomic*. Using this terminology it follows from Theorems 1.6.1 and 1.6.2 that *an experiment $\mathscr{E} = (\Omega, \mathscr{A})$ such that Ω is finite or denumerably infinite, is always purely atomic*.

It follows that σ-algebras of subsets of denumerable sets are isomorphic if and only if they contain the same number of atoms (i.e., either they contain the same finite number of atoms, or both contain a denumerably infinite number of atoms). No such simple structural theorem is known about σ-algebras of subsets of a nondenumerable space. (See Exercise E.1.7.)

The following consequence of Theorem 1.6.2 should be noted.

COROLLARY TO THEOREM 1.6.2.* *If $\mathscr{A}$ is a σ-algebra of subsets of a denumerable set Ω and $\mathscr{A}_1$ is an arbitrary (not necessarily denumerable) subset of $\mathscr{A}$, then the union (and also the intersection) of all sets belonging to $\mathscr{A}_1$ belongs to $\mathscr{A}$.*

Let $\mathscr{A}$ and $\mathscr{A}'$ be two σ-algebras of the subsets of the same finite or denumerable set Ω, and suppose that $\mathscr{A} \subseteq \mathscr{A}'$. In other words, suppose that the experiment $\mathscr{E}' = (\Omega, \mathscr{A}')$ is a refinement of the experiment $\mathscr{E} = (\Omega, \mathscr{A})$. It follows that every atom of $\mathscr{A}$ is the union of certain atoms of $\mathscr{A}'$, i.e., the partition of the set Ω into the atoms of $\mathscr{A}'$ is a *refinement* of the partition of Ω into the atoms of $\mathscr{A}$.

Let $\mathscr{E} = (\Omega, \mathscr{A})$ be an experiment such that Ω is denumerable. A system $B \subseteq \mathscr{A}$ of events is called a *separating system of events* of $\mathscr{E}$ if for any pair

* A Boolean algebra which is closed with respect to its binary operations without cardinality restrictions is called a *complete* Boolean algebra. Thus the statement of the corollary can be expressed also as follows: *A σ-algebra of subsets of a denumerable set is a complete Boolean algebra.* This is a special case of a more general theorem of Tarski (see [14] and [15], further [16] where the relation of Theorem 1.6.2 to other results on Boolean algebras is also discussed).

A_1, A_2 of different atoms of $\mathscr{A}$ there exists an event B of the system $\mathscr{B}$ such that B separates A_1 and A_2, i.e., is such that either $A_1 \subseteq B$ and $A_2 B = \emptyset$ or $A_2 \subseteq B$ and $A_1 B = \emptyset$. (See [22] and [24].) It is easy to prove the following:

THEOREM 1.6.3. *Let $\mathscr{E} = (\Omega, \mathscr{A})$ be an experiment such that Ω is a denumerable set. A subset $\mathscr{B}$ of $\mathscr{A}$ generates $\mathscr{A}$ (that is, the least σ-algebra $\sigma(\mathscr{B})$ containing $\mathscr{B}$ is identical with $\mathscr{A}$) if and only if $\mathscr{B}$ is a separating system of events of $\mathscr{E}$.*

Proof. Let us suppose that $\mathscr{B}$ is a separating system. Let $A_1, A_2, \ldots,$ denote all atoms of $\mathscr{A}$. Let for each value of $j \neq k$, $B(j, k)$ denote a set which contains A_k but does not contain A_j and is such that either $B(j, k)$ or $\overline{B(j, k)}$ belongs to $\mathscr{B}$. By the supposition that $\mathscr{B}$ is a separating system, there follows that such a set exists. But then $A_k = \prod_{j \neq k} B(j, k)$ and thus A_k belongs to the least σ-algebra $\sigma(\mathscr{B})$ generated by $\mathscr{B}$. As this holds for all values of k, we have $\sigma(\mathscr{B}) = \mathscr{A}$. Thus the condition is sufficient. To show the necessity one has to notice only that if all sets $B \in \mathscr{B}$ are such that either they contain both A_i and A_j or none of them, where A_i and A_j are different atoms of $\mathscr{A}$, the same holds for every set B^* in $\sigma(\mathscr{B})$. ∎

From Theorem 1.6.3 it follows that if Ω is a denumerable set and $\mathscr{A}$ is a σ-algebra of subsets of Ω, then $\mathscr{A}$ is *separable*; as a matter of fact, denoting by $\mathscr{B}$ the family of atoms of $\mathscr{A}$, $\mathscr{B}$ is denumerable and $\sigma(\mathscr{B}) = \mathscr{A}$. Evidently $\mathscr{B}$ is a semiring. If $r(\mathscr{B})$ denotes the least ring containing $\mathscr{B}$, i.e., the family of all finite unions of atoms, the ring $r(\mathscr{B})$ is also denumerable and of course $\mathscr{A} = \sigma(r(\mathscr{B}))$.

1.7 RANDOM MAPPINGS AND RANDOM VARIABLES

Let $\mathscr{E} = (\Omega, \mathscr{A})$ be an experiment and let $\xi = \xi(\omega)$ be a function defined on the space Ω, with values in a set X. If $\mathfrak{B}$ is any subset of X, let $\xi^{-1}(\mathfrak{B})$ denote the inverse image of $\mathfrak{B}$, that is, the set of those $\omega \in \Omega$ for which $\xi(\omega) \in \mathfrak{B}$, i.e., put

$$\xi^{-1}(\mathfrak{B}) = \{\omega : \xi(\omega) \in \mathfrak{B}\}. \tag{1.7.1}$$

Clearly we have

$$\xi^{-1}(X) = \Omega, \ \xi^{-1}(\emptyset) = \emptyset \tag{1.7.2}$$

and

$$\xi^{-1}(\mathfrak{B}_1 - \mathfrak{B}_2) = \xi^{-1}(\mathfrak{B}_1) - \xi^{-1}(\mathfrak{B}_2) \tag{1.7.3}$$

further

$$\xi^{-1}(\sum_n \mathfrak{B}_n) = \sum_n \xi^{-1}(\mathfrak{B}_n) \tag{1.7.4}$$

and

$$\xi^{-1}(\prod_n \mathfrak{B}_n) = \prod_n \xi^{-1}(\mathfrak{B}_n) \tag{1.7.5}$$

(In (1.7.3)–(1.7.5) $\mathfrak{B}_n$ $(n = 1, 2, \ldots)$ are arbitrary subsets of the set X.)

Thus $\xi(\omega)$ is a quantity, the value of which belongs to the set X and depends on the outcome of the experiment $\mathscr{E}$, that is, on chance.

Suppose that there is given a σ-algebra $\mathscr{F}$ of subsets of the set X, i.e., that $\mathscr{E}^* = (X, \mathscr{F})$ is an experiment. It follows from (1.7.2)–(1.7.4) that the set $\mathscr{A}_\xi$ of all subsets A' of Ω which can be written in the form $A' = \xi^{-1}(\mathfrak{B})$ with some $\mathfrak{B} \in \mathscr{F}$, is a σ-algebra. We call the function $\xi = \xi(\omega)$ a *random mapping** *of $\mathscr{E}$ into $\mathscr{E}^*$* if $\mathscr{A}_\xi \subseteq \mathscr{A}$, i.e., if for every $\mathfrak{B} \in \mathscr{F}$ we have $\xi^{-1}(\mathfrak{B}) \in \mathscr{A}$.

DEFINITION 1.7.1. *If $X = R$ is the set of real numbers, and $\mathscr{F}_1$ is the family of Borel sets of R, further $\xi = \xi(\omega)$ is a random mapping of $\mathscr{E}$ into $\mathscr{E}^* = (R, \mathscr{F}_1)$, we call the (real-valued) function $\xi(\omega)$ a (real-valued) random variable. Thus a random variable on the experiment $\mathscr{E} = (\Omega, \mathscr{A})$ is a real function on Ω such that for each $\mathfrak{B} \in \mathscr{F}_1$ we have $\xi^{-1}(\mathfrak{B}) \in \mathscr{A}$. The sets $\xi^{-1}(\mathfrak{B})$ where $\mathfrak{B}$ is a Borel set of the real line are called the level sets of the random variable $\xi(\omega)$.*

It follows that if $\xi_1, \xi_2, \ldots, \xi_r$ are random variables on the experiment $\mathscr{E}$ and $g(x_1, \ldots, x_r)$ is a Borel-measurable function† of the r real variables $x_1, \ldots, x_r$, then $\eta = g(\xi_1, \xi_2, \ldots, \xi_r)$ is also a random variable on the experiment $\mathscr{E}$. We shall denote random variables usually by Greek letters. If $\xi = \xi(\omega)$ is a constant on Ω, ξ is a random variable; we call a random variable which is not constant on Ω a *nondegenerated* random variable.

Let $\xi = \xi(\omega)$ be a random mapping of the experiment $\mathscr{E}$ into $\mathscr{E}^*$ and A an atom of $\mathscr{E}$. Then clearly $\xi(\omega)$ is a constant on A. Conversely, an event $A \in \mathscr{A}$ such that every random variable on the experiment $\mathscr{E}$ is constant on A is an atom of $\mathscr{E}$.

An r-dimensional random vector $\xi = (\xi_1, \xi_2, \ldots, \xi_r)$ on an experiment $\mathscr{E}$ is defined as an r-dimensional vector-valued function on Ω such that its r components are real-valued random variables on $\mathscr{E}$.

A complex-valued random variable on the experiment $\mathscr{E} = (\Omega, \mathscr{A})$ is defined as a complex-valued function on Ω such that its real and imaginary parts are real-valued random variables on $\mathscr{E}$.

* In measure theory what we call a random mapping is called an *$\mathscr{F}$-measurable function*. The σ-algebra A_ξ is called the σ-algebra generated by ξ. It is the least σ-algebra with respect to which ξ is a random mapping.

† A real-valued function $g(x_1, \ldots, x_r)$ of r real variables is called a Borel-measurable function if for any Borel subset $\mathfrak{B}$ of R the set of those points $(x_1, \ldots, x_r)$ of R^r for which the value of $g(x_1, \ldots, x_r)$ belongs to $\mathfrak{B}$, is a Borel subset of R^r. In other words, if R^r is the r-dimensional Euclidean space and $\mathscr{F}_r$ is the family of all Borel subsets of R^r, every (real-valued) random variable on the experiment $\mathscr{E}_r = (R^r, \mathscr{F}_r)$ is a Borel-measurable function.

If we perform the experiment $\mathscr{E} = (\Omega, \mathscr{A})$ but observe instead of its outcome ω only the value of the random mapping $\xi(\omega)$ of $\mathscr{E}$ into $\mathscr{E}^* = (X, \mathscr{F})$, this means that we really performed the experiment $\mathscr{E}^* = (X, \mathscr{F})$. Clearly, by performing $\mathscr{E}^*$ instead of $\mathscr{E}$ we usually lose some information. This is made evident by the remark made above, that denoting by $\mathscr{A}_\xi$ the family of the inverse images $\xi^{-1}(\mathfrak{B})$ of the sets $\mathfrak{B} \in \mathscr{F}$, then $\mathscr{A}_\xi$ is a σ-algebra and by definition $\mathscr{A}_\xi \subseteq \mathscr{A}$. Now the experiment $\mathscr{E}^*$ is clearly equivalent to the experiment $\mathscr{E}' = (\Omega, \mathscr{A}_\xi)$ and $\mathscr{E}$ is a refinement of $\mathscr{E}'$.

It is easy to see that the restriction of the random variable $\xi(\omega)$ to the nonempty set $B \in \mathscr{A}$ (i.e., the function $\xi(\omega)$ is now considered only for $\omega \in B$) is a random variable on the experiment $\mathscr{E}/B$.

A random mapping $\xi(\omega)$ into $\mathscr{E}' = (X, \mathscr{F})$ such that X is denumerable is called a *discrete random variable*. Each discrete random variable defines a partition of the basic space, namely, if $X = \{X_1, X_2, \ldots, X_n, \ldots\}$ and $\{X_n\}$ denotes the set consisting of the single element X_n, the nonempty sets among the sets $\xi^{-1}(\{X_n\})$ $(n = 1, 2, \ldots)$ form a partition of Ω.

DEFINITION 1.7.2. *We call the discrete random variables $\xi_1, \xi_2, \ldots, \xi_n$ qualitatively independent, if the corresponding partitions are qualitatively independent.*

If $\mathscr{E} = (\Omega, \mathscr{A}) = \prod \mathscr{E}_n$ where $\mathscr{E}_n = (\Omega_n, \mathscr{A}_n)$ $(n = 1, 2, \ldots)$ is a sequence of experiments, and $\xi_1, \xi_2, \ldots, \xi_n, \ldots$ are discrete random variables on $\mathscr{E}$ (such that if $\omega \in \Omega$ and $\omega = (\omega_1, \ldots, \omega_n, \ldots)$ where $\omega_n \in \Omega_n$ $(n = 1, 2, \ldots)$, then ξ_n depends only on ω_n), then the random variables ξ_n are qualitatively independent.

Let us consider the following example:

Example 1.7.1. Let $\mathscr{E} = (\Omega, \mathscr{A})$ be the experiment defined as follows: Let Ω be the interval $(0, 1)$ and let $\mathscr{A}$ be the set of all Borel subsets of Ω. Let us consider the decimal expansion of an $\omega \in \Omega$, i.e., we put for $0 \leqq \omega < 1$,

$$\omega = \sum_{n=1}^{\infty} \frac{\varepsilon_n(\omega)}{10^n}$$

where the possible values of the "digit" $\varepsilon_n(\omega)$ $(n = 1, 2, \ldots)$ are 0, 1, 2, 3, 4, 5, 6, 7, 8, 9. Clearly, $\varepsilon_n(\omega)$ $(n = 1, 2, \ldots)$ are discrete random variables on the experiment $\mathscr{E}$ with values in the set $X = (0, 1, \ldots, 9)$, and it is easy to see that they are qualitatively independent.

We shall need the following:

THEOREM 1.7.1. *If $\mathscr{E} = (\Omega, \mathscr{A})$ is an experiment and $\xi = \xi(\omega)$ is a real-valued function on Ω such that, denoting by I_x the interval $(-\infty, x)$ one has $\xi^{-1}(I_x) \in \mathscr{A}$ for every real value of x, then ξ is a random variable.*

Proof. The proof follows immediately from the definition of the family of

Borel subsets of R as the least σ-algebra of subsets of R containing all the sets I_x, and from (1.7.3) and (1.7.4). ■

If $\mathscr{E} = (\Omega, \mathscr{A})$ is an experiment and $A \in \mathscr{A}$ is an event A, let us define the function $\alpha = \alpha(\omega)$ $(\omega \in \Omega)$ as follows:

$$\alpha = \alpha(\omega) = \begin{cases} 1 & \text{if} \quad \omega \in A \\ 0 & \text{if} \quad \omega \in \bar{A} \end{cases} \tag{1.7.6}$$

Evidently α is a random variable.

DEFINITION 1.7.3. *We call the random variable α defined by* (1.7.6) *the indicator* of the event A.*

1.8 QUALITATIVE ENTROPY AND INFORMATION

Qualitative independence of algebras of events respectively of random variables can be characterized in information-theoretical terms. For this purpose we introduce the following definitions:

DEFINITION 1.8.1. *Let $\xi = (\Omega, \mathscr{A})$ be an experiment such that $\mathscr{A}$ is finite. The qualitative entropy† $h(\mathscr{A})$ of the finite algebra of events $\mathscr{A}$ (respectively of the experiment $\mathscr{E}$) is defined by*

$$h(\mathscr{A}) = \log_2 n \tag{1.8.1}$$

where n denotes the number of atoms of $\mathscr{A}$.

Remark: The qualitative entropy of an experiment may be considered as a measure of the amount of uncertainty concerning the result of the experiment. The uncertainty, of course, vanishes if the experiment is carried out and one is informed about its outcome. Thus the entropy $h(\mathscr{A})$ can also be considered as a measure of the amount of information which is obtained from the experiment $\mathscr{E}$. Clearly, if $\mathscr{E}' = (\Omega, \mathscr{A}')$ is a refinement of $\mathscr{E} = (\Omega, \mathscr{A})$, one has $h(\mathscr{A}) \leqq h(\mathscr{A}')$.

DEFINITION 1.8.2. *The qualitative mutual information $i(\mathscr{A}_1, \ldots, \mathscr{A}_r)$ of r finite algebras of events $\mathscr{A}_1, \ldots, \mathscr{A}_r$ of the same basic space Ω $(r \geqq 2)$ is defined by the formula*

$$i(\mathscr{A}_1, \ldots, \mathscr{A}_r) = \log_2 \frac{N_1 N_2 \ldots N_r}{N} \tag{1.8.2}$$

* In measure theory the indicator of an event is sometimes called its "characteristic function." We shall not use this terminology, because in probability theory the expression "characteristic function" has a quite different meaning (see Section 2.9).

† Compare with the notion of *topological entropy* introduced by Adler, Konheim and McAndrew [30].

where N_k denotes the number of atoms of $\mathscr{A}_k$ $(k=1, 2, \ldots, n)$ while N denotes the number of atoms of the least algebra $\mathscr{A}$ of subsets of Ω containing each of the algebras $\mathscr{A}_k$ $(k=1, 2, \ldots, r)$.

Remark: Thus denoting by $\mathscr{A}$ the least algebra containing the algebras $\mathscr{A}_1, \ldots, \mathscr{A}_r$, one has

$$i(\mathscr{A}_1, \ldots, \mathscr{A}_r) = h(\mathscr{A}_1) + h(\mathscr{A}_2) + \cdots + h(\mathscr{A}_r) - h(\mathscr{A}) \tag{1.8.3}$$

One evidently has

$$i(\mathscr{A}_1, \ldots, \mathscr{A}_r) \geqq 0 \tag{1.8.4}$$

i.e.,

$$h(\mathscr{A}) \leqq \sum_{k=1}^{r} h(\mathscr{A}_k) \tag{1.8.5}$$

From the above definitions we can easily deduce the following theorems:

THEOREM 1.8.1. *The finite algebras of events (belonging to the same experiment) $\mathscr{A}_1, \ldots, \mathscr{A}_r$ are qualitatively independent if and only if their qualitative mutual information $i(\mathscr{A}_1, \ldots, \mathscr{A}_r)$ is equal to zero.*

THEOREM 1.8.2. *Let $\mathscr{E}_1 = (\Omega_1, \mathscr{A}_1)$ and $\mathscr{E}_2 = (\Omega_2, \mathscr{A}_2)$ be two experiments, and let us form their product $\mathscr{E} = \mathscr{E}_1 \times \mathscr{E}_2 = (\Omega, \mathscr{A})$ (see Section 1.3). Then one has*

$$h(\mathscr{A}) = h(\mathscr{A}_1) + h(\mathscr{A}_2) \tag{1.8.6}$$

If ξ is a random variable taking on a finite number of values, the qualitative entropy $h(\xi)$ of ξ is defined as being equal to the qualitative entropy $h(\mathscr{A}_\xi)$ of the algebra $\mathscr{A}_\xi$ generated by ξ. Similarly, if $\xi_1, \ldots, \xi_r$ are random variables on the same experiment, taking on a finite number of values, the mutual qualitative information $i(\xi_1, \ldots, \xi_r)$ of these random variables is defined as being equal to $i(\mathscr{A}_{\xi_1}, \ldots, \mathscr{A}_{\xi_r})$. It follows that $i(\xi_1, \ldots, \xi_r) = 0$ if and only if the random variables $\xi_1, \ldots, \xi_r$ are qualitatively independent.

EXERCISES

E.1.1. Show that the following relations hold:

(a) $A \circ A = \varnothing$

(b) $A \circ \varnothing = A$

(c) $A \circ \Omega = \bar{A}$

(d) $A \circ B = (A + B) - AB$

(e) $(A \circ B)B = B - A$

(f) $A(B - C) = AB - AC$

(g) $AB - C = (A - C)(B - C)$

(h) $A - BC = (A - B) + (A - C)$

(i) If $A \circ B = C \circ D$, then $A \circ C = B \circ D$

E.1.2. Prove the following identities:

(a) $(A + B)(A + \bar{B}) + (\bar{A} + B)(\bar{A} + \bar{B}) = \Omega$

(b) $(A + B)(A + \bar{B})(\bar{A} + B)(\bar{A} + \bar{B}) = \varnothing$

(c) $(A + \bar{B}) \circ (\bar{A} + B) = A \circ B$

(d) $A\bar{B} \circ \bar{A}B = A \circ B$

(e) $A + B + C = (A - B) + (B - C) + (C - A) + ABC$

E.1.3. Let N be a square-free integer, i.e., the product of different primes: $N = p_1 p_2 \ldots p_r$. Let $\mathscr{A}$ denote the set of all divisors of N. Show that if for $n \in \mathscr{A}$, $m \in \mathscr{A}$ we define $n + m$ as the least common multiple and $n \cdot m$ as the greatest common divisor of n and m; further put $\bar{n} = N/n$, then we get a Boolean algebra.

E.1.4. Let S_k denote the sum of all products of k different events chosen from the sequence $A_1, \ldots, A_n$ and let P_k denote the product of all sums of k different events chosen from the sequences $A_1, \ldots, A_n$. Prove that the polynomials S_k and P_{n-k+1} are identical ($k = 1, 2, \ldots, n$; $n = 2, 3, \ldots$).

E.1.5. Let $\mathscr{A}$ denote the set of the integers $0, 1, \ldots, 2^n - 1$. Let each of these integers be represented in the binary system: $k = \sum_{i=1}^{n} \varepsilon_i(k) 2^{i-1}$ where $\varepsilon_i(k) = 0$ or 1. If $k \in \mathscr{A}$ and $l \in \mathscr{A}$, define $k + l$ by adding the corresponding binary digits of k and l (by the rules $1 + 0 = 0 + 1 = 1 + 1 = 1, 0 + 0 = 0$) and define $k \cdot l$ by multiplying the corresponding binary digits. Define $\bar{k} = \sum_{i=1}^{n} (1 - \varepsilon_i(k)) 2^{i-1}$. Show that in this way one gets a Boolean algebra.

E.1.6. Suppose we throw three red and two white dice, and dice of the same color are indistinguishable. How many atoms does the corresponding algebra of observable events contain?

E.1.7. Let Ω be a nondenumerable set and let $\mathscr{A}$ be the family of subsets A of Ω such that either A or $\bar{A}$ is denumerable. Show that (a) $\mathscr{A}$ is a σ-algebra, (b) $\mathscr{A}$ contains nondenumerably many atoms, (c) not every union of atoms belongs to $\mathscr{A}$, (d) not every $A \in \mathscr{A}$ is the union of denumerably many atoms.

E.1.8. Let Ω_{2^n} be the set $\{0, 1, \ldots, 2^n - 1\}$. Define $\varepsilon_i(k)$ as in E.1.5. Let A_i denote the set of those numbers k for which $\varepsilon_i(k) = 1$. Prove that the events $A_1, A_2, \ldots, A_n$ (a) are qualitatively independent, (b) form a minimal separating system in $\mathscr{E} = (\Omega_{2^n}, \mathscr{P}(\Omega_{2^n}))$.

E.1.9. Find the dual of Theorem 1.4.1.

E.1.10. Let $\xi_n = \xi_n(\omega)$ ($n = 1, 2, \ldots$) be a sequence of random variables on an experiment $\mathscr{E} = (\Omega, \mathscr{A})$.

(a) For each $\omega \in \Omega$ arrange the n numbers $\xi_1(\omega), \ldots, \xi_n(\omega)$ according to their order of magnitude into a nondecreasing sequence; let these ordered values be $\xi_1^*(\omega) \leqq \xi_2^*(\omega) \leqq \cdots \leqq \xi_n^*(\omega)$. Prove that $\xi_k^*(\omega)$ is a random variable ($k = 1, 2, \ldots, n$) on $\mathscr{E}$.

(b) Let the random variables ξ_n be uniformly bounded, i.e., $|\xi_n(\omega)| \leqq k$ for every $\omega \in \Omega$ and for $n = 1, 2, \ldots$. Prove that $\sup_n \xi_n(\omega)$ and $\inf_n \xi_n(\omega)$ are random variables on $\mathscr{E}$.

(c) Suppose that $\lim_{n\to\infty} \xi_n(\omega) = \eta(\omega)$ exists for all $\omega \in \Omega$. Prove that $\eta = \eta(\omega)$ is a random variable on $\mathscr{E}$.

PROBLEMS

P.1.1.* Let $T_n(n = 1, 2, \ldots)$ be the number of different partitions of a set Ω_n having n elements (i.e., according to Theorem 1.6.1 the total number of algebras of subsets of Ω_n).

(a) Show that $T_1 = 1$, $T_2 = 2$, $T_3 = 5$, $T_4 = 15$, $T_5 = 52$, $T_6 = 203$.

(b) Show that the recursion formula below holds.

$$T_{n+1} = 1 + \sum_{k=1}^{n} \binom{n}{k} T_k$$

Using this formula show that $T_{10} = 115{,}975$.

(c) Show that

$$\sum_{k=1}^{\infty} \frac{T_k x^k}{k!} = e^{e^x - 1} - 1$$

(d) Show that

$$T_n = \sum_{\sum_{k=1}^{n} k l_k = n} \frac{n!}{\prod_{k=1}^{n} k!^{l_k}\, l_k!} \quad (n = 1, 2, \ldots)$$

(e) Show that

$$T_n = \frac{1}{e} \sum_{k=1}^{\infty} \frac{k^n}{k!} \quad (n = 1, 2, \ldots)$$

(f) Let the linear functional $L(p)$ be defined on the set of all polynomials p of the variable x by putting $L(1) = 1$, $L(x(x-1)\ldots(x-k+1)) = 1$ for $k = 1, 2, \ldots$. Show that $T_n = L(x^n)$ $(n = 1, 2, \ldots)$.

(g) Show that $T_n = \sum_{r=1}^{n} S(n, r)$ where the $S(n, r)$ are the Stirling numbers of the second kind defined by the identity $x^n = \sum_{r=1}^{n} S(n, r)x(x-1)\ldots(x-r+1)$. Show that $S(n, r)$ is equal to the number of partitions of Ω_n into r nonempty parts.

(h) Show that if $S(n. r)$ is defined as in (g), then

$$\sum_{n=r}^{\infty} \frac{S(n, r)x^n}{n!} = \frac{(e^x - 1)^r}{r!} \quad (r = 1, 2, \ldots)$$

(i) Prove that the recursion formula $S(n+1, r) = S(n, r-1) + rS(n, r)$ holds $(1 \leqq r \leqq n+1; n = 1, 2, \ldots)$. (We put $S(n, 0) = 0$ for $n \geqq 1$.)

(j) Show that if $L_r(r = 1, 2, \ldots)$ is the linear functional defined on the set of

* For problems P.1.1 and P.1.2 see Rota [12] and Rényi [10].

all polynomials of the variable x by putting $L_r(x(x-1)\ldots(x-r+1))=1$, $L_r(1)=0$, $L_r(x(x-1)\ldots(x-k+1))=0$ if $k \neq r$, then $S(n,r)=L_r(x^n)$.

P.1.2. Let Q_n denote the number of all such partitions of a set Ω_n of n elements in which every subset of the partition has an odd number of elements.

(a) Show that $Q_1=1$, $Q_2=1$, $Q_3=2$, $Q_4=5$, $Q_5=12$, $Q_6=37$.

(b) Show that

$$\sum_{n=1}^{\infty} \frac{Q_n x^n}{n!} = e^{\sinh x} - 1$$

P.1.3. (a) Show that the maximal number of qualitatively independent open "intervals" (i.e., sets of points $(x_1, \ldots, x_n)$ such that $a_i < x_i < b_i$, $i = 1, 2, \ldots, n$) in the n-dimensional Euclidean space is equal to $2n$.

(b) Show that the maximal number of qualitatively independent open solid spheres (i.e., sets of points $(x_1, \ldots, x_n)$ for which $\sum_{i=1}^{n}(x_i - a_i)^2 < R^2$) in the Euclidean space of n dimensions is equal to $n+1$.

(c) Let $n(k)$ denote the maximal number of qualitatively independent open convex k-gons in the plane. Show that

$$\lim_{k\to+\infty} \frac{n(k)}{\log k} = \frac{1}{\log 2}$$

(A general explicit formula for $n(k)$ is not known. See A. Rényi, C. Rényi, Surányi [11] and as regards (b) see also Anusiak [26].)

P.1.4. (a) Let F_n denote the number of different polynomials of n events, which can be obtained by using only the operations of addition and multiplication. (For instance, $F_3 = 18$ because all such polynomials of the events, A, B, C are the following: A; B; C; AB; AC; BC; $A+B$; $B+C$; $A+C$; $A+BC$; $B+AC$; $C+AB$; $AB+AC$; $AB+BC$; $AC+BC$; ABC; $A+B+C$; $AB+AC+BC$.) Show that $F_4 = 166$, $F_5 = 7{,}579$, $F_6 = 7{,}828{,}352$. (A general formula for F_n for all values of n is not known. See Birkhoff [1]. The asymptotic behavior of F_n has been studied by Korobkov [17] and Hansel [18]; the sharpest results have been obtained by Kleitman [19].)

(b) A system of nonempty sets such that no set of the system is a subset of another set belonging to the system, is called a Sperner system. Show that the number of Sperner systems consisting of subsets of a set Ω_n having n elements is equal to the number F_n defined in (a).

(c) A function $f(x_1, x_2, \ldots, x_n)$ defined for all 2^n sequences $(x_1, x_2, \ldots, x_n)$ consisting of zeros and ones, and whose values at every point is either 0 or 1, is called a *Boolean function*. A Boolean function is called monotonic if from $x_k \leqq y_k$ $(k = 1, 2, \ldots, n)$, it follows that $f(x_1, x_2, \ldots, x_n) \leqq f(y_1, y_2, \ldots, y_n)$. Show that the total number of monotonic nonconstant Boolean functions of n variables is equal to F_n defined in (a).

P.1.5. Let Ω_n be a set having n elements. An unordered r-tuple $(A_1, A_2, \ldots, A_r)$ of different subsets of Ω_n such that $\sum_{i=1}^{r} A_i = \Omega_n$ is called a *covering* of Ω_n. Let C_n denote the total number of coverings of Ω_n. Show that $C_1 = 1$, $C_2 = 5$,

$C_3 = 109$, $C_4 = 32{,}297$. Show that the following identities are valid (see Comtet [20]).

$$2^{2^n-1} - 1 = \sum_{k=1}^{n} \binom{n}{k} C_k$$

and

$$C_n = \frac{1}{2} \sum_{k=0}^{n} (-1)^k \binom{n}{k} 2^{2^{n-k}}$$

P.1.6. Let $I(n, k)$ denote the number of k-tuples of qualitatively independent events concerning the experiment $\mathscr{E}_n = (\Omega_n, \mathscr{P}(\Omega_n))$ where $\Omega_n = \{1, 2, \ldots, n\}$ and $\mathscr{P}(\Omega_n)$ is the set of all subsets of Ω_n. Show that

$$I(n, k) = 2^k!\, S(n, 2^k)$$

where $S(n, r)$ are the Stirling numbers of the second kind defined in P.1.1(g). Notice that it follows that $I(n, k) = 0$ if $n < 2^k$, i.e., the maximal number of qualitatively independent events concerning the experiment $\mathscr{E}_n$ is $[\log_2 n]$. (The notation $[x]$ signifies the integral part of x and $\log_2 n$ the logarithm with base 2 of the number n.)

P.1.7. (a) Let $T(n, 2)$ denote the number of pairs Π_1, Π_2 of partitions of the set $\Omega_n = \{1, 2, \ldots, n\}$ such that Π_2 is a refinement of Π_1. In other words, let $T(n, 2)$ denote the number of pairs of experiments $\mathscr{E}_1 = (\Omega_n, \mathscr{A}_1)$, $\mathscr{E}_2 = (\Omega_n, \mathscr{A}_2)$ such that $\mathscr{E}_2$ is a refinement of $\mathscr{E}_1$. Using P.1.1(c) and (e) show that

$$\sum_{n=1}^{\infty} \frac{T(n, 2)x^n}{n!} = e^{(e^{(e^x-1)}-1)} - 1$$

Thus we have $T(1, 2) = 1$, $T(2, 2) = 3$, $T(3, 2) = 12$, $T(4, 2) = 60$.

(b) Let $T(n, k)$ denote the number of k-tuples $(\Pi_1, \Pi_2, \ldots, \Pi_k)$ of partitions of the set Ω such that Π_{j+1} is a refinement of Π_j for $j = 1, 2, \ldots, k - 1$. Generalizing the result under (a), show that

$$\sum_{n=1}^{\infty} \frac{T(n, k)x^n}{n!} = E_{k+1}(x) \qquad \text{for } k = 1, 2, \ldots$$

where $E_1(x) = e^x - 1$, and $E_r(x) = E_1(E_{r-1}(x))$ for $r = 2, 3, \ldots$, i.e., $E_{k+1}(x)$ is obtained by iterating the function $E_1(x)$ $k + 1$ times.

P.1.8. (a) Let $D_n^{(2)}$ denote the maximal number of *pairwise* qualitatively independent events concerning the experiment $\mathscr{E}_n = (\Omega_n, \mathscr{P}(\Omega_n))$. Show that $D_4^{(2)} = 3$, $D_6^{(2)} = 10$ and, in general, $D_{2n}^{(2)} = \frac{1}{2}\binom{2n}{n}$ for $n \geqq 2$.

Hint: Use Sperner's following theorem (see [21]): If $C_1, C_2, \ldots, C_N$ is a Sperner system of subsets of Ω_n (i.e., none of the sets C_i is a subset of another set C_j with $j \neq i$), then

$$N \leqq \binom{n}{[n/2]}$$

Recently a very elegant proof for this theorem has been given by Lubell [29], which is so short that it is worth while to reproduce it here: Let us consider all possible *maximal chains* of subsets of Ω_n, i.e., sequences of subsets A_k ($k = 1, 2, \ldots, n$) of Ω_n such that A_k has exactly k elements and $A_k \subseteq A_{k+1}$ for $k = 1, 2, \ldots, n-1$. The total number of such maximal chains is clearly equal to $n!$ and the number of maximal chains containing a subset A of Ω_n, where A consists of r elements, is equal to $r! \cdot (n-r)!$ and thus is $\geqq [n/2]!\,(n-[n/2])!$. Now if $C_1, \ldots, C_N$ is a Sperner system of subsets of Ω_n, then a maximal chain of subsets of Ω_n cannot contain more than one C_i; thus we have $N[n/2]!\,(n-[n/2])! \leqq n!$, which proves $N \leqq \binom{n}{[n/2]}$.*

It is easy to see that if the events $A_1, \ldots, A_q$ concerning the experiment Ω_n are pairwise qualitatively independent, then the sets $A_1, A_2, \ldots, A_q, \bar{A}_1, \bar{A}_2, \ldots, \bar{A}_q$ form a Sperner system and thus $2q \leqq \binom{n}{[n/2]}$. For $n = 2m$ equality is possible: an extremal system with $2q = \binom{2m}{m}$ is obtained by taking all subsets of the set $\Omega = 1, 2, \ldots, 2m$ which have m elements and do not contain the number 1.

(b) Show that $D_5^{(2)} = 4$, $D_7^{(2)} = 15$, further that in general $D_{2n+1}^{(2)} = \binom{2n}{n+1}$ for $n \geqq 2$.

Remark: It has been pointed out by Katona that the result $D_{2n+1}^{(2)} = \binom{2n}{n+1}$ can be deduced from a theorem of Erdös, Chao Ko and Rado (see [25]). An extremal system is obtained by taking all subsets of Ω_{2n+1} which have $n+1$ elements and do not contain the number 1.

(c) Let $D_n^{(r)}$ ($r \geqq 3$) denote the maximal number of subsets of Ω_n such that any r of these subsets are qualitatively independent. Show that

$$D_n^{(r)} \leqq r - 2 + 1/2\binom{[n/2^{r-2}]}{[1/2[n/2^{r-2}]]} \quad (r \geqq 3,\ n \geqq 2^{r-2}) \qquad \text{(P.1.8.1)}$$

Remark: This inequality has been found recently by Erdös and the author and has not been published previously. The exact value of $D_n^{(r)}$ for $r \geqq 3$ is not known.

The inequality (P.1.8.1) can be proved as follows: Suppose $r \geqq 3$. Let $A_1, \ldots, A_N$ be a system of subsets of Ω_n such that any r of the sets A_i are qualitatively independent, and $N > r-2$. Then the sets $A_1, \ldots, A_{r-2}$ divide the set Ω_n into 2^{r-2} nonempty subsets of the form $A_1^{\varepsilon_1} A_2^{\varepsilon_2} \ldots A_{r-2}^{\varepsilon_{r-2}}$ (where $\varepsilon_i = \pm 1$). Thus, among these sets there can be found one which has not more than $n/2^{r-2}$ elements. Let this set be denoted by B, and the number of its elements by $s \leqq [n/2^{r-2}]$. Then the sets BA_i and $B\bar{A}_i$ ($i = r-1, r, \ldots, N$) form a Sperner

* For another (probabilistic) proof see [31].

system of subsets of B and thus $2(N - r + 2) \leqq \binom{s}{[s/2]}$. As $\binom{s}{[s/2]}$ is a non-decreasing function of s, (P.1.8.1) follows. Notice that (P.1.8.1) holds for $r = 2$ also, as it reduces to $D_n^{(2)} \leqq \frac{1}{2}\binom{n}{[n/2]}$, which, as shown in (a) and (b), is valid, with equality if n is even.

P.1.9. (a) A separating system of events is called *miminal* if omitting any one of the events of the system, the remaining system is no more separating. Show that the number of events of a minimal separating system of events with respect to the experiment $\mathscr{E}_n$ (defined in P.1.8) cannot exceed $n - 1$.

(b) Let us consider separating systems of events concerning the experiment $\mathscr{E}_n$ such that each set (event) of the system contains k elements. Let $N(n, k)$ denote the minimum of the number of sets of such a system. Show that if $n \geqq k(k + 1)/2 + 1$, one has $N(n, k) = \{2(n - 1)/(k + 1)\}$ where $\{x\}$ denotes the least integer $\geqq x$. (See Rényi [22] and Katona [23].)

(c) Let $s_n(k)$ denote the total number of ordered k-tuples of events $A_1, A_2, \ldots, A_k$ concerning the experiment $\mathscr{E}_n$, such that $A_1, A_2, \ldots, A_k$ is a separating system of events in Ω_n. (The events $A_1, \ldots, A_k$ are not necessarily different.) Prove that

$$s_n(k) = \binom{2^k}{n} n!$$

Notice that this implies if $A_1, \ldots, A_k$ is a separating system in $\mathscr{E}_n$, then $k \geqq \log_2 n$ (see [24]).

P.1.10. Let $\Delta(n)$ denote the number of double partitions of the set Ω_n, i.e., the number of systems $A_1, A_2, \ldots, A_s$ of subsets of Ω_n such that each element of Ω_n is contained in exactly two of the sets A_i, which are supposed to be different. We put by definition $\Delta(0) = 1$ and $\Delta(1) = 0$. Show that $\Delta(2) = 1$, $\Delta(3) = 8$, $\Delta(4) = 80$, $\Delta(5) = 1{,}088$, $\Delta(6) = 19{,}232$, $\Delta(7) = 424{,}000$, and that the recursion formula

$$2^n \sum_{j=0}^{n} \binom{n}{j} \Delta(j) \left[\sum_{m=0}^{n-j} \frac{S(n-j, m)}{2^m}\right] = \sum_{k=0}^{n} \binom{n}{k} (-1)^{n+k}\, T_{n+k}$$

holds, where T_n is the number of partitions of a set having n elements (see P.1.1(a)) and $S(N, m)$ denote the Stirling numbers of the second kind, defined in P.1.1(g) (see Comtet [27] and Baróti [28]).

Hint: Let $\{A_1, A_2, \ldots, A_s\}$ be a double partition of the set $\Omega_n = \{1, 2, \ldots, n\}$. Then every element k of Ω_n belongs to exactly two sets of the double partition, say to A_i and A_j, where $i \neq j$. Let us call a system $\{A_1, A_2, \ldots, A_s\} = \pi_n$ of subsets of Ω_n, such that each element of Ω_n is contained in exactly two of the sets A_i $(1 \leqq i \leqq s)$, the sets A_i now being not necessarily all different, a *double covering* of Ω_n. Let $N(\pi_n) = s$ denote the number of sets of the system π_n and $R(\pi_n)$ the number of sets which occur twice* in the system π_n. It follows for

* Thus a double covering π_n is a double partition if $R(\pi_n) = 0$.

$x = 1, 2, \ldots$ that

$$\binom{x}{2}^n = \sum_{\pi_n} \frac{x(x-1)\cdots(x-N(\pi_n)+1)}{2^{R(\pi_n)}}$$

where the summation has to be extended over all double coverings π_n of Ω_n. Now use the functional L of Rota introduced in P.1.1 (f).

REFERENCES

[1] G. Birkhoff, "Lattice Theory," *Am. Math. Soc. Colloquium Publ.*, **25**, 1940.

[2] O. Frink, "Representations of Boolean Algebras," *Bull. Am. Math. Soc.*, **47**: 755–756, 1941.

[3] W. I. Glivenko, "Théorie Générale des Structures," *Actualités Scientifiques et Industrielles*, 652, Hermann, Paris, 1938.

[4] R. L. Goodstein, *Boolean Algebra*, Pergamon Press, New York, 1963.

[5] P. R. Halmos, *Measure Theory*, Van Nostrand, New York, 1950.

[6] P. R. Halmos, *Lectures on Boolean Algebras*, Van Nostrand, Princeton, 1963.

[7] D. A. Kappos, *Strukturtheorie der Wahrscheinlichkeitsfelder und -räume*, Springer, Heidelberg, 1960.

[8] A. N. Kolmogoroff, *Grundbegriffe der Wahrscheinlichkeitsrechnung*, Springer, Berlin, 1933.

[9] E. Marczewski, "Independence d'Ensembles et Prolongement de Mesures," *Colloquium Mathematicum*, **1**: 122–132, 1948.

[10] A. Rényi, "New Methods and Results of Combinatorial Analysis" (in Hungarian), *Magyar Tudományos Akadémia Matematikai és Fizikai Oztályának Közleményei*, **16**: 77–105, 159–177, 1966.

[11] A. Rényi, C. Rényi, and J. Surányi, "Sur l'Indépendence des Domaines Simples dans l'Espace Euclidien à n-Dimensions," *Colloquium Mathematicum* **2**: 130–135, 1951.

[12] G. C. Rota, "The Number of Partitions of a Set," *Am. Math. Monthly*, **71**: 498–504, 1964.

[13] A. Rényi, *Wahrscheinlichkeitsrechnung, mit einem Anhang über Informationstheorie*, VEB Deutscher Verlag der Wissenschaften, Berlin, 1962.

[14] A. Tarski, "Über additive und multiplikative Mengenkörper und Mengenfunktionen," *Comptes Rendus Soc. Sci. Lettr. Varsovie*, Cl. III **30**: 151–181, 1937.

[15] R. Sikorski, *Boolean Algebras*, 2nd ed., Academic Press, New York, 1964.

[16] H. Hanisch, W. M. Hirsch, and A. Rényi, "Measures in Denumerable Spaces," *Am. Math. Monthly*, **76**: 494–502, 1969.

[17] V. K. Korobkov, "On Monotonic Functions of the Algebra of Logic" (in Russian), *Problemi Kibernetiki*, **13**, 1965.

[18] M. G. Hansel, "Sur le Nombre des Fonctions Booléens Monotones des n Variables," *Comptes Rendus Acad. Sci.* (*Paris*), **262**: 1088, 1966.

[19] D. Kleitman, "On Dedekind's Problem: the Number of Monotone Boolean Functions," *Proc. Am. Math. Soc.* (in print).

[20] L. Comtet, "Recouvrements, Bases de Filtre et Topologies d'un Ensemble Fini," *Comptes Rendus Acad. Sci.* (*Paris*), **262**: 1091–1094, 1966.

[21] E. Sperner, "Ein satz über Untermengen einer endlichen Menge," *Math. Zeitschrift*, **27**: 544–548, 1928.

[22] A. Rényi, "On the Theory of Random Search," *Bull. Am. Math. Soc.*, **71**: 809–828, 1965.

[23] Gy. Katona, "On Separating Systems of a Finite Set," *J. Combinatorial Theory*, **1**: 174–197, 1966.

[24] A. Rényi, "On Random Generating Elements of a Finite Boolean Algebra," *Acta Sci. Math.* (*Szeged*), **22**: 71–81, 1961.

[25] P. Erdös, Chao Ko, and R. Rado, "Intersection Theorems for Systems of Finite Sets," *Quart. J. Math.* (*Oxford*), *Ser.* 2, **2**: 313–320, 1961.

[26] J. Anusiak, "On Set-Theoretically Independent Collections of Balls," *Colloquium Mathematicum*, **13**: 223–233, 1965.

[27] L. Comtet, "Birécouvrements et Birévêtements d'un Ensemble Fini," *Studia Sci. Math. Hung.*, **3**: 137–152, 1968.

[28] G. Baróti, "Calcul des Nombres des Birecouvrements et Birevêtements d'un Ensemble Fini, Employant la Méthode Fonctionelle de Rota," *Proceedings of the Colloquium on Combinatorial Mathematics, Balatonfüred, J. Bolyai Society*, Budapest (in print).

[29] D. Lubell, "A Short Proof of Sperner's Lemma," *J. Combinatorial Theory*, **1**: 299, 1966.

[30] R. L. Adler, A. G. Konheim, and M. H. McAndrew, "Topological Entropy," *Trans. Am. Math Soc.*, **114**: 309–319, 1965.

[31] A. Rényi, "Lectures on The Theory of Search," *University of North Carolina at Chapel Hill Institute of Statistics Mimeo Series No. 600.7*, 1969.

CHAPTER 2

PROBABILITY

2.1 THE INTUITIVE NOTION OF PROBABILITY

In everyday life, speaking about events depending on chance, people often say that an event will *probably* happen (or that it will probably not happen); for instance, that tomorrow there will probably be rain. Weather forecasts also often use expressions like "the probability of having rain on June 30, 1969 (in a certain area) is about 80%." Let us examine what such a statement really means. By saying that the probability of having rain tomorrow is 80% (or, what amounts to the same, 0.8) the meteorologist means that in a situation similar to that observed on the given day, there is usually rain on the next day in about 8 out of 10 cases; thus, while it is not certain that it will rain tomorrow, the *degree of certainty* of this event is 0.8. On what evidence is such a statement based? The meteorologist compares the given meteorological situation with similar situations in the past. If he says that the probability of having rain tomorrow is about 80%, this means that among the similar weather situations known to him (that is, to his computer) in the past years there was rain on the next day in 80% of all cases. Of course, such a statement depends on the period of comparison, and the criteria of similarity adopted. If the meteorologist had considered a longer or shorter period of comparison, he might have obtained another estimate—say 75%—of the probability in question. Another meteorologist might arrive at a different estimate, even if he uses the same basic data and the same period of comparison, by adopting other criteria of similarity, and thus feeding a different program into his computer to work on the same set of data. Thus, estimates of probability of an event made by different persons may be different and each such estimate is to a certain extent subjective. However, it is reasonable to assume that every random event has a definite probability even if the exact value of this probability is not always known; it is a natural assumption that every probability has an objectively determined value in spite of the fact that the estimates of this value made by different persons are to some degree subjective.

The notion of probability has been much discussed from the point of view of philosophy, especially as it is closely connected with the questions of causality and indeterminism, and with the nature of the physical laws. We do not want to go into these discussions here,* as we want to concentrate on the mathematical theory of probability. We take here the point of view that the probability of any random event has a determined value,† independently of whether this value is known to anybody exactly or only approximately or not known at all. Of course, such a statement cannot be proved; it is a basic assumption, a sort of axiom, and the question as to whether this is true or not has no exact mathematical sense. However, that this supposition is reasonable is indirectly shown through the wide practical applicability of probability theory to the description and forecasting of random phenomena.

Of course, theories of probability based on different philosophical concepts may lead in practice to the same or almost the same results. Thus, while the great success of the applications of probability theory in science and everyday life proves that the above made assumption about the existence of objective probabilities of random events is a reasonable one, these facts do not necessarily disprove the reasonability of quite different approaches to the notion of probability.

The simple example given above can be used also to discuss another important question of principle. Suppose the same meteorologist was asked a week earlier about the chances of having rain in the area in question on June 30, 1969. It is very likely that he would have produced a quite different estimate of the probability in question—say, 55%. Now which estimate is the right one? Of course the estimate given on June 29 is based on the weather up to this day, thus on much more information, and therefore it is more reliable. But this does not mean that the forecast given on June 22 is wrong. As a matter of fact, the two answers given by the meteorologist are answers to different questions. In the first case the question is: What is the probability of having rain on June 30 given the weather situation up to June 29, while in the second case the question is: What is the probability of having rain on June 30 given the weather situation up to June 22 only. As the two questions are different, both of the answers may be correct (or nearly correct) even if the two figures are far from each other.

In general, it makes sense to ask for the probability of an event A only if the conditions under which the event A may or may not occur are specified and the value of the probability depends essentially on these conditions. In other words, *every probability is in reality a conditional probability*. This

* The author has expressed his views on these questions in the form of fictitious letters of B. Pascal to P. Fermat (see [46]). It is a fact that Pascal and Fermat had a correspondence on some problems concerning games of chance, and their letters were the starting point of the development of probability theory. However, in these letters they do not discuss the notion of probability from a general point of view. The author assumed that Pascal and Fermat must have thought about this question and tried to reconstruct their thoughts in the form of a fictitious continuation of their correspondence.

† Depending on the conditions under which the occurrence or nonoccurrence of the event is considered; about this dependence more will be said later.

evident fact is somewhat obscured by the practice of omitting the explicit statement of the conditions if it is clear under which conditions the probability of an event is considered. Thus, if somebody asks about the probability of throwing 6 with a die, it is implicitly assumed that the die is a fair one. If one asks what is the probability that a card drawn at random from a pack of cards should be an ace, it is tacitly assumed that the pack is complete and well shuffled, etc. If the weather forecast concerning June 30 is issued on June 29, it is understood that the forecast is made on the basis of all data available on June 29.

This example shows that *the basic notion of probability theory should be the notion of the conditional probability of* A *under the condition* B, denoted in what follows by $P(A \mid B)$, where both A and B are random events connected with the same experiment $\mathscr{E}$. In the above example the experiment in question is the observation of the weather on each day of June 1969 in a sufficiently wide area, containing the smaller area for which the forecasts are made. Clearly not every event may be taken as a condition. The impossible event $\varnothing$ cannot be chosen for B, because $\varnothing$ never takes place by definition, and thus the condition that $\varnothing$ has taken place is never satisfied and the question does not make sense. Thus it is reasonable to suppose that the conditional probability $P(A \mid B)$ of an event A under condition B is defined for every observable event A in a given experiment $\mathscr{E} = (\Omega, \mathscr{A})$, that is, for each $A \in \mathscr{A}$ but only for B belonging to a certain subset $\mathscr{B}$ of the set $\mathscr{A}$ of observable events; we shall call the family $\mathscr{B}$ of events B which may be taken as conditions *the family of admissible conditions*.

The question arises of what assumptions should be made, in setting up the mathematical model of random phenomena, about the conditional probability $P(A \mid B)$, which in our interpretation of an experiment is a function of the two events (sets) $A \in \mathscr{A}$ and $B \in \mathscr{B}$. The assumptions (axioms) which will be introduced in the next section can be best understood through the close connection between probability and (relative) frequency. As a matter of fact, probabilities are used to forecast frequencies, and their estimates are based on observed frequencies. Therefore, the basic properties of probabilities have to reflect the properties of frequencies.

The *frequency* $\mathscr{F}(A)$ of an event A in a series of observations of repetitions* of the same experiment $\mathscr{E}$ is defined as the number of those observations at

* As regards the repetitions of the experiment $\mathscr{E}$, it is assumed that each time the *same* experiment is repeated; this implies that the same conditions are restored each time and thus the outcomes of the different experiments do not have any influence on each other. For instance, if the experiment consists in drawing a card from a pack, it is understood that each time a card is drawn it is replaced and the pack is well shuffled before the experiment is repeated, i.e., another card drawn. (Later this will be expressed by saying that the experiments are *independent* of each other. As, however, the notion of independence is used as a technical term in probability, which will be defined later, we tried to avoid the use of this term here.)

which the event A happened. The *relative frequency* $f(A)$ of the event A in such a sequence of observations is defined as the ratio of the frequency of A and of the total number of observations. Thus, if the series of observations consists of N observations, we have $f(A) = \mathscr{F}(A)/N$. Thus, $100f(A)$ gives us the percentage of those observations which led to the occurrence of the event A.

It is an experimental fact that the relative frequencies of random events in long sequences of observations are fairly near to a constant value, and this value is called the probability of the events in question. This relation between the notions of probability and relative frequency has been described very succinctly and accurately by Pólya [66] as follows: *Probability is the theoretical value of long-range relative frequency.*

The *conditional relative frequency* $f(A \mid B)$ of an event A under condition B is defined as the ratio of the number of those observations at which both A and B were observed and the number of those observations at which B was observed (if there are such). In other words, we put

$$f(A \mid B) = f(AB)/f(B) \quad \text{provided that} \quad f(B) > 0 \tag{2.1.1}$$

Thus, the conditional relative frequency of the event A with respect to the condition B is equal to the relative frequency of the event A in that subsequence of our observations which consists only of those observations which led to the occurrence of the event B. Thus we may compute $f(A \mid B)$ as follows: First we select from our observations those, if any, for which the condition B is fulfilled and disregard all the others; then we compute the relative frequency of A among the selected observations. (In case the condition B is not fulfilled for any of the observations, $f(A \mid B)$ is not defined.)

In our meteorological example, the series of observations consists in observations of the weather in a certain area. The condition B is the weather situation on the day when the forecast is made, and A is the event of having rain on the next day. Suppose that the meteorologist found 400 days in the period considered on which the weather conditions were similar to that on June 29, 1969. Suppose that among these 400 days there was rain (in the area in question) on the following day in 320 cases. Then the conditional relative frequency $f(A \mid B)$ equals $320/400 = 0.8$. Thus, if the meteorologist gives the estimate 80% for the probability of having rain on June 30, this means that he estimates this probability by the corresponding conditional relative frequency in the data for the past available to him. In doing this the meteorologist is guided by the empirical fact (supported by an enormous amount of evidence) that the conditional relative frequency of an event A under a condition B is—while depending to some extent on chance—rather stable, provided that it is computed from a fairly long sequence of observations. This means that if the meteorologist had used another (longer or shorter, but not too short) sequence of observations, he would have obtained another value for $f(A \mid B)$ but this value would have been very likely only slightly different. In general, on the basis of experience, it is reasonable to make the assumption that the conditional relative frequency $f(A \mid B)$ of an event A under condition B, if computed from a long sequence of observations, is near to a definite number and oscillates around it at random only slightly. This number is what we call conditional probability of the event A under condition B and we denote it by $P(A \mid B)$. Now, on the basis of this assumption, it is clear that we have to attribute to the conditional probabilities such properties as are possessed by conditional relative frequencies.

There are two such basic properties, each of which can be verified immediately from the definition.

(1) If A_1 and A_2 are two events which are mutually exclusive under condition B (i.e., if B takes place, then A_1 and A_2 cannot both take place simultaneously, that is, $A_1A_2B = \varnothing$), then we evidently have $\mathscr{F}((A_1 + A_2)B) = \mathscr{F}(A_1B) + \mathscr{F}(A_2B)$, and thus

$$f(A_1 + A_2 \mid B) = f(A_1 \mid B) + f(A_2 \mid B) \text{ provided that } A_1A_2B = \varnothing \quad (2.1.2)$$

(2) If A, B and C are three events, we have

$$f(A \mid BC) = \frac{\mathscr{F}(ABC)}{\mathscr{F}(BC)}, f(AB \mid C) = \frac{\mathscr{F}(ABC)}{\mathscr{F}(C)}, \quad \text{and} \quad f(B \mid C) = \frac{\mathscr{F}(BC)}{\mathscr{F}(C)}$$

provided that $\mathscr{F}(BC) > 0$, which implies $\mathscr{F}(C) > 0$, and thus

$$f(A \mid BC) = \frac{f(AB \mid C)}{f(B \mid C)} \quad (2.1.3)$$

Especially, we obtain from (2.1.3) that

$$f(A \mid B) = \frac{f(AB \mid C)}{f(B \mid C)} \quad \text{if } B \subseteq C \text{ and } \mathscr{F}(B) > 0 \quad (2.1.3')$$

because in case $B \subseteq C$ we have $BC = B$.

The conditional relative frequency $f(A \mid B)$ is evidently always a number lying in the interval $[0, 1]$ and one clearly has

$$f(B \mid B) = 1 \quad \text{if } \mathscr{F}(B) > 0 \quad (2.1.4)$$

These evident properties of conditional relative frequency form the intuitive background of the axioms of conditional probability spaces which will be introduced in the next section.

2.2 CONDITIONAL PROBABILITY SPACES

DEFINITION 2.2.1. *Let $\mathscr{E} = (\Omega, \mathscr{A})$ be an experiment and let $\mathscr{B}$ denote a subset of the σ-algebra $\mathscr{A}$. The subset $\mathscr{B}$ is called a bunch of events if it has the following properties:*

(1) *If $B_1 \in \mathscr{B}$ and $B_2 \in \mathscr{B}$, then $B_1 + B_2 \in \mathscr{B}$.*
(2) *There exists a sequence $B_n \in \mathscr{B}$ $(n = 1, 2, \ldots)$ such that $\sum_{n=1}^{\infty} B_n = \Omega$.*
(3) *The empty set $\varnothing$ does not belong to $\mathscr{B}$.*

DEFINITION 2.2.2. *A conditional probability space is a system $S = [\Omega, \mathscr{A}, \mathscr{B}, P(A \mid B)]$ where $\mathscr{E} = (\Omega, \mathscr{A})$ is an experiment, $\mathscr{B} \subseteq \mathscr{A}$ is a bunch of sets and $P(A \mid B)$ is a function defined for $A \in \mathscr{A}$ and $B \in \mathscr{B}$ which satisfies the following three conditions:*

(α) *$P(A \mid B)$ is, for any fixed $B \in \mathscr{B}$, considered as a function of A a measure on $\mathscr{A}$, i.e., a nonnegative and σ-additive set function on $\mathscr{A}$; that is, if $A_n \in \mathscr{A}$ $(n = 1, 2, \ldots)$ and $A_n A_m = \varnothing$ for $n \neq m$, we have*

$$P\left(\sum_{n=1}^{\infty} A_n \mid B\right) = \sum_{n=1}^{\infty} P(A_n \mid B) \tag{2.2.1}$$

(β) *One has for every $B \in \mathfrak{B}$*

$$P(B \mid B) = 1 \tag{2.2.2}$$

(γ) *If $B \in \mathscr{B}, C \in \mathscr{B}, B \subseteq C$, one has*

$$P(B \mid C) > 0 \tag{2.2.3}$$

and for every $A \in \mathscr{A}$, one has

$$P(A \mid B) = \frac{P(AB \mid C)}{P(B \mid C)} \tag{2.2.4}$$

If $S = [\Omega, \mathscr{A}, \mathscr{B}, P(A \mid B)]$ is a conditional probability space, every $B \in \mathscr{B}$ will be called an admissible condition and $P(A \mid B)$ the conditional probability of the event A under condition B.

Clearly the above axioms (α), (β), (γ) are counterparts of the corresponding properties of conditional relative frequencies discussed in Section 2.1. Especially, (α) corresponds to (2.1.2), (β) to (2.1.4) and (γ) to (2.1.3′).

Thus, a conditional probability space $S = [\Omega, \mathscr{A}, \mathscr{B}, P(A \mid B)]$ is defined as an experiment $\mathscr{E} = [\Omega, \mathscr{A}]$ in which a certain subset $\mathscr{B}$ of events is specified (which is a bunch of sets) and a set function $P(A \mid B)$ of the two set variables $A \in \mathscr{A}$ and $B \in \mathscr{B}$ is given which is for each fixed $B \in \mathscr{B}$ a measure* on $\mathscr{A}$

* The basic notions and theorems of measure theory used in this book are summarized in Appendix A. For a detailed treatment of measure theory see Halmos [1], Royden [2] and Neveu [3].

normed by the condition (2.2.2) and satisfying the condition (2.2.3), and these measures are related to each other by (2.2.4).

The following result will often be needed in what follows:

LEMMA 2.2.1. *A finite-valued finitely additive and nonnegative set function $\mu(A)$ defined on an algebra $\mathscr{A}$ of subsets of a set Ω is a measure (i.e., is σ-additive) on $\mathscr{A}$ if and only if for every sequence A_n of sets such that $A_n \in \mathscr{A}$, $A_{n+1} \subseteq A_n$ $(n = 1, 2, \ldots)$, and $\prod_{n=1}^{\infty} A_n = \varnothing$, one has*

$$\lim_{n \to \infty} \mu(A_n) = 0$$

Proof. Let μ satisfy the condition $\lim_{n \to \infty} \mu(A_n) = 0$ if $A_n \in \mathscr{A}$, $A_{n+1} \subseteq A_n$ $(n = 1, 2, \ldots)$ and $\prod A_n = \varnothing$. Let $B_n \in \mathscr{A}$ be a sequence of disjoint sets, and put $A_n = \sum_{k=n+1}^{\infty} B_k$. Then $\{A_n\}$ satisfies the above condition, and thus $\lim \mu(A_n) = 0$; as clearly $\mu(\sum_{k=1}^{\infty} B_k) = \sum_{k=1}^{n} \mu(B_k) + \mu(A_n)$, it follows that μ is σ-additive.

Conversely, let us suppose that μ is a measure, i.e., σ-additive; then if $A_{n+1} \subseteq A_n \in \mathscr{A}$, the sets $B_n = A_n - A_{n+1}$ are disjoint, and $\sum_{k=1}^{\infty} B_k = A_1$. Thus $\mu(A_k) = \sum_{n=k}^{\infty} \mu(B_n)$; and as the remainder sums of a convergent series tend to 0, this implies $\mu(A_k) \to 0$ if $k \to \infty$. ■

Let us deduce a few immediate consequences of the above axioms.

One has for each $B \in \mathscr{B}$ and $A \in \mathscr{A}$

$$P(A \mid B) = P(AB \mid B) \tag{2.2.5}$$

To obtain (2.2.5) put $C = B$ into (2.2.4) and use (2.2.2). Clearly, (2.2.5) implies that

$$P(A \mid B) = 0 \quad \text{if } AB = \varnothing \tag{2.2.6}$$

because, $P(A \mid B)$ being a measure in A, one has*

$$P(\varnothing \mid B) = 0 \quad \text{for } B \in \mathscr{B} \tag{2.2.6$'$}$$

One has further, for each $A \in \mathscr{A}$ and $B \in \mathscr{B}$

$$P(A \mid B) \leqq 1 \tag{2.2.7}$$

with equality standing in (2.2.7) if $B \subseteq A$; thus, especially, one has

$$P(\Omega \mid B) = 1 \tag{2.2.8}$$

Formula (2.2.7) follows easily from (2.2.5), because $P(A \mid B)$ being a measure, one has $P(A \mid B) \leqq P(A' \mid B)$ if $A \subseteq A'$; thus we have $P(A \mid B) = P(AB \mid B) \leqq P(B \mid B) = 1$.

* Note that one of the reasons why the impossible event $\varnothing$ is not admissible as a condition is that if $P(A \mid \varnothing)$ were defined, we should have the contradiction that $P(\varnothing \mid \varnothing) = 0$ by (2.2.6′) and $P(\varnothing \mid \varnothing) = 1$ by (2.2.2).

Let us mention the following consequence of axiom (γ):

$$P(A \mid BC) = \frac{P(AB \mid C)}{P(B \mid C)} \quad \text{if } B, C \text{ and } BC \text{ belong to } \mathscr{B} \tag{2.2.4'}$$

To obtain (2.2.4′) substitute BC into (2.2.4) instead of B. Thus we obtain

$$P(A \mid BC) = \frac{P(ABC \mid C)}{P(BC \mid C)} \tag{2.2.9}$$

Applying (2.2.5) to both the numerator and the denominator at the right-hand side of (2.2.9) we obtain (2.2.4′). Note that (2.2.4′) is the counterpart for conditional probabilities of the relation (2.1.3) for conditional relative frequencies.

We now prove a theorem which gives insight into the structure of conditional probability spaces.

THEOREM 2.2.1. *If $S = [\Omega, \mathscr{A}, \mathscr{B}, P(A \mid B)]$ is a conditional probability space, there exists a σ-finite* measure μ on $\mathscr{A}$ such that for each $B \in \mathscr{B}$ one has*

$$0 < \mu(B) < +\infty \tag{2.2.10}$$

and for each $A \in \mathscr{A}$ and $B \in \mathscr{B}$ one has

$$P(A \mid B) = \frac{\mu(AB)}{\mu(B)} \tag{2.2.11}$$

The measure μ is determined uniquely up to a positive constant factor.

We shall in what follows often need the following basic fact on measures:†

LEMMA 2.2.2. *If $\mu(A)$ is a σ-finite measure on an algebra $\mathscr{A}$ of subsets of a set Ω, then $\mu(A)$ can be uniquely extended to a measure on $\sigma(\mathscr{A})$.*

In order to prove our theorem we shall also need the following:

LEMMA 2.2.3. *Let $S = [\Omega, \mathscr{A}, \mathscr{B}, P(A \mid B)]$ be a conditional probability space, $A_i \in \mathscr{A}$, $B_i \in \mathscr{B}$ $(i = 1, 2)$ and $A_1 + A_2 \subseteq B_1 B_2$. Then we have*

$$P(A_1 \mid B_1) P(A_2 \mid B_2) = P(A_1 \mid B_2) P(A_2 \mid B_1) \tag{2.2.12}$$

Proof. Let us put $B_3 = B_1 + B_2$. By supposition (1) of Definition 2.2.1 we have $B_3 \in \mathscr{B}$. From (γ) we get

$$P(A_1 \mid B_1) P(A_2 \mid B_2) = \frac{P(A_1 \mid B_3)}{P(B_1 \mid B_3)} \cdot \frac{P(A_2 \mid B_3)}{P(B_2 \mid B_3)} \tag{2.2.13}$$

* A measure μ on a σ-algebra $\mathscr{A}$ of subsets of a set Ω is called σ-finite if the set Ω can be partitioned into a denumerable sequence of disjoint sets $\Omega_n \in \mathscr{A}$ $(n = 1, 2, \ldots)$ such that $\mu(\Omega_n) < +\infty$ for $n = 1, 2, \ldots$. Thus a finite measure (i.e., a measure such that $\mu(\Omega) < +\infty$) is a fortiori σ-finite. (See also Appendix A.)

† See Appendix A.

and similarly,

$$P(A_1 \mid B_2)P(A_2 \mid B_1) = \frac{P(A_1 \mid B_3)}{P(B_2 \mid B_3)} \cdot \frac{P(A_2 \mid B_3)}{P(B_1 \mid B_3)} \tag{2.2.14}$$

As the right-hand sides of (2.2.13) and (2.2.14) are equal, (2.2.12) follows. ■

Proof. Now we turn to the proof of Theorem 2.2.1. Let B_0 be an arbitrary element of $\mathscr{B}$ which will be fixed throughout the proof. Let us put for each $B \in \mathscr{B}$,

$$\mu(B) = \frac{P(B \mid B + B_0)}{P(B_0 \mid B + B_0)} \tag{2.2.15}$$

In view of (2.2.3), $\mu(B)$ as defined by (2.2.15) is positive and finite for each $B \in \mathscr{B}$, i.e., (2.2.10) is satisfied. If $A \in \mathscr{A}$ is any set such that there exists a set $B \in \mathscr{B}$, for which $A \subseteq B$, let us put

$$\mu(A) = P(A \mid B)\mu(B) \tag{2.2.16}$$

We have to show that the definition (2.2.16) of μ does not depend on the choice of B, i.e., that if $B_i \in \mathscr{B}$ $(i = 1, 2)$ and $A \subseteq B_1 B_2$, we have

$$P(A \mid B_1)\mu(B_1) = P(A \mid B_2)\mu(B_2). \tag{2.2.17}$$

This can be shown as follows: using Lemma 2.2.3 and (γ) we get, putting $B_4 = B_0 + B_1 + B_2$,

$$P(A \mid B_1)\mu(B_1) = P(A \mid B_1)\frac{P(B_1 \mid B_4)}{P(B_0 \mid B_4)} = \frac{P(A \mid B_4)}{P(B_0 \mid B_4)} \tag{2.2.18}$$

Similarly, replacing B_1 by B_2, we get

$$P(A \mid B_2)\mu(B_2) = \frac{P(A \mid B_4)}{P(B_0 \mid B_4)} \tag{2.2.19}$$

As the right-hand sides of (2.2.18) and (2.2.19) are equal, (2.2.17) follows. Thus we have shown that the definition of $\mu(A)$ by (2.2.16) is unambiguous, i.e., it does not depend on the choice of the set B provided that $A \subseteq B$. Incidentally we have shown also that the definitions of $\mu(A)$ and $\mu(B)$ by (2.2.15) and (2.2.16) are in accordance, because if $A \in \mathscr{B}$, we can choose $B = A$ in (2.2.16).

Now let $\mathscr{A}^*$ be the family of those sets $A \in \mathscr{A}$ for which there exists a $B \in \mathscr{B}$ such that $A \subseteq B$. Clearly if $A_1 \in \mathscr{A}^*$ and $A_2 \in \mathscr{A}^*$, then there exist sets $B_i \in \mathscr{B}$ such that $A_i \subseteq B_i$ $(i = 1, 2)$; as by (1) (Definition 2.2.1) we have $B_1 + B_2 \in \mathscr{B}$ and evidently $A_1 + A_2 \subseteq B_1 + B_2$, it follows that $A_1 + A_2 \in \mathscr{A}^*$. Besides this, if $A_1 \in \mathscr{A}^*$ and $A_2 \in \mathscr{A}^*$, then clearly $A_1 - A_2 \in \mathscr{A}^*$; thus $\mathscr{A}^*$ is a ring of sets. Now we show that μ is a measure on $\mathscr{A}^*$. Let $A_n \in \mathscr{A}^*$

$(n=1, 2, \ldots)$ be a sequence of pairwise disjoint sets and suppose that $\sum_{n=1}^{\infty} A_n \in \mathscr{A}^*$. Then by definition there is a $B \in \mathscr{B}$ such that $\sum_{n=1}^{\infty} A_n \subseteq B$. It follows that $A_n \subseteq B$ for $n=1, 2, \ldots$ and thus, as we have supposed that $P(A \mid B)$ is for each fixed $B \in \mathscr{B}$ a measure on $\mathscr{A}$, it follows that

$$\mu\left(\sum_{n=1}^{\infty} A_n\right) = P\left(\sum_{n=1}^{\infty} A_n \mid B\right)\mu(B) = \sum_{n=1}^{\infty} P(A_n \mid B)\mu(B) = \sum_{n=1}^{\infty} \mu(A_n)$$

Thus, μ is a measure on $\mathscr{A}^*$. By Lemma 2.2.2 the definition of μ can be uniquely extended to the least σ-algebra $\sigma(\mathscr{A}^*)$ containing $\mathscr{A}^*$ so that it will be a measure on $\sigma(\mathscr{A}^*)$. Now clearly, $\sigma(\mathscr{A}^*)$ contains every set $A \in \mathscr{A}$ which can be covered by the union of a denumerable sequence of sets $B_n \in \mathscr{B}$ $(n=1, 2, \ldots)$. But by supposition (2) of Definition 2.2.1 the space Ω itself can be covered by a denumerable sequence of sets belonging to $\mathscr{B}$: thus the same is true for each $A \in \mathscr{A}$ and $\mathscr{A} \subseteq \sigma(\mathscr{A}^*)$. As $\mathscr{A}$ itself is a σ-algebra, we have $\mathscr{A} = \sigma(\mathscr{A}^*)$. Thus we have defined the measure μ for every $A \in \mathscr{A}$, and μ is clearly σ-finite. As for each $A \in \mathscr{A}$ and $B \in \mathscr{B}$, we have $AB \subset B$, and thus $AB \in \mathscr{A}^*$, it follows from (2.2.16) and (2.2.5) that

$$P(A \mid B) = P(AB \mid B) = \frac{\mu(AB)}{\mu(B)} \tag{2.2.20}$$

To prove Theorem 2.2.1 it remains to show that μ is uniquely determined up to a positive constant factor. Suppose ν is another σ-finite measure on $\mathscr{A}$ and one has for each $A \in \mathscr{A}$ and $B \in \mathscr{B}$

$$\frac{\nu(AB)}{\nu(B)} = \frac{\mu(AB)}{\mu(B)}$$

It follows that for any fixed $B \in \mathscr{B}$, one has for $A \subseteq B$

$$\nu(A) = C(B)\mu(A) \tag{2.2.21}$$

where $C(B) = \nu(B)/\mu(B)$. Now let us consider a sequence B_k $(k=1, 2, \ldots)$ such that $B_k \in \mathscr{B}$ and $\sum_{k=1}^{\infty} B_k = \Omega$. Such a sequence exists according to (2) (Definition 2.2.1). Now let us take $B = \sum_{k=1}^{n} B_k$ in (2.2.21). We obtain

$$\nu(A) = c_n \mu(A) \tag{2.2.22}$$

for all $A \in \mathscr{A}$ for which $A \subseteq \sum_{k=1}^{n} B_k$, where c_n is a positive constant. As $\sum_{k=1}^{n} B_k \subseteq \sum_{k=1}^{n+1} B_k$, it follows that the constants c_n are all equal (namely, all are equal to $\nu(B_1)/\mu(B_1)$) because $B_1 \subseteq \sum_{k=1}^{n} B_k$ for $n=1, 2, \ldots,$ and thus there exists a positive constant c such that

$$\nu(A) = c\mu(A) \tag{2.2.23}$$

for the ring $\mathscr{A}_1$ of those sets $A \in \mathscr{A}$ for which there can be found a positive integer n such that $A \subseteq \sum_{k=1}^{n} B_k$. Thus, (2.2.23) is also valid for the least

σ-algebra $\sigma(\mathscr{A}_1)$ containing the ring $\mathscr{A}_1$ and by an argument similar to that used above it is easily seen that $\sigma(\mathscr{A}_1) = \mathscr{A}$. Thus (2.2.23) holds for all $A \in \mathscr{A}$ and the proof of Theorem 2.2.1 is complete. ∎

We introduce the following:

DEFINITION 2.2.3. *A conditional probability space $S = [\Omega, \mathscr{A}, \mathscr{B}, P(A \mid B)]$ such that $P(A \mid B) = \mu(AB)/\mu(B)$ for $A \in \mathscr{A}$ and $B \in \mathscr{B}$, where μ is a σ-finite measure on $\mathscr{A}$, is called full if the bunch $\mathscr{B}$ is identical with the family of all sets $B \in \mathscr{A}$ such that $0 < \mu(B) < +\infty$.*

It is easy to prove the following:

THEOREM 2.2.2. *Every conditional probability space can be extended to a full conditional probability space.*

Proof. Let $S = [\Omega, \mathscr{A}, \mathscr{B}, P(A \mid B)]$ be a conditional probability space. Then there exists by Theorem 2.2.1 a σ-finite measure μ such that for $A \in \mathscr{A}$, $B \in \mathscr{B}$ one has $0 < \mu(B) < +\infty$ and $P(A|B) = \mu(AB)/\mu(B)$. Let $\mathscr{B}^*$ denote the family of all sets $B \in \mathscr{A}$ such that $0 < \mu(B) < +\infty$. Clearly $\mathscr{B}^*$ is a bunch of sets and the family $\mathscr{B}^*$ does not depend on the choice of the measure μ. We have, evidently, $\mathscr{B} \subseteq \mathscr{B}^*$. If $\mathscr{B}$ is not identical with $\mathscr{B}^*$, let us extend the definition of $P(A \mid B)$ for each $B \in \mathscr{B}^* - \mathscr{B}$ by putting

$$P(A \mid B) = \frac{\mu(AB)}{\mu(B)} \quad \text{for } B \in \mathscr{B}^* - \mathscr{B} \tag{2.2.24}$$

It is easy to see that $S^* = [\Omega, \mathscr{A}, \mathscr{B}^*, P(A \mid B)]$ is a full conditional probability space. This proves Theorem 2.2.2. ∎

DEFINITION 2.2.4. *A full conditional space $S = [\Omega, \mathscr{A}, \mathscr{B}, P(A \mid B)]$ such that $P(A \mid B) = \mu(AB)/\mu(B)$ for $A \in \mathscr{A}$ and $B \in \mathscr{B}$, where μ is a σ-finite measure on $\mathscr{A}$ and $\mathscr{B}$ is the family of those sets $B \in \mathscr{A}$ for which $0 < \mu(B) < +\infty$, is called the full conditional probability space generated by the measure μ on the experiment $\mathscr{E} = [\Omega, \mathscr{A}]$, and will be denoted by $S = [\Omega, \mathscr{A}, \mu]$.*

Clearly, every σ-finite measure on the σ-algebra $\mathscr{A}$ of subsets of a set Ω, which is not identically 0, generates a full conditional probability space $S = [\Omega, \mathscr{A}, \mu]$. If μ and ν are σ-finite measures on $\mathscr{A}$, such that $\nu(A) = c\mu(A)$ for all $A \in \mathscr{A}$, where c is a positive constant (not depending on A), then we have $[\Omega, \mathscr{A}, \mu] = [\Omega, \mathscr{A}, \nu]$; thus two measures which are proportional to each other generate the same full conditional probability space.

According to Theorem 2.2.2 it is no essential restriction to deal only with full conditional probability spaces. In what follows we shall in general consider only full conditional probability spaces.

Let us now consider some examples.

Example 2.2.1. Let Ω be a nonempty finite or denumerable set, and let $\mathscr{A}$ be the set of all subsets of Ω. Let ω_n $(n = 1, 2, \ldots)$ denote the elements of the set Ω. Let p_n $(n = 1, 2, \ldots)$ be an arbitrary sequence of positive numbers, and put for each $A \in \mathscr{A}$

$$\mu(A) = \sum_{\omega_n \in A} p_n \tag{2.2.25}$$

The full conditional probability space $S = [\Omega, \mathscr{A}, \mu]$ generated by the measure μ on the experiment $\mathscr{E} = (\Omega, \mathscr{A})$ is called a *discrete conditional probability space*. Note that the family $\mathscr{B}$ of admissible conditions in this example is the set of those nonempty subsets A of Ω for which $\mu(A) < +\infty$. In case the series $\sum_{k=1}^{\infty} p_n$ is convergent, $\mathscr{B}$ contains all nonempty subsets of Ω. If the series $\sum_{n=1}^{\infty} p_n$ is divergent, $\mathscr{B}$ contains all finite nonempty subsets of Ω; if the numbers p_n have a positive lower bound, $\mathscr{B}$ consists of such sets only.

Remark: It is easy to see that Example 2.2.1 gives us essentially all possible full conditional probability spaces over a denumerable basic set Ω. As a matter of fact, it follows from Theorem 1.6.2 that in case Ω is denumerable, it is not an essential restriction to suppose that $\mathscr{A}$ consists of all subsets of Ω, because every σ-algebra of subsets of a denumerable set is isomorphic with the σ-algebra of all subsets of another denumerable set Ω', namely the set of all atoms of $\mathscr{A}$. It is easy to see further that a measure μ on the σ-algebra of all subsets of a denumerable set Ω is σ-finite if and only if all sets consisting of a single element have finite μ-measure. Thus every full conditional probability space on a denumerable set Ω is generated by a measure μ which is uniquely determined by the sequence of nonnegative numbers $p_n = \mu(\{\omega_n\})$ where ω_n $(n = 1, 2, \ldots)$ are the elements of the set Ω (the outcomes of the experiment $\mathscr{E} = (\Omega, P(\Omega))$, and $\{\omega_n\}$ denotes the set which consists of the single element ω_n. Now let Ω^* denote the set of those $\omega_n \in \Omega$ for which $p_n > 0$. Clearly the conditional probability space $S^* = [\Omega^*, \Omega^*\mathscr{A}, \mu]$ is of the type described in Example 2.2.1.

The following example is a particular case of Example 2.2.1, but it deserves special attention.

Example 2.2.2. Let Ω be a finite or denumerable set. Let $\mathscr{A}$ denote the set of all subsets of Ω, and let $\mathscr{B}$ denote the set of all nonempty finite subsets of Ω. Let $\mathscr{N}(A)$ denote the number of elements of the set A and put for each $A \in \mathscr{A}$ and $B \in \mathscr{B}$

$$P(A \mid B) = \frac{\mathscr{N}(AB)}{\mathscr{N}(B)} \tag{2.2.26}$$

Then $S = [\Omega, \mathscr{A}, \mathscr{B}, P(A \mid B)]$ is a full conditional probability space generated by the σ-finite measure $\mathscr{N}(A)$; thus it can be denoted also by $[\Omega, \mathscr{A}, \mathscr{N}]$.

Remarks: Clearly, Example 2.2.2 is that special case of Example 2.2.1 when all numbers p_n are equal to 1 (or, to the same positive constant c). The experiment $\mathscr{E} = (\Omega, \mathscr{A})$ in Example 2.2.2 can be described as follows: The experiment consists in choosing at random an element of the set Ω. The probabilities corresponding to this random choice can be characterized by saying that under the condition that the element of Ω chosen at random belongs to an arbitrary finite nonempty subset B of Ω, each element of B has the same conditional probability to be chosen.*

Example 2.2.3. Let R denote the set of real numbers and let $\mathscr{A}$ be the set of all Borel subsets of R. Let $F(x)$ be a nondecreasing real function defined on the real line R, which is everywhere continuous from the left (i.e., $F(x-0) = F(x)$). Let μ_F denote the Lebesgue-Stieltjes measure generated on $\mathscr{A}$ by the function $F(x)$. In other words, define μ_F on the semiring of all intervals $I[a, b) = \{x: a \leqq x < b\}$ $(a < b)$ by putting $\mu_F(I[a, b)) = F(b) - F(a)$. If A is a union of the disjoint intervals I_k $(k = 1, 2, \ldots, n)$, put $\mu_F(A) = \sum_{k=1}^{n} \mu_F(I_k)$ and extend the measure μ_F to the least σ-algebra containing all such sets A, i.e., to the σ-algebra of Borel subsets of R (see Appendix A). The measure μ_F generates a full conditional probability space $[R, \mathscr{A}, \mu_F]$ on the experiment $(R, \mathscr{A})$. Note that in this case every interval $I[a, b)$ such that $F(a) < F(b)$ is an admissible condition and if $I[a, b)$ is such an interval and $a \leqq c \leqq d \leqq b$, we have

$$P(I[c, d) \mid I[a, b)) = \frac{F(d) - F(c)}{F(b) - F(a)}$$

Note that if we replace the function $F(x)$ by the function $G(x) = eF(x) + f$, where e is an arbitrary positive constant and f is an arbitrary real constant, we have $\mu_G = e\mu_F$ and thus $[R, \mathscr{A}, \mu_G] = [R, \mathscr{A}, \mu_F]$.

The experiment $(R, \mathscr{A})$ can be described by saying that it consists in choosing at random a real number. The measure $\mu_F(A)$ is denoted also by $\int_A dF(x)$; thus we have for $A \in \mathscr{A}$ and every $B \in \mathscr{A}$ for which $\mu_F(B) > 0$

$$P(A \mid B) = \frac{\int_{AB} dF(x)}{\int_B dF(x)} \tag{2.2.27}$$

Example 2.2.4. Let R denote the set of real numbers, let $\mathscr{A}$ be the set of Borel subsets of R, and let $f(x)$ be a nonnegative Lebesgue-measurable function on the real line such that $\int_a^b f(x)\,dx$ is finite for every finite interval $I(a, b)$

* In case the set Ω is finite, the conditional probability space of Example 2.2.2 will later be called a *classical probability space*. See Sections 2.3 and 2.4.

$(-\infty < a < b < +\infty)$. Let $\mathscr{B}$ denote the family of those Borel sets B of the real line for which $0 < \int_B f(x)\,dx < +\infty$ and put

$$P(A \mid B) = \frac{\int_{AB} f(x)\,dx}{\int_B f(x)\,dx} \quad \text{for } A \in \mathscr{A},\ B \in \mathscr{B} \tag{2.2.28}$$

Clearly, $S = [R, \mathscr{A}, \mathscr{B}, P(A \mid B)]$ is a full conditional probability space; it is generated by the measure $\mu_f(A) = \int_A f(x)\,dx$. We may also denote S by $[R, \mathscr{A}, \mu_f]$.

Remark 1: Clearly, if we replace the function $f(x)$ by the function $g(x) = cf(x)$, where c is a positive constant, we obtain the same conditional probability space.

Remark 2: Note that if $F(x)$ is absolutely continuous and we put* $F'(x) = f(x)$, then the conditional probability space of Example 2.2.3 will be that of Example 2.2.4.

The following example is a special case of Example 2.2.4 which is of particular interest:

Example 2.2.5. Let R be the real line, let $\mathscr{A}$ be the σ-algebra of Borel subsets of R, and let $\lambda(A)$ denote the Lebesgue measure of the set $A \in \mathscr{A}$. Put for all $A \in \mathscr{A}$ and for all $B \in \mathscr{B}$ such that $0 < \lambda(B) < +\infty$,

$$P(A \mid B) = \frac{\lambda(AB)}{\lambda(B)} \tag{2.2.29}$$

In this way we get a full conditional probability space $S = [R, \mathscr{A}, \lambda]$. The experiment described by this example may be characterized by saying that we choose at random a real number so that under the condition that this real number belongs to a Borel set B, having positive and finite Lebesgue measure, it is *uniformly distributed* on this set, i.e., the conditional probability of this random number lying in a subset A of B $(A \in \mathscr{A})$ is proportional to the Lebesgue measure of A, the factor of proportionality being $\lambda(B)^{-1}$.

The above examples can easily be generalized for the Euclidean space of dimension r $(r = 2, 3, \ldots)$.

In measure theory a measure μ on a σ-algebra $\mathscr{A}$ is called *complete* if the conditions $A \in \mathscr{A}$, $\mu(A) = 0$ and $C \subseteq A$ imply $C \in \mathscr{A}$ (and thus $\mu(C) = 0$). It is known (see Appendix A) that if μ is any measure on the σ-algebra $\mathscr{A}$, and

* The derivative of an absolutely continuous function exists almost everywhere (see Appendix A). For those x for which $F'(x)$ does not exist, we put $f(x) = 0$.

$\mathscr{A}^*$ is the family of subsets of Ω of the form $A \circ C$ where $A \in \mathscr{A}$ and C is a subset of a set $A' \in \mathscr{A}$ such that $\mu(A') = 0$, then the set function μ^* defined on $\mathscr{A}^*$ by putting $\mu^*(A \circ C) = \mu(A)$ is a complete measure on $\mathscr{A}^*$. Thus, every measure can be made complete by extending its definition to the least σ-algebra $\mathscr{A}^*$ containing the σ-algebra $\mathscr{A}$ and all subsets of sets of μ-measure 0. It is useful to carry out this extension for measures generating a conditional probability space. This extension means of course that the underlying experiment $\mathscr{E}$ is replaced by another experiment $\mathscr{E}'$ which is a refinement of $\mathscr{E}$; it means that every set of outcomes which is the subset of a set of measure 0 is also considered to be an observable event. Thus the conditional probability space $S = [\Omega, \mathscr{A}, \mu]$ is replaced by the conditional probability space $S^* = [\Omega, \mathscr{A}^*, \mu^*]$; we call this the *completion* of S and we call S^* *a complete full conditional probability space*. In other words, we introduce the following:

DEFINITION 2.2.5. *A full conditional probability space $S = [\Omega, \mathscr{A}, \mu]$ generated by the σ-finite measure μ is called complete if from $A \in \mathscr{A}$, $\mu(A) = 0$ and $C \subseteq A$ it follows that $C \in \mathscr{A}$; in other words, a full conditional probability space is called complete if it is generated by a complete measure.*

2.3 PROBABILITY SPACES

Let $S = [\Omega, \mathscr{A}, \mu]$ be a full conditional probability space generated by a σ-finite measure μ over the experiment $\mathscr{E} = (\Omega, \mathscr{A})$. Two essentially different cases have to be distinguished; either μ is bounded, i.e., $\mu(\Omega) < +\infty$, or $\mu(\Omega) = +\infty$. In the first case, Ω itself is an admissible condition, thus the conditional probabilities $P(A \mid \Omega)$ with respect to Ω are defined. However, Ω means the certain event which always takes place; thus to take Ω as a condition means to take no condition at all, because the "condition" is by definition always fulfilled, i.e., whatever the outcome of the experiment may be, the event Ω always happens. Thus $P(A \mid \Omega)$ may be interpreted as the unconditional probability of the event A. We shall call $P(A \mid \Omega)$ the *probability* of the event A and denote it for the sake of brevity by $P(A)$, i.e., we put

$$P(A) = P(A \mid \Omega) \tag{2.3.1}$$

Clearly,

$$P(A) = \frac{\mu(A)}{\mu(\Omega)} \tag{2.3.2}$$

and thus

$$P(\Omega) = 1 \tag{2.3.3}$$

We now introduce the following:

DEFINITION 2.3.1. *A full conditional probability space $S = [\Omega, \mathscr{A}, \mu]$ generated by a bounded measure μ on the experiment $\mathscr{E} = (\Omega, \mathscr{A})$ is called a probability space. If $S = [\Omega, \mathscr{A}, \mu]$ is a probability space, $P(A \mid \Omega) = \mu(A)/\mu(\Omega)$ is called the probability of the event A and will be denoted by $P(A)$.*

As the measure μ is determined only up to a positive constant factor, it is no restriction to assume that $\mu(\Omega) = 1$, in which case $P(A) = \mu(A)$.

Clearly the probability $P(A)$, considered as a set function on $\mathscr{A}$, is itself a measure on $\mathscr{A}$ subject to the condition (2.3.3). Such a measure is usually called a normed measure or a probability measure; we shall also call such a measure a *probability distribution*. This terminology reflects that such a measure shows us how the probability 1 of the certain event Ω is distributed over the space Ω. Thus we introduce the following:

DEFINITION 2.3.2. *A measure $P(A)$ on a σ-algebra $\mathscr{A}$ of subsets of a set Ω, for which $P(\Omega) = 1$, is called a probability distribution over the experiment $\mathscr{E} = (\Omega, \mathscr{A})$.*

According to Definition 2.3.2 we can restate Definition 2.3.1 as follows:

DEFINITION 2.3.1′. *A probability space is an experiment over which a probability distribution is given, i.e., it is a triple $(\Omega, \mathscr{A}, P)$ where Ω is an arbitrary nonempty set, $\mathscr{A}$ is a σ-algebra of subsets of Ω and P is a measure on $\mathscr{A}$ such that $P(\Omega) = 1$.*

The notion of a probability space is due to Kolmogoroff [4].

If $S = (\Omega, \mathscr{A}, P)$ is a probability space, then the set $\mathscr{B}$ of admissible conditions is clearly the family of all events B such that $P(B) > 0$, and we have

$$P(A \mid B) = \frac{P(AB)}{P(B)} \tag{2.3.4}$$

Kolmogoroff has defined conditional probabilities by formula (2.3.4). Thus we arrive at Kolmogoroff's notion of probability space as a special case of the more general notion of a conditional probability space, namely, as a full conditional probability space generated by a bounded measure. The only difference between our approach and that of Kolmogoroff—as far as probability spaces are concerned—is that while in Kolmogoroff's theory the notion of probability is the primary notion and conditional probabilities are defined by means of ordinary probabilities by formula (2.3.4), in our approach conditional probability is the basic concept, and (ordinary) probabilities are special cases of conditional probabilites. Of course, this difference is only a methodological one. However, the important difference is that our theory of conditional probability spaces also includes conditional probability spaces which are generated by an *unbounded* measure, while these cannot be fitted into the frames of Kolmogoroff's theory, because in these conditional probability spaces ordinary probabilities cannot be defined at all.

Note that formula (2.3.1) is the counterpart of the following evident relation for frequencies: $f(A) = f(A \mid \Omega)$. This shows that ordinary probabilities correspond to relative frequencies in the same way as conditional probabilities to conditional relative frequencies; in other words, the probability of an event A is that number to which the relative frequency $f(A)$ of this event will usually be near in a long sequence of observations.

Using the notion of a probability space we can describe a conditional probability space as a compatible family of ordinary probability spaces. This can be seen as follows: Let $S = [\Omega, \mathscr{A}, \mathscr{B}, P(A \mid B)]$ be a conditional probability space. For any fixed $C \in \mathscr{B}$ put $P_C(A) = P(A \mid C)$. Clearly $(\Omega, \mathscr{A}, P_C)$ is an ordinary probability space; we shall denote this probability space by $S \mid C$ and call it the probability space obtained by restricting S to C. Now let $B \in \mathscr{B}$ be an admissible condition such that $B \subseteq C$. Then by definition we have $P_C(B) = P(B \mid C) > 0$; thus B is an admissible condition in the probability space $S \mid C$ too, and it makes sense to speak about the conditional probability of A with respect to B in the probability space $S \mid C$. Denoting this conditional probability by $P_C(A \mid B)$ we have, by (2.3.4),

$$P_C(A \mid B) = \frac{P_C(AB)}{P_C(B)} = \frac{P(AB \mid C)}{P(B \mid C)} \tag{2.3.5}$$

However, by our axiom (γ) of Definition 2.2.1 we have $P(AB \mid C)/P(B \mid C) = P(A \mid B)$; thus we obtain that

$$P_C(A \mid B) = P(A \mid B) \tag{2.3.6}$$

Thus $P_C(A \mid B)$ does not depend on C (provided that $B \subseteq C$). In other words, axiom (γ) states that if the conditional probability $P_C(A \mid B)$ is computed from the probability space $S|C$ where $C \in \mathscr{B}$, $B \subseteq C$, one should always get the same value as the value of $P(A \mid B)$ in the conditional probability space S. Thus (γ) is really a condition of compatibility of the probability spaces $S \mid C$. Thus a conditional probability space can be looked at as a family of ordinary probability spaces, which are compatible with each other in the sense that one always gets the same value for a conditional probability independently of the probability space in which this conditional probability is computed. The converse of this statement is also true, as is shown by the following:

THEOREM 2.3.1. *Let $\mathscr{E} = (\Omega, \mathscr{A})$ be an experiment. Let $\mathscr{B} \subseteq \mathscr{A}$ be a bunch of sets. Suppose that for each $C \in \mathscr{B}$ a probability space $S_C = (\Omega, \mathscr{A}, P_C)$ is given, such that $P_C(C) = 1$ and that these spaces are compatible in the sense that* *for* $B \in \mathscr{B}$, $B \subseteq C$, $C \in \mathscr{B}$ *one has* $P_C(B) > 0$ *and the ratio*

$$\frac{P_C(AB)}{P_C(B)} \tag{2.3.7}$$

does not depend on C. *Put for* $A \in \mathscr{A}$, $B \in \mathscr{B}$

$$P(A \mid B) = \frac{P_C(AB)}{P_C(B)} \tag{2.3.8}$$

where $B \subseteq C \in \mathscr{B}$. (*This ratio depends by supposition only on* A *and* B *but not on* C.) *Then* $S = [\Omega, \mathscr{A}, \mathscr{B}, P(A \mid B)]$ *is a conditional probability space and for each* $A \in \mathscr{A}$ *and* $C \in \mathscr{B}$ *one has* $P_C(A) = P(A \mid C)$.

Proof. If $A \in \mathscr{A}$, $B \in \mathscr{B}$, $C \in \mathscr{B}$, $B \subseteq C$, one gets from (2.3.8)

$$\frac{P(AB \mid C)}{P(B \mid C)} = \frac{P_C(AB)}{P_C(B)} = P(A \mid B) \tag{2.3.9}$$

Thus (γ) holds. One obtains for $A = B$, from (2.3.8),

$$P(B \mid B) = \frac{P_C(B)}{P_C(B)} = 1 \tag{2.3.10}$$

Thus (β) holds also. Finally, for a fixed $B \in \mathscr{B}$, $P_C(A)$ being by supposition a measure, $P(A \mid B)$ is a measure on $\mathscr{A}$ also, i.e., (α) holds. Thus $[\Omega, \mathscr{A}, \mathscr{B}, P(A \mid B)]$ satisfies all axioms of a conditional probability space. Finally one has $P_C(A) = P_C(AC)/P_C(C) = P(A \mid C)$. Thus Theorem 2.3.1 is proved. ■

As every conditional probability space is a compatible family of ordinary probability spaces, every result on probability spaces is at the same time a result about arbitrary conditional probability spaces, as it is valid for each of the probability spaces from which it is composed. Most results on conditional probability spaces can be formulated as results on probability spaces. We shall state and prove such results, for the sake of brevity, for probability spaces only and shall not reformulate them for conditional probability spaces, but the reader should always have in mind that these are at the same time also statements about arbitrary conditional probability spaces. There are, however, certain theorems which cannot be formulated in this way, namely those in which conditional probabilities $P(A \mid B)$ occur for an infinity of different conditions B at the same time and these are not necessarily all subsets of the same admissible condition B_0. Such theorems have been, and of course will be, stated in terms of conditional probability spaces; but whenever it is possible we shall state our theorems in terms of probability spaces.

Now we give some examples of probability spaces.

Example 2.3.1. The conditional probability space of Example 2.2.1 is a probability space if and only if the series $\sum p_k$ is convergent. In this case we can suppose, without restricting the generality, that

$$\sum_{k=1}^{\infty} p_k = 1 \tag{2.3.11}$$

Thus if Ω is a denumerable set, $\mathscr{A}$ is the σ-algebra of all subsets of Ω, and the measure $P(A)$ on $\mathscr{A}$ is defined by

$$P(A) = \sum_{\omega_k \in A} p_k \tag{2.3.12}$$

where the p_k are arbitrary positive numbers subject to the condition (2.3.11), then P is a probability distribution and $(\Omega, \mathscr{A}, P) = S$ is a probability space. We call such a space a *discrete probability space*. Especially if Ω is finite, a conditional probability space over the experiment $\mathscr{E} = (\Omega, \mathscr{A})$ is always a discrete probability space. In case Ω is finite, $\Omega = \{\omega_1, \omega_2, \ldots, \omega_N\}$ and the numbers $p_k (k = 1, 2, \ldots, N)$ are all equal and thus are all equal to $1/N$, we call the space S a *classical probability space* (see Section 2.4). This is the same as the space of Example 2.2.2 in case Ω is finite. A classical probability space can be characterized as an experiment consisting in choosing at random one out of N objects such that each object has the same probability (namely $1/N$) to be chosen.

It is convenient to introduce the following:

DEFINITION 2.3.3. *A probability distribution over the set of all subsets of a denumerable set is called a discrete probability distribution.*

Example 2.3.2. The conditional probability space of Example 2.2.3 is a probability space if and only if the function $F(x)$ is bounded. In this case (as $cF(x) + d = G(x)$ leads to the same conditional probability space as $F(x)$) it is not a restriction to suppose that

$$\lim_{x \to -\infty} F(x) = 0 \quad \text{and} \quad \lim_{x \to +\infty} F(x) = 1 \tag{2.3.13}$$

We introduce the following:

DEFINITION 2.3.4. *A nondecreasing function $F(x)$ defined on the whole real line, continuous from the left in every point and satisfying the conditions* (2.3.13), *is called a probability distribution function** *(on the real line)*.

* A probability distribution function is often called a *cumulative* probability distribution function (to distinguish it from what we call a density function and what is sometimes called distribution function); we shall not use the adjective "cumulative" in this book.

Clearly, if $F(x)$ is a probability distribution function and μ_F is the Lebesgue-Stieltjes measure obtained by means of the function F in the way described in Example 2.2.3, then μ_F is a probability distribution on the Borel sets of the real line, and in the probability space $(R, \mathscr{A}, \mu_F)$ the probability of an event $A \in \mathscr{A}$ is given by

$$P(A) = \mu_F(A) = \int_A dF(x) \tag{2.3.14}$$

Example 2.3.3. Let $F(x)$ be a probability distribution function on the real line, suppose that $F(x)$ is absolutely continuous, and put $f(x) = F'(x)$ for those x for which the derivative exists. For those x for which $F'(x)$ does not exist, put $f(x) = 0$. Then $f(x)$ is a nonnegative measureable function such that

$$\int_a^b f(x)\, dx = F(b) - F(a) \quad \text{for} \quad -\infty < a < b < +\infty \tag{2.3.15}$$

and thus, especially,

$$\int_{-\infty}^{+\infty} f(x)\, dx = 1 \tag{2.3.16}$$

We introduce the following definition:

DEFINITION 2.3.5. *A nonnegative measurable function $f(x)$ defined on the real line and satisfying the condition* (2.3.16) *is called a probability density function* (*on the real line*).

If $f(x)$ is a probability density function and we put

$$P(A) = \int_A f(x)\, dx \tag{2.3.17}$$

for each $A \in \mathscr{A}$ where $\mathscr{A}$ denotes the family of Borel subsets of the real line R, then P is a probability measure and $(R, \mathscr{A}, P)$ is a probability space.

Examples 2.3.2 and 2.3.3 can easily be generalized for Euclidean spaces for an arbitrary number of dimensions. The following example is a special case of Example 2.3.3 which is of special interest:

Example 2.3.4. Let B be a Borel subset of the real line R having finite positive Lebesgue measure $\lambda(B)$ and define the function $f(x)$ as follows:

$$f(x) = \begin{cases} 1/\lambda(B) & \text{if} \quad x \in B \\ 0 & \text{if} \quad x \notin B \end{cases} \tag{2.3.18}$$

Clearly, $f(x)$ is a probability density function and

$$P(A) = \int_A f(x)\, dx = \frac{\lambda(AB)}{\lambda(B)} \tag{2.3.19}$$

is a probability distribution over the experiment $(R, \mathscr{A})$, i.e., $(R, \mathscr{A}, P)$ is a probability space. We shall call the distribution (2.3.19) the *uniform probability distribution on the set B.*

DEFINITION 2.3.6. *A probability space is called complete if it is generated by a complete measure.*

Clearly a discrete probability space is always complete. The following example exhibits a nondiscrete complete probability space of particular interest.

Example 2.3.5. Let I denote the interval $[0, 1]$; let $\mathscr{M}$ denote the class of all Lebesgue-measurable subsets of I and let $P(A)$ denote the Lebesgue measure of the set $A \in \mathscr{M}$; then $(I, \mathscr{M}, P)$ is a complete probability space. This probability space will be referred to as *the Lebesgue probability space.*

By completing the measure (2.3.17) we obtain a complete probability space $(R, \mathscr{M}, P)$; this reduces essentially to the Lebesgue probability space if we choose for $f(x)$ the function

$$f(x)=\begin{cases}1 & \text{for } 0 \leqq x \leqq 1\\ 0 & \text{for } x < 0 \text{ and } 1 < x\end{cases}$$

i.e., the *indicator* of the interval $[0, 1]$.

One advantage of the completion of a probability space can be seen in connection with the example. If we consider the experiment $(R, \mathscr{M})$ instead of the experiment $(R, \mathscr{A})$, then if $f(x)$ is any density function and $g(x)$ is any function such that $g(x) = f(x)$ except for $x \in D$ where D is a set of Lebesgue measure 0 (i.e., if $f(x)$ and $g(x)$ are equal *almost everywhere*), then $g(x)$ is also Lebesgue-measurable. Thus it is also a density function and it generates the same probability space, because for every measurable set A we have $\int_A g(x)\,dx = \int_A f(x)\,dx$. Thus in defining a density function it is sufficient to define it almost everywhere. This is convenient for instance if the density function is obtained as the derivative $f(x) = F'(x)$ of an absolutely continuous probability distribution function F, because the derivative of such a function exists almost everywhere, but not necessarily everywhere.

In what follows we shall always suppose that the probability spaces dealt with are extended to be complete whenever this is needed, without explicitly mentioning this in every instance.

2.4 SOME REMARKS ON THE HISTORY OF PROBABILITY THEORY*

A few words should be said here about the development of the mathematical theory of probability. Some problems concerning dice and other games of

* As regards a detailed treatment of the early history of probability, see Todhunter [50] and David [51].

chance have been considered and solved by Cardano [5] and Galilei [6], but as an independent branch of knowledge probability theory started with the correspondence of Pascal and Fermat, in 1654. Their ideas have been developed further by C. Huyghens and J. Bernoulli [7], who wrote the first books on probability, further by Montmort, A. de Moivre, D. Bernoulli, Buffon, Bayes [8], and many others in the eighteenth century. The classical theory of probability was enriched by many new ideas and systematized by Laplace [9], who also wrote a highly interesting philosophical treatise on probability [10]. The definition of probability given by Laplace and accepted throughout the nineteenth century is as follows: The probability of an event A is equal to the ratio of the cases favorable for the occurrence of the event A and the total number of possible cases, provided that these cases are "equally possible." Expressed in terms of modern probability theory this definition means that Laplace had only those probability spaces in mind which we called (exactly for this reason) classical probability spaces. The objection was often raised against the definition of Laplace that it contains a vicious circle, as in his definition of probability he presupposed the notion of equally possible, i.e., equally probable cases. This objection is, however, not valid: in defining physical quantities, e.g., mass or weight, usually the equality of two such quantities is first defined (e.g., in the case of weight, by means of a balance) before defining a numerical measure for this quantity. Laplace's definition is not satisfactory from a modern point of view for other reasons: in his theory the notion of an event is an intuitive concept and is not defined at all, and besides this, his approach is too narrow.

In the nineteenth century important contributions were made to probability theory by Poisson [11], Gauss [12], Helmert [13], Lobačevsky [14], Čebishev [15], Markoff [16], Liapounoff [17], Bertrand [18], Poincaré [19], etc. It was gradually realized that probability theory could be successfully applied not only to games of chance but also to random phenomena in nature (e.g., to the kinetic theory of gases and in general to statistical mechanics) and in social life (statistics, insurance, etc.). As is well known, the notion of mathematical rigor (in fact the notion of mathematics itself) underwent a fundamental change in the nineteenth century. The axiomatic foundations of most branches of mathematics were thoroughly studied and clarified on the basis of the theory of sets and mathematical logic. It was realized that a mathematical theory has to start from clearly stated axioms, implicitly defining the basic notions, and every proof has to be based exclusively on these axioms and the rules of logic, without making use of the intuitive background of the basic notions. Classical probability theory did not correspond to these new standards, and therefore even in the first decades of this century many mathematicians did not consider probability theory as a branch of mathematics proper, but looked upon it as occupying an intermediate position between mathematics and the natural sciences (or between mathematics and philosophy). This state of affairs slowed down the development of probability theory in a period where there was a very rapid development in other branches of mathematics.

The need for an axiomatic foundation of probability theory, the necessity of developing a mathematical theory of probability which corresponds to the new standards of mathematical rigor, was clearly realized by D. Hilbert, and he included this problem into his famous list of the most important and urgent unsolved problems in mathematics, given in 1900.

The first serious attempt to solve this problem was made by von Mises [20] in 1919. While the theory which he gave was not quite satisfactory and is nowadays mainly of historical interest only, his work contributed much to center the interest of mathematicians on giving a satisfactory foundation to probability theory. This task was accomplished by Kolmogoroff in 1933; his theory of probability spaces [4] was generally accepted and since that time it has formed the basis of modern probability theory. Besides being natural and simple and satisfying modern requirements of rigor, Kolmogoroff's theory had the great advantage that, by basing the theory of probability on the theory of measure, it established the possibility of applying the results of highly developed branches of modern mathematics to probability theory. Of course, Kolmogoroff had his predecessors, among whom Borel [21], Lomnicki [22], Lévy [23], Steinhaus [24] and Jordan [25] have to be mentioned.

Kolmogoroff's theory was not only satisfactory from a purely mathematical point of view, but it also served as the solid foundation for the rapidly expanding field of applications of probability theory. Of particular importance was that it made possible a rigorous treatment of stochastic processes, which had already been studied from a naive point of view by Bachelier [26]. Nevertheless, in the course of development there arose certain problems which could not be fitted into the frames of the theory of probability spaces of Kolmogoroff. The common feature of these problems is that unbounded measures occur in them, while in Kolmogoroff's theory, probability is a bounded measure. Unbounded measures were used to compute probabilities in statistical mechanics, quantum mechanics, and also in mathematical statistics (in connection with the method of Bayes), in integral geometry, in probabilistic number theory, etc. At first glance it seems that unbounded measures can play no role in probability theory, because in view of the meaning of probability, and with respect to the connection between probability and relative frequency, a probability can never have a value greater than 1. While this is of course true, observing attentively how unbounded measures were used in the fields mentioned, either in a heuristic and unrigorous way or by passing to the limit (see Section 2.5), it turns out that such measures were used only to compute conditional probabilities as ratios of the values of an unbounded measure for two sets, the first being the subset of the second, and in this way one always gets reasonable values, i.e., values between 0 and 1 for conditional probabilities. This is the reason why the mentioned more-or-less heuristic reasonings often led to reasonable results. However, the use of unbounded measures does not fit into Kolmogoroff's theory; thus the necessity arose to generalize this theory. This has been done by the author* (see [27]–[33]). Clearly, in a theory in which unbounded measures are allowed, conditional probability has to be taken as the fundamental concept. This is natural also from a philosophical point of view. As a matter of fact, as emphasized in Section 2.1, the probability of an event depends essentially on the circumstances under which the occurrence or nonoccurrence of the event is observed. It is a commonplace to say that in reality every probability is conditional. This was realized a long time ago by many authors, among whom I mention—without aiming at completeness—Barnard [34], Copeland [35], Fréchet [36], Good [37], Jeffreys [38], Keynes [39], Koopman [40], Popper [41], Reichenbach [42]. However, these authors did not develop their theory on a measure-theoretic basis. The axiomatic theory of conditional probability spaces, as developed by the author and exposed in the present book, combines the measure-theoretic approach of Kolmogoroff with the idea of choosing conditional probability as the basic concept. In this way a theory could be obtained which contains that of Kolmogoroff as a special case and in which the above mentioned problems involving an unbounded measure can be given a rigorous mathematical foundation. In other words, in this theory probabilities obtained from unbounded measures receive full "civil rights."

It should be added that independently of the author, but some years later, having in view applications to psychology, Luce [43] also developed a theory of conditional probability spaces; he restricted himself, however, to the case when the basic space is denumerable.

We would like to emphasize that the theory of probability exposed in the present book should not be considered as one which is different from the usually accepted theory based on Kolmogoroff's notion of a probability space, but as a generalized version of the same theory. As a matter of fact, in the greater part of this book the discussion is based on the usual notion of a probability space.

It should be added that research work is going on concerning mathematical models of probability which are different in some respect from the usually accepted theory. One such direction has been mentioned already in Chapter 1. (See Kappos [71].) Recently some authors have emphasized the advantages of supposing that probability is only an additive set function (instead of σ-additivity) (see e.g., Dubbins and Savage [44]). In this book we do not go into the details of these investigations.

* After having worked out this theory the author was informed that the idea of such a generalized theory has been put forward by Kolmogoroff himself in a lecture some years earlier, but he did not publish his ideas regarding this question.

As we have seen, in Kolmogoroff's theory the probability is not necessarily defined for all subsets A of the basic set Ω, but only for sets A belonging to a certain σ-algebra $\mathscr{A}$. This restriction is usually explained by purely mathematical difficulties encountered when one wants to extend a measure μ to all subsets of the set Ω, provided that Ω is nondenumerable. (See Ulam [67].) In the present book we have taken the point of view that it is not necessarily desirable to extend the definition of $P(A)$ for a collection of sets that is as large as is mathematically possible. As has been pointed out in Chapter 1, in every real experiment the family of events which are really observable usually does not contain all logically conceivable events. From this it follows that the restriction of the definition of a probability measure $P(A)$ to a σ-algebra $\mathscr{A}$ which is not identical with the power set $\mathscr{P}(\Omega)$, may correspond to the impossibility of observation of certain events. This may be the case also if Ω is denumerable or even finite, when no mathematical difficulties arise in extending the definition of $P(A)$ to $\mathscr{P}(\Omega)$, but this may not be desirable at all in view of the impossibility of observing certain events in the real situation of which our probability space is a mathematical model. Therefore we have interpreted the σ-algebra of those events A for which the probability $P(A)$ is defined as the family of *observable* events.*

2.5 LIMITS OF CONDITIONAL PROBABILITY SPACES

First we introduce the following:

DEFINITION 2.5.1. *Let* $S = [\Omega, \mathscr{A}, \mathscr{B}, P(A \mid B)]$ *and* $S_n = [\Omega, \mathscr{A}, \mathscr{B}_n, P_n(A \mid B)]$ $(n = 1, 2, \ldots)$ *be conditional probability spaces over the same experiment* $\mathscr{E} = (\Omega, \mathscr{A})$. *Suppose that* $\mathscr{B} \subseteq \liminf_{n \to \infty} \mathscr{B}_n$, *i.e., for each* $B \in \mathscr{B}$ *we have* $B \in \mathscr{B}_n$ *except for a finite number of values of* n. *We shall say that the conditional probability spaces* S_n *tend for* $n \to +\infty$ *to the conditional probability space* S *with respect to* $\mathscr{B}$, *if*†

$$\lim_{n \to +\infty} P_n(A \mid B) = P(A \mid B) \quad \textit{for all } A \in \mathscr{A} \textit{ and } B \in \mathscr{B} \tag{2.5.1}$$

If (2.5.1) *holds, we write*

$$\lim_{n \to \infty} S_n = S(\mathscr{B}) \tag{2.5.2}$$

* As is well known, the impossibility of observing certain events is very much emphasized in quantum mechanics. However, here we have in mind not this type of unobservability, but only unobservability in the "classical" sense. In quantum mechanics the observable events do not form even an algebra of events. In this book we do not deal with the problems of probability in quantum mechanics.

† Of course, $P_n(A \mid B)$ is not necessarily defined for all n, but it is defined for all but a finite number of values of n.

Note that the limit of a sequence of conditional probability spaces is, if it exists, unique only with respect to a given bunch $\mathscr{B} \subseteq \liminf_{n\to\infty} \mathscr{B}_n$. It is possible that the sequence S_n tends to a conditional probability space $S = [\Omega, \mathscr{A}, \mathscr{B}, P(A \mid B)]$ with respect to a certain bunch $\mathscr{B}$, but to another conditional probability space $S^* = [\Omega, \mathscr{A}, \mathscr{B}^*, P^*(A \mid B)]$ with respect to another bunch $\mathscr{B}^*$. However, for every B which belongs to the intersection of the two bunches $\mathscr{B}$ and $\mathscr{B}^*$ (if this is not empty) one necessarily has $P(A \mid B) = P^*(A \mid B)$.

We shall now prove the following:

THEOREM 2.5.1. *If $S_n = [\Omega, \mathscr{A}, \mathscr{B}_n; P_n(A \mid B)]$ is a sequence of conditional probability spaces, over the same experiment $\mathscr{E} = (\Omega, \mathscr{A})$, $\mathscr{B}$ is a bunch of sets such that $\mathscr{B} \subseteq \liminf_{n\to\infty} \mathscr{B}_n$, further if the limit*

$$\lim_{n\to\infty} P_n(A \mid B) = P(A \mid B) \tag{2.5.3}$$

exists for all $A \in \mathscr{A}$ and all $B \in \mathscr{B}$, and

$$P(B \mid C) > 0 \quad \text{if } B \in \mathscr{B},\ B \subseteq C \text{ and } C \in \mathscr{B} \tag{2.5.4}$$

then $S = [\Omega, \mathscr{A}, \mathscr{B}, P(A \mid B)]$ is a conditional probability space and thus we have

$$\lim_{n\to\infty} S_n = S(\mathscr{B}) \tag{2.5.5}$$

Proof. In order to prove the theorem we have to verify that the set function $P(A \mid B)$ defined by (2.5.3) fulfills the axioms (α), (β) and (γ) of Definition 2.2.1. Axioms (β) and (γ) are easily verified, because we have

$$P(B \mid B) = \lim_{n\to\infty} P_n(B \mid B) = 1$$

Further, if $B \in \mathscr{B}$, $C \in \mathscr{B}$, $B \subseteq C$, we have, in view of (2.5.2) and (2.5.3), as (γ) holds for P_n,

$$P(A \mid B) = \lim_{n\to+\infty} P_n(A \mid B) = \lim_{n\to\infty} \frac{P_n(AB \mid C)}{P_n(B \mid C)} = \frac{P(AB \mid C)}{P(B \mid C)}$$

That $P(A \mid B)$ also satisfies (α), i.e., is a measure for each fixed $B \in \mathscr{B}$, can be shown by using the Vitali-Hahn-Saks theorem (see Section 4.3 where a simple proof of this theorem is given), according to which if μ_n is a sequence of probability distributions on an experiment $\mathscr{E} = (\Omega, \mathscr{A})$ and the limit $\lim_{n\to+\infty} \mu_n(A) = \mu(A)$ exists for *every* $A \in \mathscr{A}$, then μ is also a probability distribution on $\mathscr{A}$. As $P_n(A \mid B)$ is for each fixed B and for sufficiently large values of n a probability distribution, $P(A \mid B)$ is also a probability distribution for each $B \in \mathscr{B}$. This proves Theorem 2.5.1. ∎

Note that even if S_n is a probability space for each n and we have $\lim_{n\to+\infty} S_n = S(\mathscr{B})$ for a certain bunch $\mathscr{B} \subseteq \liminf_{n\to+\infty} \mathscr{B}_n$, it does *not* follow that S is also a probability space: it may be a conditional probability space generated by an unbounded measure. This is shown by the following examples:

Example 2.5.1. Let Ω be the set of natural numbers and let $\mathscr{A}$ be the family of all subsets of Ω. Let $\mathscr{B}_n$ be the set of all subsets of Ω containing at least one element of the set $\Omega_n = \{1, 2, \ldots, n\}$ and put

$$P_n(A \mid B) = \frac{\mathscr{N}(AB\Omega_n)}{\mathscr{N}(B\Omega_n)} \quad \text{for } B \in \mathscr{B}_n \text{ and } A \in \mathscr{A} \tag{2.5.6}$$

where $\mathscr{N}(C)$ denotes the number of elements of the finite set C. The conditional probability space $S_n = [\Omega, \mathscr{A}, \mathscr{B}_n, P_n]$ is thus the classical probability space over the first n natural numbers, considered as a conditional probability space over the experiment $\mathscr{E} = (\Omega, \mathscr{A})$. Let $\mathscr{B}$ be the set of all finite nonempty subsets of the set Ω; then clearly $\mathscr{B} \subset \liminf_{n\to+\infty} \mathscr{B}_n$. Further, we have for every $A \in \mathscr{A}$ and $B \in \mathscr{B}$

$$\lim_{n\to+\infty} P_n(A \mid B) = \frac{\mathscr{N}(AB)}{\mathscr{N}(B)} \tag{2.5.7}$$

Thus if $S = [\Omega, \mathscr{A}, \mathscr{N}]$ denotes the (full) conditional probability space of Example 2.2.2 generated over the experiment $\mathscr{E} = (\Omega, \mathscr{A})$ by the (unbounded) counting measure $\mathscr{N}$, we have $\lim_{n\to\infty} S_n = S(\mathscr{B})$. Thus the conditional probability space S is the limit (with respect to the set of all nonempty subsets of Ω) of the *ordinary* probability spaces S_n.

Example 2.5.2. Let R denote the real axis, let $\mathscr{M}$ be the family of Lebesgue-measurable subsets of R, and let S_n denote the probability space generated on the experiment $(R, \mathscr{M})$ by the uniform probability distribution over the interval $(-n, +n)$. In other words, let $\mathscr{B}_n$ denote the family of those measurable subsets of the real line whose intersection with the interval $I_n = (-n, +n)$ has a positive measure and put

$$P_n(A \mid B) = \frac{\lambda(ABI_n)}{\lambda(BI_n)} \quad \text{for } A \in \mathscr{M},\ B \in \mathscr{B}_n \tag{2.5.8}$$

where $\lambda(E)$ denotes the Lebesgue measure of the set E. Clearly $S_n = [R, \mathscr{M}, \mathscr{B}_n, P_n(A \mid B)]$ is identical to the ordinary probability space generated by the measure $\lambda(AI_n)$ on $(R, \mathscr{M})$. Let $\mathscr{B}$ denote the family of all measurable subsets of R which have a positive and finite Lebesgue measure. Then clearly, $\lim_{n\to\infty} S_n = S(\mathscr{B})$ where $S = [R, \mathscr{M}, \lambda]$ is the conditional probability space generated over the experiment $(R, \mathscr{M})$ by the Lebesgue measure λ.

Let us add a few remarks. The examples above show that the limit of a convergent sequence of ordinary conditional probability spaces may be a conditional probability space generated by an unbounded measure. It is easy to see that every conditional probability space can be obtained in this way. As a matter of fact, if $S = [\Omega, \mathscr{A}, \mathscr{B}, P(A \mid B)]$ is any conditional probability space and $C_n \in \mathscr{B}$ is a sequence of conditions such that $C_n \subseteq C_{n+1}$ and $\sum_{n=1}^{\infty} C_n = \Omega$ (such a sequence always exists according to the definition of a bunch of events), then we have $S = \lim_{n\to+\infty} S_n$ where $S_n = S \mid C_n$ is the ordinary probability space obtained by restricting S to C_n. Thus, we may look at conditional probability spaces generated by unbounded measures as limits of ordinary probability spaces. In fact, this is the only way in which conditional probability spaces generated by unbounded measures can be dealt with within the frames of Kolmogoroff's theory, and this device was often used previously. However, this manner of dealing with such spaces is rather cumbersome, and the way chosen in this book is more advantageous. The situation can be compared with that concerning generalized functions (distributions of Schwartz, see [68]). When the need for generalized functions like Dirac's delta function $\delta(x)$ arose in physics, mathematicians were inclined to deal with them by passing to the limit. For instance, the "integral" $\int f(x)\delta(x)\,dx$, defined by putting $\int f(x)\delta(x)\,dx = f(0)$, is the limit of the ordinary integrals $\int f(x)\delta_n(x)\,dx$ where, e.g.,

$$\delta_n(x) = \begin{cases} n & \text{for} \quad |x| < \dfrac{1}{2n} \\[2ex] 0 & \text{for} \quad |x| \geqq \dfrac{1}{2n} \end{cases}$$

It turned out, however, that it is possible and much more convenient to generalize the notion of a function so as to include mathematical objects like the delta function of Dirac. This has been done by Schwartz. It should be added that this is more than an analogy. As a matter of fact, while the Fourier transform of a probability distribution in the real line is an ordinary function, the Fourier transform of an unbounded measure does not exist as an ordinary function, but it may be defined as a generalized function. The Fourier transform of the Lebesgue measure over the whole real line is nothing else than the delta function of Dirac (see [32]).

Finally, we give an example which shows that even in the case when the limit of a sequence of probability spaces is a conditional probability space generated by a probability measure, this limit cannot be considered as a probability space, but only as a conditional probability space, because the limit may exist only with respect to a bunch which does not include all events having positive probability. This is the case in the following example:

Example 2.5.3. Let Ω be the set of natural numbers, $\mathscr{A}$ the set of all subsets, and $\mathscr{B}$ the set of all nonempty subsets of Ω, and put for $n = 1, 2, \ldots$

$$P_{n,k} = \begin{cases} \dfrac{1}{2^{k+1}} + \dfrac{1}{2n} & \text{for } 1 \leq k \leq n \\ \dfrac{1}{2^{k+1}} & \text{for } k > n \end{cases} \tag{2.5.9}$$

Clearly,

$$P_n(A) = \sum_{k \in A} P_{n,k} \tag{2.5.10}$$

is a probability distribution over the experiment $(\Omega, \mathscr{A})$ and thus $S_n = (\Omega, \mathscr{A}, P_n)$ is a probability space. Evidently, if $\mathscr{B}^*$ denotes the family of all finite nonempty subsets of Ω,

$$\lim_{n \to \infty} P_n(A \mid B) = \frac{P(A\,B)}{P(B)} = P(A|B) \quad \text{for } A \in \mathscr{A} \text{ and } B \in \mathscr{B}^* \tag{2.5.11}$$

where

$$P(A) = \sum_{k \in A} \frac{1}{2^k} \tag{2.5.12}$$

is a probability measure over $(\Omega, \mathscr{A})$. Thus we have

$$\lim_{n \to \infty} S_n = S^* = [\Omega, \mathscr{A}, \mathscr{B}^*, P(A \mid B)] \tag{2.5.13}$$

However, as

$$\lim_{n \to +\infty} P_{n,k} = \frac{1}{2^{k+1}} \quad \text{and} \quad \sum_{k=1}^{\infty} \frac{1}{2^{k+1}} = \frac{1}{2}$$

it follows from the *Vitali-Hahn-Saks theorem* that the limit $\lim_{n \to \infty} P_n(A)$ does not exist* for all sets $A \in \mathscr{A}$. As a matter of fact, we have

$$\lim_{n \to \infty} P_n(A) = \tfrac{1}{2}P(A) \tag{2.5.14}$$

if A is any finite set; however, we have

$$\lim_{n \to \infty} P_n(\Omega) = P(\Omega) = 1 \tag{2.5.15}$$

Thus, if $\lim_{n \to \infty} P_n(A)$ would exist for every $A \in \mathscr{A}$, it would *not* be a probability measure, which contradicts the Vitali-Hahn-Saks theorem. Thus

* See problem P.2.7, where sets A for which $\lim_{n \to \infty} P_n(A)$ does not exist are constructed.

the relation $\lim_{n\to\infty} S_n = S = (\Omega, \mathscr{A}, P)(\mathscr{B})$ does not hold. This means that while the conditional probability space S^*, which is the limit of the sequence S_n of probability spaces with respect to $\mathscr{B}^*$, is generated by the bounded measure P and thus can be extended to the (full) probability space $S = (\Omega, \mathscr{A}, P)$, nevertheless, it is the limit of the sequence S_n only as a conditional probability space.

The results of this section show that even if somebody wants to consider only ordinary probability spaces, he is sometimes forced to consider conditional probability spaces also, as these enter the scene as limits of sequences of ordinary probability spaces.

We now prove a theorem which sheds light on the examples discussed above. Before stating the theorem we introduce some definitions.

DEFINITION 2.5.3. *A sequence $S_n = (\Omega, \mathscr{A}, P_n)$ of probability spaces is called convergent to a probability space $S = (\Omega, \mathscr{A}, P)$ if $\lim_{n\to\infty} P_n(A) = P(A)$ for every $A \in \mathscr{A}$.*

DEFINITION 2.5.4. *A sequence of probability measures P_n on the experiment $(\Omega, \mathscr{A})$ is called uniformly tight with respect to the bunch $\mathscr{B} \subseteq \mathscr{A}$ if for every ε such that $0 < \varepsilon < 1$, there exists a set $B_\varepsilon \in \mathscr{B}$ such that $P_n(B_\varepsilon) \geqq 1 - \varepsilon$ for all n.*

Now we can state the following:

THEOREM 2.5.2. *Let $S_n = (\Omega, \mathscr{A}, P_n)$ be a sequence of probability spaces on the experiment $\mathscr{E} = (\Omega, \mathscr{A})$. Let $\mathscr{B} \subseteq \mathscr{A}$ be a bunch of events. Suppose that S_n (considered as the full conditional probability space $[\Omega, \mathscr{A}, \mathscr{B}, P_n]$ generated by P_n on $\mathscr{E}$) converges to a conditional probability space $S^* = [\Omega, \mathscr{A}, \mathscr{B}, \mu]$. Then μ is a bounded measure and the probability space S_n converges to the probability space $S = (\Omega, \mathscr{A}, P)$ where $P(A) = \mu(A)/\mu(\Omega)$ (in the sense of Definition 2.5.3) if and only if the measures P_n are uniformly tight with respect to the bunch $\mathscr{B}$.*

Proof. We first prove that the condition of uniform tightness is sufficient for the boundedness of μ and the convergence of S_n to S. Let us choose an arbitrary $\varepsilon > 0$ and a corresponding $B_\varepsilon \in \mathscr{B}$ such that $P_n(B_\varepsilon) \geqq 1 - \varepsilon$ for $n = 1, 2, \ldots.$ Choose an arbitrary $B \in \mathscr{B}$; then we have

$$\lim_{n\to\infty} \frac{P_n(B_\varepsilon)}{P_n(B + B_\varepsilon)} = \frac{\mu(B_\varepsilon)}{\mu(B + B_\varepsilon)}$$

As $P_n(B_\varepsilon)/P_n(B + B_\varepsilon) \geqq 1 - \varepsilon$, it follows that $\mu(B_\varepsilon)/\mu(B + B_\varepsilon) \geqq 1 - \varepsilon$, i.e., $\mu(B) \leqq \mu(B + B_\varepsilon) \leqq \mu(B_\varepsilon)/(1 - \varepsilon)$. As $B \in \mathscr{B}$ was arbitrary, it follows that

$\mu(\Omega) = \sup_{B \in \mathscr{B}} \mu(B) \leqq \mu(B_\varepsilon)/(1-\varepsilon)$, i.e., μ is bounded. We may suppose $\mu(\Omega) = 1$. Then it follows that $\mu(B_\varepsilon) \geqq 1 - \varepsilon$. Thus we get, for any $A \in \mathscr{A}$,

$$(1-\varepsilon)\mu(A) \leqq (1-\varepsilon) \lim_{n\to\infty} \frac{P_n(A)}{P_n(B_\varepsilon)} \leqq \varliminf_{n\to\infty} P_n(A)$$

$$\leqq \varlimsup_{n\to+\infty} P_n(A) \leqq \lim_{n\to+\infty} \frac{P_n(A)}{P_n(B_\varepsilon)} \leqq \frac{\mu(A)}{1-\varepsilon}$$

As $\varepsilon > 0$ can be taken arbitrarily small, it follows that $\lim_{n\to\infty} P_n(A) = \mu(A)$, i.e., S_n converges to S.

We turn now to the proof of the necessity of the condition. Let us suppose that $P_n(A) \to P(A)$ for every $A \in \mathscr{A}$ where $P(A)$ is a probability measure. We shall show that in this case the probability measures P_n are uniformly tight with respect to any bunch $\mathscr{B} \subseteq \mathscr{A}$. Let $\mathscr{B}$ be such a bunch and let $B_n \in \mathscr{B}$ be a sequence of events such that $\sum_{n=1}^{\infty} B_n = \Omega$ and put $C_N = \sum_1^N B_n$ and $D_N = \overline{C_N}$. Then by supposition $C_N \in \mathscr{B}$, $D_{N+1} \subseteq D_N$ for $N = 1, 2, \ldots$, and $\prod_{N=1}^{\infty} D_N = \varnothing$. Thus it follows by the *Vitali-Hahn-Saks theorem* that $\lim_{N\to+\infty} \sup_n P_n(D_N) = 0$, i.e., for every $\varepsilon > 0$ we can find an N such that $\sup_n P_n(D_N) < \varepsilon$, i.e., $P_n(C_N) > 1 - \varepsilon$ for every n. Thus, the probability measures P_n $(n = 1, 2, \ldots)$ are uniformly tight with respect to the bunch $\mathscr{B}$. This completes the proof of Theorem 2.5.2. ■

2.6 LINEAR INEQUALITIES AND IDENTITIES OF PROBABILITY THEORY

Let $S = (\Omega, \mathscr{A}, P)$ be a probability space. In this section we shall consider linear identities and inequalities of the form

$$\sum_{k=1}^{N} c_k P(F_k) = 0 \tag{2.6.1}$$

and

$$\sum_{k=1}^{N} c_k P(F_k) \geqq 0 \tag{2.6.2}$$

respectively, where $c_1, c_2, \ldots, c_N$ are real constants and

$$F_k = F_k(A_1, A_2, \ldots, A_n) \tag{2.6.3}$$

are certain polynomials of the events $A_1, A_2, \ldots, A_n$. We shall prove a general theorem which gives a simple and useful criterion to decide whether an identity of the type (2.6.1), or an inequality of the type (2.6.2) is valid for every probability space S and for every sequence of events $A_1, A_2, \ldots, A_n$ in S. (See Theorem 2.6.1 below.)

As an example of an identity of the type (2.6.1) we mention the following:

$$P(A+B) = P(A) + P(B) - P(AB) \tag{2.6.4}$$

where A and B are arbitrary events.

As an example of an inequality of the type (2.6.2) we mention the following:

$$P(A \circ C) \leqq P(A \circ B) + P(B \circ C) \tag{2.6.5}$$

where A, B and C are arbitrary events and $A \circ B$ denotes the symmetric difference of the events A and B (see (1.2.15)).

We have written these examples in the form in which they will be used later, but of course both can be written in the form (2.6.1) or (2.6.2), respectively, by carrying over all terms to one side of the equality, or inequality, respectively. Note that Ω itself is a polynomial of the events $A_1, \ldots, A_n$, because $\Omega = A_1 + \overline{A_1}$. If one of the polynomials F_k in (2.6.1) or (2.6.2) is equal to Ω, we call the identity (2.6.1) or the inequality (2.6.2) inhomogeneous, otherwise we call it homogeneous. Thus (2.6.4) is a homogeneous linear identity and (2.6.5) is a homogeneous linear inequality. As an example of an inhomogeneous identity we mention the following: If $B_{n,1}, B_{n,2}, \ldots, B_{n,2^n}$ are all basic functions (see Definition 1.4.1) of the events $A_1, A_2, \ldots, A_n$, we have

$$\sum_{k=1}^{2^n} P(B_{n,k}) = 1 \tag{2.6.6}$$

As an example of an inhomogeneous inequality we mention the following (which is a consequence of (2.6.4)): For any events A and B

$$P(A) + P(B) - P(AB) \leqq 1 \tag{2.6.7}$$

We shall prove the following:

THEOREM 2.6.1. *A linear identity of the form* (2.6.1) *or a linear inequality of the form* (2.6.2) *is valid for arbitrary events* $A_1, A_2, \ldots, A_n$ *in any probability space* $S = (\Omega, \mathscr{A}, P)$ *if it is valid in the* 2^n *special cases when each one of the events* $A_1, A_2, \ldots, A_n$ *is equal to* Ω *or to* $\varnothing$.

Remark: Let $S_0 = (\Omega, \mathscr{A}_0, P)$ denote the probability space such that $\mathscr{A}_0$ consists only of the two events $\varnothing$ and Ω. (We call such a probability space a *trivial probability space*.) Theorem 2.6.1 can be formulated also in the following form: *If an identity or an inequality of the form* (2.6.1) *or* (2.6.2), *respectively, is valid for the trivial probability space* S_0, *then it is valid for any probability space* S.

Proof. It is clearly sufficient to prove the assertion of the theorem for inequalities. As a matter of fact, the identity (2.6.2) is equivalent to the pair of inequalities

$$\sum_{k=1}^{N} c_k P(F_k) \geqq 0 \quad \text{and} \quad \sum_{k=1}^{N} (-c_k) P(F_k) \geqq 0 \tag{2.6.8}$$

Thus if (2.6.1) is valid when each A_k is either $\varnothing$ or Ω, then both inequalities in (2.6.8) are valid for these cases, and thus if the assertion of Theorem 2.6.1 is true for inequalities, the inequalities (2.6.8) are valid for arbitrary events $A_1, \ldots, A_n$ and (2.6.1) holds for arbitrary events, too. To prove the statement of Theorem 2.6.1 for inequalities, we shall use Theorem 1.4.1. By this theorem each of the polynomials F_k can be expressed in the form

$$F_k = \sum_{j=1}^{2^n} \varepsilon_{k,j} B_{n,j} \tag{2.6.9}$$

where $B_{n,1}, \ldots, B_{n,2^n}$ are the basic functions of the events $A_1, \ldots, A_n$ and each of the coefficients $\varepsilon_{k,j}$ is equal to 0 or to 1, and by definition $1 \cdot B_{n,j} = B_{n,j}$ and $0 \cdot B_{n,j} = \varnothing$. Thus we have, in view of (1.4.2),

$$P(F_k) = \sum_{j=1}^{2^n} \varepsilon_{k,j} P(B_{n,j}) \tag{2.6.10}$$

It follows that

$$\sum_{k=1}^{N} c_k P(F_k) = \sum_{j=1}^{2^n} d_j P(B_{n,j}) \tag{2.6.11}$$

where

$$d_j = \sum_{k=1}^{N} c_k \varepsilon_{k,j} \tag{2.6.12}$$

Now suppose that (2.6.2) holds if each of the A_k is equal either to Ω or to $\varnothing$. Clearly in this case exactly one and only one of the basic functions is equal to Ω and all others are equal to $\varnothing$. As a matter of fact, each basic function is a product of n factors such that the kth factor is either A_k or $\overline{A_k}$. Now if each A_k is either $\varnothing$ or Ω, take that basic function which contains the factor A_k for those values of k for which $A_k = \Omega$ and the factor $\overline{A_k}$ for those values of k for which $A_k = \varnothing$. This basic function will evidently be equal to Ω and all others to $\varnothing$. Conversely, if $B_{n,j}$ is any one of the 2^n basic functions, if we choose Ω as the value of A_k for those indices k for which the factor A_k occurs in $B_{n,j}$, and if we choose the value $\varnothing$ for A_k for those indices k for which $\overline{A_k}$ occurs in $B_{n,j}$, then evidently $B_{n,j} = \Omega$ and $B_{n,i} = \varnothing$ for $i \neq j$. Thus, as by supposition, formula (2.6.1) holds for all these substitutions, it follows from (2.6.11) that each of the coefficients d_j ($j = 1, 2, \ldots, 2^n$) is nonnegative. But this clearly implies by (2.6.11) that (2.6.2) holds for arbitrary events A_k. Thus our theorem is proved. ■

As applications of Theorem 2.6.1 we now prove some useful identities and inequalities.

Let $A_1, \ldots, A_n$ be arbitrary events, and let us define the sequence S_k $(k=1, 2, \ldots, n)$ as follows:

$$S_k = \sum_{1 \leqq i_1 < i_2 < \ldots < i_k \leqq n} P(A_{i_1} A_{i_2} \ldots A_{i_k}) \tag{2.6.13}$$

where the summation has to be extended over all k-tuples of different integers which can be selected from the integers $1, 2, \ldots, n$. Then the following identity holds, which is often called *Poincaré's identity*:

$$\sum_{k=1}^{n} (-1)^{k-1} S_k = P(A_1 + A_2 + \cdots + A_n) \tag{2.6.14}$$

To prove (2.6.14) by Theorem 2.6.1, we have only to verify that it holds for those 2^n special cases when each of the A_k is either Ω or $\varnothing$. Clearly, S_k is a symmetric function of the events $A_1, A_2, \ldots, A_n$; thus, if r $(r=0, 1, \ldots, n)$ among the events $A_1, \ldots, A_n$ are equal to Ω and the remaining $n-r$ to $\varnothing$, then the value of S_k depends only on r. Clearly, in this case, we have

$$S_k = \begin{cases} \binom{r}{k} & \text{if } 1 \leqq k \leqq r \\ 0 & \text{if } k > r \end{cases} \qquad (r=0, 1, \ldots, n) \tag{2.6.15}$$

Thus, to prove (2.6.14) we have to show that

$$\sum_{k=1}^{r} (-1)^{k-1} \binom{r}{k} = \begin{cases} 1 & \text{for } r=1, 2, \ldots, n \\ 0 & \text{for } r=0 \end{cases} \tag{2.6.16}$$

holds. But (2.6.16) is trivially satisfied for $r=0$, and for $r \geqq 1$ it holds because by the binomial theorem we have

$$1 - \sum_{k=1}^{r} (-1)^{k-1} \binom{r}{k} = (1-1)^r = 0$$

Thus by Theorem 2.6.1 the identity (2.6.14) holds for arbitrary events.

We now prove a series of inequalities which are closely connected with Poincaré's identity: we shall call them *Poincaré's inequalities*. With the same notations as above, these inequalities are as follows:

$$P(A_1 + A_2 + \cdots + A_n) \leqq \sum_{k=1}^{2m-1} (-1)^{k-1} S_k \quad \text{for } m=1, 2, \ldots \tag{2.6.17}$$

and

$$P(A_1 + A_2 + \cdots + A_n) \geqq \sum_{k=1}^{2m} (-1)^{k-1} S_k \quad \text{for } m=1, 2, \ldots \tag{2.6.18}$$

Just as we have reduced the proof of (2.6.14) to the proof of the identity (2.6.16) by Theorem 2.6.1, in order to prove the inequalities (2.6.17) and (2.6.18), we have to verify only the inequalities

$$1 \leqq \sum_{k=1}^{2m-1} (-1)^{k-1} \binom{r}{k} \tag{2.6.19}$$

and

$$1 \geqq \sum_{k=1}^{2m} (-1)^{k-1} \binom{r}{k} \tag{2.6.20}$$

But these are clearly valid, as it is easy to show (e.g., by Pascal's triangle) that for $l = 1, 2, \ldots,$

$$1 - \sum_{k=1}^{l} (-1)^{k-1} \binom{r}{k} = \sum_{k=0}^{l} (-1)^{k} \binom{r}{k} = (-1)^{l} \binom{r-1}{l} \tag{2.6.21}$$

Note that for $n = 2$, (2.6.14) reduces to (2.6.4), while for $m = 1$, (2.6.17) reduces to the obvious inequality

$$P(A_1 + A_2 + \cdots + A_n) \leqq \sum_{k=1}^{n} P(A_k) \tag{2.6.22}$$

Poincaré's identity has been generalized by Jordan as follows: Let $W_{n,r}$ denote the probability of the event that among the events $A_1, \ldots, A_n$ exactly r events occur; then we have

$$W_{n,r} = \sum_{k=0}^{n-r} (-1)^{k} \binom{r+k}{k} S_{r+k} \tag{2.6.23}$$

where S_k $(1 \leqq k \leqq n)$ is defined by (2.6.13) and $S_0 = 1$.

We prove (2.6.23) again by Theorem 2.6.1. This theorem clearly reduces the proof of (2.6.23) to the verification of the combinatorial identity

$$\sum_{k=0}^{n-r} (-1)^{k} \binom{r+k}{k} \binom{l}{r+k} = \begin{cases} 1 & \text{if } l = r \\ 0 & \text{if } l \neq r \end{cases} \tag{2.6.24}$$

which can be verified as follows: For $l < r$ all terms on the left-hand side of (2.6.24) are 0; for $l = r$ the term corresponding to $k = 0$ is equal to 1 and all others to 0; finally, for $r < l \leqq n$ we can transform the left-hand side of (2.6.24) as follows:

$$\sum_{k=0}^{n-r} (-1)^{k} \binom{r+k}{k} \binom{l}{r+k} = \binom{l}{r} \sum_{k=0}^{l-r} (-1)^{k} \binom{l-r}{k} = \binom{l}{r} (1-1)^{l-r} = 0$$

which proves (2.6.24) and Jordan's identity (2.6.23).

We want to add one more remark on inequalities of the type (2.6.2). As we have seen, the left-hand side of (2.6.2) is a linear function of the quantities $x_{n,k} = P(B_{n,k})$ $(k = 1, 2, \ldots, 2^n)$, where the $B_{n,k}$ denote the basic functions of the events $A_1, A_2, \ldots, A_n$. The quantities $x_{n,k}$ are subject only to the conditions of admissibility $x_{n,k} \geqq 0$ $(1 \leqq k \leqq 2^n)$, and $\sum_{k=1}^{2^n} x_{n,k} = 1$. Thus the problem of providing an inequality of the type (2.6.2) is equivalent to finding the minimum of a linear function of the variables $x_{n,k}$ subject to the admissibility conditions mentioned. Thus it is a problem of *linear programming*. Looking at these inequalities from the point of view of linear programming the statement of Theorem 2.6.1 expresses the well known fact that a linear function over a convex polyhedron takes on its minimum at certain vertices of the polyhedron. (See also Halperin [63].)

2.7 RANDOM VARIABLES ON PROBABILITY SPACES

The notion of a random variable has been introduced in Section 1.7. We have seen that a random variable ξ on an experiment $\mathscr{E} = (\Omega, \mathscr{A})$ (i.e., a real-valued function $\xi = \xi(\omega)$ defined for $\omega \in \Omega$ and measurable with respect to the σ-algebra $\mathscr{A}$) maps Ω into R and thus it maps the experiment $\mathscr{E}$ into another experiment $\mathscr{E}' = (R, \mathscr{F})$ where R is the real line and $\mathscr{F}$ is the σ-algebra of Borel subsets of R. Now if on the experiment $\mathscr{E}$ a probability measure P is given, then ξ generates a measure Q on $\mathscr{F}$, defined by the formula

$$Q(\mathfrak{B}) = P(\xi^{-1}(\mathfrak{B})) \tag{2.7.1}$$

for $\mathfrak{B} \in \mathscr{F}$. Thus, ξ maps the probability space $S = (\Omega, \mathscr{A}, P)$ on the probability space $S' = (R, \mathscr{F}, Q)$.

Let I_x denote the open interval $(-\infty, x)$ and put $F(x) = Q(I_x)$. Clearly the measure Q is uniquely defined by its values on the sets I_x, i.e., by the function $F(x)$, as its value for any $\mathfrak{B} \in \mathscr{F}$ can be expressed by the formula

$$Q(\mathfrak{B}) = \int_{\mathfrak{B}} dF(x) \tag{2.7.2}$$

(See (2.3.14).) If $F(x)$ is absolutely continuous and $F'(x) = f(x)$, then (2.7.2) can be expressed by a Lebesgue integral (instead of a Lebesgue-Stieltjes integral) in the form

$$Q(\mathfrak{B}) = \int_{\mathfrak{B}} f(x)\, dx \tag{2.7.3}$$

(See (2.3.17).) The function $F(x)$ is a probability distribution function in the sense of Definition 2.3.4, and if it is absolutely continuous, $f(x) = F'(x)$ is a (probability) density function in the sense of Definition 2.3.5. Clearly,

$$F(x) = P(\xi < x) \tag{2.7.4}$$

i.e., $F(x)$ is equal to the probability of the event that the random variable ξ takes on a value less than x.

Below we shall use the terminology given by the following:

DEFINITION 2.7.1. *The measure Q defined by* (2.7.1) *is called the probability distribution and the function $F(x)$ defined by* (2.7.4) *is the distribution function of the random variable ξ. If $F(x)$ is absolutely continuous, $f(x) = F'(x)$ is called the density function of ξ.*

It follows that to every probability distribution function in the sense of Definition 2.3.4 there exists a probability space S and a random variable ξ on S which has $F(x)$ as its distribution function.

An n-dimensional random vector was defined in Section 1.7 as an n-tuple $\xi = (\xi_1, \xi_2, \ldots, \xi_n)$ of real-valued random variables defined on the same experiment $\mathscr{E} = (\Omega, \mathscr{A})$. We have seen that an n-dimensional random vector maps the experiment $\mathscr{E}$ onto an experiment $\mathscr{E}' = (R^n, \mathscr{F}_n)$ where R^n is the n-dimensional Euclidean space and $\mathscr{F}_n$ is the σ-algebra of Borel subsets of R^n. If a probability measure P is given on $\mathscr{E}$, we define the probability measure Q on $\mathscr{E}'$ again by the formula (2.7.1) for every $\mathfrak{B} \in \mathscr{F}_n$. Thus an n-dimensional random vector maps the probability space $(\Omega, \mathscr{A}, P)$ on the probability space $(R^n, \mathscr{F}_n, Q)$. Let $I(x_1, \ldots, x_n)$ denote the set of those points $(y_1, \ldots, y_n)$ of R^n for which $y_k < x_k$ for $k = 1, 2, \ldots, n$, and put

$$F(x_1, x_2, \ldots, x_n) = Q(I(x_1, \ldots, x_n)) = P(\xi_1 < x_1, \xi_2 < x_2, \ldots, \xi_n < x_n) \tag{2.7.5}$$

The measure Q is uniquely defined by the function $F(x_1, x_2, \ldots, x_n)$, which expresses the probability of the event that the inequalities $\xi_1 < x_1, \xi_2 < x_2, \ldots, \xi_n < x_n$ are simultaneously fulfilled.

It follows from the definition that the function $F(x_1, x_2, \ldots, x_n)$ has the following properties:

(1) *It is monotonically nondecreasing and continuous from the left in each of its variables.* Let $\Delta_k(h)$ stand for the operation on functions $G = G(x_1, \ldots, x_n)$ of n variables, defined by the formula

$$\Delta_k(h)G = G(x_1, \ldots, x_{k-1}, x_k + h, \ldots, x_n) - G(x_1, x_2, \ldots, x_n) \tag{2.7.6}$$

(2) *If $h_k > 0$ for $k = 1, 2, \ldots, n$, one has*

$$\Delta_1(h_1)\, \Delta_2(h_2) \ldots \Delta_n(h_n)F \geqq 0 \tag{2.7.7}$$

(3) *If all variables $x_k (k = 1, 2, \ldots, n)$ tend simultaneously to $+\infty$, then $F(x_1, \ldots, x_n)$ tends to 1; if at least one of the variables x_k tends to $-\infty$, then $F(x_1, \ldots, x_n)$ tends to 0.*

The statements (1) and (3) are evident; (2) follows from Poincaré's identity

(2.6.14), by which the left-hand side of (2.7.6) is equal to the probability of the joint occurrence of the events

$$x_k \leqq \xi_k < x_k + h \quad (k = 1, 2, \ldots, n)$$

If the measure Q is absolutely continuous (with respect to the n-dimensional Lebesgue measure), then the function $F(x_1, x_2, \ldots, x_n)$ can be written in the form

$$F(x_1, x_2, \ldots, x_n) = \int_{-\infty}^{x_1} \int_{-\infty}^{x_2} \cdots \int_{-\infty}^{x_n} f(y_1, y_2, \ldots, y_n)\, dy_1\, dy_2 \ldots dy_n \tag{2.7.8}$$

where

$$f(x_1, x_2, \ldots, x_n) = \frac{\partial^n F}{\partial x_1 \partial x_2, \ldots, \partial x_n} \tag{2.7.9}$$

This derivative exists almost everywhere in R^n. It follows from (2.7.6) that the (measurable) function $f(x_1, \ldots, x_n)$ is nonnegative, and from (3) that

$$\int_{-\infty}^{+\infty} \int_{-\infty}^{+\infty} \cdots \int_{-\infty}^{+\infty} f(x_1, x_2, \ldots, x_n)\, dx_1\, dx_2 \ldots dx_n = 1 \tag{2.7.10}$$

We shall use the terminology given by the following:

DEFINITION 2.7.2. *If* $\xi = (\xi_1, \xi_2, \ldots, \xi_n)$ *is an n-dimensional random vector, we call the measure* Q *defined* (*for* $\mathfrak{B} \in \mathscr{F}_n$) *by* (2.7.1) *the joint probability distribution, the function* F *defined by* (2.7.5) *the joint distribution function, and* (*if* F *is absolutely continuous*) *the function* f *defined by* (2.7.9) *the joint density function of the random variables* $\xi_1, \ldots, \xi_n$.

It follows from (2.7.7) that if the joint probability distribution of the random variables $\xi_1, \xi_2, \ldots, \xi_n$ has the density function $f(x_1, x_2, \ldots, x_n)$, then for every $\mathfrak{B} \in \mathscr{F}_n$ one has

$$P((\xi_1, \xi_2, \ldots, \xi_n) \in \mathfrak{B}) = \int_{\mathfrak{B}} f(x_1, x_2, \ldots, x_n)\, dx_1\, dx_2 \ldots dx_n \tag{2.7.11}$$

From (2.7.11) it can be deduced that if the joint probability distribution of the random variables $\xi_1, \xi_2, \ldots, \xi_n$ is absolutely continuous ($n \geqq 2$), then the joint distribution of any m among them ($1 \leqq m < n$) is also absolutely continuous. The density function of the joint distribution of, e.g., $\xi_1, \xi_2, \ldots, \xi_m$ ($1 \leqq m < n$) can be obtained by integrating the function $f(x_1, x_2, \ldots, x_n)$ with respect to each of the variables $x_{m+1}, \ldots, x_n$ from $-\infty$ to $+\infty$. We shall use the terminology expressed by the following:

DEFINITION 2.7.3. *The probability distributions (distribution functions, density functions) of the single random variables ξ_k $(k = 1, 2, \ldots, n)$ are called the marginal distributions (distribution functions, density functions, respectively) of the joint distribution of the variables $\xi_1, \xi_2, \ldots, \xi_n$.*

Let $\xi_1, \xi_2, \ldots, \xi_n$ be n random variables, the joint distribution of which is absolutely continuous on a probability space S. Let $h(x_1, x_2, \ldots, x_n)$ be a Borel function of the n real variables $x_1, x_2, \ldots, x_n$. Then, clearly, $\eta = h(\xi_1, \xi_2, \ldots, \xi_n)$ is a random variable on S, and its distribution function is given by

$$P(\eta < y) = \iint \cdots \int_{h(x_1, x_2, \ldots, x_n) < y} f(x_1, x_2, \ldots, x_n)\, dx_1\, dx_2 \ldots dx_n \tag{2.7.12}$$

where f is the density of the joint distribution of the variables $\xi_1, \ldots, \xi_n$. As a matter of fact, (2.7.12) follows directly from (2.7.11).

Let us consider the special case when h is a linear function:

$$h(x_1, \ldots, x_n) = \sum_{k=1}^{n} a_k x_k \tag{2.7.13}$$

where $a_1, \ldots, a_n$ are real numbers, such that

$$\sum_{k=1}^{n} a_k^2 = 1 \tag{2.7.14}$$

Then we may interpret $\sum_{k=1}^{n} a_k \xi_k$ as the scalar product (ξ, a) of the vector ξ and the unit vector $a = (a_1, \ldots, a_n)$, that is, as the *projection* of the vector ξ on the line through the origin of the coordinate system in R^n, the direction of which is that of the vector a. Accordingly, we call the probability distributions of the random variables $\sum_{k=1}^{n} a_k \xi_k$ the (one-dimensional) projections of the joint distribution of the random variables $\xi_1, \ldots, \xi_n$.

The marginal distributions of a probability distribution in R^n are clearly its projections on the coordinate axes. Projections of probability distributions on hyperplanes of dimensions m $(1 < m < n)$ can be defined in a similar way for $n \geqq 3$.

A probability distribution in R^n $(n \geqq 2)$ is uniquely determined by the totality of its one-dimensional projections (see Cramér and Wold [52]). Under some restrictions* a probability distribution is already determined by any infinite sequence of its projections (see Rényi [53]). However, a finite number of projections are, in general, not sufficient to characterize a distribution in R^n; especially, a probability distribution in R^n $(n \geqq 2)$ is generally *not* determined by its marginal distributions (see [70]).

* For instance, if the density functions of the joint distribution vanishes outside some sphere.

2.8 RANDOM VARIABLES ON CONDITIONAL PROBABILITY SPACES

Let $S = [\Omega, \mathscr{A}, \mathscr{B}, P(A \mid B)]$ be a full conditional probability space generated by the σ-finite measure μ on the σ-algebra $\mathscr{A}$. By definition the family $\mathscr{B}$ of admissible conditions is the family of all sets $B \in \mathscr{A}$ for which $\mu(B)$ is finite and positive. As we have seen for each fixed $B \in \mathscr{B}$, putting

$$P_B(A) = P(A \mid B) = \frac{\mu(AB)}{\mu(B)}$$

$S_B = (\Omega, \mathscr{A}, P_B)$ is a probability space.

Let us now consider a random variable ξ on the experiment $\mathscr{E} = (\Omega, \mathscr{A})$ and put for each Borel set C of the real line

$$Q_B(C) = P_B(\xi^{-1}(C)) \tag{2.8.1}$$

and

$$F_B(x) = Q_B(I_x) = P_B(\xi < x) \tag{2.8.2}$$

further, if $F_B(x)$ is absolutely continuous, put

$$f_B(x) = F'_B(x) \tag{2.8.3}$$

We shall use the following terminology:

DEFINITION 2.8.1. *The measure Q_B defined by* (2.8.1) *is called the conditional distribution, $F_B(x)$ defined by* (2.8.2) *is called the conditional distribution function, and* (*if defined*) *the function $f_B(x)$ given by* (2.8.3) *is called the conditional density function of the random variable ξ with respect to the condition B.*

Conditional probability distributions (distribution functions and density functions) can be dealt with in exactly the same way, independently of whether the measure μ generating the conditional probability space S in question is a probability measure or not. As a matter of fact, as long as the condition B is fixed this is irrelevant, as we are working in the (ordinary) probability space S_B.

Concerning the case when μ is not a probability measure, i.e., $\mu(\Omega) = +\infty$, the question arises: Does a random variable ξ map the conditional probability space S on a conditional probability space on the real line? It is easy to see that this is not necessarily true. If ξ is an arbitrary random variable and C is a Borel set of the real line, then by supposition $\xi^{-1}(C)$ belongs to $\mathscr{A}$. However, we obtain a conditional probability space on the real line only if the family of Borel sets C, such that $\xi^{-1}(C)$ belongs to $\mathscr{B}$, is a bunch. This is, however, not necessarily true, it may even happen that $\xi^{-1}(C)$ does not belong to $\mathscr{B}$ for any Borel set C. This is shown by the following:

Example 2.8.1. Let R denote the real line, $\mathscr{F}_1$ the family of Borel subsets of R and λ the Lebesgue measure, and consider the full conditional probability space generated by λ on $(R, \mathscr{F}_1)$. Let the random variable ξ be defined by $\xi(x) = \sin x$. Then clearly for any Borel set C of the real line the Lebesgue measure of the set $\xi^{-1}(C)$ is either equal to 0 or to $+\infty$; thus, for no Borel set C is the set $\xi^{-1}(C)$ an admissible condition.

We introduce the following:

DEFINITION 2.8.2. *Let $S = [\Omega, \mathscr{A}, \mathscr{B}, P(A \mid B)]$ be a full conditional probability space generated by the σ-finite unbounded measure μ. A random variable ξ on S is called regular if the Borel sets C, for which $\xi^{-1}(C) \in \mathscr{B}$, form a bunch in R.*

The following example gives some insight into the notion of a regular random variable:

Example 2.8.2. Let us consider the full conditional probability space generated by the counting measure $\mathscr{N}$ on the set of all positive integers, i.e., we suppose $\Omega = \{1, 2, \ldots, n, \ldots\}$, $\mathscr{A} = \mathscr{P}(\Omega)$ and $P(A \mid B) = \mathscr{N}(AB)/\mathscr{N}(B)$ if B is a nonempty finite subset of Ω. In this case any real-valued function $\xi = \xi(n)$ is a random variable. A random variable ξ is regular if and only if for any real number x the set of those positive integers n, for which $\xi(n) = x$, is finite.

If a random variable ξ on a conditional probability space generated by an unbounded measure is regular, it defines a conditional probability space on the real line. According to Theorem 2.2.1 there exists a σ-finite measure ν which generates this conditional probability space on the real line. Clearly, $\nu(R) = +\infty$. Let us now suppose that the measure ν is finite for every finite interval $[a, b) = \{x: a \leqq x < b\}$ (which, of course, is not always the case). In this case, put

$$F(x) = \begin{cases} \nu([0, x)) & \text{for} \quad x > 0 \\ -\nu([x, 0)) & \text{for} \quad x \leqq 0 \end{cases} \tag{2.8.4}$$

Then we have for every interval $[c, d)$ such that $\nu([c, d)) > 0$ and for every subinterval $[a, b)$ of the interval $[c, d)$,

$$P(a \leqq \xi < b \mid c \leqq \xi < d) = \frac{F(b) - F(a)}{F(d) - F(c)} \tag{2.8.5}$$

In other words, in this case the conditional probability space on the real line defined by the random variable ξ is of the type described in Example 2.2.3. We call any nondecreasing function $F(x)$ for which (2.8.5) holds a *distribution function* of the random variable ξ. Note that if $F(x)$ is a distribution function of ξ, A is an arbitrary positive number and B is an arbitrary real number, then $AF(x) + B$ is also a distribution function of ξ; thus the distribution function of ξ is determined only up to a linear transformation. If the measure ν is absolutely continuous, then $F'(x) = f(x)$ exists almost everywhere, and we may write (2.8.5) in the form

$$P(a \leqq \xi < b \mid c \leqq \xi < d) = \frac{\int_a^b f(x)\,dx}{\int_c^d f(x)\,dx} \tag{2.8.6}$$

We call any nonnegative measurable function for which (2.8.6) holds a *density function* of ξ. Note that the density function of ξ is determined only up to a positive constant factor. In case the density function is defined, the conditional probability space defined by ξ on R is of the type described by Example 2.2.4.

2.9 EXPECTATIONS AND OTHER CHARACTERISTICS OF PROBABILITY DISTRIBUTIONS

Let ξ be a discrete random variable on a probability space $S = (\Omega, \mathscr{A}, P)$ taking on only a finite number of distinct values $x_1, x_2, \ldots, x_m$. In this case the probability distribution of ξ is uniquely determined by the sequence of values $p_k = P(\xi = x_k)$ $(k = 1, 2, \ldots, m)$. Let us now return for a moment to the intuitive meaning of probability, as described in Section 2.1. Suppose we repeat the experiment $\mathscr{E}$ under the same circumstances a large number of times, say, N times. Let the observed values of the random variable ξ in these experiments be $\xi_1, \xi_2, \ldots, \xi_N$. Clearly, each of the observed values ξ_i of ξ is equal to one of the numbers x_k. According to the intuitive meaning of probability we expect that the number x_k will occur in the sequence $\xi_1, \ldots, \xi_N$ approximately Np_k times. Thus we may expect that the average (arithmetic mean value)

$$\bar{\xi} = (\xi_1 + \xi_2 + \cdots + \xi_N)/N \tag{2.9.1}$$

of the observed values of ξ will be close to the value

$$E(\xi) = p_1 x_1 + p_2 x_2 + \cdots + p_m x_m \tag{2.9.2}$$

i.e., to the value which would be obtained if the actual value of the frequency of each value x_k were exactly equal to the corresponding theoretical value Np_k.

Note that $E(\xi)$ defined by (2.9.2) is equal to the Lebesgue integral of the function ξ over the space Ω with respect to the measure P, i.e.,

$$E(\xi) = \int_\Omega \xi\,dP \tag{2.9.3}$$

Thus we expect that if we make a large number of observations on a random variable, the mean value of the observed values will be near to $E(\xi)$.

This rather heuristic reasoning, carried out for a special case only, is the intuitive background of the following:

DEFINITION 2.9.1. *The expectation $E(\xi)$ of a random variable ξ on a probability space $S=(\Omega, \mathscr{A}, P)$ is defined as the integral of ξ over Ω with respect to the measure P, i.e., by the formula* (2.9.3), *provided that this integral exists as an (abstract) Lebesgue integral.*

The properties of the expectation follow immediately from the well known properties of the integral. We note especially the following properties:

(1) *The expectation is a linear operation, i.e., if ξ and η are two random variables on the same probability space such that their expectations $E(\xi)$ and $E(\eta)$ exist, further, a and b are real constants, then*

$$E(a\xi + b\eta) = aE(\xi) + bE(\eta) \tag{2.9.4}$$

(2) *If α is the indicator of an event A, one has*

$$E(\alpha) = P(A)$$

(3) *If ξ is a nonnegative random variable such that $E(\xi)$ exists, then $E(\xi) \geqq 0$.*

(4) *If ξ is a bounded random variable, then $E(\xi)$ exists.*

Note that if ξ is a discrete random variable taking on infinitely many distinct values x_k with the corresponding probabilities $P_k = P(\xi = x_k)$ $(k = 1, 2, \ldots)$, then

$$E(\xi) = \sum_{k=1}^{+\infty} P_k x_k \tag{2.9.5}$$

provided that the series on the right-hand side of (2.9.5) is *absolutely* convergent. In case the series (2.9.5) is only conditionally convergent, $E(\xi)$ is not defined. To motivate this it is enough to call attention to the fact that the labeling of the possible values of a random variable is completely arbitrary. Thus, the definition of $E(\xi)$ has to be independent of the labeling of the values x_k. Now, if the series on the right of (2.9.5) is only conditionally convergent, the sum of the series can be changed and even made divergent by rearranging its terms. Thus, in this case it would make no sense to define $E(\xi)$ by formula (2.9.5). (Of course if the series on the right of (2.9.5) is only conditionally convergent, this means that the Lebesgue integral (2.9.3) does not exist.)

Clearly, if the random variable ξ is constant with probability 1, i.e., $P(\xi = c) = 1$, then $E(\xi) = c$. It follows, using the linearity of E, that

$$E(\xi - E(\xi)) = 0 \tag{2.9.6}$$

Equation (2.9.6) shows that the expectation of a random variable is the "center" of its distribution in the sense that $E(\xi)$ is that value from which the expected deviations of the actual values will, on the average, be equal to 0. This is, of course, due to the fact that if ξ is not constant, then both positive and negative deviations are possible. However, the expectation does not give any information about the average size of these deviations. To measure to what extent the values of ξ are scattered around $E(\xi)$, the notion of the *variance* of a random variable is used.

DEFINITION 2.9.2. *Let ξ be a random variable, the expectation $E(\xi)$ of which exists. The variance $D^2(\xi)$ of ξ is defined as the expectation of the square of the deviation of ξ from its expected value, that is, by the formula*

$$D^2(\xi) = E((\xi - E(\xi))^2) \tag{2.9.7}$$

provided that this quantity exists: if not, we say that ξ has infinitely large variance. The positive square root of the variance of ξ is called the standard deviation of ξ and is denoted by $D(\xi)$.

It follows from the mentioned properties of expectations that if c is any real constant,

$$E((\xi - c)^2) = D^2(\xi) + (E(\xi) - c)^2 \tag{2.9.8}$$

Thus, the expected quadratic deviation of ξ from a number c is minimal if and only if $c = E(\xi)$. It follows from (2.9.8) for $c = 0$ that

$$E(\xi^2) = D^2(\xi) + E^2(\xi) \tag{2.9.9}$$

Let us consider the set $L_2(S)$ of those random variables ξ on a probability space S which have finite variance. The set $L_2(S)$ is a Hilbert space if the scalar product (ξ, η) is defined by

$$(\xi, \eta) = E(\xi\eta) \tag{2.9.10}$$

(which exists because of the Cauchy-Schwarz inequality). If $E(\xi) = 0$, the norm $\|\xi\|$ of ξ is equal to the standard deviation,

$$\|\xi\| = (\xi, \xi)^{1/2} = E(\xi^2)^{1/2} = D(\xi) \tag{2.9.11}$$

A sequence $\xi_n \in L_2(S)$ is said to converge to $\xi \in L_2(S)$ strongly (or in L_2-norm) if $\lim_{n\to\infty} \|\xi_n - \xi\| = 0$. This interpretation opens the way to the use of Hilbert-space methods in probability theory.

The expectation and the variance of a random variable ξ depend only on the distribution of ξ. Thus, if the distribution function $F(x)$ of ξ is given, $E(\xi)$ and $D^2(\xi)$ can be expressed by it. As a matter of fact, it follows from the definition of the Lebesgue integral that for any real Borel-measurable function $g(x)$ of the real variable x

$$E(g(\xi)) = \int_{-\infty}^{+\infty} g(x)\, dF(x) \tag{2.9.12}$$

provided that $E(g(\xi))$ exists. Thus, especially

$$E(\xi) = \int_{-\infty}^{+\infty} x \, dF(x) \tag{2.9.13}$$

and, in view of (2.9.9)

$$D^2(\xi) = \int_{-\infty}^{+\infty} x^2 \, dF(x) - \left(\int_{-\infty}^{+\infty} x \, dF(x)\right)^2 \tag{2.9.14}$$

If the distribution of ξ is absolutely continuous with density function $f(x)$, one can express $E(g(\xi))$ in the form

$$E(g(\xi)) = \int_{-\infty}^{+\infty} g(x) f(x) \, dx \tag{2.9.15}$$

Thus one has

$$E(\xi) = \int_{-\infty}^{+\infty} x f(x) \, dx \tag{2.9.16}$$

and

$$D^2(\xi) = \int_{-\infty}^{+\infty} x^2 f(x) \, dx - \left(\int_{-\infty}^{+\infty} x f(x) \, dx\right)^2 \tag{2.9.17}$$

If a and b are real constants, then clearly

$$D^2(a\xi + b) = a^2 D^2(\xi) \tag{2.9.18}$$

If ξ and η are random variables with finite expectations and variances, then by the Cauchy-Schwarz inequality

$$E^2(\xi\eta) \leqq E(\xi^2)E(\eta^2) \tag{2.9.19}$$

further, the variance of $\xi + \eta$ exists also, and

$$D^2(\xi + \eta) = D^2(\xi) + D^2(\eta) + 2(E(\xi\eta) - E(\xi)E(\eta)) \tag{2.9.20}$$

The expectation and the variance of a distribution are called *parameters* of the distribution; they characterize the distribution to some extent, though of course they do not determine it in general. If more information on a distribution is needed, one has to consider other parameters, e.g., the expectations of higher powers of the random variable, called its *moments*. The nth moment of the random variable ξ is defined as the expectation of ξ^n and denoted by $M_n(\xi)$. Thus,

$$M_n(\xi) = E(\xi^n) = \int_{-\infty}^{+\infty} x^n \, dF(x) \quad (n = 1, 2, \ldots) \tag{2.9.21}$$

If we put $M_0(\xi) = 1$ for every ξ, (2.9.21) holds for $n = 0$ also. A related notion is that of *central moments*; the nth central moment $m_n(\xi)$ of ξ is defined by

$$m_n(\xi) = E((\xi - E(\xi))^n) \tag{2.9.22}$$

Clearly $M_1(\xi)=E(\xi)$, $m_1(\xi)=0$ and $m_2(\xi)=D^2(\xi)$. (Of course, these quantities are defined only if the corresponding integrals exist.*) If the random variable ξ is, e.g., bounded, then all its moments exist. In this special case, and even under much more general conditions (see [57] and [58]), the distribution of ξ is uniquely determined by the sequence of its moments. As a matter of fact, if all the moments $M_n(\xi)$ of ξ are given ($n=1, 2, \ldots$), then the (two-sided) *Laplace-Stieltjes transform* $\psi_\xi(s)$ of the distribution function $F(x)$ (respective of the (two-sided) Laplace transform of the density function $f(x)$) of ξ can be obtained by

$$\psi_\xi(s)=\int_{-\infty}^{+\infty} e^{xs}\,dF(x)=\int_{-\infty}^{+\infty} e^{xs}f(x)\,dx=1+\sum_{n=1}^{\infty}\frac{M_n(\xi)s^n}{n!} \tag{2.9.23}$$

provided that the series on the right of (2.9.23) is convergent in some circle $|s|\leqq R$. If ξ is bounded, $|\xi|\leqq K$, then $|M_n(\xi)|\leqq K^n$, and thus the series on the right of (2.9.23) converges for every complex value of s and its sum is an entire function of s. By using one of the inversion formulas for the Laplace transformation (see [59] and [60]) one can evaluate $F(x)$ (respectively $f(x)$) for every (respectively almost every) value of x.

Definition 2.9.3. *The function* $\psi_\xi(s)$ *defined by* (2.9.23) *is called the moment-generating function of the random variable* ξ (*respectively of its probability distribution*).

In order that for a given sequence of numbers M_n ($n=0, 1, 2, \ldots$) there should exist a distribution function $F(x)$ such that $M_n=\int_{-\infty}^{+\infty} x^n\,dF(x)$ should hold ($n=0, 1, \ldots$), it is necessary and sufficient that $M_0=1$ and that the quadratic forms $\sum_{j=0}^{n}\sum_{k=0}^{n} M_{j+k}x_jx_k$ should all be positive definite ($n=1, 2, \ldots$). (Theorem of Hamburger.) According to Carleman's theorem for such a sequence M_n, in order for the moment problem to be determinate, i.e., $F(x)$ to be uniquely determined, it is sufficient that the series $\sum_{n=0}^{+\infty} 1/(\sqrt[2n]{M_{2n}})$ be divergent. (For these and further results on the moment problem see the books by Shohat and Tamarkin [57] and Akhiezer [58].)

In general, however, the moments and the Laplace transform do not necessarily exist. Therefore, in probability theory, instead of the Laplace transform one usually uses the Fourier-Stieltjes transform of $F(x)$ (respectively the Fourier transform of $f(x)$) defined by

$$\varphi_\xi(t)=\int_{-\infty}^{+\infty} e^{ixt}\,dF(x)\quad(-\infty<t<+\infty) \tag{2.9.24}$$

* It can be shown (see Appendix A) that if $M_n(\xi)$ exists, then $M_k(\xi)$ also exists for $1\leq k<n$. Thus, e.g., if $M_2(\xi)=E(\xi^2)$ exists, so does $M_1(\xi)=E(\xi)$ and $D^2(\xi)=E(\xi^2)-E^2(\xi)$.

This is called the *characteristic function* of the random variable ξ, respectively of its distribution function $F(x)$. Clearly, if the distribution is absolutely continuous with density function $f(x)$, we have

$$\varphi_\xi(t) = \int_{-\infty}^{+\infty} e^{itx} f(x)\, dx \tag{2.9.25}$$

In view of (2.9.12), the characteristic function $\varphi_\xi(t)$ of ξ can also be expressed by

$$\varphi_\xi(t) = E(e^{i\xi t}) \tag{2.9.26}$$

DEFINITION 2.9.4. *The function $\varphi_\xi(t)$, defined by* (2.9.26) *of the real variable t $(-\infty < t < +\infty)$ is called the characteristic function of the random variable ξ, respectively of its probability distribution.*

It is easy to see that $\varphi_\xi(0) = 1$, $|\varphi_\xi(t)| \leq 1$ for all real values of t, further, that $\varphi_\xi(t)$ is a uniformly continuous complex-valued function of t, and that $\varphi_\xi(-t)$ and $\varphi_\xi(t)$ are conjugate complex numbers.

It can be shown (see [54] or [32]) that the characteristic function of a probability distribution determines the distribution uniquely.

The moments of a random variable (if they exist) can be obtained by means of its characteristic function by evaluating the derivatives of this function at the point $t = 0$. As a matter of fact, one obtains

$$M_n(\xi) = i^{-n} \varphi_\xi^{(n)}(0) \quad (n = 1, 2, \ldots) \tag{2.9.27}$$

Clearly, if the Laplace transform $\psi_\xi(s)$ exists, we have

$$\varphi_\xi(t) = \psi_\xi(it) \tag{2.9.28}$$

It can be shown that the correspondence between distribution functions and characteristic functions is continuous in the following sense: *A sequence $F_n(x)$ of distribution functions $(n = 1, 2, \ldots)$ converges for $n \to \infty$ to a distribution function $F(x)$, in every point of continuity of the latter if and only if the characteristic function $\varphi_n(t) = \int_{-\infty}^{+\infty} e^{ixt}\, dF_n(x)$ converges for $n \to \infty$ to the characteristic function $\varphi(t) = \int_{-\infty}^{+\infty} e^{ixt}\, dF(x)$ for every real value of t.* (See [54] or [32].)

Let us consider some simple examples.

Example 2.9.1. Let α be the indicator of a random event A, i.e., define α by

$$\alpha = \alpha(\omega) = \begin{cases} 1 & \text{if} \quad \omega \in A \\ 0 & \text{if} \quad \omega \in \bar{A} \end{cases} \tag{2.9.29}$$

Then we have $E(\alpha) = P(A)$ and

$$D^2(\alpha) = P(A)(1 - P(A)) \tag{2.9.30}$$

The characteristic function of α is

$$\varphi_\alpha(t) = 1 + P(A)(e^{it} - 1) \tag{2.9.31}$$

Example 2.9.2. Let ξ be a nonnegative integer-valued random variable and put

$$P(\xi = k) = p_k \quad (k = 0, 1, \ldots) \tag{2.9.32}$$

Then, if $E(\xi)$, respectively $D^2(\xi)$ exists, we have

$$E(\xi) = \sum_{k=1}^{\infty} kp_k \tag{2.9.33}$$

and

$$D^2(\xi) = \sum_{k=1}^{\infty} k^2 p_k - \left(\sum_{k=1}^{\infty} kp_k\right)^2 \tag{2.9.34}$$

Let us denote by $G(z)$ the *generating function* of the sequence $\{p_k\}$, i.e., let us put for every complex number z such that $|z| \leqq 1$,

$$G(z) = \sum_{k=0}^{\infty} p_k z^k \tag{2.9.35}$$

Clearly, $G(z)$ is regular in $|z| < 1$: $G(z)$ is called the *generating function* of ξ. Denoting by $\varphi(t)$ the characteristic function of ξ, we evidently have

$$\varphi(t) = G(e^{it}) \tag{2.9.36}$$

Further, the Laplace-Stieltjes transforms $\psi(s)$ of the distribution of ξ is

$$\psi(s) = G(e^s) \tag{2.9.37}$$

Note that if the function $G(z)$ is regular in the point $z = 1$ too, then $G(e^s)$ is regular for $s = 0$ and around $s = 0$ has the power series

$$G(e^s) = \sum_{n=0}^{+\infty} \frac{M_n s^n}{n!} \tag{2.9.38}$$

where $M_n = M_n(\xi)$. In this case, $\sqrt[2n]{M_{2n}} \leqq cn$ for some number $c > 0$ and thus Carleman's condition for the determinateness of the moment problem is satisfied.

The notion of a characteristic function is also used for random vectors. Let $\xi = (\xi_1, \xi_2, \ldots, \xi_n)$ be an n-dimensional random vector; its characteristic function $\varphi_\xi(t_1, t_2, \ldots, t_n)$ is defined for any real values of $t_1, t_2, \ldots, t_n$, as follows:

$$\varphi_\xi(t_1, t_2, \ldots, t_n) = E(e^{i(\xi, t)}) \tag{2.9.39}$$

Here t stands for the vector $(t_1, t_2, \ldots, t_n)$ and (ξ, t) denotes the scalar product of the vectors ξ and t, i.e., $(\xi, t) = \sum_{k=1}^{n} \xi_k t_k$. Thus the characteristic function $\varphi_\xi(t_1, \ldots, t_n)$ of the random vector ξ is a complex-valued function of the n real variables $t_1, \ldots, t_n$. The characteristic function of a random vector $\xi = (\xi_1, \ldots, \xi_n)$ depends only on the joint distribution of the random variables $\xi_1, \ldots, \xi_n$ and determines it uniquely. (See Laha and Lukács [55].)

If the characteristic function of the random vector ξ is known, the characteristic functions of the projections of ξ can be easily obtained. As a matter of fact, if $a = (a_1, a_2, \ldots, a_n)$ is any unit vector, putting $\xi_a = (\xi, a)$, we have

$$\varphi_{\xi_a}(t) = \varphi_\xi(a_1 t, a_2 t, \ldots, a_n t) \tag{2.9.40}$$

Formula (2.9.40) also shows that conversely, if the characteristic function of ξ_a is given for every unit vector a, then the characteristic function $\varphi_\xi(t_1, \ldots, t_n)$ of the joint distribution of $\xi_1, \xi_2, \ldots, \xi_n$ can be obtained for any choice of the real numbers $t_1, t_2, \ldots, t_n$.

As a matter of fact, putting $t = \sqrt{\sum_{k=1}^{n} t_k^2}$, and if $t \neq 0$, $a_k = t_k/t$ $(k = 1, 2, \ldots, n)$ and $a = (a_1, \ldots, a_n)$, we get

$$\varphi_\xi(t_1, t_2, \ldots, t_n) = \varphi_{\xi_a}(t) \tag{2.9.41}$$

This remark contains the proof of the theorem of Cramér and Wold mentioned in Section 2.7, according to which a distribution in R^n is uniquely determined by the totality of its linear projections.

Note that as a special case of (2.9.40) we get the following expression for the characteristic functions of the marginal distributions of a distribution in R^n: If $\varphi_\xi(t_1, t_2, \ldots, t_n)$ is the characteristic function of the random vector $\xi = (\xi_1, \xi_2, \ldots, \xi_n)$, the characteristic function of ξ_k is obtained by substituting t for t_k and 0 for t_j with $j \neq k$ into $\varphi_\xi(t_1, t_2, \ldots, t_n)$.

2.10 CONDITIONAL EXPECTATIONS AND OTHER CHARACTERISTICS OF RANDOM VARIABLES

Let ξ be a random variable on the full conditional probability space $S = [\Omega, \mathscr{A}, \mathscr{B}, P(A \mid B)]$ generated by the σ-finite measure μ. The conditional expectation $E_B(\xi)$ of the random variable ξ with respect to the admissible condition $B \in \mathscr{B}$ is defined (both in the case when μ is finite and when it is not) as the expectation of ξ on the probability space $S_B = (\Omega, \mathscr{A}, P_B)$ where $P_B(A) = P(A \mid B)$ $(A \in \mathscr{A})$. Thus we have*

$$E_B(\xi) = \frac{1}{\mu(B)} \int_B \xi \, d\mu \tag{2.10.1}$$

* The notation $E(\xi \mid B)$ will also be used occasionally.

(provided that this integral exists), i.e., $E_B(\xi)$ is the mean value of $\xi = \xi(\omega)$ over the set B with respect to the measure μ. In (2.10.1) we have expressed the value of $E_B(\xi)$ by means of μ; evidently the expression on the right of (2.10.1) is not changed if μ is replaced by $c\mu$ with $c > 0$. Conditional variances, higher moments and characteristic functions are defined similarly.

The following theorem is often called the theorem of total expectation:

THEOREM 2.10.1. *Let ξ be a random variable on the conditional probability space S. Let B_n $(n = 1, 2, \ldots)$ be a finite or infinite sequence of pairwise disjoint admissible conditions such that their union $B = \sum_n B_n$ is also an admissible condition. If the conditional expectation $E_B(\xi)$ exists, then the expectations $E_{B_n}(\xi)$ $(n = 1, 2, \ldots)$ also exist and one has*

$$E_B(\xi) = \sum_n E_{B_n}(\xi) P(B_n \mid B) \tag{2.10.2}$$

The proof of Theorem 2.10.1 follows immediately from (2.10.1) and the σ-additivity of the integral as a set function.

Let us consider some simple examples.

Example 2.10.1. Let ξ be a random variable uniformly distributed on the whole real axis R. In other words, let ξ be a regular random variable on a conditional probability space S which defines the conditional probability space described by Example 2.2.5 on the real line. The easiest way to get such a random variable is to choose $S = [R, \mathscr{F}, \lambda]$ where $\mathscr{F}$ is the family of Borel sets of R and λ is the Lebesgue measure, and put $\xi(x) = x$ for $x \in R$. Let B denote the event $a \leqq \xi < b$ where a, b $(a < b)$ are arbitrary real numbers. Then we have

$$E_B(\xi) = \frac{1}{b-a} \int_a^b x\,dx = \frac{a+b}{2}$$

Further, denoting the variance of ξ under condition B by $D_B^2(\xi)$, we have

$$D_B^2(\xi) = \frac{1}{b-a} \int_a^b x^2\,dx - \left(\frac{a+b}{2}\right)^2 = \frac{(b-a)^2}{12}$$

Example 2.10.2. Let the experiment $\mathscr{E}$ consist in throwing two dice. We suppose that the dice are fair, i.e., that each of the thirty-six possible throws (ξ_1, ξ_2) $(1 \leqq \xi_1 \leqq 6,\ 1 \leqq \xi_2 \leqq 6)$ have the same probability, 1/36. Let η denote the sum of the two numbers thrown, i.e., if we throw (ξ_1, ξ_2), then $\eta = \xi_1 + \xi_2$. We want to compute (1) the expectation of η, (2) the conditional expectation of η under the condition that $\xi_1 \neq \xi_2$, (3) the conditional expectation of η under condition $\xi_1 < \xi_2$, and (4) the conditional expectation of η under the condition that it is even. These expectations can be computed as follows:

(1) Since $E(\xi_1) = E(\xi_2) = (1 + 2 + 3 + 4 + 5 + 6)/6 = 3.5$, by the linearity of the expectation, $E(\eta) = 7$.

(2) Let B denote the event $\xi_1 \neq \xi_2$, then $\bar{B}$ is the event $\xi_1 = \xi_2$. Clearly, $E_{\bar{B}}(\eta) = 7$, further, $P(B) = 5/6$ and $P(\bar{B}) = 1/6$. As by Theorem 2.10.1, $E(\eta) = E_B(\eta)P(B) + E_{\bar{B}}(\eta)P(\bar{B})$, it follows that $E_B(\eta) = 7$.

(3) By symmetry, $E(\eta \mid \xi_1 < \xi_2) = E(\eta \mid \xi_2 < \xi_1)$. Thus it follows from Theorem 2.10.1 that $E(\eta \mid \xi_1 < \xi_2) = 7$.

(4) Clearly, under the conditions that η is even, η and $14 - \eta$ have the same distribution and thus the same expectation; if C denotes the event that η is even, then $E_C(\eta) + E_C(14 - \eta) = 14$. It follows that the conditional expectation is equal to 7 in this case also.

Remark: This example shows that sometimes by using the general properties of expectation, one can determine the expectation of a random variable without effectively computing its distribution. Note that the relation $E(\xi_1) = 3.5$ can also be obtained by noticing that ξ_1 and $7 - \xi_1$ have the same distribution. Another instance is the following:

Example 2.10.3. Let the random point (ξ, η) be uniformly distributed over the whole (x, y) plane. For instance, consider the conditional probability space $S = [R^2, \mathscr{F}_2, \lambda_2]$ where R^2 is the (x, y) plane, $\mathscr{F}_2$ is the family of Borel subsets of R^2, and λ_2 is the two-dimensional Lebesgue measure, and put $\xi = \xi(x, y) = x$, $\eta = \eta(x, y) = y$. Let C be a bounded domain in the (x, y) plane, which is symmetrical with respect to the origin, i.e., is such that if $(x, y) \in C$, then $(-x, -y) \in C$. Clearly, C is an admissible condition. Show that $E_C(\xi) = E_C(\eta) = 0$. To show this it is sufficient to notice that the random vector $(-\xi, -\eta)$ has the same conditional distribution with respect to C as the random vector (ξ, η) and therefore $E_C(-\xi) = E_C(\xi)$, which implies $E_C(\xi) = 0$, and one gets $E_C(\eta) = 0$ in the same way. Similarly, if C is symmetrical with respect to the line $y = x$ (i.e., if C contains the point (x, y) it also contains the point (y, x)), then the conditional distributions of the random points (ξ, η) and (η, ξ) are the same with respect to C. Thus one has, e.g., $D_C^2(\xi) = D_C^2(\eta)$.

In particular, if C is the unit circle $x^2 + y^2 < 1$, one has

$$D_C^2(\xi) = E_C(\xi^2) = 1/2\ E_C(\xi^2 + \eta^2) = 1/2\pi \int_{x^2+y^2<1} (x^2 + y^2)\, dx\, dy$$

and thus

$$D_C^2(\xi) = \int_0^1 r^3\, dr = 1/4$$

Remark: One can obtain the same result in a straightforward way by computing the conditional density function $f_C(x)$ of ξ under condition C:

$$f_C(x) = \begin{cases} \dfrac{2\sqrt{1-x^2}}{\pi} & \text{for} \quad -1 < x < +1 \\ 0 & \text{for} \quad |x| > 1 \end{cases}$$

We give one more example of this kind:

Example 2.10.4. Let the point (ξ, η) be uniformly distributed in the plane as in Example 2.10.3. Let the domain C be the interior of a regular triangle whose barycenter is in the origin of the (x, y) plane. Show that

$$E_C(\xi) = E_C(\eta) = 0$$

This can be proved by using the fact that by rotating the plane by an angle of degree $2\pi/3$ around the origin, the domain C is mapped onto itself.

We shall often need the following two theorems:

THEOREM 2.10.2. *Let ξ be a random variable on the probability space $(\Omega, \mathscr{A}, P)$ such that $E(\xi)$ exists; let $B_n \in \mathscr{A}$ $(n = 1, 2, \ldots)$ be a sequence of events such that* $\lim_{n\to\infty} P(B_n) = 1$. *Then we have*

$$\lim_{n\to+\infty} E(\xi \mid B_n) = E(\xi) \tag{2.10.3}$$

Proof. According to a well known theorem on Lebesgue integrals (see Appendix A) if the integral $\int_\Omega \xi \, dP$ exists, the set function $\mu(A) = \int_A \xi \, dP$ is absolutely continuous with respect to the measure P, and thus, in view of $\lim_{n\to\infty} P(\bar{B}_n) = 0$, we have $\lim_{n\to\infty} P(\bar{B}_n)E(\xi \mid \bar{B}_n) = 0$. By Theorem 2.10.1 we have

$$E(\xi) = P(B_n)E(\xi \mid B_n) + P(\bar{B}_n)E(\xi \mid \bar{B}_n) \tag{2.10.4}$$

Thus it follows that (2.10.3) holds. ■

THEOREM 2.10.3. *Let ξ be a random variable with finite expectation. Then one can find a sequence ξ_n of random variables $(n = 1, 2, \ldots)$, each taking on only a finite number of values, so that*

$$\lim_{n\to\infty} E(|\xi - \xi_n|) = 0 \tag{2.10.5}$$

Proof. One can choose, e.g., the sequence ξ_n defined as follows:

$$\xi_n = \begin{cases} \dfrac{[n\xi]}{n} & \text{if} \quad |\xi| \leqq n \\ 0 & \text{if} \quad |\xi| > n \end{cases} \tag{2.10.6}$$

■

2.11 INEQUALITIES CONCERNING RANDOM VARIABLES

Here we shall prove some inequalities which will be needed in the following chapters.

THEOREM 2.11.1. *Let ξ be a random variable on the probability space $(\Omega, \mathscr{A}, P)$, and let $g(x)$ be a nonnegative and nondecreasing function of the real variable x. If the expectation $E(g(\xi))$ exists, one has for each value of x for which $g(x) > 0$*

$$P(\xi \geqq x) \leqq \frac{E(g(\xi))}{g(x)} \tag{2.11.1}$$

Proof. By definition,

$$E(g(\xi)) = \int_\Omega g(\xi)\, dP \tag{2.11.2}$$

Let A_x denote the set of those $\omega \in \Omega$ for which $\xi(\omega) < x$. As we have supposed that $g(x)$ is nonnegative and nondecreasing, it follows from (2.11.2) that

$$E(g(\xi)) \geqq \int_{\bar{A}_x} g(\xi)\, dP \geqq g(x) \int_{\bar{A}_x} dP = g(x)P(\bar{A}_x) \tag{2.11.3}$$

This proves (2.11.1). ■

Remark: While (2.11.1) is valid for every x such that $g(x) > 0$, it is trivial for those values of x for which $g(x) \leqq E(g(\xi))$ (as every probability is $\leqq 1$). The inequality (2.11.1) is called *Markov's inequality*.

The following two inequalities are consequences of Theorem 2.11.1:

THEOREM 2.11.2. *If ξ is a random variable having finite expectation $E(\xi)$ and variance $D^2(\xi)$, one has for every $\lambda > 0$*

$$P(|\xi - E(\xi)| \geqq \lambda D(\xi)) \leqq 1/\lambda^2. \tag{2.11.4}$$

Proof. We obtain (2.11.4) as a special case of (2.11.1) if we apply it to the random variable $(\xi - E(\xi))^2$ and take $g(x) = xD(\xi)$. ■

Remark: Equation (2.11.4) is called *Čebyshev's inequality*.

THEOREM 2.11.3. *Suppose that the Laplace transform $\psi_\xi(s) = E(e^{\xi s})$ of the distribution of the random variable ξ exists for $s \in I$ where I is an open interval* *on the real line containing $s = 0$. Then one has for $x > 0$*

$$P(|\xi| \geqq x) \leqq \inf_{\substack{c>0 \\ c \in I}} e^{-cx}\psi_\xi(c) + \inf_{\substack{d<0 \\ d \in I}} e^{-|d|x}\,\psi_\xi(d) \tag{2.11.5}$$

Proof. By applying (2.11.1) to the random variable ξ and to the function e^{cx}, where c is any positive number in the interval I, we obtain

$$P(\xi \geqq x) \leqq e^{-cx}\psi_\xi(c) \tag{2.11.6}$$

Similarly, applying (2.11.1) to the random variable $-\xi$ and for $g(x) = e^{cx}$, where c is a positive number such that $-c \in I$ we get

$$P(\xi \leqq -x) \leqq e^{-cx}\psi_\xi(-c) \tag{2.11.7}$$

Thus,

$$P(|\xi| \geqq x) = P(\xi \geqq x) + P(\xi \leqq -x) \leqq \inf_{\substack{c>0 \\ c \in I}} (e^{-cx}\,\psi_\xi(c)) + \inf_{\substack{c>0 \\ c \in I}} (e^{-cx}\,\psi_\xi(-c)) \tag{2.11.8}$$

which proves (2.11.5). ■

Remark: Inequality (2.11.5) is called *Bernstein's inequality.*

2.12 SOME REMARKS ON THE NOTION OF RANDOM VARIABLES

Let $S = (\Omega, \mathscr{A}, P)$ be a complete probability space (see Definition 2.3.6). Let $\xi(\omega)$ be a random variable on S and let $\xi^*(\omega)$ be a function defined on Ω such that $\xi^*(\omega) = \xi(\omega)$ except for $\omega \in Z$ where Z is a set such that $P(Z) = 0$. In this case we say that $\xi(\omega)$ and $\xi^*(\omega)$ are *almost surely** equal. It is easy to see that in this case $\xi^*(\omega)$ is also a random variable, further, $\xi(\omega)$ and $\xi^*(\omega)$ have the same probability distribution; moreover, if η is another random variable on S, the joint distribution of ξ and η is the same as that of ξ^* and η, etc. In general, any probabilistic statement about the random variable ξ also holds for ξ^*. Thus, from the point of view of probability theory, the random variables ξ and ξ^* may be regarded as identical. *In what follows we shall regard random variables which are almost surely equal as identical.* This shows again why it is advantageous in probability theory to consider complete probability spaces. If the probability space S is not complete, if ξ

* Often abbreviated as "a.s."

is a random variable on S, and if $\xi^*(\omega)$ is a function on Ω which coincides with $\xi(\omega)$ except for $\omega \in Z$ where $P(Z) = 0$, then it is not at all sure that $\xi^*(\omega)$ is also a random variable, (i.e., that it is measurable with respect to $\mathscr{A}$).

We have said that we consider two random variables which coincide with probability 1 as identical, not only because they have the same probability distribution, but also because they are, in general, indistinguishable from the point of view of probability theory. It should be emphasized that the mere fact that two random variables, ξ_1 and ξ_2, have the same distribution is by far not enough cause to consider them as identical. For instance, if η is a third random variable, the joint distribution of ξ_1 and η may be different from that of ξ_2 and η. Attempts have been made to define random variables through their distributions, but these attempts were not successful. One reason for this is that if we only know the probability distributions of two random variables, this information is not sufficient to find their joint distribution. For instance, if we draw a ball twice from an urn which contains a known number of balls marked with 1, 2, ..., N, then the numbers on the balls drawn at the first and at the second occasion have the same distribution whether we replace the first ball before drawing the second or not. However, the joint distribution of the two numbers is not the same in these two different experiments.

According to what has been said, every statement with probabilistic meaning which holds for a random variable ξ also holds for the random variable ξ^* if ξ and ξ^* are almost surely equal. Thus, for instance, we have mentioned that a bounded random variable possesses moments of all orders. Instead of this one can say that a random variable which is almost surely bounded has moments of all orders, etc.

Evidently, if a probabilistic statement on a family $\{\xi_t\}$ $(t \in T)$ of random variables is valid, it remains so if each of the random variables ξ_t is replaced by another variable ξ_t^* such that ξ_t and ξ_t^* are almost surely equal, provided the family $\{\xi_t\}$ (i.e., the set T) is *denumerable*. As a matter of fact, in this case, if Z_t denotes the exceptional set on which $\xi_t \neq \xi_t^*$, then $P(Z_t) = 0$ and the union of the sets Z_t $(t \in T)$ is also a set of probability 0. On the other hand, this is not necessarily true if the family $\{\xi_t\}$ is not denumerable. In this book, however, we shall deal only with denumerable families of random variables,* as the theory of *stochastic processes* lies outside the scope of this book; thus we will not encounter the mentioned difficulty.

We now extend our dictionary given in Section 1.1 with notions concerning probabilities and random variables.

* More exactly, we deal with a nondenumerable family of random variables only exceptionally, and even if we do so we construct these families by means of a denumerable sequence of random variables.

Language of probability theory	Language of measure theory	Notation
probability of an observable event A	measure of a measurable set A	$P(A)$ $(A \in \mathscr{A})$
random variable	measurable function	$\xi = \xi(\omega)$, $\eta = \eta(\omega)$, etc.
expectation of a random variable	Lebesgue integral of a measurable function	$E(\xi) = \int_\Omega \xi(\omega)\, dP$
conditional expectation of a random variable with respect to a condition B having positive probability	mean value of a measurable function over a set B (having positive measure)	$E(\xi \mid B) = \frac{1}{P(B)} \int_B \xi\, dP$
two random variables are almost surely equal	two measurable functions are equal almost everywhere	$\xi = \xi^*$ a.s.
random variable having a finite variance	square-integrable function	$E(\xi^2) < +\infty$
characteristic function of a probability distribution with distribution function $F(x)$ respectively density function $f(x)$	Fourier-Stieltjes transform of $F(x)$ (respectively Fourier transform of $f(x)$)	$\varphi(t) = \int_{-\infty}^{+\infty} e^{ixt}\, dF(x)$ respectively $\varphi(t) = \int_{-\infty}^{+\infty} e^{ixt} f(x)\, dx$
moment-generating function of a probability distribution with distribution function $F(x)$ (respectively density function $f(x)$)	Laplace-Stieltjes transform of $F(x)$ (respectively Laplace transform of $f(x)$)	$\psi(s) = \int_{-\infty}^{+\infty} e^{xs}\, dF(x)$ respectively $\psi(s) = \int_{-\infty}^{+\infty} e^{xs} f(x)\, dx$

EXERCISES

E.2.1. Show that the conditional probability $P(A \mid B)$ is a continuous function of both A and B in the sense that if $\lim_{n\to+\infty} A_n = A$, $\lim_{n\to+\infty} B_n = B$ where $A_n \in \mathscr{A}$, $B_n \in \mathscr{B}$ $(n = 1, 2, \ldots)$ and $B \in \mathscr{B}$, then $\lim_{n\to+\infty} P(A_n \mid B_n) = P(A \mid B)$.

E.2.2. (a) Let Ω be the set of all straight lines ω in the (x, y) plane. Let the equation of a line ω be written in the form $x \cos \varphi + y \sin \varphi = p$. The parameters φ and p $(0 \leq \varphi < \pi, 0 \leq p < +\infty)$ can be considered as the coordinates of the line ω. Thus, one gets a mapping $\xi(\omega) = (\varphi, p)$ of the space Ω onto the half strip $0 \leq \varphi < \pi$, $0 \leq p < +\infty$ of the (φ, p) plane. Let $\mathscr{A}$ be the set of the inverse images under the mapping ξ of Borel subsets of the half strip $0 \leq \varphi < \pi$, $0 \leq p < +\infty$ and let the measure μ be defined on $\mathscr{A}$ by putting $\mu(A) = \lambda_2(\xi(A))$ for $A \in \mathscr{A}$ where λ_2 is the two-dimensional Lebesgue measure in the (φ, p) plane. Show that the conditional probability space $[\Omega, \mathscr{A}, \mu]$ is invariant with respect to Euclidean motions of the plane; i.e., if in the (x, y) plane a new rectangular system of coordinates is introduced and the measure of sets of lines is defined as above, but with respect to the new system of coordinates, the resulting measure is the same as before. (See also Problem P.2.9.)

Remark: The conditional probability space $[\Omega, \mathscr{A}, \mu]$ constructed above may be interpreted as describing an experiment in which a straight line is chosen at random in the (x, y) plane with *uniform distribution*. This conditional probability space is used in integral geometry (see Problem P.2.1).

(b) Construct a conditional probability space describing the following experiment: A given equilateral triangle is thrown at random onto a plane so that the position of a marked vertex of the triangle is uniformly distributed over the whole plane and the angle formed by a marked side of the triangle and a fixed line in the plane is uniformly distributed in the interval $[0, 2\pi)$ (see Problem P.2.2).

Hint: Let the coordinates of the marked vertex in the plane be x and y and the angle which the marked side of the triangle forms with a fixed line in the plane be α. Map the positions of the triangle onto the set of points (x, y, α) with $0 \leqq \alpha < 2\pi$, $-\infty < x < +\infty$, and $-\infty < y < +\infty$, and use the three-dimensional Lebesgue measure in the space (x, y, α).

E.2.3. Show that if $P_1, P_2, \ldots, P_n, \ldots$ are probability measures on the same experiment $\mathscr{E}$ and c_n $(n = 1, 2, \ldots)$ is a sequence of nonnegative numbers such that $\sum_{n=1}^{\infty} c_n = 1$, then $P = \sum_{n=1}^{\infty} c_n P_n$ is also a probability measure on $\mathscr{E}$. (P is called the *mixture* of the probability measures P_n with *weights* c_n.)

E.2.4. Let R be the real line, let $\mathscr{F}_1$ be the family of all Borel subsets of R and put $\mu(A) = \int_A dx/|x|$ for $A \in \mathscr{F}_1$. Show that the measure μ is σ-finite, but that the conditional probability space $(R, \mathscr{F}_1, \mu)$ is not of the type described by Example 2.2.3.

Hint: If one wants to write the measure μ in the form $\mu(A) = \int_A dF(x)$, this can be done by putting $F(x) = \operatorname{sign} x \log |x|$, but $F(x)$ is not defined for $x = 0$ and while it is increasing both on the negative and the positive half line, it is not a nondecreasing function on the whole line.

E.2.5. Let R be the real line, let $\mathscr{F}_1$ be the σ-algebra of the Borel subsets of R and let $f_n(x)$ $(n = 1, 2, \ldots)$ be probability density functions on R. Suppose that $f_n(x)/f_n(1) < k$ for $-\infty < x < +\infty$ and $n = 1, 2, \ldots$, further, that the (finite) limit

$$\lim_{n \to +\infty} \frac{f_n(x)}{f_n(y)} = q(x, y)$$

exists for every real x and y. Show that in this case we have $q(x, y) = f(x)/f(y)$ where $f(x)$ is a bounded measurable function which is strictly positive. Let $\mathscr{B}$ denote the family of all Borel-measurable subsets of R having positive and finite Lebesgue measure, and put $P_n(A) = \int_A f_n(x)\, dx$ and $\mu(A) = \int_A f(x)\, dx$ for $A \in \mathscr{F}_1$. Show that the probability spaces $S_n = (R, \mathscr{F}_1, P_n)$ tend for $n \to +\infty$ with respect to $\mathscr{B}$ to the *conditional* probability space generated by the measure μ on $(R, \mathscr{F}_1)$.

E.2.6. Let $A_1, \ldots, A_n$ be arbitrary events, put $S_0 = 1$ and let S_k $(k = 1, 2, \ldots, n)$ be defined by (2.6.13). Show that the following inequalities are valid:

(a) $$\frac{S_{r+1}}{\binom{n}{r+1}} \leqq \frac{S_r}{\binom{n}{r}} \quad (r = 0, 1, \ldots, n-1) \qquad \text{(inequalities of Fréchet)}$$

(b) $$\frac{\binom{n}{r+1} - S_{r+1}}{\binom{n-1}{r}} \leqq \frac{\binom{n}{r} - S_r}{\binom{n-1}{r-1}} \quad (r = 1, 2, \ldots, n-1) \qquad \text{(inequalities of Gumbel)}$$

(c) Let $V_{n,r}$ denote the probability of the event that among the events $A_1, \ldots, A_n$ *at least* r events occur. Show that

$$V_{n,r} = \sum_{k=0}^{n-r} (-1)^k \binom{k+r-1}{k} S_{k+r} \quad (r = 1, \ldots, n)$$

Hint: Use Theorem 2.6.1.

□***E.2.7.** Let $S = [\Omega, \mathscr{A}, \mathscr{B}, P(A \mid B)]$ be a conditional probability space. Prove the following identities:

(a) If $A_k \in \mathscr{A}$, $A_k \subseteq B$ $(k = 1, 2, \ldots, n)$, further, $B \in \mathscr{B}$ and $\prod_{j=1}^{k} A_j \in \mathscr{B}$ for $k = 1, 2, \ldots, n-1$, one has

$$P\left(\prod_{k=1}^{n} A_k \mid B\right) = P(A_1 \mid B) \prod_{k=1}^{n} P\left(A_k \,\middle|\, \prod_{j=1}^{k-1} A_j\right)$$

(b) If $B_k \in \mathscr{B}$ $(k = 1, 2, \ldots, n)$, $B_j B_k = \varnothing$ for $j \neq k$, $A \in \mathscr{A}$, $A \subseteq \sum_{k=1}^{n} B_k \subseteq C \in \mathscr{B}$, one has

$$P(A \mid C) = \sum_{k=2}^{n} P(A \mid B_k) P(B_k \mid C)$$

Remark: This formula is called *the theorem of total probability*. It can be deduced from Theorem 2.10.1.

(c) If $B_k \in \mathscr{B}$ $(k = 1, 2, \ldots, n)$, $B_j B_k = \varnothing$ for $j \neq k$, $B = \sum_{k=1}^{n} B_k$, further, $A \in \mathscr{B}$, $A \subseteq B$, one has (*Bayes' formula*)

$$P(B_k \mid A) = \frac{P(A \mid B_k) P(B_k \mid B)}{\sum_{j=1}^{n} P(A \mid B_j) P(B_j \mid B)} \quad \text{for } k = 1, 2, \ldots, n$$

E.2.8. A bunch $\mathscr{B}$ of events of an experiment $\mathscr{E} = (\Omega, \mathscr{A})$ is called a *full bunch* if it has the following property:

If $B_1 \subseteq A \subseteq B_2$ where $B_1 \in \mathscr{B}$, $B_2 \in \mathscr{B}$ and $A \in \mathscr{A}$, then $A \in \mathscr{B}$.

A class $\mathscr{F}$ of events of an experiment $\mathscr{E}$ is called an *implication class*† if $\varnothing \notin \mathscr{F}$,

* The sign □ before an exercise or a problem signifies that its results will be used later in the main text, therefore the reader is advised to read it before passing to the next chapter.

† The notion of an implication class is the dual notion to that of a *hereditary class* (which contains together with every set belonging to it every subset of such a set). (See Halmos [1].)

further, whenever $A \subseteq B$, $A \in \mathscr{F}$ and $B \in \mathscr{A}$, then we also have $B \in \mathscr{F}$. Prove the following statements:

(a) The bunch of events of a full conditional probability space is a full bunch.

(b) If $\mathscr{A}$ is finite and $\mathscr{B}$ is a bunch of events concerning the experiment $(\Omega, \mathscr{A})$, then $\Omega \in \mathscr{B}$.

(c) If Ω is a finite set, a class of events concerning the experiment $(\Omega, \mathscr{A})$ is a full bunch if and only if it is an implication class.

E.2.9. Let Ω_n be a set having n elements. Show that the total number of all full bunches of events concerning the experiments $\mathscr{E}_n = (\Omega_n, \mathscr{P}(\Omega_n))$ is equal to the number of Sperner systems of subsets of Ω_n (see Problem P.1.8).

Hint: Let $\mathscr{B}$ be a full bunch. Let us call an event $B \in \mathscr{B}$ *minimal* if no proper subset of B is contained in $\mathscr{B}$. Clearly a full bunch in $\mathscr{E}_n$ is uniquely characterized by its minimal elements and these form a Sperner system.

(b) Let $B(n)$ denote the total number of bunches concerning the experiment $\mathscr{E}_n = (\Omega_n, \mathscr{P}(\Omega_n))$. Show that $B(1) = 1$, $B(2) = 4$, $B(3) = 45$.

(c) Let $B(n, k)$ denote the number of those bunches concerning the experiment $\mathscr{E}_n$ which consist of exactly k events ($1 \leq k \leq 2^n$). Let $b(n, k)$ denote the number of those bunches concerning the experiment $\mathscr{E}_n$ which are separating systems of Ω_n. Show that

$$B(n, k) = \sum_{r=1}^{k} S(n, r)b(r, k)$$

where $S(n, r)$ denote the Stirling numbers of the second kind (see Problem P.1.1).

(d) Show that $b(n, k) = 0$ if $k < n$.

(e) Show that if a bunch concerning the experiment $\mathscr{E}_n$ is a separating system of Ω_n and consists of exactly n events, then the intersection of all events of the bunch consists of a single element of Ω_n. Using this fact, show that

$$b(n, n) = \sum_{k=1}^{n} (-1)^{k-1} n(n-1) \cdots (n-k+1)[b(n-k, n) + b(n-k, n-1)]$$

(f) Show that

$$b(n, k) = B(n, k) \quad \text{if} \quad k \geqq 2^{n-1}$$

E.2.10. Let ξ be a positive random variable having an absolutely continuous distribution with density function $f(x)$ ($x > 0$). Suppose that the Laplace transform $\psi(c) = \int_0^\infty e^{ct} f(t)\, dt$ exists for $0 \leqq c < A$ and either A is the largest positive number with this property or $A = +\infty$. Put for $0 \leqq c < A$ and $x > 0$

$$f_c(x) = \frac{e^{cx} f(x)}{\psi(c)}$$

Then $f_c(x)$ is a density function of a probability distribution on the positive half line for each c with $0 \leqq c < A$ ($f_0(x) \equiv f(x)$). Prove the following statements:

(a) The expectation $E(c)$ of the probability distribution having the density function $f_c(x)$ exists for $0 \leqq c < A$ and $E(c)$ is a continuous increasing function of c.

(b) The function $h(c) = e^{-cx}\psi(c)$ is convex for each fixed value of $x > 0$ and if $\int_x^\infty f(t)\,dt > 0$, then $h(c)$ takes on its minimum for that (uniquely determined) value c^* of c for which $E(c^*) = x$ and $\min h(c) = h(c^*) < 1$.

Remark: Note that it follows from (2.11.6) that under the conditions stated in (b) one has (see Cramér [62])

$$P(\xi \geqq x) \leqq h(c^*)$$

PROBLEMS

P.2.1. (a) In Exercise E.2.2(a), let F_C denote the event that the random line ω cuts the closed convex curve C. Show that if C_1 and C_2 are convex curves such that C_1 lies in the interior of C_2 and $L(C)$ denotes the length of the curve C, then we have

$$P(F_{C_1} \mid F_{C_2}) = \frac{L(C_1)}{L(C_2)}$$

(b) Let ξ_{C_1} denote the length of the segment of the random line ω lying inside the closed convex curve C_1. Prove that if C_1 lies in the interior of C_2, $L(C)$ denotes the length of C, and $A(C)$ is the area of the domain bounded by C, then

$$E(\xi_{C_1} \mid F_{C_2}) = \frac{\pi A(C_1)}{L(C_2)} \quad \text{and} \quad E(\xi^3_{C_1} \mid F_{C_2}) = \frac{3A^2(C_1)}{L(C_2)}$$

Remark: These statements are reformulations of Crofton's theorems in integral geometry (see Blaschke [56] and Santalo [61]).

P.2.2. In Exercise E.2.2(b), let the length of the sides of the equilateral triangle T thrown at random on the plane be equal to z; suppose that on the plane on which the triangle T is thrown, a regular triangular lattice is drawn, consisting of equilateral triangles with sides of length 1. Let A_z denote the event that the triangle T covers at least one triangle of the lattice, and let B denote the event that the marked corner of T lies in a regular hexagon formed by six triangles of the lattice having a common vertex. Compute the conditional probability $P(A_z \mid B)$ where $1 \leqq z \leqq 2\sqrt{3}$.

Remark: Note that in case $z > 2\sqrt{3}$, the triangle T covers at least one lattice triangle in every position, while for $z < 1$, it, of course, can never cover a lattice triangle; thus, only the cases $1 \leqq z \leqq 2\sqrt{3}$ are of interest.

P.2.3. Let $(\Omega, \mathscr{A}, P)$ be a probability space. Prove the following statements:

(a) The set of values of $P(A)$ $(A \in \mathscr{A})$ is always a closed subset of the interval $[0, 1]$.

(b) If for every $A \in \mathscr{A}$ with $P(A) > 0$ there exists a set $B \in \mathscr{A}$ with $B \subseteq A$ and $0 < P(B) < P(A)$, the probability space $(\Omega, \mathscr{A}, P)$ is called *nonatomic*. Show that for a nonatomic probability space the set of values of $P(A)$ is the whole interval $[0, 1]$, i.e., for every x with $0 \leqq x \leqq 1$ one can find a set $A \in \mathscr{A}$ such that $P(A) = x$.

(c) Suppose $(\Omega, \mathscr{A}, P)$ is a discrete probability space; Ω is a denumerably infinite set, $\Omega = \{\omega_1, \omega_2, \ldots, \omega_n, \ldots\}$, and $\mathscr{A}$ is the σ-algebra $\mathscr{P}(\Omega)$ of all subsets of Ω. Put $P(\{\omega_k\}) = p_k$; show that if $p_n > p_{n+1} > 0$ for $n = 1, 2, \ldots$, the set of values of $P(A)$ is a perfect subset of the interval $[0, 1]$. Show further that this set is identical with the whole interval $[0, 1]$ if and only if

$$p_n \leqq \sum_{k=n+1}^{\infty} p_k \quad \text{for } n = 1, 2, \ldots$$

(d) Using the notations of (c), show that if the inequalities

$$p_n \leqq \frac{1}{r} \sum_{k=n}^{\infty} p_k \ (n = 1, 2, \ldots)$$

are satisfied, for every r-tuple of nonnegative numbers $x_1, x_2, \ldots, x_r$ such that $\sum_{j=1}^{r} x_j = 1$, one can find a partition $\Omega = \sum_{j=1}^{r} A_j$, $(A_j \in \mathscr{A}, A_j A_k = \varnothing$ if $j \neq k)$ such that $P(A_j) = x_j$ $(j = 1, 2, \ldots, r)$.

(e) Let $S = [\Omega, \mathscr{A}, \mu]$ be a full conditional probability space. Show that the set of values of $P(A \mid B)$ is not necessarily a closed set.

Hint: Let Ω be the set of positive integers, $\mathscr{A}$ the set of all subsets of Ω, and $\mu = \mathscr{N}$ the counting measure. Clearly, the set of values of $P(A \mid B)$ is the set of all rational numbers in the interval $[0, 1]$.

P.2.4. Let $[\Omega, \mathscr{A}, \mu]$ be a full conditional probability space and let $\mathscr{B}$ denote the family of admissible conditions, i.e., the family of those sets $B \in \mathscr{A}$ for which $0 < \mu(B) < +\infty$. Prove the following statements:

(a) If $A_n \in \mathscr{A}$ $(n = 1, 2, \ldots)$ and $\sum_{n=1}^{\infty} A_n \in \mathscr{B}$, then at least one of the sets A_n belongs to the family $\mathscr{B}$.

(b) Let us call a set $A \in \mathscr{A}$ which does not belong to $\mathscr{B}$ "a set of the first kind" if there exists a $B \in \mathscr{B}$ such that $B \subseteq A$; otherwise, A shall be called "of the second kind." Clearly $A \notin \mathscr{B}$ is of the first kind if and only if $\mu(A) = +\infty$ and it is of the second kind if and only if $\mu(A) = 0$. Let $\mathscr{D}$, respectively $\mathscr{N}$ denote the set of all $A \in \mathscr{A}$ for which $A \notin \mathscr{B}$ and A is of the first kind respectively of the second kind. Show that if $A \in \mathscr{D}$, there exist two disjoint sets A_1 and A_2 such that $A_i \in \mathscr{D}$ $(i = 1, 2)$ and $A = A_1 + A_2$.

(c) Show that if $A_n \in \mathscr{D}$ $(n = 1, 2, \ldots)$, there exist sets $B_n \in \mathscr{B}$ such that $B_n \subseteq A_n$ and $\sum_{n=1}^{\infty} B_n \in \mathscr{D}$.

(d) Show that $\mathscr{N}$ is a σ-ideal in $\mathscr{A}$, i.e., if $A_n \in \mathscr{N}$ $(n = 1, 2, \ldots)$, then $\sum_{n=1}^{\infty} A_n \in \mathscr{N}$; further, if $A \in \mathscr{N}$ and $C \in \mathscr{A}$, then $AC \in \mathscr{N}$.

Remark: Let $\mathscr{A}$ be a σ-algebra of subsets of a set Ω, and $\mathscr{B}$ a subset of $\mathscr{A}$. The following problem is unsolved: to give necessary and sufficient conditions on $\mathscr{B}$ in order that there should exist on $\mathscr{A}$ a σ-finite measure μ such that $\mu(B)$ is positive and finite if and only if $B \in \mathscr{B}$. Clearly the condition that $\mathscr{B}$ should be a full bunch (see Exercise E.2.8), and the above statements (a)–(d) are necessary conditions, but they are not yet sufficient. For the case when Ω is denumerable, the problem has been solved by Speakman [48].

P.2.5. Let $\Omega = \{\omega_1, \omega_2, \ldots, \omega_n, \ldots\}$ be a denumerably infinite set, $\mathscr{A}$ the σ-algebra of all subsets of Ω and μ a measure on $\mathscr{A}$ such that, putting $p_n = \mu(\{\omega_n\})$, one has $\lim_{n\to\infty} p_n = 0$ and the series $\sum_{n=1}^{\infty} p_n$ is divergent. Show that in the full conditional probability space $[\Omega, \mathscr{A}, \mu]$, the set of values of $P(A \mid B)$ is identical with the whole interval [0, 1].

P.2.6. (a) Let $F_1, F_2, \ldots, F_N$ be polynomials of the events $A_1, \ldots, A_n$ and let $C_{j,k}$ $(j, k = 1, 2, \ldots, N)$ be real coefficients. Let S_0 denote the trivial probability space (see the remark to Theorem 2.6.1) and let S_1 denote the probability space $(\Omega_1, \mathscr{A}, P_1)$ where Ω_1 consists of two elements, $\Omega_1 = \{0, 1\}$, say; $\mathscr{A}$ is the set of all four subsets of Ω_1 (that is, $\mathscr{A}$ consists of the four sets Ω_1, $\emptyset$, $A = \{0\}$ and $\bar{A} = \{1\}$) and let the probability measure P_1 be defined by putting $P_1(A) = P_1(\bar{A}) = 1/2$. The quadratic inequality

$$\sum_{j=1}^{N} \sum_{k=1}^{N} C_{j,k} P(F_j) P(F_k) \geqq 0 \tag{P.2.6.1}$$

is called *exact* if equality holds in (P.2.6.1) for the trivial probability space S_0 (i.e., if all A_k are equal to Ω or to $\emptyset$). Show that an exact quadratic inequality is valid for every probability space if it is valid on the probability space S_1. (See [47].)

(b) Let S_k $(k = 0, 1, \ldots, n)$ be defined as in Exercise E.2.6. By using the result under (a), prove the quadratic inequalities

$$kS_k \geqq S_{k-1}(S_1 - k + 1) \qquad (k = 1, 2, \ldots, n)$$

(c) Using (a), prove the identity

$$P^2(A + B) + P^2(AB) = P^2(A) + P^2(B) + 2P(A\bar{B})P(\bar{A}B)$$

(d) Generalize the result (a) for cubic inequalities.

Remark: For other nonlinear inequalities see [64] and [65].

P.2.7. (a) Let Ω denote the set of positive integers, Ω_n the set $\{1, 2, \ldots, n\}$, and $\mathscr{N}$ the counting measure. Let A be any subset of Ω. If the limit $d(A) = \lim_{n\to\infty} \mathscr{N}(A\Omega_n)/n$ exists, its value $d(A)$ is called the *density* of the set A. Clearly, if A_1 and A_2 are disjoint sets, both of which have a density, then the set $A_1 + A_2$ also has a density and one has $d(A_1 + A_2) = d(A_1) + d(A_2)$. Show that nevertheless, the family of all subsets A of Ω which possess a density is *not* an algebra of sets by constructing two sets A_1 and A_2 such that both possess a density but neither their union nor their intersection has a density. (This is only one of the reasons why density cannot be interpreted as a probability measure.)

Hint: Let A_1 be the set of even numbers. Let A_2 be defined as follows: A_2 consists of all even numbers lying in the intervals $[2^{2k}, 2^{2k+1})$ and of all odd numbers lying in the intervals $[2^{2k+1}, 2^{2k+2})$ $(k = 0, 1, 2, \ldots)$. Then both A_1 and A_2 have densities, but the sets $A_1 + A_2$ and A_1A_2 do not possess densities.

(b) Show that in Example 2.5.3 the limit $\lim_{n\to\infty} P_n(A)$ exists if and only if the density $d(A)$ of A exists, and for such sets

$$\lim_{n\to+\infty} P_n(A) = \frac{P(A) + d(A)}{2}$$

P.2.8. Let the axiom (γ) in Definition 2.2.2 be replaced by the following weaker axiom (γ^*): If $B \in \mathscr{B}$, $C \in \mathscr{B}$, $B \subseteq C$, and *if* (2.2.3) holds, then (2.2.4) holds for every $A \in \mathscr{A}$. Let us call the system $[\Omega, \mathscr{A}, \mathscr{B}, P(A \mid B)]$ for which this weaker set of axioms holds a *generalized conditional probability space*. Show that a generalized conditional probability space is not necessarily generated by a single σ-finite measure.*

Remark: In his earlier papers ([27]–[29]) the author first introduced the notion of a generalized conditional probability space; in reference [33] this notion was restricted and only conditional probability spaces were considered (in reference [33] these were called simple conditional probability spaces). As regards the structure of generalized conditional probability spaces, see Császár [45].

P.2.9. Let $S = [\Omega, \mathscr{A}, \mathscr{B}, P(A \mid B)]$ be a full conditional probability space generated by the measure μ and let G be a group of one-to-one transformations of the set Ω into itself which are measurable with respect to $\mathscr{A}$, i.e., for each set $A \in \mathscr{A}$ and for each transformation $T \in G$, the set $TA = \{T\omega : \omega \in A\}$ belongs to $\mathscr{A}$. The conditional probability space S is called *invariant* with respect to the group G if for each $T \in G$ there exists a positive constant $c(T)$ such that for all $A \in \mathscr{A}$ one has $\mu(TA) = c(T)\mu(A)$ (which implies $P(TA \mid TB) = P(A \mid B)$ for $A \in \mathscr{A}$, $B \in \mathscr{B}$).

A probability space $S = (\Omega, \mathscr{A}, P)$ is called invariant with respect to a group G of measurable one-to-one transformations of Ω if for every $T \in G$ and every $A \in \mathscr{A}$ one has $P(TA) = P(A)$. (See [49].)

(a) Prove that the classical probability space can be characterized as a finite (discrete) probability space which is invariant with respect to the group of all permutations of the set of outcomes of the experiment.

(b) Prove that the Lebesgue probability space (see Example 2.3.5) is invariant with respect to the group of translations mod 1 of the interval [0, 1].

(c) Show that the full conditional probability space $[R^k, \mathscr{F}_k, \lambda_k]$, where R^k is the k-dimensional Euclidean space, $\mathscr{F}_k$ is the set of all Borel subsets of R^k, and λ_k is the k-dimensional Lebesgue measure, is invariant with respect to the group of all linear transformations mapping R^k onto itself. If T is such a transformation,

$$T(x_1, \ldots, x_k) = (y_1, \ldots, y_k)$$

where

$$y_j = \sum_{h=1}^{k} C_{j,h} x_h + d_j \quad (j = 1, 2, \ldots, k)$$

then $c(T)$ is equal to the absolute value of the determinant of the nonsingular matrix $(C_{j,h})$.

(d) Let R^+ be the set of all positive numbers, $\mathscr{F}^+$ the family of Borel subsets of R^+; let the measure μ be defined by

$$\mu(A) = \int_A \frac{dx}{x} \quad \text{for } A \in \mathscr{F}^+$$

* If measures the values of which are not numbers but elements of a non-Archimedean field are admitted, then generalized conditional probability spaces may also be generated by a single measure (see Krausz [69]).

Show that the full conditional probability space $[R^+, \mathcal{F}^+, \mu]$ is invariant with respect to the group of all transformations $Tx = cx^\alpha$ where c is an arbitrary positive number and α is an arbitrary real number different from 0.

(e) Show that the measure μ defined in Exercise E.2.2(a) on the set of all straight lines of the plane is the only measure which is invariant with respect to the group of all Euclidean motions of the plane.

P.2.10. Let $A_1, A_2, \ldots, A_n$ be arbitrary events, and put

$$C_k = \sum A_{i_1} A_{i_2}, \ldots, A_{i_k} \quad (k = 1, 2, \ldots n)$$

where the summation has to be extended over all $\binom{n}{k}$ k-tuples $(i_1, i_2, \ldots, i_k)$ $(1 \leqq i_1 < i_2 < \cdots < i_k \leqq n)$. Show that

$$\sum_{k=1}^{n} P(C_k) = \sum_{k=1}^{n} P(A_k) \tag{P.2.10.1}$$

and

$$\prod_{k=1}^{n} P(C_k) \leqq \prod_{k=1}^{n} P(A_k) \tag{P.2.10.2}$$

Hint: To prove (P.2.10.1) let α_k denote the indicator of the event A_k and put $\gamma = \sum_{k=1}^{n} \alpha_k$. Also note that $\sum_{k=1}^{n} P(A_k) = \sum_{k=1}^{n} E(\alpha_k)$ and (using the linearity of the expectation)

$$\sum_{k=1}^{n} P(C_k) = \sum_{k=1}^{n} k \cdot P(C_k \bar{C}_{k+1}) = E(\gamma) = \sum_{k=1}^{n} E(\alpha_k)$$

Equation (P.2.10.1) can also be proved by the use of Theorem 2.6.1.

The inequality (P.2.10.2) follows for $n = 2$ from the identity (2.6.4) and the fact that $P(AB) \leqq \min[P(A), P(B)]$. For $n \geqq 3$ notice that if A_k and A_h are replaced by $A_k \cdot A_h$ and $A_k + A_h$, the left-hand side of (P.2.10.2) remains unchanged and the right-hand side cannot increase. By a finite number of such replacements and by relabeling the events we can achieve that $A_j \supseteq A_{j+1}$ for $j = 1, 2, \ldots, n - 1$, in which case $C_j = A_j$ for $j = 1, 2, \ldots, n$ and there is equality in (P.2.10.2).

Remark: At the "M. Schweitzer competition" for university students held in Hungary in 1968 it was proposed as one of the problems to prove the inequality (P.2.10.2). The above given proof was found by L. Lovász, one of the participants of the competition.

REFERENCES

[1] P. R. Halmos, *Measure Theory*, Van Nostrand, New York, 1950.
[2] H. L. Royden, *Real Analysis*, Macmillan, New York, 1963.
[3] J. Neveu, *Mathematical Foundations of the Calculus of Probability*, Holden-Day, Inc., 1965.
[4] A. N. Kolmogoroff, *Foundations of the Theory of Probability*, Chelsea, New York, 1950.

[5] G. Cardano, *The Book on Games of Chance*, Holt, Rinehart & Winston, New York, 1961.

[6] G. Galilei, *Le Opere*, **14**: 243–246, Firenze, 1855.

[7] J. Bernoulli, *Ars Coniectandi*, Basel, 1713.

[8] T. Bayes, "An Essay Towards Solving a Problem in the Doctrine of Chances," *Phil. Trans.*, **53**: 370–418, 1763.

[9] P. S. Laplace, *Théorie Analytique des Probabilités*, 1795.

[10] P. S. Laplace, *A Philosophical Essay on Probabilities*, Dover, New York, 1951.

[11] S. D. Poisson, *Recherches sur la Probabilité de Judgements*, Paris, 1837.

[12] C. F. Gauss, *Theoria Combinationis Observationum Erroris Minimus Obnoxiae*, Göttingen, 1821.

[13] R. Helmert, "Über die Wahrscheinlichkeit der Potenzsummen der Beobachtungsfehler und über einige damit im Zusammenhang stehende Fragen," *Z. Math. Phys.*, **21**: 192–219, 1876.

[14] N. I. Lobačevsky, "Sur la Probabilité des Résultats Moyens Tirés Des Observations Repetées, *J. Reine Angew. Math.*, **24**: 164–170, 1842.

[15] P. L. Čebishev, *Teoria Veroiatnostei*, Moscow, 1936.

[16] A. A. Markoff, *Wahrscheinlichkeitsrechnung*, Teubner, Leipzig, 1912.

[17] A. M. Liapounoff, *Selected Papers* (in Russian), Moscow, 1948.

[18] J. Bertrand, *Calcul des Probabilités*, Gauthier-Villars, Paris, 1889.

[19] H. Poincaré, *Calcul des Probabilités*, Carré-Naud, Paris, 1912.

[20] R. von Mises, *Wahrscheinlichkeitsrechnung*, Deuticke, Leipzig, 1931. (See also R. von Mises, *Mathematical Theory of Probability and Statistics*, edited and complemented by Hilda Geiringer, Academic Press, New York, 1964.)

[21] É. Borel, "Les Probabilités Dénombrables et Leurs Applications Arithmétiques," *Rend. Circ. Mat. Palermo*, **27**: 247–271, 1909.

[22] A. Lomnicki, "Nouveaux Fondements du Calcul des Probabilitiés," *Fund. Math.*, **4**: 34–71, 1923.

[23] P. Lévy, *Calcul des Probabilités*, Gauthier-Villars, Paris, 1925.

[24] H. Steinhaus, "Les Probabilités Denombrables et Leur Rapport à la Théorie de la Mesure," *Fund. Math*, **4**: 286–310, 1923.

[25] C. Jordan, "On Probability," *Proc. Phys. Math. Soc. Japan*, **7**: 96–109, 1925.

[26] L. Bachelier, *Calcul des Probabilités*, Gauthier-Villars, Paris, 1912.

[27] A. Rényi, "On a New Axiomatic Theory of Probability," *Acta Math. Acad. Sci. Hung.*, **6**: 285–335, 1955.

[28] A. Rényi, "On Conditional Probability Spaces Generated by a Dimensionally Ordered Set of Measures," *Teoria Veroiatnostei*, **1**: 61–71, 1956.

[29] A. Rényi, "A New Deduction of Maxwell's Law of Velocity Distribution," *Isv. Mat. Inst. Sofia*, **2**: 45–53, 1957.

[30] A. Rényi, "Quelques Remarques sur les Probabilités Des Événements Dépendantes," *J. Math. Pures Appl.*, **37**: 393–398, 1958.

[31] A. Rényi, *Valószinüségszámitás* (in Hungarian), Tankönyvkiadó, Budapest, 1954.

[32] A. Rényi, *Wahrscheinlichkeitsrechnung, mit einem Anhang über Informationstheorie*, VEB Deutscher Verlag der Wissenschaften, Berlin, 1962.

[33] A. Rényi, "Sur les Espaces Simples Des Probabilités Conditionnelles," *Ann. Inst. H. Poincaré*, **1**: 3–19, 1964.

[34] G. A. Barnard, "Statistical Inference," *J. Royal Stat. Soc., Ser. B.*, **11**: 115–139, 1949.

[35] A. H. Copeland, "Postulates for the Theory of Probability," *Am. J. Math.*, **63**: 741–762, 1941.

[36] M. Fréchet, *Généralités sur les Probabilités*, Gauthier-Villars, Paris, 1950.

[37] I. J. Good, *Probability and Weighing of Evidence*, Griffin, London, 1950.

[38] H. Jeffreys, *Theory of Probability*, Clarendon Press, Oxford, 1948.

[39] J. M. Keynes, *A Treatise on Probability Theory*, MacMillan, London, 1952.

[40] B. O. Koopman, "The Axioms and Algebra of Intuitive Probability," *Ann. Math.*, **41**: 269–292, 1940.

[41] K. Popper, *The Logic of Scientific Discovery*, Harper & Row, New York, 1965.

[42] H. Reichenbach, *Wahrscheinlichkeitslehre*, Sijthoff, Leiden, 1935.

[43] R. D. Luce, *Individual Choice Behavior*, Wiley, New York, 1959.

[44] L. E. Dubins and L. O. Savage, *How to Gamble if You Must*, McGraw-Hill, New York, 1965.

[45] A. Császár, "Sur la Structure des Espaces de Probabilité Conditionnelle," *Acta Math. Acad. Sci. Hung.*, **6**: 337–361, 1955.

[46] A. Rényi, *Briefe über die Wahrscheinlichkeit*, VEB Deutscher Verlag der Wissenschaften, 1969.

[47] J. Galambos and A. Rényi, "On Quadratic Inequalities in the Theory of Probability," *Studia Sci. Math. Hung.*, **3**: 351–358, 1968.

[48] J. M. O. Speakman, "An Algebraic Characterization of Convergence Ideals," *J. London Math. Soc.*, **44**: 26–30, 1968.

[49] H. Hadwiger, "Zur Axiomatik der innermathematischen Wahrscheinlichkeitstheorie," *Mitteilungen Schweizer Versicherungsmath.*, **58**: 151–165, 1958.

[50] I. Todhunter, *A History of the Mathematical Theory of Probability*, MacMillan, London, 1865.

[51] F. N. David, *Games, Gods and Gambling*, Griffin, London, 1962.

[52] H. Cramér and H. Wold, "Some Theorems on Distribution Functions," *J. London Math. Soc.*, **11**: 290–294, 1936.

[53] A. Rényi, "On Projections of Probability Distributions," *Acta Math. Acad. Sci. Hung.*, **3**: 131–142, 1952.

[54] E. Lukács, *Characteristic Functions*, Griffin, London, 1960.

[55] E. Lukács and R. G. Laha, *Applications of Characteristic Functions in Probability Theory*, Griffin, London, 1964.

[56] W. Blaschke, *Vorlesungen über Integralgeometrie*, 3rd edition, VEB Deutscher Verlag der Wissenschaften, Berlin, 1955.

[57] J. A. Shohat and J. D. Tamarkin, *The Problem of Moments*, Am. Math. Soc., Math. Surveys No. 1, New York, 1943.

[58] N. I. Akhiezer, *The Classical Moment Problems*, Oliver and Boyd, Edinburgh, 1965.

[59] D. V. Widder, *The Laplace Transform*, Princeton University Press, Princeton, 1941.

[60] G. Doetsch, *Theorie und Anwendung der Laplace-Transformation*, Springer, Berlin, 1937.

[61] L. A. Santalo, "Introduction to Integral Geometry," *Actualités Scientifiques et Industrielles*, No. 1198, Hermann, Paris, 1953.

[62] H. Cramér, "Sur un Nouveau Théorème-Limite de la Théorie des Probabilités," *Actualités Scientifiques et Industrielles*, No. 735, Hermann, Paris, 1938.

[63] T. Halperin, "Best Possible Inequalities for the Probability of a Logical Function of Events," *Am. Math. Monthly*, **72**: 343–359, 1965.

[64] P. Erdös, J. Neveu, and A. Rényi, "An Elementary Inequality Between the Probabilities of Events," *Math. Scandinavica*, **13**: 99–104, 1963.

[65] S. Zubrzycki, "Les Inégalités Entre les Moments des Variables Aléatoires Équivalentes," *Studia Math.*, **14**: 232–242, 1954.

[66] G. Pōlya, *Mathematics and Plausible Reasoning*, Vol. II: 58, Princeton University Press, Princeton, 1954.

[67] S. M. Ulam, "Zur Masstheorie in der allgemeinen Mengenlehre," *Fund. Math.*, **16**: 141–150, 1930.

[68] L. Schwartz, *Théorie des distributions*, Actualités Scientifiques et Industrielles, No. 1191, Paris, Hermann, 1950.

[69] P. H. Krauss, "Representation of Conditional Probability Measures on Boolean Algebras," *Acta Math. Acad. Sci. Hung.*, **19**: 229–241, 1968.

[70] A. Heppes, "On the Determination of Probability Distributions of More Dimensions by Their Projections," *Acta Math. Acad. Sci. Hung.*, **7**: 403–410, 1956.

[71] D. A. Kappos, *Strukturtheorie der Wahrscheinlichkeitsfelder und-räume*, Springer, Heidelberg, 1960.

CHAPTER 3

INDEPENDENCE

3.1 INDEPENDENCE OF TWO EVENTS

DEFINITION 3.1.1. *Let $S=(\Omega, \mathscr{A}, P)$ be a probability space. The events $A \in \mathscr{A}$ and $B \in \mathscr{A}$ are called independent** if*

$$P(AB)=P(A)P(B) \tag{3.1.1}$$

Remark: If A and B are independent events, we shall also say that A is independent from B (or conversely). Clearly if A is an event such that $P(A)=1$, then A is independent from every event B, because $P(AB)= P(B)=P(A)P(B)$. Similarly, if A is an event such that $P(A)=0$, then A is independent from every event B, because $P(AB)=0=P(A)P(B)$. Thus, Ω and $\varnothing$ are independent from every event $B \in \mathscr{A}$. If A and B are independent and $P(B)>0$, then it follows from (3.1.1) that

$$P(A \mid B)=\frac{P(AB)}{P(B)}=\frac{P(A)P(B)}{P(B)}=P(A) \tag{3.1.2}$$

Thus, *if A and B are independent and $P(B)>0$, the conditional probability of A, under the condition that B happened, is equal to the unconditional probability of A*; i.e., the chances of A are not changed by the fact that the event B has taken place. Thus the definition of independence of two events is in agreement with the intuitive meaning of the word "independence": two events are called independent if the occurrence of one of them has no influence on the occurrence of the other. It is easy to show that if A and B are independent, then $\bar{A}$ and $\bar{B}$ are also independent. As a matter of fact, for any events A and B, in view of $P(A\bar{B})+P(AB)=P(A)$, we have

$$P(A\bar{B})-P(A)P(\bar{B})=-[P(AB)-P(A)P(B)] \tag{3.1.3}$$

Thus, if A and B are independent, so are A and $\bar{B}$, and similarly $\bar{A}$ and B. It follows further from (3.1.3) that

$$P(\bar{A}\bar{B})-P(\bar{A})P(\bar{B})=P(AB)-P(A)P(B) \tag{3.1.4}$$

* The term "stochastically independent" is sometimes used in order to avoid misunderstanding, as the term "independence" has other meanings in other branches of mathematics (e.g., "linear independence").

Thus, if A and B are independent, so are $\bar{A}$ and $\bar{B}$. Thus if A and B are independent, then any one of the events $\varnothing$, Ω, A, or $\bar{A}$ is independent from any one of the events $\varnothing, \Omega, B, \bar{B}$. The events $\varnothing, \Omega, A$, and $\bar{A}$ form an algebra of events, wh ch is a subalgebra of the σ-algebra $\mathscr{A}$. (Such an algebra is called a *four-element algebra.*) This leads to the following generalization of Definition 3.1.1:

DEFINITION 3.1.2. *Let $S=(\Omega, \mathscr{A}, P)$ be a probability space. Two σ-algebras $\mathscr{A}_1$ and $\mathscr{A}_2$ which are both subalgebras of $\mathscr{A}$ are called independent if every event $A \in \mathscr{A}_1$ is independent from every event $B \in \mathscr{A}_2$.*

Clearly, the independence of the algebras $(\varnothing, \Omega, A, \bar{A})$ and $(\varnothing, \Omega, B, \bar{B})$ is equivalent to the independence of the events A and B. Evidently, if the events A_i $(i=1, 2, \ldots)$ mutually exclude each other, and each A_i is independent from the event B, then the sum $A=\sum_i A_i$ is also independent from B, because

$$P(AB)=\sum_i P(A_iB)=\sum_i P(A_i)P(B)=P(A)\cdot P(B) \tag{3.1.5}$$

Thus it follows that if the σ-algebras $\mathscr{A}_1$ and $\mathscr{A}_2$ are both subalgebras of $\mathscr{A}$ and the experiments $(\Omega, \mathscr{A}_1)$ and $(\Omega, \mathscr{A}_2)$ are purely atomic (see Section 1.6), the atoms of the experiment $(\Omega, \mathscr{A}_1)$ being A_i $(i=1, 2, \ldots)$ and those of the experiment $(\Omega, \mathscr{A}_2)$ being B_j $(j=1, 2, \ldots)$, then the σ-algebras $\mathscr{A}_1$ and $\mathscr{A}_2$ are independent if and only if each A_i is independent from each B_j.

If A_1 and A_2 are independent from B and $A_2 \subseteq A_1$, then $A_1\bar{A}_2$ is also independent from B. (However, in this statement the condition $A_2 \subseteq A_1$ cannot be omitted. See Example 3.2.1.)

In view of (3.1.3) and (3.1.4) it is natural to measure the dependence between two σ-algebras $\mathscr{A}_1$ and $\mathscr{A}_2$ which are both subalgebras of the σ-algebra $\mathscr{A}$ by the quantity $\delta(\mathscr{A}_1, \mathscr{A}_2)$, defined by

$$\delta(\mathscr{A}_1, \mathscr{A}_2)=4 \sup_{\substack{A \in \mathscr{A}_1 \\ B \in \mathscr{A}_2}} [P(AB)-P(A)P(B)] \tag{3.1.6}$$

According to this definition, $\mathscr{A}_1$ and $\mathscr{A}_2$ are independent if and only if $\delta(\mathscr{A}_1, \mathscr{A}_2)=0$. It is easy to see that for any two σ-algebras $\mathscr{A}_1$ and $\mathscr{A}_2$, we have

$$0 \leqq \delta(\mathscr{A}_1, \mathscr{A}_2) \leqq 1 \tag{3.1.7}$$

This can be seen from the identity

$$P(AB)-P(A)P(B)=P(AB)P(\bar{A}\bar{B})-P(A\bar{B})P(\bar{A}B) \tag{3.1.8}$$

From this, in view of $AB \cdot \bar{A}\bar{B}=\varnothing$, it follows that for any two events A and B one has $P(AB)+P(\bar{A}\bar{B}) \leqq 1$, and therefore

$$4[P(AB)-P(A)P(B)] \leqq 1 \tag{3.1.9}$$

Formula (3.1.9) is an equality if and only if $P(AB) = P(\bar{A}\bar{B}) = 1/2$, in which case $P(A\bar{B}) = P(\bar{A}B) = 0$, and therefore $P(A \circ B) = 0$; i.e., the sets A and B are identical up to a set of probability zero and $P(A) = P(B) = 1/2$. Thus, if the intersection of the σ-algebras $\mathscr{A}_1$ and $\mathscr{A}_2$ contains a set of probability 1/2, one has $\delta(\mathscr{A}_1, \mathscr{A}_2) = 1$.

Now let ξ and η be two random variables on the probability space $S = (\Omega, \mathscr{A}, P)$. Let $\mathscr{A}_\xi$ denote the σ-algebra of all sets $\xi^{-1}(\mathfrak{B})$ where $\mathfrak{B}$ is an arbitary Borel subset of the real line. We shall call $\mathscr{A}_\xi$ *the σ-algebra generated (or spanned) by the random variable* ξ. By definition $\mathscr{A}_\xi$ is a subalgebra of $\mathscr{A}$. We now introduce the following:

DEFINITION* 3.1.3. *Two random variables ξ and η on a probability space $(\Omega, \mathscr{A}, P)$ are called independent if the σ-algebras $\mathscr{A}_\xi$ and $\mathscr{A}_\eta$ generated by these random variables are independent.*

Example 3.1.1. Let us throw a fair die twice. The experiment is described by the probability space $S = (\Omega, \mathscr{A}, P)$ where the set Ω consists of the thirty-six elements (i, j) $(i, j = 1, 2, 3, 4, 5, 6)$, $\mathscr{A} = \mathscr{P}(\Omega)$, and the probability distribution P is uniform on Ω. If ξ denotes the number obtained at the first throw and η the number obtained at the second throw, the random variables ξ and η are independent.

THEOREM 3.1.1. *Let $S = (\Omega, \mathscr{A}, P)$ be a probability space, and let $\mathscr{A}_1$ and $\mathscr{A}_2$ be two algebras of sets which are subalgebras of $\mathscr{A}$. Let $\sigma(\mathscr{A}_i)$ denote the least σ-algebra containing $\mathscr{A}_i$ $(i = 1, 2)$. If each $A \in \mathscr{A}_1$ is independent from each $B \in \mathscr{A}_2$, then the σ-algebras $\sigma(\mathscr{A}_1)$ and $\sigma(\mathscr{A}_2)$ are independent.*

Proof. Let B be any fixed element of $\mathscr{A}_2$. The measures $\mu_1(A) = P(AB)$ and $\mu_2(A) = P(A)P(B)$ coincide for $A \in \mathscr{A}_1$, therefore, by Lemma 2.2.2, they also coincide on $\sigma(\mathscr{A}_1)$, i.e., every $A \in \sigma(\mathscr{A}_1)$ is independent from B. Let us now fix $A \in \sigma(\mathscr{A}_1)$. As the two measures $\nu_1(B) = P(AB)$ and $\nu_2(B) = P(A)P(B)$ coincide for $B \in \mathscr{A}_2$, they coincide for $B \in \sigma(\mathscr{A}_2)$; thus, every $B \in \sigma(\mathscr{A}_2)$ is independent from A. As A is an arbitary element of $\sigma(\mathscr{A}_1)$, our theorem is proved. ■

Clearly, the trivial σ-algebra $\mathscr{A}_0$ consisting only of the sets $\varnothing$ and Ω is independent from every other σ-algebra. Expressed otherwise, a constant random variable is independent from every other random variable.

If ξ_i is a discrete random variable taking on the different values $a_{i1}, a_{i2}, \ldots, a_{in}, \ldots$ $(i = 1, 2)$ and A_{ij} denotes the set of those $\omega \in \Omega$ for which $\xi_i(\omega) = a_{ij}$, then ξ_1 and ξ_2 are independent if and only if each A_{1j} is independent from each A_{2k} $(j, k = 1, 2, \ldots)$.

In general, we have the following:

* The independence of random mappings is defined similarly.

Corollary of Theorem 3.1.1. *Two random variables ξ and η are independent if for every real x and y the events $\xi^{-1}(I_x)$ and $\eta^{-1}(I_y)$ are independent.**

Proof. The sets I_x $(-\infty < x < +\infty)$ generate the Borel sets and thus the sets $\xi^{-1}(I_x)$ and $\eta^{-1}(I_y)$ generate $\mathscr{A}_\xi$ and $\mathscr{A}_\eta$, respectively. ■

Example 3.1.2. Let the probability space $S = (\Omega, \mathscr{A}, P)$ describe the experiment consisting of choosing at random, with uniform distribution, a point (x, y) in the unit square $0 \leqq x < 1, 0 \leqq y < 1$ of the (x, y) plane. In other words, let Ω be the set of points $\omega = (x, y)$ $(0 \leqq x < 1, 0 \leqq y < 1)$ of the plane, let $\mathscr{A}$ be the set of all Borel subsets of Ω, and let the probability $P(C)$ of every event $C \in \mathscr{A}$ be equal to the two-dimensional Lebesgue measure of the set C. Let the random variables ξ and η be defined as follows: If $\omega = (x, y)$ $(0 \leqq x < 1, 0 \leqq y < 1)$, we put $\xi(\omega) = x$ and $\eta(\omega) = y$, i.e., ξ and η are the coordinates of the random point $\omega = (\xi, \eta)$. Then the random variables ξ and η are independent. This follows immediately from the corollary of Theorem 3.1.1. Note that the independence of every $A \in \mathscr{A}_\xi$ from every $B \in \mathscr{A}_\eta$ can also be shown directly (without using Theorem 3.1.1), e.g., by Fubini's theorem.

Now let us turn to the notion of independence in conditional probability spaces. We introduce the following:

Definition 3.1.4. *If $S = [\Omega, \mathscr{A}, \mathscr{B}, P(A \mid B)]$ is a conditional probability space, the events $A \in \mathscr{A}$ and $B \in \mathscr{A}$ are called independent with respect to the condition $C \in \mathscr{B}$ if*

$$P(AB \mid C) = P(A \mid C)P(B \mid C) \tag{3.1.10}$$

Evidently, if S is generated by the measure μ and $A + B \subseteq C$, then (3.1.10) is equivalent to

$$\mu(C)\mu(AB) = \mu(A)\mu(B) \tag{3.1.11}$$

Thus, if A and B are independent with respect to some condition C such that $A + B \subseteq C$, and $D \in \mathscr{B}$ is another set such that $A + B \subseteq D$ and $\mu(D) = \mu(C)$, then A and B are also independent with respect to D.

Example 3.1.3. Let $S = [R, \mathscr{A}, \lambda]$ be the conditional probability space of Example 2.2.5, i.e., let R be the real line, let $\mathscr{A}$ be the family of Borel subsets of R, and let λ be the Lebesgue measure. Let $A \in \mathscr{A}$ and $B \in \mathscr{A}$ be any two sets such that $\lambda(A)$, $\lambda(B)$, and $\lambda(AB)$ are all positive and finite. It is easy to see that in this case one can always find conditions C under which A and B are independent. To show this it is sufficient to remark that

$$\frac{\lambda(A)\lambda(B)}{\lambda(AB)} = \lambda(A + B) + \frac{\lambda(A\bar{B})\lambda(\bar{A}B)}{\lambda(AB)} \geqq \lambda(A + B) \tag{3.1.12}$$

* I_x and I_y denote the intervals $(-\infty, x)$ and $(-\infty, y)$, respectively.

Thus, one can always find a measurable set C such that $A + B \subseteq C$ and

$$\lambda(C) = \frac{\lambda(A)\lambda(B)}{\lambda(AB)} \tag{3.1.13}$$

and A and B will be independent with respect to such a C.

Example 3.1.4. Let Ω be the set of natural numbers, let $\mathscr{A} = \mathscr{P}(\Omega)$ be the family of all subsets of Ω, and let $S = [\Omega, \mathscr{A}, \mathscr{N}]$ be the conditional probability space generated by the counting measure $\mathscr{N}$ on the experiment $(\Omega, \mathscr{A})$. Let A_p denote the set of all natural numbers divisible by the natural number p. Let B_n denote the set $\{1, 2, \ldots, n\}$. It is easy to see that if p_1 and p_2 are relatively prime integers and n is divisible by $p_1 p_2$, then the events A_{p_1} and A_{p_2} are independent with respect to the condition B_n. To show this, first we remark that by definition if p is a divisor of n,

$$P(A_p \mid B_n) = \frac{\mathscr{N}(A_p B_n)}{\mathscr{N}(B_n)} = \frac{n/p}{n} = \frac{1}{p} \tag{3.1.14}$$

Now, clearly, a number n is divisible by the relatively prime integers p_1 and p_2 if and only if it is divisible by their product $p_1 p_2$; thus we have $A_{p_1} \cdot A_{p_2} = A_{p_1 p_2}$ and therefore, in view of (3.1.14),

$$P(A_{p_1} A_{p_2} \mid B_n) = P(A_{p_1 p_2} \mid B_n) = \frac{1}{p_1 p_2} = P(A_{p_1} \mid B_n) P(A_{p_2} \mid B_n) \tag{3.1.15}$$

That is, A_{p_1} and A_{p_2} are independent with respect to B_n. Thus, for instance, if we choose at random one of the first 10,000 natural numbers, the event that this number will be even is independent from the event that it will be divisible by five.

The independence of σ-algebras, respectively of random variables, with respect to a condition C in a conditional probability space, is defined in the same manner as that of events.

3.2 INDEPENDENCE OF SEQUENCES OF EVENTS

It is possible that two events A_1 and A_2 are independent from an event B, but that their product $A_1 A_2$ is not independent from B. That is shown by the following:

Example 3.2.1. Let us throw twice with a fair coin. Let A_1 denote the event that we have thrown a head at the first throw and A_2 the event that we have thrown a head at the second throw. Let A_3 denote the event that the results of the two throws are identical, i.e., we have thrown either a head both times or a tail both times. Then clearly $P(A_i) = 1/2$ for $i = 1, 2, 3$, and $P(A_1A_2) = P(A_1A_3) = P(A_2A_3) = 1/4$. Thus, any two of the events A_1, A_2, A_3 are independent; however, we have

$$P((A_1A_2)A_3) = 1/4 \neq 1/8 = P(A_1A_2)P(A_3)$$

i.e., the event A_1A_2 is not independent from the event A_3. Note that $P(A_1\bar{A}_2A_3) = 0$; thus $A_1\bar{A}_2$ is not independent from A_3 either, because $P(A_1\bar{A}_2)P(A_3) = 1/8$.

However, for certain sequences of events it is true that not only two of them are independent but, besides this, any two products formed from these events which do not both contain the same factor are independent. This is shown by the following:

Example 3.2.2. Let us draw cards from a pack of fifty-two cards so that the card drawn is always replaced in the pack and the pack is thoroughly shuffled before drawing the next card. Let A_k denote the event that we have drawn an ace at the kth drawing $(k = 1, 2, \ldots)$. Then it is easy to see that if the natural numbers $i_1, \ldots, i_r; j_1, \ldots, j_s$ are all different, then the events $A_{i_1}A_{i_2}\cdots A_{i_r}$ and $A_{j_1}A_{j_2}\cdots A_{j_s}$ are independent. As a matter of fact, the probability of drawing any prescribed sequence of n cards is clearly the same, namely 52^{-n} and $P(A_k) = 1/13$. Therefore, if $1 \leqq i_1 < i_2 < \cdots < i_r$, we have

$$P(A_{i_1}A_{i_2}\cdots A_{i_r}) = \frac{1}{13^r} = \prod_{n=1}^{r} P(A_{i_n})$$

Thus, if the numbers $i_1, \ldots, i_r; j_1, \ldots, j_s$ are all different, we have

$$P(A_{i_1}\cdots A_{i_r})P(A_{j_1}\cdots A_{j_s}) = \frac{1}{13^{r+s}} = P(A_{i_1}\cdots A_{i_r}A_{j_1}\cdots A_{j_s})$$

This example suggests that it is useful to introduce the following:

Definition 3.2.1. *Let $S = (\Omega, \mathscr{A}, P)$ be a probability space. The events $A_r \in \mathscr{A}$ $(r = 1, 2, \ldots, n)$ are called independent* if for any k-tuple of different integers $1 \leqq i_1 < i_2 < \cdots < i_k \leqq n$ $(k = 2, 3, \ldots, n)$, we have*

$$P(A_{i_1}A_{i_2}\cdots A_{i_k}) = \prod_{n=1}^{r} P(A_{i_n}) \tag{3.2.1}$$

The infinite sequence of events A_k $(k = 1, 2, \ldots)$ will be called independent if the events $A_1, A_2, \ldots, A_n$ are independent for each $n = 2, 3, \ldots$.

Let us mention some immediate consequences of Definition 3.2.1. Any subsequence of a finite or infinite sequence of independent events is again

* Independent events are sometimes called "mutually independent" to emphasize that they are not only pairwise independent, but also in the sense of Definition 3.2.1.

a sequence of independent events. Evidently (3.2.1) contains $2^n - n - 1$ conditions; thus, if $n = 2$, we have only one condition, namely $P(A_1A_2) = P(A_1)P(A_2)$, i.e., for $n = 2$, Definition 3.2.1 reduces to Definition 3.1.1. If $n = 3$, we have four conditions: The events A, B and C are independent if we have

$$P(AB) = P(A)P(B) \qquad P(BC) = P(B)P(C)$$

$$P(AC) = P(A)P(C) \qquad P(ABC) = P(A)P(B)P(C)$$

If the first condition holds, the last can be replaced by $P(ABC) = P(AB)P(C)$, i.e., three events are independent if any two of them are independent and the product of two of them is independent from the third. If a sequence of events $A_1, A_2, \dots$ is such that any two of them are independent, we shall call these events *pairwise* independent; if any three are independent, we call them *independent by three*, etc. It follows from what has been said in Section 3.1 that if the events $A_1, A_2, \dots, A_n$ are independent, and for any event A we put $A^1 = A$ and $A^{-1} = \bar{A}$, and if $\delta_1, \delta_2, \dots, \delta_n$ is any sequence every element of which is equal to $+1$ or to -1, the events $A_1^{\delta_1}, A_2^{\delta_2}, \dots, A_n^{\delta_n}$ are also independent, and thus we have

$$P\left(\prod_{k=1}^{n} A_k^{\delta_k}\right) = \prod_{k=1}^{n} P(A_k^{\delta_k}) = \prod_{k=1}^{n} \left[\frac{(1+\delta_k)}{2} P(A_k) + \frac{(1-\delta_k)}{2}(1 - P(A_k))\right] \tag{3.2.2}$$

It follows that if $A_1, \dots, A_n$ are independent events and the probabilities $P(A_k) = p_k$ $(k = 1, 2, \dots, n)$ are given, then the probability of every polynomial of the events $A_1, \dots, A_n$ is determined thereby. As a matter of fact, in view of (3.2.2), given the probabilities $p_k = P(A_k)$ $(1 \leqq k \leqq n)$ we can compute the probability of every basic function of the events $A_1, \dots, A_n$ and, as by Theorem 1.4.1 every polynomial is the sum of basic functions which are disjoint events, it follows that we can compute the probability of every polynomial of the events A_k by means of the numbers $p_1, \dots, p_n$, i.e., these numbers determine uniquely the measure P on the least algebra of sets containing the sets $A_1, \dots, A_n$.

For instance, if the events A_k are independent and $P(A_k) = p_k$ $(k = 1, 2, \dots, n)$, one has

$$P(A_1 + A_2 + \cdots + A_n) = 1 - \prod_{k=1}^{n} (1 - p_k)$$

Similarly, as for $n = 2$, we generalize the notion of independence for algebras respectively for random variables.

DEFINITION 3.2.2. *A finite or infinite sequence of σ-algebras $\mathcal{A}_k$ $(k = 1, 2, \dots)$ of events of a probability space $S = (\Omega, \mathcal{A}, P)$ is called independent if any* *sequence of events* A_k *such that* $A_k \in \mathscr{A}_k$ $(k = 1, 2, \ldots)$ *is a sequence of independent events in the sense of Definition* 3.2.1. *A sequence* ξ_k $(k = 1, 2, \ldots)$ *of random variables is called independent if the corresponding σ-algebras* A_{ξ_k} $(k = 1, 2, \ldots)$ *generated by these random variables are independent.**

It is easy to see, in view of the continuity of measures, that if $A_1, A_2, \ldots$ is an infinite sequence of independent events, then we have

$$P\left(\prod_{n=1}^{\infty} A_n\right) = \prod_{n=1}^{\infty} P(A_n) \tag{3.2.3}$$

in the sense that if the probability of the product $\prod_{n=1}^{\infty} A_n$ is positive, then the infinite product on the right-hand side of (3.2.3) is convergent and (3.2.3) holds, while if $P(\prod_{n=1}^{\infty} A_n) = 0$, then the partial products of the infinite product on the right-hand side of (3.2.3) tend to 0.†

Example 3.2.3. Let S be the Lebesgue probability space (see Example 2.3.4). Let $R_n(x)$ be the nth Rademacher function $(n = 0, 1, 2, \ldots)$ defined by

$$R_n(x) = \operatorname{sgn} \sin(2^n \pi x) \quad (0 \leqq x < 1; n = 0, 1, \ldots) \tag{3.2.4}$$

The functions $R_n(x)$ considered as random variables on the probability space S are independent. As a matter of fact, $R_n(x)$ takes on the value $+1$ in each interval $2k/2^n < x < (2k + 1)/2^n$ and the value -1 in each interval $(2k + 1)/2^n < x < (2k + 2)/2^n$ $(k = 0, 1, \ldots, 2^{n-1} - 1)$; thus,**

$$P(R_n(x) = +1) = P(R_n(x) = -1) = 1/2 \quad (n = 1, 2, \ldots) \tag{3.2.5}$$

Now, if $\delta_1, \delta_2, \ldots, \delta_n$ is any sequence, each element of which is either $+1$ or -1, it is easy to see that the conditions $R_k(x) = \delta_k$ $(k = 1, 2, \ldots, n)$ hold in a subinterval $(l/2^n, (l + 1)/2^n)$ of length $1/2^n$ of the interval $[0, 1)$. Thus, for each such sequence δ_k of $(+1)$'s and (-1)'s we have

$$P(R_k(x) = \delta_k ; 1 \leqq k \leqq n) = \prod_{k=1}^{n} P(R_k(x) = \delta_k) \tag{3.2.6}$$

From this it follows that the random variables $R_1(x), \ldots, R_n(x), \ldots$ are independent; $R_0(x)$ is identically equal to 1, thus independent from every random variable.

Theorem 3.1.1 can be generalized for $n > 2$ as follows:

THEOREM 3.2.1. *If* $S = (\Omega, \mathscr{A}, P)$ *is a probability space and* $\mathscr{A}_k$ $(k = 1, 2, \ldots)$ *is a finite or infinite sequence of algebras of events which are subalgebras of* $\mathscr{A}$,

* Independent random variables are sometimes called "mutually independent" to emphasize that more has been supposed than their pairwise independence.

† Such an infinite product is not called convergent; an infinite product is called convergent only if its partial products tend to a number which is different from 0.

** $R_n(x)$ takes the value 0 in the points $x = k/2^n$ $(k = 0, 1, \ldots, 2^n)$, but as the Lebesgue measure of a point is 0, we may neglect these points.

and if each sequence of events A_k ($k = 1, 2, \ldots$) such that $A_k \in \mathscr{A}_k$ is independent, then the σ-algebras $\sigma(\mathscr{A}_k)$ ($k = 1, 2, \ldots$) are also independent.

The proof of Theorem 3.2.1 is essentially the same as that of Theorem 3.1.1 and therefore can be omitted.

THEOREM 3.2.2. *Let $(\Omega, \mathscr{A}, P)$ be a probability space. Suppose that the σ-algebras $\mathscr{A}_1, \mathscr{A}_2, \ldots, \mathscr{A}_r$ ($r \geqq 3$, $\mathscr{A}_i \subseteq \mathscr{A}$) are independent. Let $\sigma(\mathscr{A}_1, \mathscr{A}_2)$ denote the least σ-algebra containing both $\mathscr{A}_1$ and $\mathscr{A}_2$. Then the σ-algebras $\sigma(\mathscr{A}_1, \mathscr{A}_2), \mathscr{A}_3, \ldots, \mathscr{A}_r$ are also independent.*

Proof. According to Theorem 3.2.1 it is sufficient to show that, denoting by $\mathscr{A}_{1,2}$ the least algebra of sets containing both $\mathscr{A}_1$ and $\mathscr{A}_2$, the algebras $\mathscr{A}_{1,2}, \mathscr{A}_3, \ldots, \mathscr{A}_r$ are independent. As a matter of fact, according to Theorem 1.3.1, $\sigma(\mathscr{A}_1, \mathscr{A}_2)$ is the least σ-algebra containing the algebra $\mathscr{A}_{1,2}$. Now let C be any set in $\mathscr{A}_{1,2}$. According to Theorem 1.3.1, C can be represented in the form

$$C = \sum_{k=1}^{n} A_k B_k \tag{3.2.7}$$

where $A_k \in \mathscr{A}_1$, $B_k \in \mathscr{A}_2$ ($k = 1, 2, \ldots, n$) and the sets A_k ($k = 1, 2, \ldots, n$) form a partition of Ω. It follows that

$$P(C) = \sum_{k=1}^{n} P(A_k B_k) = \sum_{k=1}^{n} P(A_k) P(B_k) \tag{3.2.8}$$

Now choose in $\mathscr{A}_j$ an arbitrary set D_j for $j = 3, 4, \ldots, r$. It follows that

$$P(CD_3 \cdots D_r) = P(C) \prod_{j=3}^{r} P(D_j) \tag{3.2.9}$$

Thus $C, D_3, \ldots, D_r$ are independent. This proves our theorem. ■

Remark: Let $\mathscr{A}_1, \mathscr{A}_2, \ldots, \mathscr{A}_n, \ldots$ be an infinite sequence of independent σ-algebras, $\mathscr{A}_n \subseteq \mathscr{A}$ for $n = 1, 2, \ldots$. It follows from Theorem 3.2.2 combined with Theorem 3.1.1 that the σ-algebra $\mathscr{A}_1$ is independent from the least σ-algebra $\sigma(\mathscr{A}_2, \mathscr{A}_3, \ldots)$ containing all the σ-algebras $\mathscr{A}_k$, $k \geqq 2$.

Example 3.2.4. Let the experiment $\mathscr{E} = (\Omega, \mathscr{A})$ consist in choosing at random a point in the unit "cube" $0 \leqq x_k < 1$ ($k = 1, 2, \ldots, r$) of the r-dimensional Euclidean space, with uniform distribution, i.e., let $P(A)$ for each measurable subset A of Ω be equal to the r-dimensional Lebesgue measure of the set A. Let the random variables ξ_k ($k = 1, 2, \ldots, r$) be defined as follows: If $\omega = (x_1, x_2, \ldots, x_r)$, put $\xi_k(\omega) = x_k$ ($k = 1, 2, \ldots, r$). The random variables $\xi_1, \xi_2, \ldots, \xi_r$ are independent; this can be shown in exactly the same way as the corresponding statement in Example 3.1.2 of which Example 3.2.4 is the natural generalization.

DEFINITION 3.2.5. *A finite or infinite sequence A_n $(n=1, 2, \ldots)$ of events in a conditional probability space is called independent with respect to a condition B if one has for each k-tuple $i_1 < i_2 < \cdots < i_k$ $(k=2, 3, \ldots)$ of natural numbers*

$$P(A_{i_1} \cdot A_{i_2} \cdots A_{i_k} \mid B) = \prod_{j=1}^{k} P(A_{i_j} \mid B)$$

Example 3.2.5. Let $p_1, p_2, \ldots, p_r$ be any sequence of pairwise relatively prime natural numbers. The events $A_{p_1}, A_{p_2}, \ldots, A_{p_r}$ are independent with respect to the condition B_n if n is divisible by the product $p_1 p_2 \cdots p_r$; here A_p and B_n have the same meaning as in Example 3.1.4.

3.3 CONSTRUCTION OF A PROBABILITY MEASURE WITH RESPECT TO WHICH QUALITATIVELY INDEPENDENT EVENTS ARE INDEPENDENT

Let $S = (\Omega, \mathscr{A}, P)$ be a probability space and let $A_1, A_2, \ldots, A_n$ be a sequence of independent events in S. Suppose that putting

$$P(A_k) = p_k \tag{3.3.1}$$

we have

$$0 < p_k < 1 \quad (k=1, 2, \ldots, n) \tag{3.3.2}$$

It follows from (3.2.2) that each of the 2^n events

$$A_1^{\delta_1} A_2^{\delta_2} \cdots A_n^{\delta_n} \quad (\delta_k = \pm 1; k=1, 2, \ldots, n)$$

has a positive probability; thus none of these sets can be empty, i.e., the sets $A_1, A_2, \ldots, A_n$ are qualitatively independent in the sense of Section 1.5. Thus the qualitative independence of a sequence of events $A_1, \ldots, A_n$ concerning an experiment $\mathscr{E}$ is a necessary condition for the existence of a probability P over $\mathscr{E}$ such that the events $A_1, \ldots, A_n$ should be independent in the probability space $(\mathscr{E}, P)$ and have prescribed probabilites $P(A_k) = p_k$ lying strictly between 0 and 1. We shall now prove that this necessary condition is at the same time sufficient, in the sense that the following result holds:

THEOREM 3.3.1. *Let $A_1, A_2, \ldots, A_n$ be a sequence of qualitatively independent subsets of a set Ω. Let $\mathscr{A}$ denote the least algebra of subsets of $\mathscr{A}$ containing the sets $A_1, \ldots, A_n$. Let $p_1, p_2, \ldots, p_n$ be an arbitrary sequence of numbers such that $0 < p_k < 1$ $(k=1, 2, \ldots, n)$. Then there exists a uniquely determined probability P on the experiment $(\Omega, \mathscr{A})$ such that the events $A_1, \ldots, A_n$ are independent in the probability space $(\Omega, \mathscr{A}, P)$ and $P(A_k) = p_k$ $(k=1, 2, \ldots, n)$.*

Proof. As $\mathscr{A}$ consists of all polynomials of the events $A_1, \ldots, A_n$, define $P(A)$ for each basic function $A_1^{\delta_1} A_2^{\delta_2} \cdots A_n^{\delta_n}$ by putting

$$P\left(\prod_{k=1}^{n} A_k^{\delta_k}\right) = \prod_{k=1}^{n} \left[\frac{(1+\delta_k)}{2} p_k + \frac{(1-\delta_k)}{2} (1-p_k)\right] \tag{3.3.3}$$

If the canonical representation of an event A is $A = \sum_{j=1}^{r} B_{h_j}$ where $B_{h_1}, \ldots, B_{h_r}$ are different basic functions of the events $A_1, \ldots, A_n$, put $P(A) = \sum_{j=1}^{r} P(B_{h_j})$. In this way a probability measure P is defined over the experiment $(\Omega, \mathscr{A})$. Now clearly, the canonical representation of the event $A_{i_1} A_{i_2} \cdots A_{i_s}$ $(1 \leqq i_1 < i_2 < \cdots < i_s \leqq n)$ is obtained by taking the sum of those basic functions $A_1^{\delta_1} A_2^{\delta_2} \cdots A_n^{\delta_n}$ for which $\delta_{i_1} = \delta_{i_2} = \cdots = \delta_{i_s} = 1$ and the other δ_j's are arbitrary. It follows that

$$P(A_{i_1} A_{i_2} \cdots A_{i_s}) = p_{i_1} p_{i_2} \cdots p_{i_s} \prod_{j \neq i_h} (p_j + (1 - p_j)) = p_{i_1} p_{i_2} \cdots p_{i_s} \quad (3.3.4)$$

Especially, $P(\Omega) = 1$, and for $s = 1$, we have

$$P(A_i) = p_i \quad (i = 1, 2, \ldots, n) \quad (3.3.5)$$

Thus for $1 \leqq i_1 < i_2 < \cdots < i_s \leqq n$ $(s = 2, 3, \ldots, n)$,

$$P(A_{i_1} A_{i_2} \cdots A_{i_s}) = \prod_{h=1}^{s} P(A_{i_h}) \quad (3.3.6)$$

That is, the events $A_1, \ldots, A_n$ have the prescribed probabilities and are independent. ■

Remark: Note that we have proved implicitly that the set of conditions (3.2.2) is equivalent to the set of conditions (3.2.1).

As (3.2.2) contains 2^n equations, while (3.2.1) contains only $2^n - n - 1$ equations, clearly, from the set of equations (3.2.2), $n + 1$ suitably chosen equations may be omitted. For instance, if we omit from the set of equations (3.2.2) that in which all the numbers $\delta_1, \ldots, \delta_n$ are equal to -1 and those in which all but one of the numbers $\delta_1, \ldots, \delta_n$ are equal to -1, the remaining $2^n - n - 1$ equations imply those omitted and thus are sufficient to ensure the independence of the events $A_1, \ldots, A_n$. However, it is not true that any $2^n - n - 1$ of the equations (3.2.2) imply the remaining $n + 1$. For instance, if $n = 3$, we have $2^n - n - 1 = 4$, but the four equations (3.2.2) in which $\delta_1 = +1$ are not sufficient to deduce the other four. (See Exercise E.3.6.)

The following example shows that Theorem 3.3.1 cannot be generalized for an infinite sequence of events such that any finite number of these events is qualitatively independent.

Example 3.3.1. Let Ω denote the set of all infinite sequences $\omega = (\omega_1, \ldots, \omega_n, \ldots)$ where each ω_n is either 0 or 1 and only a finite number of the ω_n's are equal to 1. Let A_n denote the set of sequences $(\omega_1, \ldots, \omega_n, \ldots)$ such that $\omega_n = 1$. Then the sets $A_1, A_2, \ldots, A_N$ are qualitatively independent for every $N \geqq 2$. However, there does not exist a probability measure P on the least σ-algebra $\mathscr{A}$ containing all the sets A_n and such that the events A_n are independent and $P(A_n) = 1/2$ for $n = 1, 2, \ldots$. As a matter of fact, suppose that such a measure does exist; let B_N denote the event that

$\omega_n = 0$ for all $n \geq N$. Then clearly,

$$P(B_N) \leq P(\bar{A}_{N+1}\bar{A}_{N+2} \cdots \bar{A}_{N+k}) = 1/2^k \quad \text{for every } k \geq 1$$

and thus $P(B_N) = 0$ for every N. Thus it would follow that $P(\sum_{N=1}^{\infty} B_N) = 0$. As $\Omega = \sum_{N=1}^{\infty} B_N$ and $P(\Omega) = 1$, we have arrived at a contradiction.

To obtain a generalization of Theorem 3.3.1 for an infinite sequence of events, the appropriate notion is that of *strong qualitative independence.*

DEFINITION 3.3.1. *An infinite sequence $A_1, \ldots, A_n, \ldots$ of subsets of a set Ω is called a sequence of strongly qualitatively independent sets, if none of the sets $\prod_{n=1}^{\infty} A_n^{\delta_n}$ is empty, where each δ_n is either $+1$ or -1. An infinite sequence of σ-algebras $\mathscr{A}_n$ of subsets of Ω is called strongly qualitatively independent if each sequence $A_n \in \mathscr{A}_n$ of sets is strongly qualitatively independent when $A_n \neq \varnothing$ and $A_n \neq \Omega$ $(n = 1, 2, \ldots)$.*

Remark: Note that the set of all possible products $\prod_{n=1}^{\infty} A_n^{\delta_n}$ has the power of the continuum; thus if none of these sets is empty, as they are clearly pairwise disjoint, the set Ω has to be at least of the power of the continuum. Thus there cannot exist an infinite sequence of strongly qualitatively independent subsets of a denumerable set. However, there can exist infinitely many independent events in a probability space $(\Omega, \mathscr{A}, P)$ with a denumerable basic set Ω, but the probabilities of such independent events cannot be prescribed arbitrarily (see Problem P.3.4).

We shall now prove the following theorem due to Marczewski [2]:

THEOREM 3.3.2. *Let $A_1, A_2, \ldots, A_n, \ldots$ be a strongly qualitatively independent sequence of subsets of a set Ω; let $\mathscr{A}$ denote the least σ-algebra of subsets of Ω containing all the sets A_n. Let p_n $(n = 1, 2, \ldots)$ be an arbitrary prescribed sequence of numbers such that $0 < p_n < 1$ $(n = 1, 2, \ldots)$. Then there exists a uniquely determined probability P over the experiment $\mathscr{E} = (\Omega, \mathscr{A})$ such that the events A_n $(n = 1, 2, \ldots)$ are independent in the probability space $S = (\Omega, \mathscr{A}, P)$ and $P(A_n) = p_n$ $(n = 1, 2, \ldots)$.*

Proof. Let $\mathscr{A}_0^{(n)}$ denote the family of all polynomials of the events $A_1, \ldots, A_n$. Clearly, $\mathscr{A}_0^{(n)}$ is an algebra of events. Let $\mathscr{A}_0$ be the union of the algebras $\mathscr{A}_0^{(n)}$ $(n = 1, 2, \ldots)$. Evidently $\mathscr{A}_0$ is also an algebra of events.* Let us first define the set function $P_n(A)$ for $A \in \mathscr{A}_0^{(n)}$ $(n = 1, 2, \ldots)$, as in the proof of Theorem 3.3.1, so that $P_n(A_k) = p_k$ for $k = 1, 2, \ldots, n$, and the events $A_1, A_2, \ldots, A_n$ are independent in the probability space $(\Omega, \mathscr{A}_0^{(n)}, P_n)$. Evidently, $\mathscr{A}_0^{(n)} \subseteq \mathscr{A}_0^{(m)}$ for $n < m$. It is easy to see that if $n < m$, the measure $P_m(A)$ coincides with $P_n(A)$ on $\mathscr{A}_0^{(n)}$.

* But not a σ-algebra!

Thus, if we define $P(A)$ for every $A \in \mathscr{A}_0$ by putting $P(A) = P_n(A)$ for any value of n such that $A \in \mathscr{A}_0^{(n)}$, then this definition is not ambiguous and the set function $P(A)$ is such that $P(A_k) = p_k$ for $k = 1, 2, \ldots$, and we also have $P(\Omega) = 1$. We show first that the set function $P(A)$, defined in the above way for $A \in \mathscr{A}_0$, is finitely additive. Clearly, this is true for every n on the algebra $\mathscr{A}_0^{(n)}$ as $P(A) = P_n(A)$ on $\mathscr{A}_0^{(n)}$; but if A and B are any two elements of $\mathscr{A}_0$, then $A \in \mathscr{A}_0^{(n)}$ and $B \in \mathscr{A}_0^{(m)}$ for some n and m. Suppose $n < m$; as $\mathscr{A}_0^{(n)} \subseteq \mathscr{A}_0^{(m)}$, both A and B belong to $\mathscr{A}_0^{(m)}$, and thus, if $AB = \emptyset$, it follows that $P(A + B) = P(A) + P(B)$. Thus, $P(A)$ is finitely additive on $\mathscr{A}_0$. Now we prove that $P(A)$ is a measure on $\mathscr{A}_0$. To show this we have to prove that if C_n $(n = 1, 2, \ldots)$ is a sequence of sets belonging to $\mathscr{A}_0$ which are pairwise disjoint and $C = \sum_{n=1}^{\infty} C_n$ also belongs to $\mathscr{A}_0$, then we have

$$P(C) = \sum_{n=1}^{\infty} P(C_n)$$

This is a consequence of the following:

LEMMA 3.3.1. *Let A_k $(k = 1, 2, \ldots)$ be an infinite sequence of subsets of a set Ω and suppose that the sets A_k are strongly qualitatively independent. Let $\mathscr{A}_0$ denote the algebra of sets generated by the sets A_k, i.e., the family of all polynomials of the sets A_k $(k = 1, 2, \ldots)$. If C_n $(n = 1, 2, \ldots)$ is a sequence of pairwise disjoint sets all belonging to $\mathscr{A}_0$, their sum $\sum_{n=1}^{\infty} C_n$ belongs to $\mathscr{A}_0$ if and only if all but a finite number of the sets C_n are equal to the empty set $\emptyset$.*

Proof. Suppose $C = \sum_{n=1}^{\infty} C_n$ where $C_n \neq \emptyset$, $C_n \in \mathscr{A}_0$ $(n = 1, 2, \ldots)$, $C_n C_m = \emptyset$ if $n \neq m$, and $C \in \mathscr{A}_0$. Then $\bar{C} \in \mathscr{A}_0$ also; thus it follows, putting $C_0 = \bar{C}$, that

$$\Omega = \sum_{n=0}^{\infty} C_n \tag{3.3.7}$$

where the sets C_n $(n = 0, 1, \ldots)$ are disjoint and all belong to $\mathscr{A}_0$. Thus, each C_n belongs to some $\mathscr{A}_0^{(s_n)}$. Replacing each C_n in (3.3.7) by its canonical representation in $\mathscr{A}_0^{(s_n)}$ we obtain that

$$\Omega = \sum_{m=1}^{\infty} B_m \tag{3.3.8}$$

where each B_m is a basic function of the sets $A_1, A_2, \ldots, A_{r(m)}$ where $r(m)$ is a positive integer and we may suppose that $B_n B_m = \emptyset$ if $n \neq m$. Clearly, each B_m contains either the factor A_1 or the factor $\bar{A}_1$; thus, at least one of A_1 and $\bar{A}_1$ is a factor of infinitely many sets B_m. Suppose $A_1^{\varepsilon_1}$ (where ε_1 is $+1$ or -1) is a factor of infinitely many sets B_m. Among these sets B_m the set $A_1^{\varepsilon_1}$ itself cannot occur, because each B_m which contains the factor $A_1^{\varepsilon_1}$ is a subset of $A_1^{\varepsilon_1}$, and we have supposed that the sets B_m are pairwise disjoint. Thus, every set B_m containing the factor $A_1^{\varepsilon_1}$ contains either A_2 or $\overline{A}_2$. Thus, either A_2 or $\overline{A}_2$ is contained as a factor in infinitely many B_m's. Suppose that there are infinitely many B_m's containing the factor $A_2^{\varepsilon_2}$; the set $A_1^{\varepsilon_1}A_2^{\varepsilon_2}$ itself cannot be among them, because it would contain all the others as subsets. Thus, every one of the infinitely many sets B_m which contain the factor $A_1^{\varepsilon_1}A_2^{\varepsilon_2}$ also contain either A_3 or $\overline{A}_3$ as a factor. This argument can be continued infinitely. Thus, there exists an infinite sequence ε_k $(k = 1, 2, \ldots)$, each element of which is either $+1$ or -1, such that the following two statements hold: (a) For each natural number N the factor $\prod_{k=1}^{N} A_k^{\varepsilon_k}$ is contained in infinitely many sets B_m. (b) None of the sets $\prod_{k=1}^{N} A_k^{\varepsilon_k}$ $(N = 1, 2, \ldots)$ occurs among the sets B_m. By supposition, the set $\prod_{k=1}^{\infty} A_k^{\varepsilon_k} = A_\infty$ is not empty; let ω_0 denote one of its elements. Let m be any fixed number and $B_m = A_1^{\delta_1} \cdots A_{r(m)}^{\delta_{r(m)}}$. Now clearly, $\omega_0 \in B_m$ if and only if $\delta_k = \varepsilon_k$ $(k = 1, 2, \ldots, r(m))$. But this is not the case because of the property (b) of the sequence ε_k. Thus, it follows that $\omega_0 \notin B_m$ for $m = 1, 2, \ldots$, i.e., $\omega_0 \notin \sum_{m=1}^{\infty} B_m$, but this contradicts (3.3.8). This contradiction proves our lemma. ∎

It follows from Lemma 3.3.1 that every finitely additive and nonnegative set function on the algebra $\mathscr{A}_0$ is also σ-additive on $\mathscr{A}_0$, i.e., is a measure, and thus by Lemma 2.2.1 it can be extended to a measure on $\mathscr{A} = \sigma(\mathscr{A}_0)$. Thus, $P(A)$ can be extended in a unique way to a measure on $\sigma(\mathscr{A}_0)$, satisfying all requirements of Theorem 3.3.2, which is thus proved.* ∎

Example 3.3.2. Let p_n $(n = 1, 2, \ldots)$ be an arbitrary sequence of numbers such that $0 < p_n < 1$ $(n = 1, 2, \ldots)$. Let Ω be the set of all infinite sequences of zeros and ones; let A_n denote the set of those sequences $\omega = (\omega_1, \omega_2, \ldots, \omega_n, \ldots)$ for which $\omega_n = 1$. Clearly, each infinite product $\prod_{n=1}^{\infty} A_n^{\delta_n}$ where $\delta_n = +1$ or $\delta_n = -1$ is a nonempty set, namely, the set consisting of the single point $\omega = (\omega_1, \ldots, \omega_n, \ldots)$ where $\omega_n = (1 + \delta_n)/2$ $(n = 1, 2, \ldots)$. Thus, the conditions of Theorem 3.3.2 are satisfied, and if $\mathscr{A}$ denotes the least σ-algebra of subsets of the set Ω containing the sets A_n, then there exists on $\mathscr{A}$ a measure P such that the A_n's are independent and $P(A_n) = p_n$. Evidently, $\mathscr{A}$ is identical with the least σ-algebra containing all subsets of Ω defined by a finite number of conditions of the form $\omega_k = \varepsilon_k$, where $\varepsilon_k = 1$ or 0. Thus, if Ω_k is for each k the set $\{0, 1\}$ and A_k is the four-element algebra of all subsets of Ω_k, the experiment $(\Omega, \mathscr{A})$ is the *product of the experiments* $(\Omega_k, \mathscr{A}_k)$.

Example 3.3.3. Let Ω be the unit square $0 \leqq x \leqq 1, 0 \leqq y \leqq 1$. Let $I(a, b)$ $(0 \leqq a < b \leqq 1)$ denote the subset of Ω which consists of the points (x, y) for which $a < x < b, 0 \leq y \leq 1$, further, of the points (x, y) such that

* For another proof of Theorem 3.3.2, see Problem P.3.3.

$x = a$, $0 \leqq y < 1/2$ and the points (x, y) such that $x = b$, $1/2 \leqq y \leqq 1$. If p_n is any sequence of numbers such that $0 < p_n < 1$ ($n = 1, 2, \ldots$), put

$$A_1 = I(0, p_1)$$

$$A_2 = I(0, p_1p_2) + I(p_1, p_1 + p_2(1 - p_1))$$

$$A_3 = I(0, p_1p_2p_3) + I(p_1p_2, p_1p_2 + p_1(1 - p_2)p_3) + I(p_1, p_1 + p_2(1 - p_1)p_3) + I(p_1 + p_2(1 - p_1), p_1 + p_2(1 - p_1) + p_3(1 - p_1)(1 - p_2))$$

We define the sets A_n ($n \geqq 4$) similarly.

In other words, A_n consists of 2^{n-1} rectangles with base on the interval $[0, 1]$ and height 1, which are half-closed, that is, they possess their two horizontal sides, the lower half of the left-hand side, and the upper half of the right-hand side. The set A_{n+1} is obtained by dividing the base of each of the 2^{n-1} rectangles whose union is A_n and of each of the 2^{n-1} rectangles whose union is $\overline{A}_n$ into two subintervals, so that their lengths have the ratio $p_{n+1} : (1 - p_{n+1})$. From each pair of such subintervals we take the first (i.e., the one from the left side) and make it the base of a half-closed rectangle of the type $I(a, b)$; A_{n+1} is the union of the 2^n rectangles thus obtained.

Clearly the sets A_n are strongly qualitatively independent. The least σ-algebra $\mathscr{A}$ of subsets of Ω which contains all the sets A_n clearly contains the family of all subsets A of Ω such that $A = A' \times I_2$ where A' is a Borel subset of the interval $[0, 1]$, and I_2 is the interval $[0, 1]$ (i.e., A consists of those points (x, y) of the unit square for which $x \in A'$ and $0 \leqq y \leqq 1$). If $A = A' \times I_2$, we put

$$P(A) = \lambda(A')$$

where $\lambda(A')$ is the linear Lebesgue measure of the subset A' of the interval $0 \leqq x \leqq 1$. Thus we have $P(A_n) = p_n$ ($n = 1, 2, \ldots$).

Note that after having constructed the probability measure P on $\mathscr{A}$, we may identify the event $A \in \mathscr{A}$, such that $A = A' \times I_2$, with its base A' and the set Ω itself with the interval $I_1 = \{x: 0 \leqq x \leqq 1\}$, i.e., replace the probability space $(\Omega, \mathscr{A}, P)$ by the Lebesgue probability space. Note, however, that if A'_n denotes the projection of the set A_n to the x axis, the sets A'_n are not strongly qualitatively independent, because among the sets $\prod_{n=1}^{\infty} (A'_n)^{\delta_n}$ there are (denumerably) infinitely many empty sets. However, if we add to the interval I_1 denumerably many points, by replacing each of the points x which are end points of subintervals of the set A'_n by two points (x^+ and x^-), and if such a point is the right-hand end point of a subinterval of A'_n then one of the two new points, say, x^- belongs to A'_n and x^+ belongs to, $\overline{A}'_n$ (and for a left-hand end point x, conversely $x^+ \in A'_n$ and $x^- \in \overline{A}'_n$), then the modified sets A'_n are strongly qualitatively independent and Theorem 3.3.2 can be applied.

Theorem 3.3.2 can also be formulated in the following way: *Let $\mathscr{A}_n$ denote the four-element algebra consisting of the sets $\varnothing$, Ω, A_n and $\bar{A}_n$; let a measure P_n be given on each of these algebras by putting $P_n(A_n) = p_n$, $P_n(\bar{A}_n) = 1 - p_n$ (and, of course, $P_n(\varnothing) = 0$ and $P_n(\Omega) = 1$). Then, if the algebras $\mathscr{A}_n$ are strongly qualitatively independent, there exists a unique common extension P of the measures P_n to the least σ-algebra $\mathscr{A}$ containing all algebras $\mathscr{A}_n$ $(n = 1, 2, \ldots)$ such that in the probability space $(\Omega, \mathscr{A}, P)$ the algebras $\mathscr{A}_n$ are independent.*

In this form the theorem admits generalization, which is due to Banach [1]; before formulating his theorem we first prove the following:

THEOREM 3.3.3. *Let $(\Omega, \mathscr{A}_1, P_1)$ and $(\Omega, \mathscr{A}_2, P_2)$ be two probability spaces with the same basic set Ω. Suppose that the σ-algebras $\mathscr{A}_1$ and $\mathscr{A}_2$ are qualitatively independent, i.e., if $A_i \in \mathscr{A}_i$, $A_i \neq \varnothing$ $(i = 1, 2)$ then $A_1A_2 \neq \varnothing$. Let $\mathscr{A}$ denote the least σ-algebra containing both $\mathscr{A}_1$ and $\mathscr{A}_2$. Then there exists a probability measure P on $\mathscr{A}$ which is a common extension of both P_1 and P_2, i.e., $P(A) = P_i(A)$ if $A \in \mathscr{A}_i$ $(i = 1, 2)$, and for which the σ-algebras $\mathscr{A}_1$ and $\mathscr{A}_2$ are independent, i.e., if $A_i \in \mathscr{A}_i$ $(i = 1, 2)$ we have $P(A_1A_2) = P(A_1)P(A_2)$.*

For the proof of Theorem 3.3.3 we need the following lemma:

LEMMA 3.3.2. *Let the σ-algebras $\mathscr{A}_1$ and $\mathscr{A}_2$ of subsets of the same set Ω be qualitatively independent. If $A_i \in \mathscr{A}_i$, $A_i \neq \varnothing$, $B_i \in \mathscr{A}_i$ $(i = 1, 2)$, and $A_1A_2 \subseteq B_1B_2$, then $A_i \subseteq B_i$ $(i = 1, 2)$.*

Proof. Suppose $A_1 \subseteq B_1$ does not hold; then clearly $A_1\bar{B}_1 \neq \varnothing$. As $A_1\bar{B}_1 \in \mathscr{A}_1$, we have $A_1\bar{B}_1A_2 \neq \varnothing$, which contradicts our assumption that $A_1A_2 \subseteq B_1B_2$. ■

COROLLARY TO LEMMA 3.3.2. *If $A_1A_2 = B_1B_2 \neq \varnothing$ where $A_i \in \mathscr{A}_i$, $B_i \in \mathscr{A}_i$ $(i = 1, 2)$, then $A_1 = B_1$ and $A_2 = B_2$.*

Proof (of Theorem 3.3.3). Let $\mathscr{A}_0$ denote the least algebra of sets containing both $\mathscr{A}_1$ and $\mathscr{A}_2$. Each $D \in \mathscr{A}_0$ can be represented in the form

$$D = \sum_{k=1}^{n} A_k B_k \tag{3.3.9}$$

where $A_k \in \mathscr{A}_1$, $B_k \in \mathscr{A}_2$, and the sets A_k form a partition of Ω (see Theorem 1.3.1). We may also suppose that the sets B_k are different, because this can be achieved by uniting certain terms. Now we show that under these conditions the representation of $D \in \mathscr{A}_0$ in the form (3.3.9) is unique, i.e., that the following lemma is valid:

LEMMA 3.3.3. *Let $\mathscr{A}_1$ and $\mathscr{A}_2$ be two qualitatively independent σ-algebras of subsets of the set Ω, and let $\mathscr{A}_0$ denote the least algebra containing both $\mathscr{A}_1$ and*

$\mathscr{A}_2$. Then every element D of $\mathscr{A}_0$ has a unique representation in the form (3.3.9) where $A_k \in \mathscr{A}_1$ and $B_k \in \mathscr{A}_2$ $(k = 1, 2, \ldots, n)$; further, the A_k form a partition of Ω and the B_k are different from each other.

Proof. Let $D' \in \mathscr{A}_0$ have the representation

$$D' = \sum_{j=1}^{m} A'_j B'_j \tag{3.3.9'}$$

where $A'_j \in \mathscr{A}_1$, $B'_j \in \mathscr{A}_2$, the A'_j form a partition of Ω and the B'_j are different from each other. We have to show that if $D = D'$, then—after appropriate relabeling of the terms on the right of (3.3.9′)—one has $A'_k = A_k$ and $B'_k = B_k$ for $k = 1, 2, \ldots, n$ and $m = n$. It can be shown (as in the proof of Theorem 1.3.1) that

$$D \circ D' = \sum_{k=1}^{n} \sum_{j=1}^{m} A_k A'_j (B_k \circ B'_j) \tag{3.3.10}$$

Thus, if $D = D'$, then $D \circ D' = \varnothing$ and for all those pairs (k, j) for which $A_k A'_j \neq \varnothing$, we must have (the σ-algebras $\mathscr{A}_1$ and $\mathscr{A}_2$ being qualitatively independent) $B_k \circ B'_j = \varnothing$, i.e., $B_k = B'_j$. As the B'_j are by supposition different from each other, this can hold only for one value of j for a given value of k. It follows that, relabeling the A'_j and B'_j, one has $m = n$, $A_k = A'_k$, and $B_k = B'_k$ $(k = 1, 2, \ldots, n)$, which proves Lemma 3.3.3. ∎

Now let us put, for every set $D \in \mathscr{A}_0$ having the representation (3.3.9),

$$P(D) = \sum_{k=1}^{n} P_1(A_k) P_2(B_k) \tag{3.3.11}$$

Thus the set function $P(D)$ is defined for every $D \in \mathscr{A}_0$. It is easy to see that $P(D)$ is additive. To prove Theorem 3.3.3 it is sufficient to show that P is a measure (i.e., is σ-additive) on the algebra $\mathscr{A}_0$, i.e., in view of Lemma 2.2.1, that if $D_n \in \mathscr{A}_0$, $D_{n+1} \subseteq D_n$ $(n = 1, 2, \ldots)$, and $\prod_{n=1}^{\infty} D_n = \varnothing$, then $\lim_{n\to+\infty} P(D_n) = a > 0$ $(a \leqq 1)$ is impossible. Suppose that there would exist such a sequence D_n. If $D_n = \sum_{k=1}^{i_n} A_k^{(n)} B_k^{(n)}$, where the sets $A_k^{(n)}$ $(k = 1, 2, \ldots, i_n)$ form a partition of Ω, then by definition

$$P(D_n) = \sum_{k=1}^{i_n} P_1(A_k^{(n)}) P_2(B_k^{(n)}) \tag{3.3.12}$$

Let E_n denote the set of those indices k for which

$$P_2(B_k^{(n)}) > a/2 \tag{3.3.13}$$

and put $C_n = \sum_{k \in E_n} A_k^{(n)}$. Clearly,

$$a \leqq P(D_n) \leqq a/2(1 - P_1(C_n)) + P_1(C_n)$$

and thus

$$P_1(C_n) \geqq a/(2 - a) > 0 \tag{3.3.14}$$

Now we show that $C_{n+1} \subseteq C_n$. Take an $i \in E_{n+1}$. Let F_i denote the set of those values of j, for which

$$A_i^{(n+1)} A_j^{(n)} \neq \emptyset$$

As $D_{n+1} = \sum_{i=1}^{i_{n+1}} A_i^{(n+1)} B_i^{(n+1)} \subseteq \sum_{i=1}^{i_n} A_i^{(n)} B_i^{(n)} = D_n$, multiplying both sides by $A_j^{(n)}$, where $j \in F_i$, we get

$$\emptyset \neq A_i^{(n+1)} A_j^{(n)} B_i^{(n+1)} \subseteq A_j^{(n)} B_j^{(n)}$$

It follows by Lemma 3.3.2 that $B_i^{(n+1)} \subseteq B_j^{(n)}$ and thus if $P_2(B_i^{(n+1)}) > a/2$, then we also have $P_2(B_j^{(n)}) > a/2$ for each $j \in F_i$, and thus $j \in E_n$. It follows that $A_i^{(n+1)} = \sum_{j \in F_i} A_j^{(n)} A_i^{(n+1)} \subseteq C_n$ for $i \in E_{n+1}$, i.e., $C_{n+1} \subseteq C_n$. As P_1 is by supposition a measure, it follows from (3.3.14) that $\prod_{n=1}^{\infty} C_n \neq \emptyset$.

Now let ω_0 be any element of $\prod_{n=1}^{\infty} C_n$; for each n there exists one and only one value of h_n such that $\omega_0 \in A_{h_n}^{(n)}$, and thus for this sequence we have $A^* = \prod_{n=1}^{\infty} A_{h_n}^{(n)} \neq \emptyset$. Now let us consider the sets $B_{h_n}^{(n)}$. As by definition $A_{h_{n+1}}^{(n+1)} A_{h_n}^{(n)} \neq \emptyset$, it follows, as above,

$$B_{h_{n+1}}^{(n+1)} \subseteq B_{h_n}^{(n)}$$

As by definition $P_2(B_{h_n}^{(n)}) > a/2$ and P_2 is a measure, it follows that $B^* = \prod_{n=1}^{\infty} B_{h_n}^{(n)} \neq \emptyset$.

Now, evidently $A^* \in \mathscr{A}_1$ and $B^* \in \mathscr{A}_2$ and as we have shown that A^* and B^* are not empty, it follows that $A^*B^* \neq \emptyset$. However, $A^*B^* \subseteq D_n$ for $n = 1, 2, \ldots$, and thus it follows that $\prod_{n=1}^{\infty} D_n \neq \emptyset$, which contradicts our assumption. This proves Theorem 3.3.3. ∎

Remark: The statement of Theorem 3.3.3 can be generalized immediately for an arbitrary finite number of qualitatively independent σ-algebras.

Now we pass to the proof of the above mentioned theorem of Banach:

THEOREM 3.3.4. *Let $\mathscr{A}_n$ $(n = 1, 2, \ldots)$ be a denumerably infinite sequence of strongly qualitatively independent σ-algebras of subsets of the set Ω and let P_n be a probability measure on $\mathscr{A}_n$ $(n = 1, 2, \ldots)$. Let $\mathscr{A}$ denote the least σ-algebra containing all the σ-algebras $\mathscr{A}_n$ $(n = 1, 2, \ldots)$. Then there exists on $\mathscr{A}$ a uniquely determined probability measure P such that $P(A) = P_n(A)$ if $A \in \mathscr{A}_n$ $(n = 1, 2, \ldots)$, further, the σ-algebras $\mathscr{A}_n$ are independent with respect to P.*

Proof. Let $\mathscr{A}^{(n)}$ $(n = 1, 2, \ldots)$ denote the least σ-algebra of subsets of Ω containing the σ-algebras $\mathscr{A}_1, \mathscr{A}_2, \ldots, \mathscr{A}_n$. Using Theorem 3.3.3 repeatedly, we can define a measure $P^{(n)}$ on $\mathscr{A}^{(n)}$ for each value of n, such that $P^{(n)}(A) = P_k(A)$ if $A \in \mathscr{A}_k$ and $k \leqq n$ and the σ-algebras $\mathscr{A}_1, \ldots, \mathscr{A}_n$ are independent with respect to $P^{(n)}$.

As $\mathscr{A}^{(n)} \subseteq \mathscr{A}^{(n+1)}$, $P^{(n+1)}$ is an extension of $P^{(n)}$, i.e., $P^{(n+1)}(A) = P^{(n)}(A)$ if $A \in \mathscr{A}^{(n)}$.

Now let $\mathscr{A}_0$ denote the least *algebra* of subsets of Ω which contains all σ-algebras $\mathscr{A}_n$. Each $A \in \mathscr{A}_0$ belongs to some $\mathscr{A}^{(N)}$ (and thus also to $\mathscr{A}^{(M)}$ for $M > N$). Let us define the set function P on $\mathscr{A}_0$ by putting $P(A) = P^{(N)}(A)$ if $A \in \mathscr{A}^{(N)}$. Then $P(A)$ is finitely additive, and $P(A) = P_k(A)$ if $A \in \mathscr{A}_k$, further,

$$P(A_1 \cdots A_r) = \prod_{k=1}^{r} P(A_k) \quad \text{if } A_k \in \mathscr{A}_k \ (k = 1, 2, \ldots, r)$$

Thus it is sufficient to show that P is a measure on $\mathscr{A}_0$.

Suppose that there exists a sequence $\{D_n\}$ such that $D_n \in \mathscr{A}_0$, $D_{n+1} \subseteq D_n$ and $P(D_n) \geqq a > 0$ $(n = 1, 2, \ldots)$, further, $\prod_{n=1}^{\infty} D_n = \emptyset$. By supposition D_n can be written in the form

$$D_n = \sum_{k=1}^{r_{1,n}} A_{1,k}^{(n)} \cdot B_{1,k}^{(n)}$$

where $A_{1,k}^{(n)} \in \mathscr{A}_1$, the $A_{1,k}^{(n)}$ $(k = 1, 2, \ldots, r_n)$ form a partition of Ω, and $B_{1,k}^{(n)} \in \mathscr{A}_0^{(2)}$ where $\mathscr{A}_0^{(2)}$ denotes the least algebra containing all σ-algebras $\mathscr{A}_2, \mathscr{A}_3, \ldots$.

Now we can repeat the argument used in proving Theorem 3.3.3; we obtain that there exists a sequence $l_n^{(1)}$ such that

$$A_1^* = \prod_{n=1}^{\infty} A_{1,l_n^{(1)}}^{(n)} \neq \emptyset, \quad B_{1,l_{n+1}^{(1)}}^{(n+1)} \subseteq B_{1,l_n^{(1)}}^{(n)} \quad \text{and} \quad P(B_{1,l_n^{(1)}}^{(n)}) \geqq a/2$$

Especially for those values of n for which $D_n \in \mathscr{A}_1$, we have $A_1^* \subseteq D_n$. Now we can put

$$B_{1,l_n^{(1)}}^{(n)} = \sum_{k=1}^{r_{2,n}} A_{2,k}^{(n)} B_{2,k}^{(n)}$$

where $A_{2,k}^{(n)} \in \mathscr{A}_2$, the $A_{2,k}^{(n)}$ form a partition of Ω, and $B_{2,k}^{(n)} \in \mathscr{A}_0^{(3)}$ where $\mathscr{A}_0^{(3)}$ denotes the least algebra containing all the σ-algebras $\mathscr{A}_3, \mathscr{A}_4, \ldots$. By the same argument it follows that there exists a sequence $l_n^{(2)}$ such that $A_2^* = \prod_{n=1}^{\infty} A_{2,l_n^{(2)}}^{(n)} \neq \emptyset$, $B_{2,l_n^{(2)}}^{(n)} \supseteq B_{2,l_{n+1}^{(2)}}^{(n+1)}$, and $P(B_{2,l_n^{(2)}}^{(n)}) \geqq a/4$. We have further that $A_1^* A_2^* \subseteq D_n$ for those values of n for which $D_n \in \mathscr{A}^{(2)}$. Continuing this process we can construct an infinite sequence of sets A_r^* such that $\emptyset \neq A_r^* \in \mathscr{A}_r$ $(r = 1, 2, \ldots)$ and $A_1^* A_2^* \cdots A_r^* \subseteq D_n$ for those n for which $D_n \in \mathscr{A}^{(r)}$.

By our supposition, the σ-algebras $\mathscr{A}_r$ are strongly qualitatively independent; thus $A^* = \prod_{r=1}^{\infty} A_r^* \neq \emptyset$. Evidently, $A^* \subseteq D_n$ for every n. As a matter of fact, for each n, $D_n \in \mathscr{A}^{(N)}$ for an appropriate value of N. As it follows from the construction that $\prod_{k=1}^{N} A_k^* \subseteq D_n$ if $D_n \in \mathscr{A}^{(N)}$, it follows that $A^* \subseteq D_n$ for every n. Thus, $\prod_{n=1}^{\infty} D_n \neq \emptyset$, which contradicts our assumption. This contradiction proves Theorem 3.3.4. ■

Remark: This proof is different from (and simpler than) the original proof of the theorem given by Banach [1]. Banach proved a still more general result by his method; our method could be used also in the most general case.

3.4 PRODUCT SPACES

We now deduce from Theorem 3.3.3, as a special case, the following important theorem on the existence of product measures:

THEOREM 3.4.1. *Let* $(\Omega_i, \mathcal{A}_i, P_i)$ $(i = 1, 2)$ *be two arbitrary probability spaces. Let* $\Omega = \Omega_1 \times \Omega_2$ *denote the product of the sets* Ω_1 *and* Ω_2 *(i.e., the set of all ordered pairs* (ω_1, ω_2) *with* $\omega_i \in \Omega_i$ $(i = 1, 2)$*; let* $\mathcal{A}$ *denote the least* σ*-algebra of subsets of* Ω *containing all sets* $A_1 \times A_2$ *where* $A_i \in \mathcal{A}_i$ $(i = 1, 2)$*; then there exists a measure* P *on* $\mathcal{A}$ *such that* $P(A_1 \times A_2) = P_1(A_1)P_2(A_2)$ *if* $A_i \in \mathcal{A}_i$ $(i = 1, 2)$ *and this measure is unique.*

Proof. Let $\mathcal{A}_1^*$ respectively $\mathcal{A}_2^*$ denote the σ-algebra of all subsets of Ω which can be written in the form $A_1 \times \Omega_2$ respectively $\Omega_1 \times A_2$, where $A_1 \in \mathcal{A}_1$ respectively $A_2 \in \mathcal{A}_2$, and put $P_1^*(A_1 \times \Omega_2) = P_1(A_1)$ and $P_2^*(\Omega_1 \times A_2) = P_2(A_2)$. Clearly, $\mathcal{A}_1^*$ and $\mathcal{A}_2^*$ are qualitatively independent, and thus Theorem 3.3.3 can be applied. ■

The σ-algebras $\mathcal{A}_1^*$ and $\mathcal{A}_2^*$ are evidently independent with respect to the probability space $(\Omega, \mathcal{A}, P)$.

Remark: Theorem 3.4.1 is usually proved in a different way, using Fubini's theorem; our proof is more elementary.

Theorem 3.4.1 can be immediately generalized for any finite number of qualitatively independent algebras.

Similarly, as a special case of Theorem 3.3.4, we obtain the corresponding result about the product of an infinite sequence of probability spaces.

THEOREM 3.4.2. *Let* $S_n = (\Omega_n, \mathcal{A}_n, P_n)$ *be an arbitrary infinite sequence of probability spaces.** *Let* $\mathcal{E} = (\Omega, \mathcal{A})$ *denote the product of the experiments* $\mathcal{E}_n = (\Omega_n, \mathcal{A}_n)$ $(n = 1, 2, \ldots)$*, i.e., let* Ω *be the set of all infinite sequences* $\omega = (\omega_1, \omega_2, \ldots, \omega_n, \ldots)$ *where* $\omega_n \in \Omega_n$ $(n = 1, 2, \ldots)$ *and* $\mathcal{A}$ *is the least* σ*-algebra containing all the sets* A_n' *defined as follows:* A_n' *is the set of all sequences* $(\omega_1, \omega_2, \ldots)$ *the nth term* ω_n *of which belongs to a set* $A_n \in \mathcal{A}_n$ *and all other terms* ω_k $(k \neq n)$ *are arbitrary. Let* $\mathcal{A}_n'$ *denote the* σ*-algebra of such sets* A_n' $(n = 1, 2, \ldots)$*. Then there exists a unique probability measure* P *on the experiment* $(\Omega, \mathcal{A})$ *such that if* $A' \in \mathcal{A}_n'$ *and* $A' = \{\omega : \omega_n \in A\}$ *where*

* Some (or all) of the sets Ω_n may be identical, but even in this case it is convenient to distinguish between them.

$A \in \mathscr{A}_n$, then $P(A') = P_n(A)$ $(n = 1, 2, \ldots)$ and the σ-algebras $\mathscr{A}'_n$ are independent in the probability space $S = (\Omega, \mathscr{A}, P)$.

Proof. To prove Theorem 3.4.2 it is sufficient to show that the σ-algebras $\mathscr{A}'_n$ $(n = 1, 2, \ldots)$ are strongly qualitatively independent, and thus the conditions of Theorem 3.3.4 are fulfilled. This is easily shown. Suppose $A'_n \in \mathscr{A}'_n$ and $A'_n \neq \varnothing$ $(n = 1, 2, \ldots)$. Then there exists in the σ-algebra $\mathscr{A}_n$ a set A_n such that A'_n is the set of all sequences $\omega = (\omega_1, \omega_2, \ldots)$ for which $\omega_n \in A_n$. As A'_n is not empty, A_n is not empty either. Let ω^*_n be any element of A_n. Then evidently, putting $\omega^* = (\omega^*_1, \ldots, \omega^*_n, \ldots)$, we have $\omega^* \in A'_n$ $(n = 1, 2, \ldots)$, i.e., $\omega^* \in \prod_{n=1}^{\infty} A'_n$, which proves that $\prod_{n=1}^{\infty} A'_n$ is not empty. ■

Remark 1: Let A_k be an arbitrary set belonging to $\mathscr{A}_k$ for $k = 1, 2, \ldots, n$. The subset $A = A_1 \times A_2 \times \cdots \times A_n \times \Omega_{n+1} \times \cdots$ of the product space Ω (consisting of all points $\omega = (\omega_1, \omega_2, \ldots)$ such that $\omega_k \in A_k$ for $k = 1, 2, \ldots, n$ while ω_k is arbitrary for $k > n$) is called a *cylinder set*. The cylinder sets of Ω form a semiring $\mathscr{C}$. The measure P is defined on $\mathscr{C}$ by putting for $A = A_1 \times A_2 \times \cdots \times A_n \times \Omega_{n+1} \times \cdots$, $P(A) = \prod_{k=1}^{n} P_k(A_k)$. The definition of P is extended to the least algebra $\mathscr{B}$ containing $\mathscr{C}$ in an obvious manner, so that it should be finitely additive. Clearly, $\mathscr{A}$ is the least σ-algebra containing the algebra $\mathscr{B}$. Thus, to prove Theorem 3.4.2, in view of Lemma 2.2.2, one has to show only that the set function defined above is a measure on $\mathscr{B}$. This is usually proved directly, while here we deduced it as a consequence of the more general Theorem 3.3.4. As regards further generalizations, see [3] and [4].

DEFINITION 3.4.2. *The probability space $S = (\Omega, \mathscr{A}, P)$ constructed according to Theorem 3.4.2 is called the product of the probability spaces $S_n = (\Omega_n, \mathscr{A}_n, P_n)$ and is denoted by $S = \prod_{n=1}^{\infty} S_n$.*

Note that Theorem 3.3.2 is a consequence of Theorem 3.4.2. As a matter of fact, if Ω is any set, A_n is a sequence of strongly qualitatively independent subsets of Ω, and $\mathscr{A}$ is the least σ-algebra containing the sets A_n $(n = 1, 2, \ldots)$, then the nonempty sets $\prod_{n=1}^{\infty} A_n^{\delta_n}$ where $\delta_n = +1$ or $\delta_n = -1$ are all atoms of the σ-algebra $\mathscr{A}$. The set of these atoms can be mapped on the set X of all sequences of zeros and ones by mapping the set $\prod_{n=1}^{\infty} A_n^{\delta_n}$ onto the sequence $(X_1, X_2, \ldots)$ where $X_n = (1 + \delta_n)/2$. In this way Theorem 3.3.2 can be obtained from Theorem 3.4.2.

Remark 2: In what follows we shall often prove theorems about infinite sequences of independent random variables having prescribed distributions. It follows from Theorem 3.4.2 that these theorems are not empty, i.e., there always exists a probability space on which independent random variables having the prescribed distributions can be defined: as a matter of fact, such a probability space can always be constructed as a product space.

Example 3.4.1. We want to construct a probability space which describes the compound experiment consisting of an infinite sequence of independent repetitions of an experiment $(\Omega_0, \mathscr{A}_0, P_0)$ such that $\mathscr{A}_0$ is a four-element algebra* consisting of the four events $\varnothing$, Ω, A, and $\bar{A}$, and $P(A)=p$, $P(\bar{A})=q$ where $0<p<1$ and $p+q=1$. One construction can be obtained from Example 3.3.2 by putting $p_n=p$ for $n=1, 2, \ldots$. Another construction is as follows: Let Ω be the unit square $0 \leqq x \leqq 1, 0 \leqq y \leqq 1$, and let A_n $(n=1, 2, \ldots)$ be the set which is the union of the rectangles $A_{n,k}$ $(k=0, 1, \ldots, 2^{n-1}-1)$ where $A_{n,k}$ denotes the set of those points (x, y) for which $2k/2^n < x < (2k+1)/2^n$, $0 \leqq y \leqq 1$ and, further, of the points (x, y) with $x = 2k/2^n$, $0 \leqq y < 1/2$, and the points (x, y) with $x=(2k+1)/2^n$, $1/2 \leqq y \leqq 1$. It is easy to see that the sets A_n $(n=1, 2, \ldots)$ are strongly qualitatively independent; thus, by Theorem 3.3.2 we can find a measure on the least σ-algebra $\mathscr{A}$ of subsets of the unit square containing all sets A_n, such that the sets A_n $(n=1, 2, \ldots)$ are independent with respect to this measure and have the probability p. Let us denote the measure thus obtained by $P^{(p)}$. It is easy to see that $\mathscr{A}$ contains the family of all subsets of the unit square which are of the form $A \times I_2$ where A is a Borel subset of the interval $0 \leqq x \leqq 1$ and I_2 is the interval $0 \leqq y \leqq 1$. Note that now (in contrast with Example 3.3.2) for each value of p $(0<p<1)$ we have used the *same* sequence of qualitatively independent sets A_n, but obtained for each p another measure $P^{(p)}$ on the σ-algebra $\mathscr{A}$ (while in Example 3.3.2 we have for different values of p used different sequences of sets A_n, but the measure was always the same). It will be shown in Chapter 4 that the measures $P^{(p)}$ are pairwise orthogonal, i.e., for every pair of numbers $p_1 \neq p_2$ $(0<p_1<1, 0<p_2<1)$ there exists a set $A \in \mathscr{A}$ such that $P^{(p_1)}(A)=1$ and $P^{(p_2)}(A)=0$. Note that if $p=1/2$ and $B=A \times I_2$ where A is a Borel subset of the interval $0 \leqq x \leqq 1$ and I_2 is the interval $0 \leqq y \leqq 1$, then $P^{(1/2)}(B)$ is equal to the Lebesgue measure of the set A.

However, the simplest way of constructing a probability space describing the mentioned sequence of experiments is that based on Theorem 3.4.2. Let S_n be, for each value of n, the probability space $(\Omega_0, \mathscr{A}_0, P_0)=S_0$. The direct product $S=\prod_{n=1}^{\infty} S_n$ of these probability spaces, constructed according to Theorem 3.4.2, satisfies our requirements. Clearly, $S=(\Omega, \mathscr{A}, P)$ where Ω is the set of all sequences of zeros and ones and if A_n denotes the set of all sequences $(X_1, X_2, \ldots)$ of zeros and ones such that $X_n=1$, then A_n means that the event A has taken place in the nth repetition of the experiment S_0. One has $P(A_n)=p$ $(n=1, 2, \ldots)$ and the events A_n $(n=1, 2, \ldots)$

* Such a sequence of experiments is called a Bernoulli sequence (see Chapter 4).

are independent. Thus, if $i_1, i_2, \ldots, i_k, j_1, j_2, \ldots, j_l$ are different positive integers, we have clearly,

$$P(A_{i_1}A_{i_2}\cdots A_{i_k}\bar{A}_{j_1}\bar{A}_{j_2}\cdots\bar{A}_{j_l}) = p^k q_l \tag{3.4.1}$$

Further, if ν_n denotes the frequency of occurrence of the event A during the first n repetitions of the experiment S_0, we have

$$P(\nu_n = k) = \binom{n}{k} p^k q^{n-k} \qquad (k = 0, 1, \ldots, n) \tag{3.4.2}$$

Definition 3.4.2. *The probability distribution* $P_{n,k} = \binom{n}{k} p^k q^{n-k}$ *(where* $0 < p < 1$ *and* $q = 1 - p$*) on the set of integers* $k = 0, 1, \ldots, n$*, is called the binomial distribution of order* n *and of parameter* p*.*

3.5 INDEPENDENT RANDOM VARIABLES

In this section we will prove some general theorems concerning the evaluation of the distribution, expectation, variance, etc., of functions of independent random variables. It will turn out that the solution of all these questions is much simpler for independent variables than in the general case.

Theorem 3.5.1. *Let* $\xi_1, \xi_2, \ldots, \xi_n$ *be independent random variables (see Definition 3.2.2). Let* $F_k(x)$ *denote the distribution function of* ξ_k $(k = 1, 2, \ldots, n)$*. Then the joint distribution function of the random variables* $\xi_1, \xi_2, \ldots, \xi_n$ *is*

$$F(x_1, x_2, \ldots, x_n) = F_1(x_1)F_2(x_2)\cdots F_n(x_n) \tag{3.5.1}$$

If the distribution of ξ_k *is absolutely continuous with density function* $f_k(x)$ $(k = 1, 2, \ldots, n)$*, then the density function of the joint distribution of the random variables* $\xi_1, \ldots, \xi_n$ *is*

$$f(x_1, x_2, \ldots, x_n) = f_1(x_1)f_2(x_2)\cdots f_k(x_k) \tag{3.5.2}$$

The proof of Theorem 3.5.1 follows immediately from the definitions.

In what follows we shall need the following theorems:

Theorem 3.5.2. *Let* $\xi_1, \xi_2, \ldots, \xi_n$ *be independent real or vector-valued random variables, further, let* $g_1(x), g_2(x), \ldots, g_n(x)$ *be Borel-measurable real or vector-valued functions, such that if* ξ_j *is an* r_j*-dimensional vector-valued random variable* $(r_j \geqq 1)$*, then* g_j *is a function of* r_j *variables. Then the random variables* $g_1(\xi_1), g_2(\xi_2), \ldots, g_n(\xi_n)$ *are independent.*

Proof. The proof follows immediately from the remark that $\mathscr{A}_{g_j(\xi_j)} \subseteq \mathscr{A}_{\xi_j}$ $(j = 1, 2, \ldots, n)$. ■

THEOREM 3.5.3. *Let* $\xi_1, \xi_2, \ldots, \xi_n$ *be independent random variables and* $2 \leq k < n$. *Let* η *be the k-dimensional random vector having the components* $\xi_1, \xi_2, \ldots, \xi_k$. *Then* $\eta, \xi_{k+1}, \ldots, \xi_n$ *are independent.*

Proof. We have to prove that the σ-algebras $\mathscr{A}_\eta, \mathscr{A}_{\xi_{k+1}}, \ldots, \mathscr{A}_{\xi_n}$ are independent. This, however, follows from Theorem 3.2.2. ■

THEOREM 3.5.4. *Let* $\xi_1, \xi_2, \ldots, \xi_n$ *be independent random variables,* $2 \leq k < n$ *and let* $G(x_1, \ldots, x_k)$ *be a Borel-measurable function of k variables. Put* $\zeta = G(\xi_1, \ldots, \xi_k)$. *Then the random variables* $\zeta, \xi_{k+1}, \ldots, \xi_n$ *are independent.*

Proof. The statement of the theorem follows immediately by combining Theorem 3.5.2 and 3.5.3, as ζ is a function of the random vector $\eta = (\xi_1, \ldots, \xi_k)$. ■

THEOREM 3.5.5. *Let* ξ *and* η *be independent random variables having finite expectations. Then* $E(\xi\eta)$ *also exists and one has*

$$E(\xi\eta) = E(\xi)E(\eta) \tag{3.5.3}$$

Remark: Two random variables ξ and η for which $E(\xi)$, $E(\eta)$ and $E(\xi\eta)$ exist and (3.5.3) holds are called *uncorrelated*. Thus, *independent random variables having finite expectation are uncorrelated*. This terminology is made clear by the following:

DEFINITION 3.5.1. *If* ξ *and* η *are two random variables having finite expectations and finite positive variances, the correlation coefficient* $R(\xi, \eta)$ *is defined by the formula*

$$R(\xi, \eta) = \frac{E(\xi\eta) - E(\xi)E(\eta)}{D(\xi)D(\eta)}$$

If $R(\xi, \eta) = 0$, ξ *and* η *are called uncorrelated.*

An alternative expression for $R(\xi, \eta)$ is

$$R(\xi, \eta) = E\left(\frac{\xi - E(\xi)}{D(\xi)}\right)\cdot\left(\frac{\eta - E(\eta)}{D(\eta)}\right)$$

It follows from (2.9.19), when applied to $\xi^* = (\xi - E(\xi))/D(\xi)$ and $\eta^* = (\eta - E(\eta))/D(\eta)$ in view of $E(\xi^{*2}) = E(\eta^{*2}) = 1$, that

$$|R(\xi, \eta)| \leq 1$$

for any pair of random variables ξ and η, with equality if and only if ξ^* and η^* are almost surely equal.

Proof. It is sufficient to prove (3.5.3) for bounded random variables ξ and η, because the general case can be reduced to this special case by using Theorem 2.10.2. We may also suppose that η is a discrete random variable, because the general case can be reduced to this by means of Theorem 2.10.3. If η is discrete, let the values which it takes on with positive probability be $y_1, y_2, \ldots$; let A_n denote the event $\eta = y_n$ $(n = 1, 2, \ldots)$. Thus we have, by Theorem 2.10.1,

$$E(\xi\eta) = \sum_n E(\xi\eta \mid A_n)P(A_n) \tag{3.5.4}$$

It follows from the independence of the variables ξ and η that the conditional distribution of ξ under the condition A_n is the same as the unconditional distribution of ξ. Thus, $E(\xi \mid A_n) = E(\xi)$. As η is, under condition A_n, equal to y_n (i.e., is constant on A_n), we have $E(\xi\eta \mid A_n) = y_n E(\xi \mid A_n)$. Thus, (3.5.3) follows from (3.5.4) for the special case considered. As the general case can be reduced to this special case (as pointed out above), Theorem 3.5.5 is therewith proved. ■

THEOREM 3.5.6. *If $\xi_1, \xi_2, \ldots, \xi_n$ are pairwise uncorrelated random variables with finite variance, then*

$$D^2(\xi_1 + \xi_2 + \cdots + \xi_n) = D^2(\xi_1) + D^2(\xi_2) + \cdots + D^2(\xi_n) \tag{3.5.5}$$

In particular, the variance of the sum of pairwise independent random variables is equal to the sum of their variances.

Proof. Using the formula (2.9.9), we obtain for arbitrary random variables $\xi_1, \xi_2, \ldots, \xi_n$ with finite variances,

$$D^2(\xi_1 + \xi_2 + \cdots + \xi_n) = \sum_{k=1}^{n} D^2(\xi_k) + 2 \sum_{1 \leq j < k \leq n} [E(\xi_j \xi_k) - E(\xi_j)E(\xi_k)] \tag{3.5.6}$$

(Note that for $n = 2$, (3.5.6) reduces to (2.9.20).) Thus, (3.5.5) holds if the random variables ξ_k are pairwise uncorrelated. As pairwise independent variables are, by Theorem 3.5.5., uncorrelated, Theorem 3.5.6 is proved. ■

THEOREM 3.5.7. *Let ξ and η be independent random variables having the distribution functions $F(x)$ and $G(x)$, respectively. Then the distribution function $H(x)$ of $\xi + \eta$ is given by*

$$H(x) = \int_{-\infty}^{+\infty} F(x - y)dG(y) \tag{3.5.7}$$

If the distribution of ξ is absolutely continuous with density function $f(x) = F'(x)$, then the distribution of $\xi + \eta$ is also absolutely continuous with the density function

$$h(x) = \int_{-\infty}^{+\infty} f(x - y)dG(y) \tag{3.5.8}$$

If the distribution of η is absolutely continuous also, with density function $G'(x) = g(x)$, then

$$h(x) = \int_{-\infty}^{+\infty} f(x-y)g(y)dy \tag{3.5.9}$$

Proof. Equation (3.5.7) follows from Theorem 3.5.1 as

$$P(\xi + \eta < x) = \int_{z+y<x} dF(z)dG(y) = \int_{-\infty}^{+\infty} F(x-y)dG(y)$$

Equation (3.5.8) is an immediate consequence of (3.5.7) and (3.5.9) of (3.5.8). ■

DEFINITION 3.5.2. *The probability distribution of the sum of two independent random variables is called the convolution of the distribution of the terms. Accordingly, the distribution function $H(x)$ defined by (3.5.7) is called the convolution of the distribution functions $F(x)$ and $G(x)$. Similarly, the density function $h(x)$ defined by (3.5.9) is called the convolution of the density functions $f(x)$ and $g(x)$. The convolution of two distribution functions $F(x)$ and $G(x)$ respective of two density functions $f(x)$ and $g(x)$ shall be denoted by $F(x) * G(x)$ respective of $f(x) * g(x)$.*

THEOREM 3.5.8. *The convolution operation is commutative and associative, i.e., if $F_1(x)$, $F_2(x)$, $F_3(x)$ are distribution functions, then*

$$F_1(x) * F_2(x) = F_2(x) * F_1(x) \tag{3.5.10}$$

and

$$F_1(x) * [F_2(x) * F_3(x)] = [F_1(x) * F_2(x)] * F_3(x) \tag{3.5.11}$$

Similarly, for density functions,

$$f_1(x) * f_2(x) = f_2(x) * f_1(x) \tag{3.5.12}$$

and

$$f_1(x) * [f_2(x) * f_3(x)] = [f_1(x) * f_2(x)] * f_3(x) \tag{3.5.13}$$

Proof. The commutativity and associativity of convolution follows immediately from the commutativity and associativity of the addition of real numbers. For instance, if ξ_1, ξ_2, ξ_3 are independent random variables, then by Theorem 3.5.4, ξ_1 is independent from $\xi_2 + \xi_3$ and also $\xi_1 + \xi_2$ is independent from ξ_3, and thus, if $F_i(x) = F_i$ is the distribution function of ξ_i $(i = 1, 2, 3)$, both $(F_1 * F_2) * F_3$ and $F_1 * (F_2 * F_3)$ are equal to the distribution function of $\xi_1 + \xi_2 + \xi_3$. ■

In view of Theorem 3.5.8, we may write convolution products of any number of "factors" without brackets. The nth convolution power of a distribution function $F(x)$ is defined as the distribution function of a sum of n

independent random variables, each having the distribution function $F(x)$. The nth convolution power of the distribution function $F(x)$ will be denoted by F^{*n}; similar notation will be used for convolution powers of density functions.

The following theorem is an obvious generalization of the second statement of Theorem 3.5.5:

THEOREM 3.5.9. *If $\xi_1, \xi_2, \ldots, \xi_n$ are independent random variables with finite expectations, then $\xi_1\xi_2 \cdots \xi_n$ also has finite expectation, and*

$$E(\xi_1\xi_2 \cdots \xi_n) = \prod_{k=1}^{n} E(\xi_k) \tag{3.5.14}$$

Proof. For the proof it is sufficient to point out that if the variables ξ_1, ξ_2, ξ_n are independent, then by Theorem 3.5.4 each is independent from the product of the others and thus Theorem 3.5.9 follows from Theorem 3.5.5 by induction. ■

THEOREM 3.5.10. *Let $\xi_1, \xi_2, \ldots, \xi_n$ be independent random variables. Then we have, putting $\zeta_n = \xi_1 + \xi_2 + \cdots + \xi_n$,*

$$\varphi_{\zeta_n}(t) = \prod_{k=1}^{n} \varphi_{\xi_k}(t) \tag{3.5.15}$$

i.e., the characteristic function of the sum of independent random variables is equal to the product of the characteristic functions of the terms.

Proof. To prove (3.5.15) we first show that the statement of Theorem 3.5.9 is also valid for complex-valued random variables. As a matter of fact, if $\xi_1 + i\eta_1$ and $\xi_2 + i\eta_2$ are independent complex-valued random variables (i.e., the random vectors (ξ_1, η_1) and (ξ_2, η_2) are independent), then we have

$$\begin{aligned} E((\xi_1 + i\eta_1)(\xi_2 + i\eta_2)) &= E(\xi_1\xi_2 - \eta_1\eta_2) + iE(\xi_1\eta_2 + \xi_2\eta_1) \\ &= E(\xi_1 + i\eta_1)E(\xi_2 + i\eta_2) \end{aligned}$$

The same holds, of course, for the product of an arbitrary number of complex-valued independent random variables. Now,

$$\varphi_{\zeta_n}(t) = E(e^{i\zeta_n t}) = E\left(\prod_{k=1}^{n} e^{i\xi_k t}\right)$$

Thus the statement of Theorem 3.5.10 follows from Theorem 3.5.9. ■

Remark: The converse of Theorem 3.5.10 is not true: $\varphi_{\xi+\eta}(t) = \varphi_\xi(t)\varphi_\eta(t)$ does not imply that ξ and η are independent. However, the following statement is true:

THEOREM 3.5.11. *A necessary and sufficient condition for the independence of two random variables ξ and η is that for every real number c the relation*

$$\varphi_{\xi+c\eta}(t) = \varphi_{\xi}(t)\varphi_{c\eta}(t) \tag{3.5.16}$$

should hold.

Proof. It follows from (3.5.16) that for every pair (a, b) of real numbers we have

$$\varphi_{a\xi+b\eta}(t) = \varphi_{a\xi}(t)\varphi_{b\eta}(t) \tag{3.5.17}$$

Equation (3.5.17) implies that the distribution of every projection $a\xi + b\eta$ of the vector (ξ, η) is the same as if ξ and η were independent. By the theorem of Cramer and Wold (see Section 2.7) it follows that ξ and η are in fact independent. The necessity of (3.5.16) is evident because if ξ and η are independent, so are ξ and $c\eta$. ■

Example 3.5.1. Let δ_k $(k = 1, 2, \ldots, n)$ be independent random variables, each taking on the values 1 and 0 with probability p respectively q $(0 < p < 1, q = 1 - p)$. Then $\nu_n = \delta_1 + \delta_2 + \cdots + \delta_n$ has binomial distribution of order n and parameter p (see Section 3.4). Let us compute $E(\nu_n)$ and $D(\nu_n)$. We have $E(\delta_k) = 1 \cdot p + 0 \cdot q = p$ and $D^2(\delta_k) = E(\delta_k^2) - E^2(\delta_k) = E(\delta_k) - E^2(\delta_k) = p - p^2 = pq$. Thus, from the additivity of the expectation and from Theorem 3.5.6, we get

$$E(\nu_n) = np \tag{3.5.18}$$

and

$$D^2(\nu_n) = npq \tag{3.5.19}$$

Of course, (3.5.18) and (3.5.19) could also be obtained by direct computation, or by means of (2.9.27), taking into account that the characteristic function of ν_n is, with respect to Theorem 3.5.10,

$$\varphi_{\nu_n}(t) = (pe^{it} + q)^n$$

Thus, $E(\nu_n) = -i\varphi_{\nu_n}'(0) = np$ and $E(\nu_n^2) = -\varphi_{\nu_n}''(0) = n^2p^2 + npq$ and thus, $D^2(\nu_n) = npq$.

3.6 INDEPENDENCE AND ORTHOGONALITY

DEFINITION 3.6.1. *A sequence ξ_n $(n = 1, 2, \ldots)$ of random variables on a probability space $S = (\Omega, \mathscr{A}, P)$ is called an orthonormal system, if* each ξ_n*

* If η and ζ are two random variables such that $\eta \in L_2(S)$, $\zeta \in L_2(S)$, and $E(\eta\zeta) = 0$, then η and ζ are called *orthogonal*. "Orthonormal" is an abbreviation for "orthogonal and normed."

belongs to the Hilbert space $L_2(S)$ (*i.e.*, $E(\xi_n^2)$ *exists*) *and one has*

$$E(\xi_n^2)=1 \qquad (n=1,2,\ldots) \tag{3.6.1}$$

and

$$E(\xi_n\xi_m)=0 \qquad \textit{for } n\neq m, \quad n,m=1,2,\ldots \tag{3.6.2}$$

An orthonormal system $\{\xi_n\}$ *is called complete, if there does not exist a random variable* $\eta\in L_2(S)$ *such that adding* η *to the system* $\{\xi_n\}$ *it remains orthonormal. In other words, the orthonormal system* $\{\xi_n\}$ *is complete if from* $\eta\in L_2(S)$, $E(\eta\xi_n)=0$ *for* $n=1,2,\ldots$, *it follows that* $\eta=0$ *almost surely. If* $\{\xi_n\}$ *is an orthonormal system on* S *and* η *is an arbitrary random variable belonging to* $L_2(S)$, *the sequence*

$$c_n=E(\eta\xi_n) \qquad (n=1,2,\ldots) \tag{3.6.3}$$

is called the sequence of Fourier coefficients of η *and the series*

$$\sum_{n=1}^{\infty} c_n\xi_n \tag{3.6.4}$$

is called the Fourier series of η *with respect to the system* $\{\xi_n\}$.

The following basic fact will be needed in what follows:

LEMMA 3.6.1. *If* $\{\xi_n\}$ *is an orthonormal system on the probability space* S, η *is a random variable belonging to* $L_2(S)$, *and* c_n *is defined by* (3.6.3), *then one has**

$$\sum_{n=1}^{\infty} c_n^2 \leqq E(\eta^2) \tag{3.6.5}$$

This implies

$$\lim_{n\to\infty} c_n=0 \tag{3.6.6}$$

If the orthonormal system $\{\xi_n\}$ *is complete, then†*

$$\sum_{n=1}^{\infty} c_n^2 = E(\eta^2) \tag{3.6.7}$$

and the partial sums of the series (3.6.4) *converge strongly in* $L_2(S)$ *to* η, *i.e., putting*

$$\eta_n=\sum_{k=1}^{n} c_k\xi_k \qquad (n=1,2,\ldots) \tag{3.6.8}$$

one has

$$\lim_{n\to\infty} E((\eta-\eta_n)^2)=0 \tag{3.6.9}$$

Proof. Clearly,

$$0\leqq E((\eta-\eta_n)^2)=E(\eta^2)-\sum_{k=1}^{n} c_k^2 \tag{3.6.10}$$

* Formula (3.6.5) is called Bessel's inequality.

† Formula (3.6.7) is called Parseval's relation.

which implies (3.6.5). By the Riesz-Fischer theorem (see Appendix B) the series (3.6.4) converges in L_2-norm to a random variable $\eta^* \in L_2$,

$$E((\eta - \eta^*)^2) = E(\eta^2) - \sum_{k=1}^{\infty} c_k^2 \tag{3.6.11}$$

Now, $\eta - \eta^*$ is orthogonal to every ξ_n. Thus, if $\{\xi_n\}$ is complete, $\eta^* = \eta$ almost surely and (3.6.7) and (3.6.9) follow from (3.6.11) and (3.6.10). ■

If $\{\xi_n\}$ is a sequence of pairwise independent random variables all belonging to $L_2(S)$, $E(\xi_n) = 0$ for $n \geqq 2$, and $E(\xi_n^2) = 1$ for $n = 1, 2, \ldots$, then the system $\{\xi_n\}$ is orthonormal, because by Theorem 3.5.5 we have $E(\xi_n \xi_m) = E(\xi_n)E(\xi_m) = 0$ for $n \neq m$. If the random variables ξ_n are not only pairwise independent, but mutually independent, then even more is true, namely the following:

Theorem 3.6.1. *If the random variables ξ_n are independent ($n = 1, 2, \ldots$), $E(\xi_n^2) = 1$ and $E(\xi_n) = 0$ for $n = 1, 2, \ldots$, then all the products $\xi_{k_1}\xi_{k_2} \cdots \xi_{k_r}$ ($1 \leqq k_1 < k_2 < \cdots < k_r$; $r = 1, 2, \ldots$) belong to $L_2(S)$ and they form, together with the constant* 1, *an orthonormal system.*

Proof. It follows from Theorem 3.5.9 that

$$E((\xi_{k_1}\xi_{k_2} \cdots \xi_{k_r})^2) = \prod_{j=1}^{r} E(\xi_{k_j}^2) = 1 \tag{3.6.12}$$

If we take any two nonidentical products $\xi_{k_1}\xi_{k_2} \cdots \xi_{k_r}$ ($k_1 < k_2 < \cdots < k_s$) and $\xi_{l_1}\xi_{l_2} \cdots \xi_{l_s}$ ($l_1 < l_2 < \cdots < l_s$), then it follows from Theorem 3.5.9 and from our supposition that $E(\xi_n) = 0$ for $n \geqq 1$, that their product has expectation 0; we have further, $E(1^2) = 1$ and $E(1 \cdot \xi_{k_1}\xi_{k_2} \cdots \xi_{k_r}) = 0$. ■

Corollary to Theorem 3.6.1. *Let $\{\xi_n\}$ be an orthonormal system of independent random variables on a probability space S, consisting of at least three random variables. Then the system $\{\xi_n\}$ cannot be complete.*

Proof. Among the values $E(\xi_n)$, one at most, can be different from 0, because if, e.g., $E(\xi_1) \neq 0$ and $E(\xi_2) \neq 0$, then $E(\xi_1\xi_2) = E(\xi_1)E(\xi_2) \neq 0$, i.e., ξ_1 and ξ_2 would not be orthogonal. Thus we can find two elements of the sequence $\{\xi_n\}$, say ξ_2 and ξ_3 such that $E(\xi_2) = E(\xi_3) = 0$. According to Theorem 3.6.1, the random variable $\xi_2\xi_3$ is orthogonal to all ξ_n and is also normed, i.e., $E((\xi_2\xi_3)^2) = 1$. Thus, either $\xi_2\xi_3$ is contained in the sequence $\{\xi_n\}$, or the system $\{\xi_n\}$ is not complete. The first case is, however, impossible, as $\xi_2\xi_3$ is certainly not identical with ξ_2 or ξ_3, because, e.g., $\xi_2\xi_3 = \xi_2$ would imply $E(\xi_2^2\xi_3) = 1$, which contradicts $E(\xi_3) = 0$. Now, if $\xi_2\xi_3 = \xi_4$, say, we have $E(\xi_2\xi_3\xi_4) = E(\xi_2)E(\xi_3\xi_4) = 0$; on the other hand, as $\xi_4 = \xi_2\xi_3$, we have $E(\xi_2\xi_3\xi_4) = E(\xi_2^2\xi_3^2) = 1$, which is a contradiction. Thus the system $\{\xi_n\}$ is not complete. ■

Remark: The supposition that the system $\{\xi_n\}$ contains at least three random variables is necessary in the formulation of the above corollary. For instance, if $S = (\Omega_2, \mathscr{A}, P)$ where Ω_2 is the two-element set $\Omega_2 = \{1, 2\}$, $\mathscr{A}$ is the four-element algebra $\mathscr{A} = (\varnothing, \Omega, \{1\}, \{2\})$ and $P(\{1\}) = P(\{2\}) = 1/2$, and we define ξ_1 and ξ_2 by putting $\xi_1(1) = \xi_1(2) = 1$, $\xi_2(1) = 1$, and $\xi_2(2) = -1$, then ξ_1 and ξ_2 form a complete orthonormal system, and are independent.

Example 3.6.1. Let S be the Lebesgue probability space, and consider the *Rademacher functions*

$$R_n(x) = \operatorname{sgn}(\sin 2^n \pi x) \qquad (0 \leqq x \leqq 1, n = 1, 2, \ldots) \tag{3.6.13}$$

As shown in Example 3.2.3, the Rademacher functions are independent random variables on S. Let us now define the functions $W_n(x)$ $(0 \leqq x \leqq 1;$ $n = 0, 1, 2, \ldots)$ as follows:

$$W_0(x) \equiv 1 \qquad \text{for } 0 \leqq x \leqq 1 \tag{3.6.14}$$

Further, if the representation of $n \geqq 1$ in the binary system is $n = 2^{k_1} + 2^{k_2} + \cdots + 2^{k_r}$ where $0 \leqq k_1 < k_2 < \cdots < k_r$ are integers, put

$$W_n(x) = R_{k_1+1}(x) R_{k_2+1}(x) \cdots R_{k_r+1}(x) \tag{3.6.15}$$

The functions $W_n(x)$ $(n = 0, 1, \ldots)$ are called the *Walsh functions*. It follows from Theorem 3.6.1 that the Walsh functions form an orthonormal system on the Lebesgue probability space. We now show that this system is complete. Let x be a real number, $0 \leqq x < 1$, which is not a binary rational number. Let the binary expansion of x be

$$x = \sum_{k=1}^{\infty} \frac{\varepsilon_k(x)}{2^k} \qquad (\varepsilon_k(x) = 0 \text{ or } 1) \tag{3.6.16}$$

Then we have

$$\varepsilon_k(x) = \frac{1 - R_k(x)}{2} \tag{3.6.17}$$

i.e., $R_k(x) = +1$ or $R_k(x) = -1$, according to whether $\varepsilon_k(x) = 0$ or $\varepsilon_k(x) = 1$. Let $i_{n,m}(x)$ denote the indicator function of the interval $(m/2^n, (m+1)/2^n)$, where m and n are nonnegative integers and $0 \leqq m < 2^n$. Let the binary expansion of $m/2^n$ be

$$\frac{m}{2^n} = \sum_{k=1}^{n} \frac{\delta_k}{2^k} \qquad (\delta_k = 0 \text{ or } 1, k = 1, 2, \ldots, n) \tag{3.6.18}$$

then $i_{n,m}(x)$ can be written in the form

$$i_{n,m}(x) = \prod_{j=1}^{n} \left(\frac{1 + (1 - 2\delta_j) R_j(x)}{2} \right) \tag{3.6.19}$$

if x is not a binary rational number.

Multiplying out we see that $i_{n,m}(x)$ is a linear combination of certain Walsh functions.

$$i_{n,m}(x) = \sum_{l=0}^{2^n-1} a_{n,m,l}\, W_l(x) \tag{3.6.20}$$

It follows that if $f = f(x) \in L_2(S)$ is such a function that

$$E(fW_n) = 0 \qquad \text{for } n = 0, 1, \dots \tag{3.6.21}$$

then we have

$$\int_{m/2^n}^{(m+1)/2^n} f(x)\, dx = 0 \qquad \text{for } 0 \leqq m < 2^n;\ n = 1, 2, \dots \tag{3.6.22}$$

and thus

$$\int_0^{(m+1)/2^n} f(x)\, dx = 0 \qquad (0 \leqq m < 2^n;\ n = 1, 2, \dots) \tag{3.6.23}$$

Thus, putting $F(x) = \int_0^x f(t)\, dt$, we get $F(r) = 0$ for every binary rational number r in $(0, 1)$. The function $F(x)$, being the indefinite integral of an integrable function, is continuous, thus $F(x) = 0$ for all x in $(0, 1)$ and therefore $f(x) = 0$ for almost all x. This proves that the system of Walsh functions is complete.

In view of the corollary to Theorem 3.6.1 the Walsh functions cannot be independent. However, it can be shown that they are pairwise independent. As each Walsh function $W_n(x)$ $(n \geqq 1)$ takes on only the values $+1$ and -1, to show this it is sufficient to prove that denoting by A_n the set of those x $(0 \leqq x \leqq 1)$ for which $W_n(x) = 1$, the sets A_n are pairwise independent. As $P(A_n) = 1/2$ for $n \geqq 1$, we have to show only that for $n \neq m$ and $n, m \geqq 1$, $P(A_n A_m) = 1/4$. Now we have, using the orthogonality of the Walsh functions,

$$P(A_n A_m) = \int_0^1 \left(\frac{1 + W_n(x)}{2}\right)\left(\frac{1 + W_m(x)}{2}\right) dx = \frac{1}{4} \quad \text{if } n \neq m\ (n, m \geqq 1)$$

which proves our assertion. Note that the Walsh functions are not independent by three: as a matter of fact, for instance, the Walsh functions $W_1(x) = R_1(x)$, $W_2(x) = R_2(x)$, and $W_3(x) = R_1(x)R_2(x)$ are not independent, as their product is equal to 1 (except for the points 0, 1/4, 1/2, 3/4, 1).

Example 3.6.2. Let $S_N = (\Omega_N, \mathscr{A}, P)$ be the classical probability space with N elements, i.e., let Ω_N be the set $\{1, 2, \dots, N\}$, let $\mathscr{A} = \mathscr{P}(\Omega_N)$, and let P be the uniform probability measure on $\mathscr{A}$. Suppose that N is a squarefree integer, then $N = p_1 p_2 \dots p_s$, where $p_1, p_2, \dots, p_s$ are different prime numbers.

Let $r_p(n)$ denote the remainder of the integer n modulo p and define the random variables $\xi_k = \xi_k(n)$ $(k = 1, 2, \ldots, s)$ on S_N as follows:

$$\xi_k(n) = \frac{r_{p_k}(n) - (p_k - 1)/2}{\sqrt{(p_k^2 - 1)/12}} \tag{3.6.24}$$

Then the random variables satisfy the conditions of Theorem 3.6.1. They are independent, $E(\xi_k) = 0$, and $E(\xi_k^2) = 1$ $(k = 1, 2, \ldots)$.

We now introduce the following:

DEFINITION* 3.6.2. *A sequence ξ_n $(n = 1, 2, \ldots)$ of independent random variables on a probability space S is called saturated with respect to independence if there does not exist a random variable η on S such that $\eta, \xi_1, \xi_2, \ldots, \xi_n, \ldots$ are mutually independent random variables, and η is not constant almost everywhere.*

Example 3.6.3. The Rademacher functions $R_n(x)$ $(n = 1, 2, \ldots)$ form a saturated sequence of independent random variables on the Lebesgue probability space S. As a matter of fact, if there would exist a measurable function $f(x)$ on S such that the random variables $f(x), R_1(x), \ldots, R_n(x), \ldots$ would be independent, then if $g(x)$ is any Borel-measurable function, the random variables $g(f(x)), R_1(x), \ldots, R_n(x), \ldots$ would also be independent by Theorem 3.5.2. Taking for $g(x)$ any bounded function, it would follow that $g(f(x)) - \int_0^1 g(f(u))\,du$ is orthogonal to all Walsh functions, and thus $g(f(x))$ is equal to a constant almost everywhere.

Example 3.6.4. The random variables $\xi_1, \ldots, \xi_r$ in Example 3.6.2 form a saturated system of independent random variables on S_N. This can be shown as follows: The values $\xi_1(n), \ldots, \xi_r(n)$ determine the number n $(1 \leqq n \leqq N)$ uniquely. If $\eta(n)$ would be such a random variable that the random variables $\eta, \xi_1, \ldots, \xi_r$ would be mutually independent, then the distribution of η under the condition that the values of $\xi_1, \ldots, \xi_r$ are fixed would be the same as the unconditional distribution of η. As, however, there is only one n (with $1 \leqq n \leqq N$) for which $\xi_1, \ldots, \xi_r$ take on given values, it follows that under the condition that the values of $\xi_1, \ldots, \xi_r$ are fixed, η is certainly constant. But this implies that η is constant everywhere.

In Example 3.6.4 we have proved that a system of random variables is saturated by showing that the values $\xi_k(n)$ $(k = 1, 2, \ldots, r)$ of these random variables determine n uniquely. Note that the Rademacher functions have the property that the sequence $R_n(x)$ $(n = 1, 2, \ldots)$ determines x uniquely.

Let us introduce the following:

DEFINITION 3.6.3. *Let $S = (\Omega, \mathscr{A}, P)$ be a probability space and ξ_n $(n = 1, 2, \ldots)$ be a sequence of random variables on S. We say that the sequence $\{\xi_n\}$*

* See [10].

separates the points of Ω *if there exists a subset* Z *of* Ω *such that* $P(Z)=0$; *and if* $\omega_1 \notin Z$, $\omega_2 \notin Z$, *and* $\omega_1 \neq \omega_2$, *then there exists at least one value of* n *such that* $\xi_n(\omega_1) \neq \xi_n(\omega_2)$.

Using this terminology we can state that the *Rademacher functions separate the points of the Lebesgue probability space.*

Another (closely related) property of the Rademacher functions is the following: The least σ-algebra $\mathscr{A}_0$ of subsets of the interval $[0, 1]$ with respect to which all the Rademacher functions are measurable, is the σ-algebra of all Borel subsets of $[0, 1]$. As a matter of fact, $\mathscr{A}_0$ contains all intervals (r_1, r_2) $(0 \leq r_1 < r_2 \leq 1)$ with binary rational end points r_1 and r_2; thus, $\mathscr{A}_0$ contains all intervals $[a, b]$ $(0 \leq a < b \leq 1)$ and $\mathscr{A}_0$ contains all Borel subsets of $[0, 1]$. We now introduce the following:

DEFINITION 3.6.4. *A finite or infinite sequence* ξ_n $(n=1, 2, \ldots)$ *of random variables on a probability space* $S=(\Omega, \mathscr{A}, P)$ *is called a spanning system if, denoting by* $\mathscr{A}_0$ *the least* σ*-algebra of subsets of* Ω *with respect to which all the random variables are measurable, for every* $A \in \mathscr{A}$ *there exists an* $A^* \in \mathscr{A}_0$ *such that* $P(A \circ A^*)=0$.

The following theorem shows the connection between the notion of a saturated system of independent random variables and that of spanning systems:

THEOREM 3.6.2. *If* ξ_n $(n=1, 2, \ldots)$ *is a sequence of independent random variables on a probability space* S *which is a spanning system, then the sequence* ξ_n *is saturated with respect to independence.*

Proof. Suppose that contrary to the statement of the theorem there exists a random variable η, which is not almost surely constant, and for which the random variables $\eta, \xi_1, \xi_2, \ldots, \xi_n, \ldots$ are independent. Then there exists a Borel set $\mathfrak{B}$ of the real line such that, putting $A=\eta^{-1}(\mathfrak{B})$, one has $P(A)=p$ and $A \in \mathscr{A}$, where $0<p<1$. By supposition, A is independent from every set belonging to the least σ-algebra $\mathscr{A}_0$ with respect to which the random variables ξ_n $(n=1, 2, \ldots)$ are measurable. As, by supposition, to every set $A \in \mathscr{A}$ there exists a set $A^* \in \mathscr{A}_0$ such that $P(A \circ A^*)=0$, A has to be independent from itself, which implies $P^2(A)=P(A)$, i.e., $p=0$ or $p=1$, in contradiction to our supposition $0<p<1$. This contradiction shows that the system $\{\xi_n\}$ is saturated with respect to independence. ■

Remark: Theorem 3.6.2 implies that the Rademacher functions form a saturated system and also that the random variables ξ_{p_k} $(1 \leq k \leq r)$ in Example 3.6.4 form a saturated system.

The converse of Theorem 3.6.2 is not true: there exist saturated systems of independent random variables which do not form a spanning system. This is shown by the following:

Example 3.6.5. Let S_3 be the classical probability space on the three-element basic space $\Omega_3=\{1, 2, 3\}$. Let the random variable ξ be defined by $\xi(1)=1$, $\xi(2)=\xi(3)=0$. Then ξ, in itself, forms a saturated system with respect to independence, that is, there does not exist a nonconstant random variable η on S_3 which is independent from ξ, because if ξ and η are independent, the conditional distribution of η under condition $\xi=1$ is the same as its unconditional distribution. As, however, ξ takes on the value 1 for $\omega=1$ only, η is constant under condition $\xi=1$; thus it is constant on S_3. On the other hand, ξ is not spanning $\mathscr{A}$, because $\mathscr{A}_\xi$ consists only of the four sets $\varnothing$, Ω_3, $\{1\}$, and $\{2, 3\}$, and thus is not identical with $\mathscr{A}$ which consists of all eight subsets of Ω_3.

We have mentioned two properties of the Rademacher functions: that they form a separating system and also a spanning system. Now we will show that, in general, for an arbitrary sequence of random variables each of these properties implies the other if Ω is a complete separable metric space, and $\mathscr{A}$ is the family of all Borel subsets of Ω (i.e., the least σ-algebra containing all open spheres of Ω).

THEOREM 3.6.3. *If $S=(\Omega, \mathscr{A}, P)$ is a probability space such that Ω is a complete separable metric space and $\mathscr{A}$ is the σ-algebra of Borel subsets of Ω, then every finite or infinite sequence ξ_n $(n=0, 1, 2, \dots)$ of random variables on S which separates the points of Ω is a spanning system, and conversely.*

Proof. We may suppose, without restricting the generality, that $0 \leqq \xi_n \leqq 1$. Let the binary expansion of $\xi_n(\omega)$ be*

$$\xi_n(\omega)=\sum_{k=1}^{\infty} \frac{\varepsilon_{n,k}(\omega)}{2^k} \qquad (n=0, 1, 2, \dots) \tag{3.6.25}$$

and define the random variable $\xi(\omega)$ as follows:

$$\xi(\omega)=\sum_{j=1}^{\infty} \frac{\delta_j(\omega)}{2^j} \tag{3.6.26}$$

where

$$\delta_{(2k+1)2^n}(\omega)=\varepsilon_{n,k}(\omega) \qquad (k, n=0, 1, 2, \dots) \tag{3.6.27}$$

It is easy to see that if the random variables ξ_n separate the points of Ω, so does the single random variable ξ, and conversely. Further, the σ-algebra $\mathscr{A}_\xi$ generated by ξ is identical with that generated by the sequence $\{\xi_n\}$. Thus, it is sufficient to prove Theorem 3.6.3 for a single random variable ξ.

Now we prove first that if ξ separates the points of Ω, then it spans the full σ-algebra $\mathscr{A}$. We may suppose that $\xi(\omega_1) \neq \xi(\omega_2)$ for all pairs $\omega_1 \in \Omega$,

* For binary rational numbers we choose a finite expansion.

$\omega_2 \in \Omega$ with $\omega_1 \neq \omega_2$. Now, according to a theorem of Kuratowski (see [20] and Parthasarathy [21]), if a Borel-measurable function in a complete separable metric space maps the space into another complete separable metric space so that different points have different maps, then the map of the whole space is a Borel set. Thus, if $\xi(A)$ denotes the set of values taken on by $\xi(\omega)$ in a set $A \in \mathscr{A}$, then $\xi(A)$ is a Borel subset of the real line. Thus, $A = \xi^{-1}(\xi(A)) \in \mathscr{A}_\xi$, i.e., $\mathscr{A}_\xi$ is identical to $\mathscr{A}$. To prove the second part of the theorem, we may suppose that $\mathscr{A}_\xi = \mathscr{A}$. Now, if $\xi(\omega_1) = \xi(\omega_2)$, then clearly, for every Borel set $\mathfrak{B}$ of the real line either both ω_1 and ω_2 are contained in $\xi^{-1}(\mathfrak{B})$ or none of them (if $\omega_1 \neq \omega_2$). Thus, a set $A \in \mathscr{A}$ such that $\omega_1 \in A$ but $\omega_2 \notin A$ cannot be contained in $\mathscr{A}_\xi$; however, there is an open sphere around ω_1, which does not contain ω_2, which, being open, belongs to $\mathscr{A}$. This contradicts our supposition that $\mathscr{A}_\xi = \mathscr{A}$ and proves Theorem 3.6.3. ■

In view of Theorem 3.6.2 we obtain the following:

Corollary to Theorem 3.6.3. *If the random variables ξ_n $(n = 1, 2, \ldots)$ on a probability space $S = (\Omega, \mathscr{A}, P)$, where Ω is a complete separable metric space, are independent and separate the points of Ω, then the system $\{\xi_n\}$ is saturated with respect to independence.*

If, instead of the Rademacher functions, we take an arbitrary sequence $\{\xi_n\}$ of independent random variables on a probability space S such that $E(\xi_n) = 0$ and $E(\xi_n^2) = 1$ for $n = 1, 2, \ldots$, even if the sequence $\{\xi_n\}$ is saturated with respect to independence, we cannot expect that the set of all products $\xi_{k_1} \xi_{k_2} \cdots \xi_{k_r}$ $(1 \leqq k_1 < k_2 < \cdots < k_r)$ (together with the constant 1) should be a complete system, as in the case of the Rademacher functions. As a matter of fact, it is easy to see that the random variable $\eta = \xi_1^2 - E(\xi_1^3)\xi_1 - 1$ is orthogonal to all the products $\xi_{k_1} \xi_{k_2} \cdots \xi_{k_r}$ $(1 \leqq k_1 < k_2 < \cdots < k_r)$ and also to the constant 1. Thus, if η is not identically 0 (as in the case of Rademacher functions), the system $\{\xi_{k_1} \xi_{k_2} \cdots \xi_{k_r}\}$ is certainly not complete. However, all products of nonnegative integral powers of the random variables ξ_n, i.e., all the random variables $\xi_1^{a_1} \xi_2^{a_2} \cdots \xi_n^{a_n}$ $(a_k \geqq 0$ integer, $1 \leqq k \leqq n$; $n = 1, 2, \ldots)$, together with the constant 1, will be complete in many cases.* It has been conjectured by H. Steinhaus that this is always true. Note that as $[R_n(x)]^{2k+1} = R_n(x)$ and $[R_n(x)]^{2k} = 1$, the set of all products of powers of the Rademacher functions is identical with the set of all Walsh functions; thus, this is one instance in which the mentioned conjecture is true. Another instance is the single random variable $\xi(x) = x$ on the Lebesgue probability space, which itself forms a system which is saturated with respect to independence according to Theorem 3.6.2, and the system $1, x, x^2, \ldots, x^n, \ldots$ is, as is well known, complete. A third example is the

* This system is of course not orthogonal, but can be orthogonalized by the well known procedure of E. Schmidt, see [13].

system of Example 3.6.2: If $f(n)$ is any function defined for $1 \leqq n \leqq N = p_1 p_2 \cdots p_s$ (where $p_1, \ldots, p_s$ are different prime numbers) such that one has

$$\sum_{n=1}^{N} f(n) \prod_{k=1}^{s} r_{p_k}^{a_k}(n) = 0 \qquad \text{for any integers } a_k \geqq 0 \ (k = 1, 2, \ldots, s)$$

it follows that for any integers b_k with $0 \leqq b_k \leqq p_k - 1$, one has

$$\sum_{n=1}^{N} f(n) \left(\prod_{k=1}^{s} z_k^{b_k} \right)^n = 0 \qquad \text{where } z_k = e^{2\pi i/p_k} \ (1 \leqq k \leqq s)$$

Thus the polynomial

$$\sum_{n=1}^{N} f(n) x^n$$

of degree N vanishes at the N points

$$x = \prod_{k=1}^{s} z_k^{b_k} \qquad (0 \leqq b_k \leqq p_k - 1, \quad 1 \leqq k \leqq s)$$

and at $x = 0$, i.e., $f(n)$ is identically equal to 0. Nevertheless, the conjecture of Steinhaus is not true without any restriction, as is shown by Example 3.6.5. All powers of the random variable ξ in this example are orthogonal to the random variable η defined by $\eta(1) = 0$, $\eta(2) = 1$, $\eta(3) = -1$. Another similar example is the following:

Example 3.6.6. Let the random variable $\xi(x)$ on the Lebesgue probability space be defined as follows:

$$\xi(x) = \begin{cases} 2x & \text{for } 0 \leqq x \leqq 1/2 \\ 1 & \text{for } 1/2 \leqq x \leqq 1 \end{cases} \tag{3.6.28}$$

Then $\xi(x)$ forms, in itself, a system which is saturated with respect to independence, but all powers of ξ are orthogonal to any random variable η which is equal to 0 in the interval $(0, 1/2)$ and the integral of which over the interval $(1/2, 1)$ vanishes.

However, the following theorem is valid (see [11], [12], [19], [42] and [43]).

THEOREM 3.6.4. *A necessary and sufficient condition on a system $\{\xi_n\}$ of bounded random variables on the probability space $S = (\Omega, \mathscr{A}, P)$ (where Ω is a complete separable metric space and $\mathscr{A}$ is the family of Borel subsets of Ω), in order that the system of all products of powers of these random variables should be complete in $L_2(S)$, is that the ξ_n should form a spanning system.*

In view of Theorem 3.6.3, we get the following:

COROLLARY TO THEOREM 3.6.4. *Let ξ_n $(n \geqq 1)$ be bounded random variables on the probability space $S = (\Omega, \mathscr{A}, P)$ where Ω is a complete separable metric space and $\mathscr{A}$ is the family of Borel subsets of Ω. A necessary and sufficient* *condition for the system of all products of powers of the random variables* ξ_n *to be complete in* $L_2(S)$ *is that the* ξ_n *should separate the points of* Ω.

A proof of Theorem 3.6.4 is given in Chapter 5.

Example 3.6.7. Let an arbitrary sequence of numbers p_n ($n = 1, 2, \ldots$) be given such that $0 < p_n \leqq 1/2$. Construct the subsets A_n of the interval (0, 1) as follows: Let A_1 be the interval $(0, p_1)$; let A_2 be the union of the intervals $(0, p_1p_2)$ and $(p_1, p_1 + (1 - p_1)p_2)$. Continue this construction so that if $A_1, \ldots, A_n$ are already constructed, then they define a partition of the interval (0, 1) into 2^n nonempty intervals. Divide each of these intervals into two subintervals so that their lengths have the ratio $p_{n+1} : (1 - p_{n+1})$ and let A_{n+1} denote the union of the left-hand part of each of the mentioned 2^n intervals. (Note that in case $p_n = 1/2$ for all n, A_n is the set on which $R_n(x) = 1$.) The sets A_n are clearly independent in the Lebesgue probability space and $P(A_n) = p_n$. Let α_n denote the indicator of the set A_n; then the random variables α_n ($n = 1, 2, \ldots$) are independent. We now show that the system $\{\alpha_n\}$ is saturated with respect to independence if and only if $\sum_{n=1}^{\infty} p_n = +\infty$. Let us first consider the case when $\sum_{n=1}^{\infty} p_n < +\infty$. In this case almost all x $(0 < x < 1)$ belong to a finite number of the sets A_k only. If we denote by $B(i_1, i_2, \ldots, i_r)$ the set of those x which belong to the sets $A_{i_1}, A_{i_2}, \ldots, A_{i_r}$ but to no other set A_n, then we have

$$P(B(i_1, i_2, \ldots, i_r)) = \prod_{j=1}^{r} \frac{p_{i_j}}{1 - p_{i_j}} \cdot \prod_{k=1}^{\infty} (1 - p_k) \tag{3.6.29}$$

Further, the sets $B(i_1, i_2, \ldots, i_r)$ are disjoint, and

$$\sum P(B(i_1, \ldots, i_r)) = 1 \tag{3.6.30}$$

Now each $B(i_1, \ldots, i_r)$ can be decomposed into two Borel sets, $B'(i_1, \ldots, i_r)$ and $B''(i_1, \ldots, i_r)$ such that

$$P(B'(i_1, \ldots, i_r)) = 1/2P(B(i_1, \ldots, i_r))$$

As a matter of fact, the measure of the intersection of the set $B(i_1, \ldots, i_r)$ with the interval $(0, y)$ is a continuous nondecreasing function of y, increasing from 0 to $P(B(i_1, \ldots, i_r))$. Thus, there exists a value of y for which this measure is equal to $1/2P(B(i_1, \ldots, i_r))$. Denoting by A the union of all the sets $B'(i_1, \ldots, i_r)$, A is a Borel set such that $P(A) = 1/2$ and the sets $A, A_1, A_2, \ldots, A_n, \ldots$ are mutually independent. Thus, if α denotes the indicator of the set A, the random variables $\alpha, \alpha_1, \alpha_2, \ldots, \alpha_n, \ldots$ are mutually independent. In case $\sum_{n=1}^{\infty} p_n < +\infty$, the system $\{\alpha_n\}$ is not saturated with respect to independence. Let us turn now to the case $\sum_{n=1}^{\infty} p_n = +\infty$. In this case, the 2^n sets $A_1^{\delta_1} \cdots A_n^{\delta_n}$ $(\delta_k = \pm 1, k = 1, 2, \ldots, n)$ form a partition of the interval (0, 1) into 2^n subintervals such that the measure of each is $\leqq \prod_{k=1}^{n}(1 - p_k) = \Delta_n$. It follows from the condition $\sum_{n=1}^{\infty} p_n = +\infty$ that

$\lim_{n\to+\infty}\Delta_n = 0$. If A would be a measurable set such that $p(A) = p$ $(0 < p < 1)$ and the sets $A, A_1, A_2, \ldots$ would be mutually independent, let $F(x)$ denote the measure of the intersection of A and the interval $(0, x)$. It follows that $F(x) = px$ would hold on an everywhere dense denumerable set, and thus—as $F(x)$ is continuous—for every x. However, such a set A cannot exist. As a matter of fact, such a set A would be independent from every interval $[a, b]$ $(0 \leqq a < b \leqq 1)$ and, by Theorem 3.1.1, from every Borel subset of $(0, 1)$ and thus from itself too, which is impossible.

That such a set A cannot exist, follows also from the so-called Vitali-Lebesgue theorem, according to which if A is a measurable set of positive measure, then for almost all points x of A, one has

$$\lim_{h\to 0} \frac{P(AI_{x-h,x+h})}{2h} = 1 \tag{3.6.31}$$

where $I_{x-h,x+h}$ denotes the interval $(x - h, x + h)$; i.e., the set A has density 1 at almost all of its points, thus its density cannot be equal to p $(0 < p < 1)$ everywhere (see Saks [15] and also Knopp [22]).

Note that implicitly we have also shown that the least σ-algebra containing all the sets A_n $(n = 1, 2, \ldots)$ is, in the case $\sum_{n=1}^{\infty} p_n = +\infty$, identical with the family of all Borel subsets of the interval $(0, 1)$ and thus the α_n form, in this case, a spanning set. Accordingly, by Theorem 3.6.4, the system of all products $\alpha_{i_1} \ldots \alpha_{i_r}$ is complete in $L_2(0, 1)$. This can also be shown directly by the above argument, because there exists an everywhere dense set R in $(0, 1)$ such that the indicator of the interval $(0, r)$ is for each $r \in R$ a linear combination of some functions $\alpha_{i_1} \ldots \alpha_{i_r}$. Thus, if $f(x)$ is orthogonal to all products $\alpha_{i_1} \ldots \alpha_{i_r}$, then $\int_0^r f(x)\, dx = 0$ for all $r \in R$, and thus $f(x) = 0$ almost everywhere.

As regards the theory of orthonormal systems, see Alexits [14]. Concerning further connections between independence and orthogonality see Révész [13]. As regards other related questions and further connections between the notion of independence and certain problems of analysis and number theory, see the highly inspiring book [23] by Kac.

3.7 INDEPENDENCE AND ERGODIC THEORY

Let $S = (\Omega, \mathscr{A}, P)$ be a probability space, let $T = T\omega$ be a mapping of the set Ω onto itself, and denote for each $A \in \mathscr{A}$ by $T^{-1}A$ the set of those $\omega \in \Omega$ for which $T\omega \in A$.

Definition 3.7.1. *The mapping $T = T\omega$ of Ω onto itself such that for each $A \in \mathscr{A}$ one has $T^{-1}A \in \mathscr{A}$ and further*

$$P(T^{-1}A) = P(A) \tag{3.7.1}$$

is called a measure-preserving transformation of the probability space $S = (\Omega, \mathscr{A}, P)$; *the probability measure* P *is called invariant under the transformation* T; *the system* $(\Omega, \mathscr{A}, P, T)$ *is called a dynamic system.*

Remark: Note that it has not been supposed that T should be a one-to-one transformation: If T is a one-to-one measure-preserving transformation of the probability space S and if TA denotes the set into which the set A is mapped by T, then for each $A \in \mathscr{A}$ one has $TA \in \mathscr{A}$ and $P(TA) = P(A)$, i.e., in this case T^{-1} is also a measure-preserving transformation of S.

We denote by $T^n\omega$ $(n = 1, 2, \ldots)$ the element of Ω obtained by applying transformation T n times, i.e., $T^1\omega = T\omega$, $T^n\omega = T(T^{n-1}\omega)$ (for $n \geqq 2$) and we denote by $T^{-n}A$ the set of those $\omega \in \Omega$ for which $T^n\omega \in A$ $(n = 2, 3, \ldots)$. We put $T^0\omega = \omega$ for all $\omega \in \Omega$.

Example 3.7.1. Let S denote the Lebesgue probability space, let T_a be the transformation obtained by adding to x the number a mod 1 $(0 < a < 1)$, i.e., putting

$$T_a x = \begin{cases} x + a & \text{for } 0 \leqq x < 1 - a \\ x + a - 1 & \text{for } 1 - a \leqq x < 1 \end{cases}$$

T_a is a measure-preserving one-to-one transformation of S, further, $T_a^{-1} = T_{1-a}$.

Example 3.7.2. Let S denote the Lebesgue probability space, and define the transformation T by $T_x = (2x)$ where (y) denotes the fractional part of y (i.e., $(y) = y$ for $0 \leqq y < 1$, and $(y) = y - 1$ for $1 \leqq y < 2$). Then T is a measure-preserving transformation of S, which is, however, not one-to-one (each y $(0 \leqq y < 1)$ has two inverse images under T, namely $y/2$ and $(1 + y)/2$). It is easy to show that if $R_n(x)$ denotes the nth Rademacher function, one has

$$R_n(Tx) = R_{n+1}(x) \qquad (n = 1, 2, \ldots) \tag{3.7.2}$$

Thus, if A is the set $(0, 1/2)$, i.e., the set on which $R_1(x) = 1$, then $T^{-1}A$ is the set on which $R_2(x) = 1$, i.e., the union of the intervals $(0, 1/4)$ and $(1/2, 3/4)$. In general, $T^{-n}A$ is the set on which $R_{n+1}(x) = 1$ $(n = 1, 2, \ldots)$.

Example 3.7.3. Let $S = (\Omega, \mathscr{A}, P)$ be a probability space, and let S^* denote the probability space obtained by forming the product of denumerably infinite copies of S. Thus $S^* = (\Omega^*, \mathscr{A}^*, P^*)$ where Ω^* is the set of all sequences $(\omega_1, \omega_2, \ldots, \omega_n, \ldots)$ where $\omega_n \in \Omega$, $\mathscr{A}^*$ is the least σ-algebra of subsets of Ω^* containing the cylinder sets of Ω^* and P^* is the product measure constructed according to Theorem 3.4.2. Let the transformation T of Ω^* be defined as follows: If $\omega^* = (\omega_1, \omega_2, \ldots, \omega_n, \ldots)$, then $T\omega^* = (\omega_2, \omega_3, \ldots, \omega_{n+1}, \ldots)$. Note that T is not one-to-one if Ω has more than one element. It follows by definition, that if $A = A_1 \times A_2 \times \cdots \times A_n \times \cdots$ where $A_n \in \mathscr{A}$, then $T^{-1}A = \Omega \times A_1 \times A_2 \times \cdots \times A_{n-1} \times \cdots$.

Thus if A is a cylinder set,

$$P^*(T^{-1}A) = P^*(A)$$

It follows that $P^*(T^{-1}A) = P^*(A)$ for all $A \in \mathscr{A}^*$, i.e., that T preserves the measure P^*. If $A \in \mathscr{A}$ is any set such that $P(A) = p$ $(0 < p < 1)$ and $B = A \times \Omega \times \Omega \times \cdots$, the set $T^{-n}B$ consists of all $\omega^* = (\omega_1, \omega_2, \ldots, \omega_n, \ldots)$ such that $\omega_n \in A$, and thus one has for $0 \leqq n_1 < n_2 < \cdots < n_k$ $(k = 1, 2, \ldots)$,

$$P^*(T^{-n_1}B \cdot T^{-n_2}B \cdots T^{-n_k}B) = p^k$$

Thus the sets $T^{-n}B$ $(n = 0, 1, \ldots)$ are independent.

The transformation T is called the *Bernoulli shift* and the dynamic system consisting of the infinite product space $S^* = (\Omega^*, \mathscr{A}^*, P^*)$ and the shift T is called a (one-sided) *Bernoulli scheme*.*

The above two examples suggest the following question: If in a probability space $S = (\Omega, \mathscr{A}, P)$ there is given an infinite sequence A_n $(n = 1, 2, \ldots)$ of independent events such that $P(A_n) = p$ $(0 < p < 1; n = 1, 2, \ldots)$, under what additional conditions does there exist a measure-preserving transformation T of S such that $T^{-n}A_1 = A_{n+1}$ $(n = 0, 1, 2, \ldots)$?

A sufficient (though not necessary) condition is given by the following:

THEOREM 3.7.1. *Let $S = (\Omega, \mathscr{A}, P)$ be a probability space, and let $A_n \in \mathscr{A}$ $(n = 1, 2, \ldots)$ be a sequence of independent events such that $P(A_n) = p$ where $0 < p < 1$ for $n = 1, 2, \ldots$ and $\mathscr{A}$ is the least σ-algebra containing all events A_n $(n = 1, 2, \ldots)$. If the events A_n are strongly qualitatively independent, there exists a measure-preserving transformation T of S such that*

$$T^{-n}A_1 = A_{n+1} \qquad (n = 0, 1, 2, \ldots) \tag{3.7.3}$$

Proof. Let $\delta = (\delta_1, \delta_2, \ldots, \delta_n, \ldots)$ be an arbitrary sequence, each element of which is either $+1$ or -1. Let us put

$$B(\delta) = \prod_{k=1}^{\infty} A_k^{\delta_k} \tag{3.7.4}$$

By supposition none of the sets $B(\delta)$ is empty; thus we can choose for each $B(\delta)$ an element $\omega(\delta) \in B(\delta)$. Let us now define the transformation $T\omega$ of Ω as follows: for each $\delta = (\delta_1, \delta_2, \ldots)$ put $\delta' = (\delta_2, \delta_3, \ldots)$ and for each δ put

$$T\omega = \omega(\delta') \quad \text{for } \omega \in B(\delta) \tag{3.7.5}$$

* Some authors call the system (S^*, T) a Bernoulli scheme only in the special case when A is a finite algebra. The adjective "one-sided" is used to distinguish the concept from that of a two-sided Bernoulli scheme which differs from the one-sided one in that a product space is considered which is infinite in both directions.

It is easy to see that for each value of N and for each sequence $\delta_1, \delta_2, \ldots, \delta_N$, where $\delta_k = +1$ or $= -1$ ($k = 1, 2, \ldots, N$), one has

$$T^{-1}\left(\prod_{k=1}^{N} A_k^{\delta_k}\right) = \prod_{k=1}^{N} A_{k+1}^{\delta_k} \tag{3.7.6}$$

In particular, one has

$$T^{-1}A_n = A_{n+1} \tag{3.7.7}$$

and thus, by induction, we get

$$T^{-n}A_1 = A_{n+1} \tag{3.7.8}$$

Now, by supposition,

$$P\left(\prod_{k=1}^{N} A_k^{\delta_k}\right) = p^{1/2\Sigma_{k=1}^{N}(1+\delta_k)} \cdot q^{1/2\Sigma_{k=1}^{N}(1-\delta_k)} = P\left(\prod_{k=1}^{N} A_{k+1}^{\delta_k}\right) \tag{3.7.9}$$

Thus, if $\mathscr{A}_0$ denotes the family of all sets of the form $\prod_{k=1}^{N} A_k^{\delta_k}$ ($N=1,2,\ldots$) for each $A \in \mathscr{A}_0$ one has $T^{-1}A \in \mathscr{A}_0$ and

$$P(T^{-1}A) = P(A) \tag{3.7.10}$$

Clearly, $\mathscr{A}_0$ is a semiring. It follows that if $r(\mathscr{A}_0)$ denotes the least ring of subsets of Ω containing $\mathscr{A}_0$, then $T^{-1}A \in \mathscr{A}$ and (3.7.10) holds for $A \in r(\mathscr{A}_0)$. As, by supposition, $\mathscr{A}$ is the least σ-algebra containing $r(\mathscr{A}_0)$, it follows that (3.7.10) holds for every $A \in \mathscr{A}$. Thus T is a measure-preserving transformation of the probability space S. In view of (3.7.8), Theorem 3.7.1 is proved. ■

Remark 1: The transformation T constructed in the above proof is one-to-one if and only if each $B(\delta)$ consists of a single element only i.e., if the sets A_n separate the points of Ω. If the sets $B(\delta)$ contain more than one point, but for each δ, $B(\delta)$ and $B(\delta')$ have the same cardinal number, the construction can be modified so that T becomes one-to-one.

Remark 2: Example 3.7.2 shows that the condition in Theorem 3.7.1, that the events A_n should be strongly qualitatively independent, is not necessary.

Notice, however, that if each binary rational point r of the interval (0, 1) is replaced by a couple of points r^-, r^+ as in Example 3.3.3, then the conditions of Theorem 3.7.1 are satisfied. The conditions of Theorem 3.7.1 are evidently satisfied in the case of the Bernoulli scheme of Example 3.7.3.

Remark 3: The idea of the construction given in the proof of Theorem 3.7.1 is essentially due to Doob, (see [29], pp. 455–457). Doob considers a more general problem,* but he constructs T as a set transformation, not a point transformation.

* General stationary processes.

The idea of the proof of Theorem 3.7.1 can be used to prove the following more general result:

THEOREM 3.7.2. *Let $S = (\Omega, \mathscr{A}, P)$ be a probability space and let ξ_n $(n = 1, 2, \ldots)$ be a sequence of independent random variables on S, all having the same discrete distribution. Let $\mathscr{A}_n$ denote the σ-algebra generated by ξ_n $(n = 1, 2, \ldots)$ and suppose that the σ-algebras $\mathscr{A}_n$ are strongly qualitatively independent. Suppose that $\mathscr{A}$ is the least σ-algebra containing all the σ-algebras $\mathscr{A}_n$ $(n = 1, 2, \ldots)$. Then there exists a measure-preserving transformation T of S such that*

$$\xi_1(T^n\omega) = \xi_{n+1}(\omega) \quad \textit{for all } \omega \in \Omega \textit{ and for } n = 1, 2, \ldots \tag{3.7.11}$$

The proof of Theorem 3.7.2 follows step by step that of Theorem 3.7.1 and thus it can be omitted.

Notice that if in Theorem 3.7.1, $\alpha_n(\omega)$ is the indicator of the set A_n, then (3.7.8) can be written in the form below.

$$\alpha_1(T^n\omega) = \alpha_{n+1}(\omega) \tag{3.7.8$'$}$$

This shows that Theorem 3.7.2 contains Theorem 3.7.1 as a special case.

DEFINITION 3.7.2. *A measurable transformation T of a probability space $S = (\Omega, \mathscr{A}, P)$ is called ergodic if $T^{-1}A = A$, $A \in \mathscr{A}$ implies that $P(A) = 0$ or $P(A) = 1$.*

Remark: A set $A \in \mathscr{A}$ such that $T^{-1}A = A$ is called *invariant* with respect to T. Thus T is ergodic if it does not admit nontrivial invariant sets.

Example 3.7.4. The transformation T_a of Example 3.7.1 is ergodic if and only if a is an irrational number. This can be shown as follows: If a is rational, $a = p/q$ say, where $0 < p < q$ and A_β is the union of the intervals $(k/q, (k + \beta)/q)$ $(k = 0, 1, \ldots, q - 1; 0 < \beta < 1)$, then $T^{-1}A_\beta = A_\beta$ and $P(A_\beta) = \beta$. On the other hand, if a is irrational, the ergodicity of T_a can be deduced from the fact, well known from number theory, that the points $T_a^n 0 = (na)$ are everywhere dense in (0, 1) (see Billingsley [17]).

Example 3.7.5. Let S be the Lebesgue probability space and let U_β $(\beta > 1)$ denote the transformation of the interval [0, 1) defined by

$$U_\beta x = (\beta x) \qquad (0 \leqq x < 1) \tag{3.7.12}$$

where (βx) denotes the fractional part of βx.

It is easy to see that if q is an integer, $q \geqq 2$, then U_q is a measure-preserving ergodic transformation of the Lebesgue probability space. If $\beta > 1$ is not an integer, then U_β does not preserve the Lebesgue measure. It can be shown, however, (see [24]) that there exists a (unique) measure Q_β on the measurable subsets of the interval [0, 1) which is equivalent to the Lebesgue measure* and which is preserved under U_β (see also [25] and [28]).

For instance, if $\beta = (\sqrt{5}+1)/2$, this measure is defined for each measurable subset A of the interval [0, 1) by

$$Q_\beta(A) = \int_A \rho(x)\,dx \tag{3.7.13}$$

where

$$\rho(x) = \begin{cases} \dfrac{3\sqrt{5}+5}{10} & \text{for} \quad 0 \leqq x < \dfrac{\sqrt{5}-1}{2} \\[2ex] \dfrac{5+\sqrt{5}}{10} & \text{for} \quad \dfrac{\sqrt{5}-1}{2} < x < 1 \end{cases}$$

It can be shown (see [24]) that the transformation U_β is ergodic for every $\beta > 1$ (with respect to the measure Q_β as well as with respect to the Lebesgue measure). The measure Q_β has been determined explicitly by Parry ([26] and [27]) and by Cigler [44] for every $\beta > 1$.

The transformations U_β have been studied in connection with a certain generalization of the q-ary expansion of real numbers. It can be shown (see [24]) that for each $\beta > 1$ every real number x ($0 \leqq x < 1$) admits there presentation

$$x = \sum_{k=1}^{\infty} \frac{\varepsilon_k(x, \beta)}{\beta^k} \tag{3.7.14}$$

where the "digits" $\varepsilon_k(x, \beta)$ take on the values $0, 1, \ldots, [\beta]$; $\varepsilon_k(x, \beta)$ can be expressed by the formula

$$\varepsilon_k(x, \beta) = [\beta U_\beta^{k-1} x] \qquad \text{for } 0 \leqq x < 1,\ k = 1, 2, \ldots. \tag{3.7.15}$$

If $\beta = q \geqq 2$ where q is an integer, (3.7.14) is the well known q-ary expansion of x. Thus, for $q = 10$, (3.7.14) is nothing else than the decimal expansion of x. In this case, the $\varepsilon_n(x, q)$ ($n = 1, 2, \ldots$) are independent random variables on the Lebesgue probability space. This is not true concerning $\varepsilon_n(x, \beta)$ if β is not an integer (neither with respect to the Lebesgue measure nor with respect to the measure Q_β, which is invariant under the transformation U_β). For instance, in the case $\beta = (\sqrt{5}+1)/2$ if $\varepsilon_n(x) = 1$, then one has necessarily $\varepsilon_{n+1}(x) = 0$.

* Two measures on the same σ-algebra are called equivalent if each is absolutely continuous with respect to the other. See Appendix A.

DEFINITION 3.7.3. *A measurable transformation T of a probability space $S = (\Omega, \mathscr{A}, P)$ is called mixing* if for every $A \in \mathscr{A}$ and $B \in \mathscr{A}$, one has*

$$\lim_{n \to +\infty} P(T^{-n}A \cdot B) = P(A)P(B) \tag{3.7.16}$$

Remark: The above definition can be interpreted as follows: T is mixing if for each $A \in \mathscr{A}$, the set $T^{-n}A$ is for $n \to +\infty$ asymptotically independent from every $B \in \mathscr{A}$. Clearly, (3.7.16) implies that if $P(A) > 0$, then $T^{-n}A$ has a nonempty intersection with each $B \in \mathscr{A}$ for which $P(B) > 0$, if n is sufficiently large. Thus, the transformation T^{-1} really "mixes" the points of Ω rather thoroughly. This explains why such a transformation is called mixing.

We will now prove the following:

THEOREM 3.7.3. *A mixing transformation is ergodic.*†

Proof. Suppose that T is mixing and A is an invariant set, i.e., $T^{-1}A = A$. It follows that $T^{-n}A = A$ for $n = 1, 2, \ldots$. Thus, by (3.7.16), one has for the invariant set A and for every $B \in \mathscr{A}$,

$$P(AB) = P(A)P(B) \tag{3.7.17}$$

In other words, A is independent from every $B \in \mathscr{A}$. Choosing $B = A$, it follows that $P(A) = P^2(A)$, i.e., $P(A) = 0$ or $P(A) = 1$; thus, T is ergodic. ■

Example 3.7.6. Let us consider the Bernoulli scheme $(\Omega^*, \mathscr{A}^*, P^*, T)$ described in Example 3.7.3, i.e., let $(\Omega^*, \mathscr{A}^*, P^*)$ be the product of an infinite sequence of exemplars of the probability space $S = (\Omega, \mathscr{A}, P)$ and let T be the shift transformation in S^*. We shall show that T is mixing. To prove this we need the following:

LEMMA 3.7.1. *Let P_n $(n = 1, 2, \ldots)$ be a sequence of probability measures on the experiment $(\Omega, \mathscr{A})$. Let $\mathscr{A}_0$ be a semiring of subsets of Ω containing Ω, which generates $\mathscr{A}$ (i.e., $\sigma(\mathscr{A}_0) = \mathscr{A}$). Suppose that*

$$\lim_{n \to +\infty} P_n(A) = Q(A) \tag{3.7.18}$$

for each $A \in \mathscr{A}_0$ where $Q(A)$ is a probability measure on $\mathscr{A}$.

* Mixing transformations are sometimes called "strongly mixing" to distinguish from "weakly mixing." T is called weakly mixing if for every $A \in \mathscr{A}$ and $B \in \mathscr{A}$, one has

$$\lim_{N \to +\infty} \frac{1}{N} \sum_{k=1}^{N} |P(T^{-k}A \cdot B) - P(A)P(B)| = 0$$

† It can be shown in the same way that even weak mixing implies ergodicity.

Put for $A \in \mathscr{A}$,

$$M(A) = \sup_n P_n(A) \tag{3.7.19}$$

and suppose that for every sequence of sets B_n *such that* $B_n \in \mathscr{A}$ *and* $B_{n+1} \subseteq B_n$ $(n = 1, 2, \ldots)$ *and* $\prod_{n=1}^{\infty} B_n = \varnothing$, *one has**

$$\lim_{n \to +\infty} M(B_n) = 0 \tag{3.7.20}$$

Then (3.7.18) *holds for all* $A \in \mathscr{A}$.

Proof. Let $\mathscr{B}$ denote the family of those sets $A \in \mathscr{A}$ for which (3.7.18) holds. Clearly, $\mathscr{B}$ is an algebra. We prove that $\mathscr{B}$ is a σ-algebra. Let A_k $(k = 1, 2, \ldots)$ be a sequence of disjoint sets all belonging to $\mathscr{B}$, and put $A = \sum_{k=1}^{\infty} A_k$, $C_N = \sum_{k=1}^{N} A_k$ $(N = 1, 2, \ldots)$. Then we have

$$0 \leqq P_n(A) - P_n(C_N) \leqq M(A\bar{C}_N) \quad \text{for } N = 1, 2, \ldots \tag{3.7.21}$$

As, clearly, $A\bar{C}_{N+1} \subset A\bar{C}_N$ and $\prod_{N=1}^{\infty} A\bar{C}_N = \varnothing$, it follows that

$$\lim_{N \to \infty} M(A\bar{C}_N) = 0$$

thus, $M(A\bar{C}_N) < \varepsilon$ if $N \geqq N_0(\varepsilon)$. As $C_N \in \mathscr{B}$, it follows that

$$Q(C_N) \leqq \varliminf_{n \to +\infty} P_n(A) \leqq \varlimsup_{n \to +\infty} P_n(A) \leqq Q(C_N) + \varepsilon \tag{3.7.22}$$

As Q is, by supposition, a measure,

$$Q(A) - \varepsilon \leqq Q(C_N) \leqq Q(A) \tag{3.7.23}$$

if $N \geqq N_1(\varepsilon)$. Thus we obtain

$$Q(A) - \varepsilon \leqq \varliminf_{n \to +\infty} P_n(A) \leqq \varlimsup_{n \to +\infty} P_n(A) \leqq Q(A) + \varepsilon \tag{3.7.24}$$

As (3.7.24) holds for every $\varepsilon > 0$, it follows that

$$\lim_{n \to +\infty} P_n(A) = Q(A) \tag{3.7.25}$$

Thus $\mathscr{B}$ is a σ-algebra, and it contains $\sigma(\mathscr{A}_0) = \mathscr{A}$, which was to be proved. ■

Now we return to the Bernoulli scheme $(\Omega^*, \mathscr{A}^*, P^*, T)$. Let $\mathscr{A}_0$ denote the family of all cylinder sets in Ω^*. Let us choose an $A \in \mathscr{A}_0$; then $A = A_1 \times A_2 \times \cdots \times A_r \times \Omega \times \Omega \times \cdots$ where $A_k \in \mathscr{A}$ for $k = 1, 2, \ldots, r$.

Let us also choose for B a set belonging to $\mathscr{A}_0$, say, $B = B_1 \times B_2 \times \cdots \times B_s \times \Omega \times \Omega \cdots$. We have

$$T^{-n}A = \overset{1}{\Omega} \times \overset{2}{\Omega} \times \cdots \times \overset{n}{\Omega} \times A_1 \times \cdots \times A_r \times \Omega \times \cdots$$

* A set function $M(A)$ having the property postulated is called *continuous on the empty set*.

Thus, if $n \geqq s$, $T^{-n}A$ and B are independent. Thus (3.7.16) holds for P^* if both A and B are cylinder sets. We can now apply Lemma 3.7.1 to the sequence of measures $P_n(B) = P(T^{-n}A \cdot B)$ where A is fixed, the condition (3.7.20) being satisfied because $M(B) = \sup_n P(T^{-n}A \cdot B) \leqq P(B)$. It follows that (3.7.16) holds for P^*, for $A \in \mathscr{A}_0$ and for all $B \in \mathscr{A}$. Now let us fix a $B \in \mathscr{A}$ and apply Lemma 3.7.1 to the sequence of measures $P_n(A) = P(T^{-n}A \cdot B)$. The condition (3.7.20) is again satisfied, because

$$M(A) = \sup_n P(T^{-n}A \cdot B) \leqq \sup_n P(T^{-n}A) = P(A)$$

It follows that (3.7.16) holds for P^* for every $A \in \mathscr{A}$ and $B \in \mathscr{A}$. Thus, the shift T is mixing and therefore by Theorem 3.7.1 it is also ergodic.

Example 3.7.7. Let us consider the measure-preserving transformation T in Theorem 3.7.1. One can show essentially in the same way as in Example 3.7.6 that T is mixing and therefore ergodic.

Example 3.7.8. Let us consider the transformation T_a of the Lebesgue probability space defined in Example 3.7.1 and suppose that a is irrational. We shall show that T_a (while being ergodic as pointed out in Example 3.7.4) is not mixing. Let both A and B be equal to the interval (0, 1/2). As the sequence (na) $(n = 1, 2, \ldots)$ is everywhere dense in (0, 1), we can select a sequence $n_1 < n_2 < \cdots < n_k < \cdots$ of integers such that $\lim_{k \to +\infty}(n_k a) = 0$. It follows that $\lim_{k \to \infty} P(T_a^{-n_k}A \cdot A) = 1/2$ while $P(A)P(A) = 1/4$, which shows that T is not mixing. Thus, mixing is really a stronger property than ergodicity.

For a more detailed study of ergodicity see Halmos [16] and Jacobs [18].

3.8 INDEPENDENCE AND INFORMATION

DEFINITION 3.8.1. *Let ξ be a discrete random variable taking on a finite number of different values $x_1, x_2, \ldots, x_N$ with the corresponding positive probabilities $P(\xi = x_k) = p_k$ $(k = 1, 2, \ldots, N)$. The entropy $H(\xi)$ of ξ is defined by the formula**

$$H(\xi) = \sum_{k=1}^{N} p_k \log_2 \frac{1}{p_k} \tag{3.8.1}$$

Remark: Clearly $H(\xi) \geqq 0$, with equality if and only if $N = 1$, i.e., if ξ is (almost surely) constant. The entropy $H(\xi)$ of ξ can be interpreted as a measure of the amount of uncertainty present with respect to the value of

* The logarithm with base 2 is denoted by $\log_2$.

the random variable ξ, before observing the actual value of ξ. Another possible interpretation is that $H(\xi)$ is a measure of the amount of information received when the actual value of ξ is observed. Equation (3.8.1) is called *Shannon's formula.* As $H(\xi)$ depends only on the distribution $\mathscr{P}=\{p_1, p_2, \ldots, p_N\}$ of ξ, we shall also use the notation

$$H[\mathscr{P}] = \sum_{k=1}^{N} p_k \log_2 \frac{1}{p_k} \tag{3.8.1'}$$

If $f(x)$ is any function defined on the set $\{x_1, \ldots, x_N\}$ such that $f(x_i) \neq f(x_j)$ if $i \neq j$, then $P(f(\xi) = f(x_k)) = P(\xi = x_k) = p_k$, and

$$H(f(\xi)) = H(\xi) \tag{3.8.2}$$

More generally, one has the following:

LEMMA 3.8.1. *If $f(x)$ is an arbitrary function, one has*

$$H(f(\xi)) \leqq H(\xi) \tag{3.8.3}$$

equality taking place in (3.8.3) *if and only if $f(x_i) \neq f(x_j)$ for $i \neq j$.*

Proof. Let the different values of $f(\xi)$ be $y_1, y_2, \ldots, y_r$ and let us denote by E_j the set of those values of k for which $f(x_k) = y_j$ $(j = 1, 2, \ldots, r)$. Put $q_j = P(f(\xi) = y_j) = \sum_{k \in E_j} p_k$. Then we have

$$H(\xi) - H(f(\xi)) = \sum_{j=1}^{r} q_j H[\mathscr{P}_j] \tag{3.8.4}$$

where $\mathscr{P}_j$ denotes the probability distribution $\{p_k/q_j;\ k \in E_j\}$. As $H[P_j] \geqq 0$, with equality if and only if E_j consists of a single element only, the inequality (3.8.3) follows. ■

Remark: Lemma 3.8.1 shows that if $\mathscr{A}_\xi$ denotes the algebra generated by ξ, then $H(\xi)$ depends only on the probability space $(\Omega, \mathscr{A}_\xi, P)$. Thus, instead of the entropy of ξ, respectively the entropy of the distribution $\mathscr{P}$ of ξ, we may speak about the *entropy of the algebra* $\mathscr{A}_\xi$. We denote the entropy of an algebra $\mathscr{A}$ of events by $H\{\mathscr{A}\}$. The statement of Lemma 3.8.1 can also be expressed as follows: *If the algebra $\mathscr{A}'$ is a refinement of the algebra $\mathscr{A}$, one has*

$$H\{\mathscr{A}\} \leqq H\{\mathscr{A}'\} \tag{3.8.5}$$

The meaning of (3.8.5) is: *If an experiment is refined, more information is obtained.*

DEFINITION 3.8.2. Let $\mathscr{P} = \{p_1, p_2, \ldots, p_N\}$ and let $\mathscr{Q} = \{q_1, q_2, \ldots, q_N)$ be two finite probability distributions containing the same number $N \geqq 2$ of positive terms. The *divergence** $D(\mathscr{P}, \mathscr{Q})$ of the distribution $\mathscr{P}$ from the distribution $\mathscr{Q}$ is defined by the formula

$$D(\mathscr{P}, \mathscr{Q}) = \sum_{k=1}^{N} p_k \log_2 \frac{p_k}{q_k} \tag{3.8.6}$$

Remark: The divergence $D(\mathscr{P}, \mathscr{Q})$ may be interpreted as a measure of the discrepancy of the two distributions $\mathscr{P}$ and $\mathscr{Q}$; it is asymmetric, i.e., in general $D(\mathscr{P}, \mathscr{Q}) \neq D(\mathscr{Q}, \mathscr{P})$.

Notice that the definition of $D(\mathscr{P}, \mathscr{Q})$ depends on the labeling of the terms of the distributions $\mathscr{P}$ and $\mathscr{Q}$. More exactly, it depends on how a one-to-one mapping between the terms of $\mathscr{P}$ and $\mathscr{Q}$ is established by the labeling of the terms of these distributions. Notice that the kth term on the right of (3.8.6) is $\geqq 0$ or <0, according to whether $p_k \geqq q_k$ or $p_k < q_k$. Nevertheless, we have the following:

Lemma 3.8.2. *For any pair of distributions $\mathscr{P}$ and $\mathscr{Q}$ for which it is defined, one has*

$$D(\mathscr{P}, \mathscr{Q}) \geqq 0 \tag{3.8.7}$$

There is equality in (3.8.7) *if and only if $p_k = q_k$ for $k = 1, 2, \ldots, N$.*

Proof. Using the elementary inequality $\ln(1+x) \leqq x$ for $x > -1$, equality taking place only for $x = 0$, one obtains

$$-D(\mathscr{P}, \mathscr{Q}) \ln 2 = \sum_{k=1}^{N} p_k \ln\left(1 + \frac{q_k - p_k}{p_k}\right) \leqq \sum_{k=1}^{N} (q_k - p_k) = 0$$

This proves (3.8.7). ∎

Corollary 1 of Lemma 3.8.2. *The entropy of a discrete random variable ξ, taking on N different values, is maximal if and only if $p_k = 1/N$ for $k = 1, 2, \ldots, N$; in this case, $H(\xi) = \log_2 N$.*

Proof. Let $\mathscr{P}_N$ denote the distribution $\{p_1, p_2, \ldots, p_N\}$ and let U_N denote the uniform distribution $\{1/N, 1/N, \ldots, 1/N\}$ then one has

$$\log_2 N - H(\xi) = D(\mathscr{P}_N, U_N) \geqq 0 \quad ∎ \tag{3.8.8}$$

Remark: According to the mentioned interpretation of entropy the uncertainty concerning the value of a random variable ξ, capable of taking on N different values, is maximal if all these values are equiprobable.

* Also called the I-divergence of $\mathscr{P}$ from $\mathscr{Q}$. The symmetric quantity $D(\mathscr{P}, \mathscr{Q}) + D(\mathscr{Q}, \mathscr{P})$ is called the J-divergence of $\mathscr{P}$ and $\mathscr{Q}$.

COROLLARY 2 OF LEMMA 3.8.2. *If $\mathscr{P}=\{p_1, \ldots, p_N\}$ is a probability distribution and $\mathscr{Q}=\{q_1, \ldots, q_N\}$ is a set of N positive numbers such that $\sum_{k=1}^{N} q_k \leqq 1$, then*

$$\sum_{k=1}^{N} p_k \log_2 \frac{p_k}{q_k} \geqq 0$$

Proof. Put $\lambda=\sum_{k=1}^{N} q_k$ and $q_k'=q_k/\lambda$. Then $\mathscr{Q}'=\{q_1', \ldots, q_N'\}$ is a probability distribution and therefore by Lemma 3.8.1 $D(\mathscr{P}, \mathscr{Q}') \geqq 0$. Thus,

$$\sum_{k=1}^{N} p_k \log_2 \frac{p_k}{q_k} = \log_2 \frac{1}{\lambda} + D(\mathscr{P}, \mathscr{Q}') \geqq 0 \quad \blacksquare$$

DEFINITION 3.8.3. *The mutual information $I(\xi, \eta)$ of the random variables ξ and η, each taking on only a finite number of different values, is defined by*

$$I(\xi, \eta) = H(\xi) + H(\eta) - H((\xi, \eta)) \tag{3.8.9}$$

where $H((\xi, \eta))$ denotes the entropy of the random vector (ξ, η).

Remark: Let the possible different values of ξ and η be $x_1, x_2, \ldots, x_N$ and $y_1, y_2, \ldots, y_M$, respectively, and let the joint distribution of ξ and η be

$$r_{j,k} = \mathscr{P}(\xi = x_j, \eta = y_k) \quad (1 \leqq j \leqq N, 1 \leqq k \leqq M) \tag{3.8.10}$$

Let us put

$$p_j = \mathscr{P}(\xi = x_j) \quad (1 \leqq j \leqq N) \tag{3.8.11}$$

and

$$q_k = P(\eta = y_k) \quad (1 \leqq k \leqq M) \tag{3.8.12}$$

Then we evidently have

$$\sum_{k=1}^{M} r_{j,k} = p_j \quad (1 \leqq j \leqq N) \tag{3.8.13}$$

and

$$\sum_{j=1}^{N} r_{j,k} = q_k \quad (1 \leqq k \leqq M) \tag{3.8.14}$$

Thus,

$$H(\xi) = \sum_{j=1}^{N} p_j \log_2 \frac{1}{p_j} = \sum_{j=1}^{N} \sum_{k=1}^{M} r_{j,k} \log_2 \frac{1}{p_j} \tag{3.8.15}$$

and similarly,

$$H(\eta) = \sum_{k=1}^{M} q_k \log_2 \frac{1}{q_k} = \sum_{j=1}^{N} \sum_{k=1}^{M} r_{j,k} \log_2 \frac{1}{q_k} \tag{3.8.16}$$

Further,

$$H((\xi, \eta)) = \sum_{j=1}^{N} \sum_{k=1}^{M} r_{j,k} \log_2 \frac{1}{r_{j,k}} \tag{3.8.17}$$

Thus we obtain*

$$I(\xi, \eta) = \sum_{j=1}^{N} \sum_{k=1}^{M} r_{j,k} \log_2 \frac{r_{j,k}}{p_j \cdot q_k} \tag{3.8.18}$$

This means that if we denote by $\mathscr{R}$ the joint distribution of the random variables ξ and η, and by $\mathscr{P}\mathscr{Q}$ the distribution $\{p_j q_k\}$ (i.e., $\mathscr{P}\mathscr{Q}$ would be the joint distribution of the random variables ξ and η if they were independent, which is not supposed here), then

$$I(\xi, \eta) = D(\mathscr{R}, \mathscr{P}\mathscr{Q}) \tag{3.8.19}$$

From (3.8.19) and Lemma 3.8.2 we get the following:

THEOREM 3.8.1. *For any pair ξ, η of random variables one has*

$$I(\xi, \eta) \geqq 0 \tag{3.8.20}$$

There is equality in (3.8.20) *if and only if ξ and η are independent.*

Remark: In information theory, $I(\xi, \eta)$ is interpreted as the amount of information concerning ξ obtained if η is observed. As, by definition, $I(\eta, \xi) = I(\xi, \eta)$, this means that the amount of information on ξ obtained by observing η is equal to the amount of information on η obtained by observing ξ, and this amount of information is always nonnegative and is 0 if and only if ξ and η are independent. Thus, two random variables are independent if and only if by observing one of them we get no information on the other at all. Theorem 3.8.1 throws new light on the meaning of independence: *it could even serve as the definition of independence.*

Let us introduce the notation

$$p_{j|k} = P(\xi = x_j \mid \eta = y_k) = \frac{r_{j,k}}{q_k} \tag{3.8.21}$$

$$q_{k|j} = P(\eta = y_k \mid \xi = x_j) = \frac{r_{j,k}}{p_j} \tag{3.8.22}$$

and denote by $\mathscr{P}_k$ the distribution $\{p_{j|k}\,;\, 1 \leqq j \leqq N\}$ and by $\mathscr{Q}_j$ the distribution $\{q_{k|j}\,;\, 1 \leqq k \leqq M\}$. Thus $\mathscr{P}_k$ is the conditional distribution of ξ under the condition $\eta = y_k$ and $\mathscr{Q}_j$ is the conditional distribution of η under the condition $\xi = x_j$. Let us put

$$H(\xi \mid \eta) = \sum_{k=1}^{M} q_k H[\mathscr{P}_k] \tag{3.8.23}$$

* $0 \log_2 0 = 0$ by definition.

and

$$H(\eta \mid \xi) = \sum_{j=1}^{N} p_j H[\mathcal{Q}_j] \tag{3.8.24}$$

Then $H(\xi \mid \eta)$ and $H(\eta \mid \xi)$, being the averages of nonnegative numbers, are nonnegative.

The quantity $H(\xi \mid \eta)$ respectively $H(\eta \mid \xi)$ is interpreted *as the (average) conditional entropy of* ξ *given* η (respectively of η given ξ). One evidently has

$$H(\xi \mid \eta) = H((\xi, \eta)) - H(\eta) \tag{3.8.25}$$

Note that it follows from Lemma 3.8.1 that

$$H(\xi \mid \eta) \geqq 0 \tag{3.8.26}$$

because η is a function of (ξ, η) and thus $H(\eta) \leqq H((\xi, \eta))$. We get from (3.8.25),

$$I(\xi, \eta) = H(\xi) - H(\xi \mid \eta) = H(\eta) - H(\eta \mid \xi). \tag{3.8.27}$$

Formula (3.8.27) can be interpreted as follows: *The information on* ξ *obtained by observing* η *is equal to the decrease of uncertainty concerning* ξ *obtained by observing* η *(and conversely).*

It follows from (3.8.26) that

$$H(\xi \mid \eta) \leqq H(\xi) \tag{3.8.28}$$

Thus we obtain the following:

COROLLARY TO THEOREM 3.8.1. *By observing* η, *the uncertainty concerning* ξ *cannot increase; it remains unchanged if and only if* ξ *and* η *are independent, otherwise it decreases definitely.*

A third way of expressing Theorem 3.8.1 is the following: *For any pair of random variables* ξ *and* η, *one has*

$$H((\xi, \eta)) \leqq H(\xi) + H(\eta) \tag{3.8.29}$$

with equality standing in (3.8.29) *if and only if* ξ *and* η *are independent.*

Formula (3.8.27) can be written in the form

$$I(\xi, \eta) \leqq H(\xi) \tag{3.8.30}$$

In (3.8.30) we have equality if and only if ξ is constant (with probability 1) under the condition that the value of η is given, i.e., if ξ is a function of η. Especially, we have

$$I(\xi, \xi) = H(\xi) \tag{3.8.31}$$

Thus the entropy of a random variable ξ is equal to the amount of information contained in the value of the random variable ξ concerning itself, i.e., to the total amount of information in the value of ξ, in accordance with our previous alternative interpretation of the meaning of $H(\xi)$.

The unit of information is called a *bit*. Thus the amount of information contained in the value of a random variable taking on two different values—say the values 0 and 1—with probability 1/2, is equal to one bit.

As $I(\xi, \eta)$ is symmetric, it follows from (3.8.30) that

$$I(\xi, \eta) \leqq \min(H(\xi), H(\eta)) \tag{3.8.32}$$

Theorem 3.8.1 shows that there is an intimate connection between the notion of independence and information.* Now let us consider how this connection can be extended for more than two random variables.

We will prove the following:

THEOREM 3.8.2. *Let $\xi_1, \xi_2, \ldots, \xi_n$ be arbitrary random variables on the same probability space, each ξ_k taking on only a finite number of different values. Then one has, denoting by $H((\xi_1, \xi_2, \ldots, \xi_n))$ the entropy of the joint distribution of the variables $\xi_1, \xi_2, \ldots, \xi_n$,*

$$H((\xi_1, \xi_2, \ldots, \xi_n)) \leqq \sum_{k=1}^{n} H(\xi_k) \tag{3.8.33}$$

equality standing in (3.8.33) *if and only if $\xi_1, \xi_2, \ldots, \xi_n$ are mutually independent.*

Proof. Let $x_{k,j}$ $(j = 1, 2, \ldots, N_k)$ be all different values taken on with positive probabilities by ξ_k and put $p_{k,j} = P(\xi_k = x_{k,j})$. Let us denote by $\mathscr{P}_k$ the distribution $\{p_{k,1}, p_{k,2} \ldots, p_{k,N_k}\}$ and by $\mathscr{P}_1\mathscr{P}_2 \cdots \mathscr{P}_n$ the probability distribution having the terms $p_{1,j_1}p_{2,j_2} \cdots p_{n,j_n}$. Let $\mathscr{R}$ denote the joint distribution of the variables $\xi_1, \ldots, \xi_n$, i.e., put $\mathscr{R} = \{r(j_1, j_2, \ldots, j_n)\}$ where $r(j_1, j_2, \ldots, j_n) = P(\xi_1 = x_{1,j_1}, \ldots, \xi_n = x_{n,j_n})$. Then we have

$$H((\xi_1, \ldots, \xi_n)) - \sum_{k=1}^{n} H(\xi_k) = \sum_{j_1=1}^{N_1} \cdots \sum_{j_n=1}^{N_n} r(j_1, \ldots, j_n) \log_2 \frac{r(j_1, \ldots, j_n)}{p_{1,j_1} \cdots p_{n,j_n}}$$

and thus

$$H((\xi_1, \ldots, \xi_n)) - \sum_{k=1}^{n} H(\xi_k) = D(\mathscr{R}, \mathscr{P}_1\mathscr{P}_2 \cdots \mathscr{P}_n) \geqq 0 \tag{3.8.34}$$

with equality standing in (3.8.34) if and only if $\mathscr{R} = \mathscr{P}_1\mathscr{P}_2 \cdots \mathscr{P}_n$, i.e., if $\xi_1, \xi_2, \ldots, \xi_n$ are mutually independent. ■

DEFINITION 3.8.4. *Let us put*

$$I(\xi_1, \ldots, \xi_n) = H((\xi_1, \ldots, \xi_n)) - \sum_{k=1}^{n} H(\xi_k) \tag{3.8.35}$$

and call $I(\xi_1, \ldots, \xi_n)$ the mutual information of the random variables $\xi_1, \ldots, \xi_n$.

* It is instructive to compare this connection with that between qualitative independence and qualitative information (see Section 1.8).

Remark: Clearly, Definition 3.8.4 reduces to Definition 3.8.3 for $n=2$.

COROLLARY 1 TO THEOREM 3.8.2. *The random variables $\xi_1, \xi_2, \ldots, \xi_n$ are mutually independent if and only if their mutual information equals* 0.

One can express the statement of Theorem 3.8.2 in terms of the mutual information of pairs of random variables, e.g., as follows:

COROLLARY 2 TO THEOREM 3.8.2. *The random variables are mutually independent if and only if for each k $(1 \leqq k \leqq n-1)$ one has*

$$I((\xi_1, \ldots, \xi_k), \xi_{k+1}) = 0 \tag{3.8.36}$$

i.e., if by observing the variables $\xi_1, \xi_2, \ldots, \xi_k$ we get no information on ξ_{k+1} $(1 \leqq k \leqq n-1)$.

Proof. We evidently have for $k \geqq 2$,

$$I(\xi_1, \ldots, \xi_{k+1}) - I(\xi_1, \ldots, \xi_k) = I((\xi_1, \xi_2, \ldots, \xi_k), \xi_{k+1}) \tag{3.8.37}$$

Summing (3.8.37) for $k=2, 3, \ldots, n-1$ and adding $I(\xi_1, \xi_2)$ to both sides we get

$$I(\xi_1, \ldots, \xi_n) = \sum_{k=1}^{n-1} I((\xi_1, \ldots, \xi_k), \xi_{k+1}) \tag{3.8.38}$$

from which our statement follows. ■

Remark: Clearly $I(\xi_1, \ldots, \xi_n)$ is invariant under any permutation of the random variables $\xi_1, \ldots, \xi_n$, while the terms on the right-hand side of (3.8.38) change if the ξ_k are permuted. Thus, e.g., for $n=3$ we get from (3.8.38) three different decompositions:

$$\begin{aligned} I(\xi_1, \xi_2, \xi_3) &= I(\xi_1, \xi_2) + I((\xi_1, \xi_2), \xi_3) \\ &= I(\xi_1, \xi_3) + I((\xi_1, \xi_3), \xi_2) \\ &= I(\xi_2, \xi_3) + I((\xi_2, \xi_3), \xi_1) \end{aligned}$$

For $n=4$ we get similarly twelve decompositions of the type (3.8.38) for $I(\xi_1, \xi_2, \xi_3, \xi_4)$. However, in case $n \geqq 4$ there exist still other decompositions of $I(\xi_1, \ldots, \xi_n)$ into the sum of $n-1$ mutual informations of pairs of random variables; for instance, for $n=4$ we have besides the twelve decompositions of the type (3.8.38), also three decompositions* of the type

$$I(\xi_1, \xi_2, \xi_3, \xi_4) = I((\xi_1, \xi_2), (\xi_3, \xi_4)) + I(\xi_1, \xi_2) + I(\xi_3, \xi_4) \tag{3.8.39}$$

Thus for $n=4$ we have $12 + 3 = 15$ decompositions of $I(\xi_1, \xi_2, \xi_3, \xi_4)$ into the sum of three mutual informations of pairs of random variables. It can

* The other two such decompositions are obtained from (3.8.39) by interchanging ξ_2 with ξ_3, respectively with ξ_4, on the right-hand side.

be shown that for every $n \geqq 3$ the total number of decompositions of $I(\xi_1, \ldots, \xi_n)$ into the sum of $n-1$ mutual informations is equal to $(2n-3)!! = 1 \cdot 3 \cdot 5 \cdots (2n-3)$. This can be shown as follows: Any decomposition of $I(\xi_1, \ldots, \xi_n)$ can be obtained as follows: one splits the set $\{1, 2, \ldots, n\}$ into two disjoint nonempty subsets; after this, one splits those subsets which contain more than one element into two disjoint nonempty subsets; one continues this process until possible. For instance, (3.8.39) corresponds to the following splitting:

$$\{1, 2, 3, 4\}$$

$$\{1, 2\} \qquad \{3, 4\}$$

$$\{1\} \quad \{2\} \quad \{3\} \quad \{4\}$$

To every such splitting process there corresponds an oriented rooted tree having n end points such that from every point which is not an end point there start two outgoing edges. We call such trees "binary trees."* If the end points are labeled by the numbers $1, 2, \ldots, n$, then to every vertex there corresponds a uniquely determined subset of the set $\{1, 2, \ldots, n\}$ such that if the vertex P is joined by outgoing edges with the vertices P' and P'' the sets corresponding to P' and P'' are disjoint and their union is the set corresponding to P. It is easy to show by induction that the total number of such labeled trees is equal to $(2n-3)!!$. Now, to every such tree there corresponds a uniquely determined decomposition of $I(\xi_1, \ldots, \xi_n)$. For instance, to the tree

5

(3.8.40)

1 2 3 4

* Binary trees are used in information theory for another quite different purpose: to describe minimal binary prefix codes. Suppose we have an alphabet consisting of n "letters" and we let correspond to each letter a sequence of zeros and ones called "code words"; we call the set of these code words a binary prefix code if it has the following properties: (1) All code words are different, (2) no code word is the initial segment of another code word. A binary prefix code is called minimal if by omitting any digit from any code words, the resulting set of zero-one sequences is no more a binary prefix code. Now clearly, to every minimal binary prefix code consisting of n code words there corresponds a binary tree having n end points, and conversely. For example, to the tree (3.8.40) there corresponds the code 1, 011, 010, 001, 000.

there corresponds the decomposition

$$I(\xi_1, \xi_2, \xi_3, \xi_4, \xi_5) = I((\xi_1, \xi_2, \xi_3, \xi_4), \xi_5) + I((\xi_1, \xi_2), (\xi_3, \xi_4)) + I(\xi_1, \xi_2) + I(\xi_3, \xi_4) \tag{3.8.41}$$

To each decomposition in question there corresponds a criterion of independence. For instance, to (3.8.40) there corresponds the criterion that the random variables $\xi_1, \ldots, \xi_5$ are independent if and only if each of the four mutual informations on the right-hand side of (3.8.41) vanishes. We will now prove the following:

THEOREM 3.8.3. *Let $\xi_1, \xi_2, \ldots, \xi_n$ be independent random variables on the probability space S, each taking on a finite number of values, and let ζ be an arbitrary random variable on S, taking on a finite number of values. Then one has*

$$\sum_{k=1}^{n} I(\xi_k, \zeta) \leqq I((\xi_1, \ldots, \xi_n), \zeta) \tag{3.8.42}$$

Proof. It is sufficient to prove (3.8.42) for $n = 2$, as the general case follows by induction. Thus we have to show only that if ξ and η are independent random variables, then

$$I(\xi, \zeta) + I(\eta, \zeta) \leqq I((\xi, \eta), \zeta) \tag{3.8.43}$$

Let x, y, and z run over all values taken on by ξ, η, and ζ, respectively, with positive probability. Then one has by definition,

$$I((\xi, \eta), \zeta) - I(\xi, \zeta) - I(\eta, \zeta) = D(\mathscr{P}, \mathscr{Q})$$

where $\mathscr{P}$ is the joint distribution of ξ, η, and ζ, i.e., $\mathscr{P} = \{P(\xi = x, \eta = y, \zeta = z)\}$ and $\mathscr{Q}$ is the distribution* having the terms

$$Q(x, y, z) = \frac{P(\xi = x, \zeta = z)P(\eta = y, \zeta = z)}{P(\zeta = z)}$$

Therefore, (3.8.43) follows by Lemma 3.8.2. This proves Theorem 3.8.3. ■

Remark: The meaning of Theorem 3.8.3 can be expressed as follows: The informations concerning ζ obtained by observing the *independent* random variables $\xi_1, \xi_2, \ldots, \xi_n$ do not overlap. Notice that (3.8.43) is not true if the independence of ξ and η is not supposed. For instance, if $\xi = \eta$, then $I((\xi, \eta), \zeta) = I(\xi, \zeta) = I(\eta, \zeta)$ and thus $I(\xi, \zeta) + I(\eta, \zeta) = 2I((\xi, \eta), \zeta)$.

Example 3.8.1. Let $\xi_1, \xi_2, \ldots, \xi_n$ be independent, identically distributed random variables, taking on each a finite number N of different values

* $\mathscr{Q}$ would be the joint distribution of ξ, η, and ζ if ξ and η were conditionally independent under every condition $\zeta = z$.

$x_1, x_2, \ldots, x_N$ and let us put $\zeta_n = \xi_1 + \xi_2 + \cdots + \xi_n$. Let us consider the sequence $H(\zeta_n)$. Evidently,

$$H(\zeta_n) - H(\zeta_{n-1}) = I(\xi_1, \zeta_n) \geqq 0 \tag{3.8.44}$$

Thus, $H(\zeta_n)$ ($n = 1, 2, \ldots,$) is an increasing sequence. Now the possible values of ζ_n are all of the form $k_1x_1 + \cdots + k_Nx_N$ where $k_1, k_2, \ldots, k_N$ are nonnegative integers and $k_1 + k_2 + \cdots + k_N = n$. Thus the number of possible values of ζ_n cannot exceed n^N and thus, by Corollary 1 of Lemma 3.8.2,

$$H(\zeta_n) \leqq N \log_2 n \tag{3.8.45}$$

On the other hand, it follows from Theorem 3.8.3 that

$$I(\xi_1, \zeta_n) \leqq \frac{I((\xi_1, \ldots, \xi_n), \zeta_n)}{n} = \frac{H(\zeta_n)}{n} \tag{3.8.46}$$

Thus it follows that

$$I(\xi_1, \zeta_n) \leqq \frac{N \log_2 n}{n} \tag{3.8.47}$$

which implies

$$\lim_{n \to +\infty} I(\xi_1, \zeta_n) = 0 \tag{3.8.48}$$

(For another proof of (3.8.48) see Richter and Worm [41].) The meaning of (3.8.48) is that ζ_n becomes, in the limit for $n \to \infty$, independent of ξ_1 (and similarly of any ξ_k with k fixed).

The notion of mutual information can be extended for arbitrary random variables, as follows:

DEFINITION 3.8.6. *If ξ and η are arbitrary random variables or random vectors on the probability space S, we define the mutual information $I(\xi, \eta)$ by*

$$I(\xi, \eta) = \sup I(f(\xi), g(\eta)) \tag{3.8.49}$$

where the supremum is taken over all Borel-measurable functions f and g, taking on a finite number of distinct values only.*

It follows immediately that the statement of Theorem 3.8.1 remains valid for arbitrary random variables.

The formula (3.8.9) cannot be generalized for arbitrary random variables, because if ξ has a continuous distribution then one has, in general, $\sup H(f(g)) = +\infty$ if the upper bound is taken for all Borel functions f taking on a finite number of values.† As a matter of fact, if the distribution function $F(x)$ of

* Of course the supremum may be equal to $+\infty$.

† If ξ is a discrete random variable taking on denumerably infinitely many values x_k ($k = 1, 2, \ldots$) with the corresponding probabilities $p_k = P(\xi = x_k)$, then we may put $H(\xi) = \sum_{k=1}^{+\infty} p_k \log_2 1/p_k$, provided that this series is convergent.

ξ is continuous, we can find for each n real numbers, $x_{n,k}$ $(k = 1, 2, \dots, n-1)$ such that $F(x_{n,k}) = k/n$ $(1 \leq k \leq n-1)$. If we put $f_n(x) = k$ for $x_{n,k} \leq x < x_{n,k+1}$ $(k = 1, 2, \dots, n-2)$, $f_n(x) = 0$ for $x < x_{n,1}$, and $f_n(x) = n-1$ for $x \geq x_{n,n-1}$, then $f(\xi)$ takes on each of the values $0, 1, \dots, n-1$ with probability $1/n$, and thus $H(f_n(\xi)) = \log_2 n$. However, formula (3.8.18) can be extended as follows: If the joint distribution of ξ and η is absolutely continuous with density function $h(x, y)$, and $f(x)$ and $g(y)$ denote the density functions of ξ and η, then one has

$$I(\xi, \eta) = \int_{-\infty}^{+\infty} \int_{-\infty}^{+\infty} h(x, y) \log_2 \frac{h(x, y)}{f(x)g(y)}\, dx\, dy \tag{3.8.50}$$

provided that the integral on the right-hand side of (3.8.50) is convergent; otherwise, $I(\xi, \eta) = +\infty$ (see [32]–[34]).

As regards the role of the notions of entropy, divergence and information in information theory and statistics, see [35]–[39].

3.9 SUFFICIENT FUNCTIONS

We shall now introduce an important notion which is closely connected to the notion of independence, and which is capable of being defined in terms of information.

Definition 3.9.1. *Let ξ and η be arbitrary random variables or random vectors on a probability space S such that $I(\xi, \eta)$ is finite. Let $g(x)$ be a Borel-measurable function. The random variable $g(\xi)$ is called a sufficient function of ξ for η if one has*

$$I(g(\xi), \eta) = I(\xi, \eta) \tag{3.9.1}$$

Remark: The definition of sufficiency can be expressed in words as follows: *$g(\xi)$ is sufficient for η if $g(\xi)$ contains all information on η furnished by ξ.*

We shall prove the following:

Theorem 3.9.1. *Let ξ and η be random variables taking on only a finite number of values $x_1, \dots, x_N$ and $y_1, \dots, y_M$, respectively. Let $g(x)$ be an arbitrary function. We have $g(\xi)$ sufficient for η if and only if the conditional probability distribution of η, under the condition that the value of ξ is fixed, depends on the value of $g(\xi)$ only, i.e., if one has*

$$P(\eta = y_k \mid \xi = x_i) = P(\eta = y_k \mid \xi = x_j) \quad \textit{whenever} \quad g(x_i) = g(x_j) \tag{3.9.2}$$

or—expressed otherwise—if ξ and η are independent under the condition that the value of $g(\xi)$ is fixed, i.e., for any z such that $P(g(\xi) = z) > 0$, one has

$$P(\xi = x_j, \eta = y_k \mid g(\xi) = z) = P(\xi = x_j \mid g(\xi) = z) P(\eta = y_k \mid g(\xi) = z) \tag{3.9.3}$$

Proof. Let us put $\mathscr{P}=\{p_j\}$ where $p_j=P(\xi=x_j)$, $\mathscr{Q}=\{q_k\}$ where $q_k=P(\eta=y_k)$, and $\mathscr{R}=\{r_{j,k}\}$ where $r_{j,k}=P(\xi=x_j, \eta=y_k)$. Let $z_1, z_2, \ldots, z_s$ denote the different values taken on by $g(x)$ on the set $x_1, \ldots, x_N$; and suppose that $g(x_j)=z_l$ for $j \in E_l$ where $E_1, E_2, \ldots, E_s$ is a partition of the set $\{1, 2, \ldots, N\}$. Let us put further $h(x_j)=l$ if $j \in E_l$ and

$$t_{l,k}=P(g(\xi)=z_l, \eta=y_k) \quad \text{and} \quad t_l=P(g(\xi)=z_l)$$

Then the numbers

$$u_{j,k}=\frac{t_{h(x_j),k}\cdot p_j}{t_{h(x_j)}} \tag{3.9.4}$$

form a probability distribution $\mathscr{U}$ and one has

$$I(\xi, \eta)-I(g(\xi), \eta)=D(\mathscr{R}, \mathscr{U}) \tag{3.9.5}$$

and thus $I(\xi, \eta)=I(g(\xi), \eta)$ if and only if $\mathscr{R}=\mathscr{U}$; i.e., if

$$\frac{t_{h(x_j),k}\cdot p_j}{t_{h(x_j)}}=r_{j,k} \tag{3.9.6}$$

that is, if and only if

$$\frac{r_{i,k}}{p_i}=\frac{r_{j,k}}{p_j} \quad \text{whenever} \quad g(x_i)=g(x_j) \tag{3.9.7}$$

Thus (3.9.2) is necessary and sufficient for (3.9.1). As further, if $g(x_j)=z_l$, then

$$\begin{aligned} P(\xi=x_j, \eta=y_k \mid g(\xi)=z_l) &= \frac{P(\xi=x_j, \eta=y_k)}{P(g(\xi)=z_l)} \\ &= P(\xi=x_j \mid g(\xi)=z_l)P(\eta=y_k \mid \xi=x_j) \end{aligned}$$

Thus, (3.9.3) is equivalent to $P(\eta=y_k \mid \xi=x_j)=P(\eta=y_k \mid g(\xi)=z_l)$ if $g(x_j)=z_l$. Thus (3.9.3) is equivalent to (3.9.2). This proves Theorem 3.9.1. ■

Theorem 3.9.1 shows how the notion of sufficiency is related to that of independence and of information. For further connections between these notions see [37]–[39].

Example 3.9.1. Suppose we have two dice: the first being fair, i.e., such that the probability of each of the sides equals 1/6, the other being loaded, such that the side marked with 6 has probability $p>1/6$ and the other five sides have equal probabilities $(1-p)/5$. Suppose that the two dice look quite similar. Let us now choose one of the dice at random (each having the probability 1/2 to be chosen) and make n throws with it. Let the results be $\xi_1, \xi_2, \ldots, \xi_n$. Let the random variable η be equal to 0 if the fair dice has been chosen and let η be 1 if the loaded one has been chosen. We are interested in the value of η, but we cannot observe η directly; we can observe only the variables $\xi_1, \xi_2, \ldots, \xi_n$ and we want to get information on η from these observations. We show that in this case it is sufficient to count how many times we have thrown a 6, all other information in the observations $\xi_1, \xi_2, \ldots, \xi_n$ being irrelevant as far as η is concerned. In other words, if $g(x)$ is a function such that $g(6)=1$ and $g(j)=0$ for $j=1, 2, 3, 4, 5$ and we put $\zeta_n = \sum_{k=1}^{n} g(\xi_k)$, then ζ_n (considered as a function of the vector $(\xi_1, \xi_2, \ldots, \xi_n)$) is sufficient for η. As a matter of fact, if $x_1, x_2, \ldots, x_n$ is any sequence consisting of the numbers 1, 2, 3, 4, 5, 6 and $z = \sum_{k=1}^{n} g(x_k)$, we have

$$P(\xi_k = x_k \text{ for } 1 \leqq k \leqq n \text{ and } \eta = 0) = \frac{1}{2 \cdot 6^n}$$

and

$$P(\xi_k = x_k \text{ for } 1 \leqq k \leqq n \text{ and } \eta = 1) = \frac{1}{2} p^z \left(\frac{1-p}{5}\right)^{n-z}$$

Thus we have, for $i=0$ and $i=1$,

$$P(\eta = i \mid \xi_k = x_k \text{ for } 1 \leqq k \leqq n) = P\left(\eta = i \mid \zeta_n = \sum_{k=1}^{n} g(x_k)\right)$$

and thus by Theorem 3.8.3 ζ_n is a sufficient function of the random vector $(\xi_1, \ldots, \xi_n)$ for η.

3.10 MARKOV CHAINS

In this section we shall introduce the important notion of Markov chains, which is closely related to that of independence. The most simple and convenient way to define this notion is in terms of information.

DEFINITION 3.10.1. *A sequence* $\{\eta_n\}$ $(n = 1, 2, \ldots)$ *of random variables on a probability space S, each taking on a finite number of values, is called a* (*discrete-parameter*) *Markov chain if one has*

$$I((\eta_1, \eta_2, \ldots, \eta_n), \eta_{n+1}) = I(\eta_n, \eta_{n+1}) \quad \text{for } n = 1, 2, \ldots \tag{3.10.1}$$

Remark: One can formulate the definition of a Markov chain in words as follows: *The sequence of random variables* $\{\eta_n\}$ $(n = 1, 2, \ldots)$ *is a Markov chain if* η_n *contains all information on* η_{n+1} *which is present in the vector* $\eta_1, \eta_2, \ldots, \eta_n$. In other words, $\{\eta_n\}$ $(n = 1, 2, \ldots)$ *is a Markov chain if* η_n (*considered as a function of* $\eta_1, \eta_2, \ldots, \eta_n$) *is sufficient for* η_{n+1} *for* $n = 1, 2, \ldots$.

This remark, combined with Theorem 3.9.1 leads to the following:

Theorem 3.10.1. *The sequence of random variables η_n, each taking on a finite number of values only, is a Markov chain if and only if the conditional distribution of η_{n+1}, given the values of $\eta_1, \eta_2, \ldots, \eta_n$, depends only on η_n.*

Remark: Usually the statement of Theorem 3.10.1 is taken as the definition of a Markov chain. Theorem 3.10.1 expresses the fact that the definition given above (which has the advantage that it expresses the intuitive meaning of the concept more clearly) is equivalent to the usual one for sequences of random variables η_n such that each η_n takes on a finite number of different values only.

We shall now introduce the following:

Definition 3.10.2. *Let η_n be a Markov chain, each η_n taking on only a finite number of values. The functions*

$$p_k(y \mid x) = P(\eta_{k+1} = y \mid \eta_k = x) \quad (k = 1, 2, \ldots) \tag{3.10.2}$$

(defined for those values of x for which $P(\eta_k = x) > 0$) are called the transition probabilities of the Markov chain $\{\eta_n\}$. If $p_k(y \mid x)$ does not depend on k, the Markov chain $\{\eta_n\}$ is called homogeneous.

It follows from Theorem 3.10.1 that if $\{\eta_n\}$ is a Markov chain and x_k is any number such that $P(\eta_k = x_k) > 0$ $(k = 1, 2, \ldots, n)$, one has

$$P(\eta_k = x_k \text{ for } 1 \leqq k \leqq n) = P(\eta_1 = x_1) \prod_{k=1}^{n-1} p_k(x_{k+1} \mid x_k) \tag{3.10.3}$$

Especially if the Markov chain $\{\eta_n\}$ is homogeneous, i.e., $p_k(y \mid x) = p(y \mid x)$ for $k = 1, 2, \ldots$, one has

$$P(\eta_k = x_k \text{ for } 1 \leqq k \leqq n) = P(\eta_1 = x_1) \prod_{k=1}^{n-1} p(x_{k+1} \mid x_k) \tag{3.10.3$'$}$$

Let us consider a homogeneous Markov chain $\{\eta_n\}$ such that the set of values of each η_n is the same set $\{x_1, x_2, \ldots, x_N\}$. Let us put

$$p_{j,k} = p(x_k \mid x_j)$$

and let us consider the matrix $\Pi = (p_{j,k})$ in which the kth element of the jth row is $p_{j,k}$ $(j, k = 1, 2, \ldots, N)$. Clearly, the matrix Π has nonnegative elements, and all its row sums are equal to 1. Such a matrix is called a *stochastic matrix*. It follows easily from (3.10.3′) that if we put

$$p_{j,k}^{(n)} = P(\eta_{n+1} = x_k \mid \eta_1 = x_j)$$

and denote the matrix $(p_{j,k}^{(n)})$ by Π_n, then

$$p_{j,k}^{(n+1)} = \sum_{l=1}^{N} p_{j,l}^{(n)} p_{l,k}$$

and therefore,

$$\Pi_n = \Pi^n \quad (n = 1, 2, \ldots) \tag{3.10.4}$$

Equation (3.10.4) shows that the joint distribution of any of the variables of a homogeneous Markov chain $\{\eta_n\}$ is uniquely determined by the matrix Π of transition probabilities and the initial distribution $w_j = P(\xi_1 = x_j)$ $(j = 1, 2, \ldots, N)$.

As, by Lemma 3.8.1, for any random variables $\eta_1, \eta_2, \ldots$, one has

$$I(\eta_n, \eta_{n+1}) \leqq I((\eta_r, \eta_{r+1}, \ldots, \eta_n), \eta_{n+1}) \leqq I((\eta_1, \ldots, \eta_n), \eta_{n+1})$$

it follows immediately from the definition that if $\eta_1, \eta_2, \ldots$ is a Markov chain, then $\eta_r, \eta_{r+1}, \ldots$ is also a Markov chain. Thus, by Theorem 3.10.1, if $\{\eta_n\}$ is a Markov chain, one has for every $r < n$, if y_j denotes any element of the set $\{x_1, \ldots, x_N\}$ $(r \leqq j \leqq n)$,

$$P(\eta_{n+1} = y_{n+1} \mid \eta_j = y_j \text{ for } r \leqq j \leqq n) = P(\eta_{n+1} = y_{n+1} \mid \eta_n = y_n) \tag{3.10.5}$$

It follows from (3.10.3) that

$$P(\eta_k = y_k \text{ for } r \leqq k \leqq n) = P(\eta_r = y_r) \prod_{k=r}^{n-1} p_k(y_{k+1} \mid y_k) \tag{3.10.6}$$

We conclude that if $d \geqq 2$ and $n \geqq 2$,

$$\begin{aligned} &P(\eta_j = y_j, n+1 \leqq j \leqq n+d \mid \eta_j = y_j, j \leqq n) \\ &\qquad = P(\eta_j = y_j, n+1 \leqq j \leqq n+d \mid \eta_n = y_n) \end{aligned} \tag{3.10.7}$$

From (3.10.7) one easily deduces the following:

THEOREM 3.10.2. *If $\{\eta_n\}$ is a Markov chain and y_n is a number such that $P(\eta_n = y_n) > 0$, then the random vectors $(\eta_1, \ldots, \eta_{n-1})$ and $(\eta_{n+1}, \ldots, \eta_{n+d})$ are independent under the condition $\eta_n = y_n$ $(n = 2, 3, \ldots)$. Conversely, if a sequence of random variables $\{\eta_n\}$ (each taking on a finite number of values) has the mentioned property, it is a Markov chain.*

Remark: The statement of Theorem 3.10.2 can be expressed in a very suggestive way if one interprets the index n as time and η_n as the state at time n of a system Σ undergoing random changes of states in time. Using this terminology one can say that *the random changes of state of a system are described by a Markov chain if under the condition that the present state of the system is given, its past and its future are independent.*

Example 3.10.1. Let ξ_n $(n = 1, 2, \ldots)$ be a sequence of independent random variables, each taking on a finite number of values only. Put $\eta_n = \xi_1 + \xi_2 + \cdots + \xi_n$. It follows immediately from Theorem 3.10.2

that $\{\eta_n\}$ is a Markov chain. Such a Markov chain is called an *additive Markov chain*. If the random variables ξ_n are identically distributed, the Markov chain $\{\eta_n\}$ is homogeneous.

This example can be generalized as follows:

Example 3.10.2. Let $\{\xi_n\}$ be a sequence of independent random variables, each taking on the values $x_1, x_2, \ldots, x_N$ only. Let $f(x, y)$ be an arbitrary function and define the random variables η_n $(n = 0, 1, \ldots)$ as follows: Let η_0 be an arbitrary random variable, and put $\eta_n = f(\xi_n, \eta_{n-1})$ for $n = 1, 2, \ldots$. Then $\{\eta_n\}$ is a Markov chain; if the ξ_n are identically distributed, $\{\eta_n\}$ is a homogeneous Markov chain.

This method of constructing Markov chains is fairly general; an arbitrary homogeneous Markov chain with a finite number of states can be obtained in this way. As a matter of fact, if we want to construct in this way a homogeneous Markov chain with a finite number of states, we may suppose without restricting the generality that the set of values of the variables η_n is the set $\{1, 2, \ldots, N\}$. Let the transition probabilities of the chain $\{\eta_n\}$ be

$$p_{j,k} = P(\eta_{n+1} = k \mid \eta_n = j) \quad (j, k = 1, 2, \ldots, N)$$

Let $\xi_1, \xi_2, \ldots, \xi_n, \ldots$ be independent, identically distributed random N-dimensional vectors such that the possible values of the ξ_n are all possible sequences of integers $(k_1, k_2, \ldots, k_N)$ $(1 \leqq k_i \leqq N, i = 1, 2, \ldots, N)$ and for each $n \geqq 1$ and every such sequence $(k_1, k_2, \ldots, k_N)$ one has

$$P(\xi_n = (k_1, k_2, \ldots, k_N)) = p_{1,k_1} p_{2,k_2} \cdots p_{N,k_N}$$

Define the function $f(x, y)$ as follows: x runs over all N-dimensional vectors, y runs over the integers $1, 2, \ldots, N$, and $f(x, y)$ is equal to the yth component of the vector x. Let η_1 be an arbitrary random variable with possible values $1, 2, \ldots, N$ and put $\eta_n = f(\xi_n, \eta_{n-1})$ for $n = 2, 3, \ldots$. Then $\{\eta_n\}$ is a homogeneous Markov chain having the prescribed transition probability matrix $(p_{j,k})$, because

$$P(\eta_n = k \mid \eta_{n-1} = j) = P(f(\xi_n, j) = k) = p_{j,k}$$

Example 3.10.3. The random variables $\varepsilon_k(\beta, x)$ $(k = 1, 2, \ldots)$ in Example 3.7.5 (as random variables on the Lebesgue probability space) form a Markov chain if $\beta = (\sqrt{5} + 1)/2$, with the transition probabilities

$$P(\varepsilon_{k+1}(\beta, x) = 0 \mid \varepsilon_k(\beta, x) = 1) = 1$$

$$P(\varepsilon_{k+1}(\beta, x) = 0 \mid \varepsilon_k(\beta, x) = 0) = 1/\beta$$

$$P(\varepsilon_{k+1}(\beta, x) = 1 \mid \varepsilon_k(\beta, x) = 0) = 1 - 1/\beta$$

As regards the general theory of Markov chains see Chung [40].

EXERCISES

E.3.1. (a) Let

$$x = \cfrac{1}{b_1(x) + \cfrac{1}{b_2(x) + \cfrac{1}{b_3(x) + \cdots}}}$$

where the $b_k(x)$ are positive integers, be the continued fraction expansion of the real numbers x $(0 < x < 1)$. Let $\mathscr{A}_n$ denote the least σ-algebra of subsets of the interval $(0, 1)$ with respect to which $b_n(x)$ is measurable. Show that the σ-algebras $\mathscr{A}_n$ are strongly qualitatively independent.

(b) Define the transformation Tx $(0 < x < 1)$ as follows:

$$Tx = \left(\frac{1}{x}\right)$$

where (y) denotes the fractional part of y. Show that Tx preserves the probability measure P defined for every measurable subset of the interval $(0, 1)$ by

$$P(A) = \frac{1}{\ln 2} \int_A \frac{dx}{1 + x}$$

Show that if $b_n(x)$ $(n = 1, 2, \ldots)$ denotes the nth "digit" of the continued fraction expansion of x, then $b_{n+1}(x) = b_1(T^n x)$ for $n = 1, 2, \ldots$ and $P(b_n(x) = k) = \log_2(1 + 1/k(k + 2))$ $(k = 1, 2, \ldots)$.

Remark: It can be shown that T is mixing.

E.3.2. Let the random point (x, y) be uniformly distributed in the unit circle $x^2 + y^2 < 1$. Show that the polar coordinates ρ and φ of the point (x, y) $(\rho = \sqrt{x^2 + y^2}$, $\varphi = \operatorname{arctg} y/x)$ are independent random variables.

E.3.3. Verify the identities (2.6.14), (2.6.23), the inequalities (2.6.17), (2.6.18), and the inequalities (a) and (b) of E.2.6 for the special case when the events $A_1, A_2, \ldots, A_n$ are independent.

E.3.4. Let $S = [\Omega, \mathscr{A}, \mathscr{N}]$ denote the conditional probability space where Ω is the set of all positive integers, $\mathscr{A}$ the set of all subsets of Ω, and $\mathscr{N}$ the counting measure. Using the fact that if A_p denotes the event that an integer chosen at random is divisible by the prime number p, and if $p_1, p_2, \ldots, p_r$ denote different primes, the events $A_{p_1}, A_{p_2}, \ldots, A_{p_r}$ are independent with respect to the condition $B_n = \{1, 2, \ldots, n\}$ provided that n is divisible by all the primes $p_1, p_2, \ldots, p_r$ (see Examples 3.1.4 and 3.2.5), show that denoting by C_n the event that an integer chosen at random is relatively prime to n, one has

$$P(C_n \mid B_n) = \prod_{p \mid n} \left(1 - \frac{1}{p}\right)$$

where p runs over all prime divisors of n.

Remark: Note that we have obtained a probabilistic proof of Euler's classical formula,

$$\varphi(n) = n\prod_{p/n}\left(1-\frac{1}{p}\right)$$

for the number $\varphi(n)$ of positive integers less than n and relatively prime to n.

E.3.5. Let us carry out a sequence of independent experiments $(\Omega_0, \mathscr{A}_0, P_k)$ where $\mathscr{A}_0$ is the four-element algebra $(\varnothing, \Omega, A, \bar{A})$ and $P_k(A) = p_k$ $(k = 1, 2, \ldots)$, i.e., form the product $S = (\Omega, \mathscr{A}, P)$ of the probability spaces $S_k = (\Omega_0, \mathscr{A}_0, P_k)$ $(k = 1, 2, \ldots)$. Let ν_n denote the frequency of the occurrence of the event A during the first n experiments and put

$$\Pi_{n,k} = P(\nu_n = k) \quad (k = 0, 1, \ldots, n; n = 1, 2, \ldots)$$

Show that

$$\Pi_{n,k}^2 \geqq \Pi_{n,k-1}\Pi_{n,k+1} \quad (k = 1, 2, \ldots, n-1; n = 2, 3, \ldots)$$

E.3.6. What is the necessary and sufficient condition for a subset of $2^n - n - 1$ equations of the system of equations (3.2.2) to imply the remaining $n + 1$ equations of this system?

Hint: The sum of all the 2^{n-1} numbers $P(A_1^{\delta_1}A_2^{\delta_2}\cdots A_n^{\delta_n})$ $(\delta_i = \pm 1)$ for which $\delta_k = 1$ is equal to $P(A_k)$ $(k = 1, 2, \ldots, n)$ and the sum of all the 2^n numbers $P(A_1^{\delta_1}A_2^{\delta_2}\cdots A_n^{\delta_n})$ is equal to 1. Thus, if we consider the numbers $P(A_k)$ $(k = 1, 2, \ldots, n)$ as given and the 2^n numbers $P(A_1^{\delta_1}A_2^{\delta_2}\cdots A_n^{\delta_n})$ as unknowns, these unknowns satisfy $n + 1$ linear equations; if we choose $2^n - n - 1$ out of the equations (3.2.2), we get for the 2^n unknowns a system of 2^n linear equations; these equations determine the unknowns uniquely if and only if the (zero-one) matrix of the coefficients of this system of equations is nonsingular.

□**E.3.7.** Consider the probability space $(\Omega, \mathscr{A}, P)$ constructed in Example 3.3.7, where Ω is the set of all sequences $\omega = \{\omega_1, \ldots, \omega_n, \ldots\}$ of zeros and ones, $\mathscr{A}$ is the least σ-algebra of subsets of Ω containing all the sets A_k where A_k is the set of those sequences $\omega = \{\omega_1, \omega_2, \ldots, \omega_n, \ldots\}$ for which $\omega_k = 1$, the measure P is determined by the conditions that $P(A_k) = p$ $(0 < p < 1)$, and the events A_k $(k = 1, 2, \ldots)$ are independent.

(a) Let us denote for each $\omega = (\omega_1, \omega_2, \ldots, \omega_n, \ldots)$ those values of n for which $\omega_n = 1$, in increasing order of magnitude by $\tau_1(\omega) < \tau_2(\omega) < \cdots < \tau_k(\omega) < \cdots$. Prove that $\tau_k = \tau_k(\omega)$ is for each value of k $(k = 1, 2, \ldots)$ a random variable and determine its probability distribution.

Hint: The event $B_n^{(k)} = \{\omega\colon \tau_k(\omega) = n\}$ can be written in the form $B_n^{(k)} = \sum A_{i_1}\ldots A_{i_{k-1}}\bar{A}_{j_1}\cdots\bar{A}_{j_{n-k}}A_n$, where the summation is extended over all $(k-1)$-tuples of integers $(i_1, i_2, \ldots, i_{k-1})$; among the integers $1, 2, \ldots, n-1$ and $j_1, \ldots, j_{n-k}$ denote those of the integers $1, 2, \ldots, n-1$ which are not among the $i_1, i_2, \ldots, i_{k-1}$. Thus we have $B_n^{(k)} \in \mathscr{A}$ and

$$P(\tau_k = n) = \binom{n-1}{k-1}p^k q^{n-k}$$

where $q = 1 - p$; $n = k, k + 1, \ldots$; $k = 1, 2, \ldots$.

Remark: The distribution of τ_k is called the *negative binomial distribution of order k and parameter p*.

(b) Prove that the sequence τ_k is almost surely infinite, i.e., if Z is the set of those ω for which the sequence τ_k is finite, one has $P(Z) = 0$.

(c) Prove that the random variables $\tau_1, \tau_2 - \tau_1, \tau_3 - \tau_2, \ldots, \tau_{k+1} - \tau_k, \ldots$ are independent and identically distributed with negative binomial distribution of order 1 and parameter p, i.e.,

$$P(\tau_{k+1} - \tau_k = n) = pq^{n-1} \quad \text{for } n = 1, 2, \ldots$$

□**E.3.8.** Let $S_0 = (\Omega_0, \mathscr{A}_0, P_0)$ be a probability space such that Ω_0 is a set consisting of r elements ($r \geqq 2$), $\Omega_0 = \{\omega_1, \omega_2, \ldots, \omega_r\}$, $\mathscr{A} = \mathscr{P}(\Omega_0)$, and P_0 is defined by the conditions that for $A_i = \{\omega_i\}$ one has $P_0(A_i) = p_i$ ($i = 1, 2, \ldots, r$) where $0 < p_i < 1$ and $\sum_{i=1}^r p_i = 1$. Let $S = (\Omega, \mathscr{A}, P)$ denote the product of denumerably many exemplars of S_0. Thus, S describes the sequence of independent repetitions of an experiment having r possible outcomes with the corresponding probabilities $p_1, p_2, \ldots, p_r$. Let $\xi_n(i)$ ($i = 1, 2, \ldots, r$) denote the number of those experiments among the first n, the outcome of which was the event A_i. Show that $\xi_n(1), \ldots, \xi_n(r)$ are, for each n, random variables on S and determine the joint probability distribution of these random variables.

Hint: Using the well known formula for the number of permutations with repetition, one obtains

$$P(\xi_n(i) = k_i \text{ for } 1 \leqq i \leqq r) = \frac{n!}{k_1!\, k_2! \cdots k_r!} p_1^{k_1} p_2^{k_2} \cdots p_r^{k_r}$$

for every set of nonnegative integers $k_1, k_2, \ldots, k_r$ such that $k_1 + k_2 + \cdots + k_r = n$.

Remark: The joint probability distribution of the random variables $\xi_n(1), \ldots, \xi_n(r)$ is called the r-dimensional *multinomial distribution of order n with parameters* $p_1, p_2, \ldots, p_r$. (For $r = 2$ we get as a special case the binomial distribution.)

E.3.9. Show that if T is a mixing transformation of the probability space S, and ξ and η are two random variables belonging to $L_2(S)$, then one has

$$\lim_{n \to +\infty} E(\xi(T^n\omega)\eta(\omega)) = E(\xi)E(\eta) \tag{E.3.9.1}$$

Hint: Equation (E.3.9.1) holds by definition if ξ and η are both indicators of arbitrary events. Thus (E.3.9.1) holds if ξ and η are random variables taking on only a finite number of values. Approximating arbitrary random variables $\xi, \eta \in L_2(S)$ by discrete ones in L_2-norm (see Appendix B), it follows that (E.3.9.1) holds for arbitrary random variables belonging to $L_2(S)$.

E.3.10. Let ξ be a positive integer-valued random variable, such that $E(\xi) = A$ where $A > 1$ is a fixed number. Show that under these conditions $H(\xi)$ is maximal if and only if ξ has a negative binomial distribution (see Exercise E.3.7) of order 1 and parameter $p = 1/A$.

Hint: According to Lemma 3.8.2, if $p_k = P(\xi = k)$ and $q_k = pq^{k-1}$ ($k = 1, 2, \ldots$)

where $p = 1/A$, one has

$$0 \leqq \sum_{k=1} p_k \log_2 \frac{p_k}{q_k} = \frac{1}{p}\left(p \log_2 \frac{1}{p} + q \log_2 \frac{1}{q}\right) - H(\xi)$$

On the other hand,

$$\sum_{k=1}^{\infty} q_k \log_2 \frac{1}{q_k} = \frac{1}{p}\left(p \log_2 \frac{1}{p} + q \log_2 \frac{1}{q}\right)$$

It follows that

$$H(\xi) \leqq \sum_{k=1}^{\infty} q_k \log_2 \frac{1}{q_k}$$

with equality if and only if $p_k = q_k$ for $k = 1, 2, \ldots$.

PROBLEMS

P.3.1. Let B be a polynomial of the events $A_1, A_2, \ldots, A_n$. Show that one can always find real coefficients $C(i_1, i_2, \ldots, i_k)$ such that

$$P(B) = C_0 + \sum_{k=1}^{n} \sum_{(i_1, \ldots i_k)} C(i_1, \ldots, i_k) P(A_{i_1} A_{i_2} \cdots A_{i_k})$$

where the inner summation runs over all possible choices of the k-tuple $(i_1, i_2, \ldots i_k)$ of integers from the integers $1, 2, \ldots, n$ and the coefficients $C(i_1, \ldots, i_k)$ do not depend on the events $A_1, \ldots, A_n$ but only on the functional dependence of B on these events. Show that the coefficients $C(i_1, \ldots, i_k)$ can be determined by considering the special case when the events $A_1, \ldots, A_n$ are independent.

□**P.3.2.** Show that if $A_1, A_2, \ldots, A_n, \ldots$ are arbitrary events in a probability space S such that the series $\sum_{n=1}^{\infty} P(A_n)$ is convergent, then $P(\limsup_{n \to +\infty} A_n) = 0$, i.e., with probability 1 only a finite number of the events A_n occur simultaneously. Show that if the events A_n are independent, one has $P(\limsup_{n\to\infty} A_n) = 0$ or $P(\limsup_{n \to +\infty} A_n) = 1$, according to whether the series $\sum_{n=1}^{\infty} P(A_n)$ is convergent or divergent. (Lemma of Borel-Cantelli.)

Remark: Compare with Example 3.3.1 and Exercise E.3.7(b).

P.3.3. A pair $G = (X, Y)$ where X is a set and Y is a subset of the set X^2 of all unordered pairs (x_1, x_2) of different elements of the set X, is called a *graph* (see [6], [7] and [8]) the elements of the set X are called the *vertices*, and the elements of Y are the *edges* of the graph G. The edge (x_1, x_2) is said to *connect* the vertices x_1 and x_2. A finite or infinite sequence of edges $(x_1, x_2), (x_2, x_3), (x_3, x_4), \ldots,$ where the vertices $x_1, x_2, x_3, \ldots$ are all different, is called a *path*, x_1 is the *starting point* of the path, and in case the path $(x_1, x_2), (x_2, x_3), \ldots, (x_{n-1}, x_n)$ is finite, the vertex x_n is the *end point* of the path. A finite sequence of edges $(x_1, x_2), (x_2, x_3), \ldots, (x_{n-1}, x_n)$ such that the vertices $x_1, x_2, \ldots, x_{n-1}$ are all different, but $x_n = x_1$, is called a *cycle*. A graph G is called *connected* if for every pair x_1, x_2 of its vertices $(x_2 \neq x_1)$ there exists in G a finite path of which x_1 is the starting point and

x_2 is the end point. A graph is called a *tree* if it is connected and does not contain any cycles. The *valency* (or *degree*) $V(x)$ of a vertex x of a graph G is defined as the number of edges of G that connect x with another vertex x' of G. Show that if G is a graph which contains infinitely many points, if it is a tree, and if the valency of each of the vertices of G is finite, then G contains an infinite path. Using this theorem of graph theory, give an alternative proof for Theorem 3.3.2.

Hint: Suppose D_n ($n = 1, 2, \ldots$) is a sequence of sets in the algebra $\mathscr{A}_0$ of all polynomials of the sets $A_1, A_2, \ldots$ such that $D_n \neq \varnothing$, $D_{n+1} \subseteq D_n$ and $\prod_{n=1}^{\infty} D_n = \varnothing$, where the sets A_n are supposed to be strongly qualitatively independent. Then each D_n can be expressed as

$$D_n = \sum_{k=1}^{N_n} B_{n,k}$$

where the sets $B_{n,k}$ are different basic polynomials of the sets $A_1, A_2, \ldots, A_{r_n}$ and are pairwise disjoint. Clearly, each $B_{n,k}$ is a subset of one and only one of the sets $B_{n-1,j}$. We can evidently suppose that $D_1 = \Omega$. Now let us construct a graph G as follows: Let the vertices of G be the sets $B_{n,k}$ ($n \geqq 1, k \geqq 1$), and let us connect $B_{n,k}$ and $B_{n-1,j}$ by an edge if and only if $B_{n,k} \subseteq B_{n-1,j}$. It is easy to see that G is a tree in which each vertex has finite valency; thus if none of the sets D_n is empty, then by the theorem mentioned above, G contains an infinite path. Taking the product of the sets which are vertices of this path, we get a set $A = \prod_{k=1}^{\infty} A_k^{\delta_k}$ such that $A \subseteq D_n$ for each n; as by supposition the set A is not empty, we obtain that $\prod_{n=1}^{\infty} D_n$ is not empty, in contradiction with our hypothesis. This proves that every additive set function on $\mathscr{A}_0$ is a measure.

P.3.4. Put $((x)) = \min(x, 1 - x)$ for $0 \leqq x \leqq 1$. Prove that if Ω is a denumerably infinite set, $\mathscr{A} = \mathscr{P}(\Omega)$, and p_n ($n = 0, 1, 2, \ldots$) is an infinite sequence of numbers such that $0 < p_n < 1$, one can find a sequence of events $A_n \in \mathscr{A}$ and a measure P on $\mathscr{A}$ such that $P(A_n) = p_n$ and the events A_n ($n = 0, 1, 2, \ldots$) are independent with respect to the probability space $(\Omega, \mathscr{A}, P)$, if and only if $\sum_{n=0}^{\infty} ((p_n)) < +\infty$.

Hint: (1) *Necessity*. Suppose there exists a sequence $\{A_n\}$ of events and a measure P with the required properties. Evidently,

$$\max_{\delta_i = \pm 1} P(A_0^{\delta_0} A_1^{\delta_1} \cdots A_N^{\delta_N}) \leqq \prod_{n=0}^{N} (1 - ((p_n)))$$

Now let the elements of Ω be denoted by ω_k ($k = 0, 1, \ldots$) and put

$$q_k = P(\{\omega_k\})$$

As every ω_k is contained in one of the sets $A_0^{\delta_0} A_1^{\delta_1} \cdots A_N^{\delta_N}$ for each N, it follows that

$$\max_k q_k \leqq \prod_{n=0}^{N} (1 - ((p_n)))$$

If the series $\sum_{n=0}^{\infty} ((p_n))$ is divergent, then $\lim_{N \to \infty} \prod_{n=0}^{N} (1 - ((p_n))) = 0$ and thus $\max_k q_k = 0$, which is impossible. This proves the necessity of the condition.

(2) *Sufficiency*. We may suppose that $0 < p_n \leqq 1/2$ (because we can replace A_n by $\bar{A}_n$ if $p_n > 1/2$; we know that if the events A_n are independent, and some

of the A_n are replaced by the contrary event $\bar{A}_n$, the independence remains valid). If $0 < p_n \leqq 1/2$, then $((p_n)) = p_n$ and the condition $\sum_{n=0}^{\infty} ((p_n)) < +\infty$ means simply that $\sum_{n=0}^{\infty} p_n < +\infty$. Define the sequence q_k by

$$\sum_{k=0}^{\infty} q_k z^k = \prod_{n=0}^{+\infty} (1 + p_n(z^{2^n} - 1)) \tag{P.3.4.1}$$

As we supposed, $\sum_{n=1}^{\infty} p_n < +\infty$, the infinite product on the right-hand side of (P.3.4.1) is convergent for $|z| \leqq 1$, and represents an analytic function of z (regular for $|z| < 1$ and continuous for $|z| \leqq 1$). It follows that $\sum_{n=0}^{\infty} q_k = 1$. Clearly, $q_k \geqq 0$, so $\{q_k\}$ is a probability distribution. Now let A_n be defined as follows: Let us represent each nonnegative integer k in the binary system as

$$k = \sum_j \varepsilon_j(k) 2^j$$

where $\varepsilon_j(k) = 0$ or $=1$. Let A_n be the set of those ω_k for which $\varepsilon_n(k) = 1$. It is easy to see that

$$q_k = \prod_{n=0}^{+\infty} [p_n \varepsilon_n(k) + (1 - p_n)(1 - \varepsilon_n(k))]$$

and thus

$$P(A_n) = \sum_{\varepsilon_n(k)=1} q_k = p_n \prod_{j \neq n} [p_j + (1 - p_j)] = p_n$$

Similarly, we get

$$P(A_{n_1} A_{n_2} \ldots A_{n_r}) = p_{n_1} p_{n_2} \ldots p_{n_r} \qquad \text{if } n_1 < n_2 < \cdots < n_r$$

Thus the events A_n have the prescribed probabilities and are independent. This proves the sufficiency.

Remark: Note that the nonexistence of a measure P with respect to which the sets A_n defined in Example 3.3.1 are independent and have the probability 1/2 is a special case of the necessity of the condition $\sum p_n < +\infty$, the set Ω in Example 3.3.1 being denumerable.

P.3.5. A sequence $\{\xi_n\}$ of random variables on a probability space S is called *quasi-orthonormal** if putting $a_{n,m} = E(\xi_n \xi_m)$, the quadratic form $\sum_{n=1}^{\infty} \sum_{m=1}^{\infty} a_{n,m} x_n x_m$ is bounded, i.e., there exists a constant $K > 0$ such that

$$\left| \sum_{n=1}^{\infty} \sum_{m=1}^{\infty} a_{n,m} x_n x_m \right| \leqq K \sum_{n=1}^{\infty} x_n^2 \tag{P.3.5.1}$$

for any sequence $\{x_n\}$ of real numbers such that $\sum_{n=1}^{\infty} x_n^2 < +\infty$. Show that if $\{\xi_n\}$ is a quasi-orthonormal sequence and $\eta \in L_2(S)$, then putting

$$c_n = E(\eta \xi_n) \tag{P.3.5.2}$$

one has

$$\sum_{n=1}^{\infty} c_n^2 \leqq K \cdot E(\eta^2) \tag{P.3.5.3}$$

* See Boas [30]. Concerning an application of the inequality (P.3.5.3) to the proof of a generalization of the "large sieve" of Linnik, see [31],

Hint: For every $N \geq 1$

$$0 \leq E\left(\left[\eta - \frac{1}{K}\sum_{n=1}^{N} c_n \xi_n\right]^2\right) \leq E(\eta^2) - \frac{1}{K}\sum_{n=1}^{N} c_n^2$$

Remark: Inequality (P.3.5.3) shows that the quasi-orthonormality of the sequence $\{\xi_n\}$ implies that $\lim_{n\to+\infty} c_n = 0$.

P.3.6. Let $\mathscr{A}_1$ and $\mathscr{A}_2$ be qualitatively independent σ-algebras of subsets of a set Ω. Let $\mathscr{A}_0$ denote the least algebra of subsets of Ω containing both $\mathscr{A}_1$ and $\mathscr{A}_2$. Show that there exist two experiments $(H_1, \mathscr{B}_1)$ and $(H_2, \mathscr{B}_2)$ with the following properties: Denote by H the product space $H_1 \times H_2$, by $\mathscr{B}_1'$ the σ-algebra of subsets of H consisting of the sets $B_1 \times H_2$ where $B_1 \in \mathscr{B}_1$, similarly, by $\mathscr{B}_2'$ the algebra of subsets of H consisting of the sets $H_1 \times B_2$ where $B_2 \in \mathscr{B}_2$, and finally, by $\mathscr{B}_0'$ the least algebra of subsets of H containing both $\mathscr{B}_1'$ and $\mathscr{B}_2'$. Then $\mathscr{B}_0'$ is isomorphic to $\mathscr{A}_0$ and under this isomorphism $\mathscr{B}_i'$ is isomorphic to $\mathscr{A}_i$ $(i = 1, 2)$.

Hint: Use Lemma 3.3.3.

Remark: We have deduced Theorem 3.4.1 from Theorem 3.3.3 as a special case; the above statement shows, however, that one can also deduce Theorem 3.3.3 from Theorem 3.4.1.

P.3.7. A class $\mathscr{C}$ of subsets of a set Ω is called *compact* (see Marczewski [5]) if for each sequence $C_n \in \mathscr{C}$ $(n = 1, 2, \ldots)$ such that $\prod_{n=1}^{N} C_n \neq \varnothing$ for $N = 1, 2, \ldots$, one has $\prod_{n=1}^{\infty} C_n \neq \varnothing$.

(a) Prove that if an algebra $\mathscr{A}$ of subsets of a set Ω is compact, then every finitely additive nonnegative set function P on $\mathscr{A}$ such that $P(\Omega) = 1$ is σ-additive on $\mathscr{A}$.

Hint: Show that if $A = \sum_{n=1}^{\infty} A_n$, where $A \in \mathscr{C}$ and $A_n \in \mathscr{C}$ $(n = 1, 2, \ldots)$, and the sets A_n are disjoint, then all but a finite number of the sets A_n are empty, because otherwise, putting $B_n = \sum_{k=1}^{n} A_k$, we would have $A\overline{B}_n \in \mathscr{C}$, $\prod_{n=1}^{N} A\overline{B}_n \neq \varnothing$, and $\prod_{n=1}^{\infty} A\overline{B}_n = \varnothing$, which is impossible because we have supposed that $\mathscr{C}$ is compact.

(b) Show that the least algebra containing an infinite sequence of strongly qualitatively independent events is compact.

Hint: Apply the same reasoning as in the proof of Theorem 3.3.2 (or in Problem P.3.3).

Remark: The statements (a) and (b) together imply Theorem 3.3.2, and thus can be regarded as a generalization of Theorem 3.3.2.

P.3.8. Let C denote the Cantor ternary set (also called *Cantor's discontinuum*), i.e., the set of those real numbers y in the interval $(0, 1)$ which can be represented in the form

$$y = \sum_{n=1}^{\infty} \frac{\delta_n}{3^n}$$

where $\delta_n = 0$ or $\delta_n = 2$ for $n = 1, 2, \ldots$. Let C_n ($n = 1, 2, \ldots$) denote the subset of those elements of the Cantor set for which $\delta_n = 2$.

(a) Show that the sets C_n ($n = 1, 2, \ldots$) are strongly qualitatively independent.

(b) Let $\mathscr{B}$ denote the least σ-algebra of subsets of C containing all the sets C_n ($n = 1, 2, \ldots$). Let Q be the measure (which exists according to Theorem 3.3.1) with respect to which the events C_n are independent and have the prescribed probabilities $Q(C_n) = p_n$ ($n = 1, 2, \ldots$). Let Ω be the set obtained from the set of real numbers in the interval $(0, 1)$ by replacing each binary rational number x by a couple of points x^- and x^+. Let $(\Omega, \mathscr{A}, P)$ be the probability space defined in Example 3.3.2 corresponding to the sequence p_n. Let the random variable $\xi(x)$ be defined on this probability space as follows: If x is not a binary rational number, then x has a unique binary expansion $x = \sum_{k=1}^{\infty} \varepsilon_k/2^k$ ($\varepsilon_k = 0$ or 1). In this case, put $\xi(x) = \sum_{k=1}^{\infty} 2\varepsilon_k/3^k$. If x is a binary rational number, then it has both a finite and an infinite binary expansion: $x = \sum_{k=1}^{s} \varepsilon_k/2^k$ and $x = \sum_{k=1}^{\infty} \varepsilon_k'/2^k$, where $\varepsilon_s = 1$, $\varepsilon_k' = \varepsilon_k$ for $k < s$, and $\varepsilon_s' = 0$, $\varepsilon_k' = 1$ for $k > s$.

In this case, put $\xi(x^-) = \sum_{k=1}^{s} 2\varepsilon_k/3^k$ and $\xi(x^+) = \sum_{k=1}^{\infty} 2\varepsilon_k'/3^k$. Show that ξ maps the probability space $(\Omega, \mathscr{A}, P)$ onto the probability space $(C, \mathscr{B}, Q)$ defined above.

P.3.9. Let $\xi_1, \xi_2, \ldots, \xi_n, \ldots$ be an infinite sequence of independent random variables with finite expectations $E(\xi_n)$ such that the infinite product $\prod_{n=1}^{\infty} E(\xi_n)$ is convergent.

(a) Show by a counterexample that it does not follow that $\prod_{n=1}^{\infty} \xi_n$ is almost surely convergent.

Hint: If $P(\xi_n = 3/2) = P(\xi_n = 1/2) = 1/2$ for $n = 1, 2, \ldots$, then $E(\xi_n) = 1$ and thus the conditions are satisfied. However, as $E(\sqrt{\xi_n}) = (1 + \sqrt{3})/2\sqrt{2} < 1$, it follows by Theorem 2.11.1 that if $(1 + \sqrt{3})/2\sqrt{2} < c < 1$, we have

$$P(\xi_1\xi_2 \ldots \xi_n \geqq c^{2n}) = P(\sqrt{\xi_1}\sqrt{\xi_2} \ldots \sqrt{\xi_n} \geqq c^n) \leqq \left(\frac{1 + \sqrt{3}}{2\sqrt{2}\, c}\right)^n$$

which implies that $\lim_{n\to\infty} \prod_{k=1}^{n} \xi_k = 0$ with probability 1.

(b) Show that if we add the condition that $0 \leqq \xi_n < 1$, then it follows that $\prod_{n=1}^{\infty} \xi_n$ is a random variable such that

$$E\left(\prod_{n=1}^{\infty} \xi_n\right) = \prod_{n=1}^{\infty} E(\xi_n)$$

Hint: Use the Lebesgue bounded convergence theorem.

P.3.10. Let ξ and η be two arbitrary random variables.

(a) Show that if both ξ and η are bounded and one has

$$E(\xi^n\eta^m) = E(\xi^n)E(\eta^m) \qquad (n, m = 1, 2, \ldots) \tag{P.3.10.1}$$

then ξ and η are independent.

Remark: If ξ and η are independent, so are ξ^n and η^m for every value of n and m and thus, by Theorem 3.5.2, (P.3.10.1) holds. Thus, for bounded random variables (P.3.10.1) is necessary and sufficient for the independence of ξ and η.

Hint: It follows from our suppositions that

$$E(e^{i(\xi u+\eta v)}) = E(e^{i\xi u})E(e^{i\eta v})$$

for all real values of u and v.

(b) Let ξ and η be discrete random variables, ξ taking on N different values $x_1, x_2, \ldots, x_N$ and η taking on M different values $y_1, y_2, \ldots, y_M$. Show that if the relation

$$E(\xi^n \eta^m) = E(\xi^n)E(\eta^m) \qquad (1 \leqq n \leqq N-1,\ 1 \leqq m \leqq M-1)$$

holds, then ξ and η are independent (theorem of Kantorowitch [6]).

Hint: It follows from the suppositions that if $P(x)$ is any polynomial of degree $N-1$ and $Q(x)$ is any polynomial of degree $M-1$, then

$$E(P(\xi)Q(\eta)) = E(P(\xi))E(Q(\eta))$$

Choose for $P(x)$ that polynomial $P_k(x)$ of degree $N-1$ for which $P_k(x_k) = 1$ and $P_k(x_j) = 0$ for $j \neq k$ $(1 \leqq j \leqq N)$, and for $Q(x)$ the polynomial $Q_h(x)$ of degree $M-1$ for which $Q_h(y_h) = 1$ and $Q_h(y_i) = 0$ for $i \neq h$ $(1 \leqq i \leqq M)$. As we have $E(P_k(\xi)) = P(\xi = x_k)$, $E(Q_h(\eta)) = P(\eta = y_h)$, and $E(P_k(\xi)Q_h(\eta)) = P(\xi = x_k,\ \eta = y_h)$, it follows that $P(\xi = x_k,\ \eta = y_h) = P(\xi = x_k)P(\eta = y_h)$. As k $(1 \leqq k \leqq N)$ and h $(1 \leqq h \leqq M)$ can be arbitrarily chosen it follows that ξ and η are independent.

Remark: For $N = M = 2$ the above statement reduces to the definition of independence of two events.

(c) Generalize the statement (b) for more than two random variables.

REFERENCES

[1] S. Banach, "On Measures in Independent Fields," *Studia Mathematica*, **10**: 159–181, 1948.

[2] E. Marczewski, "Independence d'Ensembles et Prolongement de Mesures," *Colloquium Mathematicum*, **1**: 122–132, 1948.

[3] E. Marczewski, "Ensembles Indépendants et Leurs Applications à la Théorie de la Mesure," *Fundamenta Mathematicae*, **35**: 13–28, 1948.

[4] E. Marczewski, "Measures in Almost Independent Fields," *Fundamenta Mathematicae*, **38**: 217–229, 1951.

[5] E. Marczewski, "On Compact Measures," *Fundamenta Mathematicae*, **40**: 113–124, 1953.

[6] L. W. Kantorowitch, "Sur une Problème de Steinhaus," *Fundamenta Mathematicae*, **14**: 266–270, 1929.

[7] O. Ore, *Theory of Graphs*, American Math. Soc., Providence, 1962.

[8] C. Berge, *The Theory of Graphs and its Applications*, Methuen, London, 1962.

[9] F. Hausdorff, *Set Theory*, 2nd edition, Chelsea, New York, 1957.

[10] H. Steinhaus, "La Théorie et les Applications des Fonctions Indépendantes au Sens Stochastique," Colloque Consacré à la Théorie des Probabilités, Part V, *Actualites Scientifiques et Industrielles*, No. 738, pp. 57–73, Hermann, Paris, 1938.

[11] A. Rényi, "On a Conjecture of H. Steinhaus," *Annales de la Soc. Polon. Math.*, **25**: 279–287, 1952.

[12] R. F. Grundy, "Complete Systems in L_2 and a Theorem of Rényi," *Michigan Mathematical J.*, **2**: 161–167, 1965.

[13] P. Révész, *The Laws of Large Numbers*, Akadémiai Kiadó, Budapest, 1967.

[14] G. Alexits, *Convergence Problems of Orthogonal Series*, Akadémiai Kiadó, Budapest, 1961.

[15] S. Saks, *Theory of the Integral*, Stechert, New York, 1937.

[16] P. Halmos, *Lectures on Ergodic Theory*, The Mathematical Society of Japan, Tokyo, 1956.

[17] P. Billingsley, *Ergodic Theory and Information*, Wiley, New York, 1965.

[18] K. Jacobs, *Neuere Methoden und Ergebnisse der Ergodentheorie*, Ergebnisse der Mathematik und ihrer Grenzgebiete, Springer, Nr. 29.

[19] L. Pukánszky and A. Rényi, "On the Approximation of Measurable Functions," *Publicationes Mathematicae (Debrecen)*, **2**: 146–149, 1951.

[20] K. Kuratowski, *Topologie*, I, Warsaw, 1933.

[21] K. R. Parthasarathy, *Probabilistic Measures on Metric Spaces*, Academic Press, New York, 1967.

[22] K. Knopp, "Mengentheoretische Behandlung einiger Probleme der Diophantischen Approximationen und der transfiniten Wahrscheinlichkeiten," *Math. Annalen*, **95**: 409–426, 1926.

[23] M. Kac, *Statistical Independence in Probability Analysis and Number Theory*, Carus Mathematics Monographs, No. 12, Wiley, New York, 1959.

[24] A. Rényi, "Representations for Real Numbers and Their Ergodic Properties, *Acta Math. Acad. Sci. Hung.*, **8**: 477–493, 1957.

[25] V. A. Rochlin, "Exact Endomorphisms of Lebesgue Spaces," *Isvestia Akad. Nauk SSSR*, **25**: 499–530, 1961.

[26] W. Parry, "On the β-Expansion of Real Numbers," *Acta Math. Acad. Sci. Hung.*, **11**: 401–416, 1960.

[27] W. Parry, "Representations for Real Numbers," *Acta Math. Acad. Sci. Hung.*, **15**: 95–105, 1964.

[28] A. O. Gelfond, "On a General Property of Number Systems" (in Russian), *Isvestia Akad. Nauk SSSR*, **23**: 809–814, 1959.

[29] J. L. Doob, *Stochastic Processes*, Wiley, New York, 1953.

[30] R. P. Boas, Jr., "A General Moment Problem," *Am. J. Math.*, **63**: 361–370, 1941.

[31] A. Rényi, "New Version of the Probabilistic Generalization of the Large Sieve," *Acta Math. Acad. Sci. Hung.*, **10**: 217–226, 1959.

[32] A. N. Kolmogoroff, "Theorie der Nachrichtenübermittlung," *Arbeiten zur Informationstheorie, I*, pp. 91–116, VEB Deutscher Verlag der Wissenschaften, Berlin, 1957.

[33] I. M. Gelfand and A. M. Jaglom, "Über die Berechnung der Menge an Information über eine zufällige Funkton, die in einer anderen zufälligen Funktion enthalten ist," *Arbeiten zur Informationstheorie, II*, pp. 7–56, VEB Deutscher Verlag der Wissenschaften, Berlin, 1958.

[34] I. M. Gelfand, A. N. Kolmogoroff and A. M. Jaglom, "Zur allgemeinen Definition der Information," *Arbeiten zur Informationstheoreie, II*, pp. 57–60, VEB Deutscher Verlag der Wissenschaften, Berlin, 1958.

[35] A. Feinstein, *Foundations of Information Theory*, McGraw-Hill, New York, 1958.

[36] J. Wolfowitz, *Coding Theorems of Information Theory*, 2nd edition, Springer, Berlin, 1964.

[37] A. Rényi, "On the Amount of Information Concerning an Unknown Parameter in a Sequence of Observations," *Publ. Math. Inst. Hung. Acad. Sci.*, **9**: 617–626, 1964.

[38] S. Kullbach, *Information Theory and Statistics*, Wiley, New York, 1959.

[39] A. Rényi, "Statistics and Information Theory," *Studia Sci. Math. Hung.*, **2**: 249–256, 1967.

[40] K. L. Chung, *Markov Chains with Stationary Transition Probabilities*, 2nd edition, Springer, Berlin, 1967.

[41] W. Richter and D. Worm, "Ein Informationstheoretischer Zugang zu Null-Eins-Gesetz," *Rev. Roumaine de Math. Pures et Appliquées*, **13**: 251 260, 1968.

[42] C. Ryll-Nardzewski, "Remarque sur un Théorème de A. Rényi," *Colloquium Mathematicum*, **2**: 319–320, 1951.

[43] D. Waterman, "On a Problem of Steinhaus," *Studia Sci. Math. Hung.* (in print).

[44] J. Cigler, "Ziffernverteilung in ϑ-adischen Brüchen," *Math. Zeitschrift*, **75**: 8–13, 1961.

CHAPTER 4

THE LAWS OF CHANCE

4.1 THE NATURE OF LAWS OF CHANCE

The notion of a "law of chance" seems at first sight rather paradoxical, because chance is by definition unpredictable. To speak about the laws of chance is thus like speaking about the "structure of chaos" or about "patterns of irregularity." Yet there are laws of chance, as there is a certain structure in chaos, and there exist patterns of irregularity.* As a matter of fact, the main aim in probability theory is to detect and to describe the laws of chance, and thereby predict as much as possible concerning random mass phenomena. Roughly speaking one can describe the nature of laws of chance as follows: While random phenomena are by their very nature unpredictable in detail, they are nevertheless predictable "in the large," to a certain extent. It is a basic experimental fact that such prediction is possible. The typical example for this is the following: Take a box containing a certain amount of gas. This gas consists of an enormous number of molecules, which move around, collide and hit the walls of the container in a random manner. If one looks at the erratic movement of particles of dust in a ray of sunlight shining into a dark room,† one gets some idea of the random movement of the molecules; however, one should realize that the dust particles which we see are "giants" compared with the molecules, and the impact of a very large number of molecules is needed to give a dust particle a visible displacement. Thus the movement of the molecules is much more erratic and unpredictable than the visible movement of the dust particles.

* These play an important part in modern art.

† This has been described already by Lucretius in his *De Rerum Natura* [49] as follows:

"Observe what happens when sunbeams are admitted into a building and shed light on its shadowy places. You will see a multitude of tiny particles mingling in a multitude of ways in the empty space within the light of beam as though contending in everlasting conflict rushing into battle rank upon rank with never a moment's pause in a rapid sequence of unions and disunions. From this you may picture what it is for the atoms to be perpetually tossed about in the illimitable void....Those small compound bodies that are least removed from the impetus of the atoms are set in motion by the impact of their invisible blows and in turn cannon against slightly larger bodies. So the movement mounts up from the atoms and gradually emerges to the level of our senses so that those bodies are in motion that we see in sunbeams, moved by blows that remain invisible." (See [49], p. 63.)

Nevertheless, some well defined and predictable laws emerge from the random movements of the molecules. If, for instance, the temperature is increased, keeping the volume unchanged, the pressure increases proportionately, according to the law of Boyle-Mariotte and Gay-Lussac. Such laws are valid *not in spite* of the basic randomness of the phenomena concerned at the molecular or atomic level, but *just because* of this randomness: such laws are consequences of the random nature of these phenomena, they are laws of chance.

Probability theory furnishes a mathematical model of random events: in this model to the real laws of chance there correspond mathematical theorems concerning a large number of independent (or weakly dependent) events or random variables. It is convenient to formulate these theorems as limit theorems for the case when the number of independent events or random variables tends to infinity. In this chapter we shall deal with such limit theorems.

The laws of chance are in a certain sense all "laws of large numbers." However, it is customary in probability theory to call "laws of large numbers" only those particular laws of chance which show that the mean value of a large number of independent (or weakly dependent) random variables is in some sense "almost" constant. Another important group of laws of chance are the *limit distribution theorems* showing that the distribution of certain functions (e.g., normed sums) of a large number of independent (or weakly dependent) random variables has a probability distribution which is fairly independent of the distribution of the single terms.

Besides the laws of large numbers there are many other types of laws of chance describing the more delicate regularities of chance fluctuations, like the laws of the iterated logarithm, the arc sine law, etc.

In this book we do not aim at formulating and proving the limit theorems in question in the greatest possible generality. There are special monographs available dealing with certain groups of limit theorems: as regards limit distribution theorems see, e.g., Gnedenko and Kolmogoroff [1], as regards the laws of large numbers, see Révész [2], as regards the fluctuation theory of random walk, see Spitzer [3], as regards "big deviations," see Ibrahimov and Linnik [4], etc. Our aim is only to make the reader familiar with the basic types of laws of chance.

In order to present these laws we first have to discuss the different notions of convergence of random variables and of distributions, in terms of which the laws of chance can be formulated. This will be done in the next two sections.

4.2 TYPES OF CONVERGENCE OF SEQUENCES OF RANDOM VARIABLES

In this section we shall define several types of convergence of sequences of random variables and discuss their properties and interrelations.

DEFINITION 4.2.1. *A sequence* $\{\xi_n\}$ $(n=1, 2, \ldots)$ *of random variables on a conditional probability space* $S=[\Omega, \mathscr{A}, \mathscr{B}, P(A \mid B)]$ *is said to converge in probability to a random variable* ξ *on* S, *if for every* $B \in \mathscr{B}$ *and every* $\varepsilon > 0$, *one has*

$$\lim_{n\to+\infty} P(|\xi_n - \xi| > \varepsilon \mid B) = 0 \tag{4.2.1}$$

If (4.2.1) *holds, we write*

$$\xi_n \Rightarrow \xi \qquad \textit{for } n \to \infty \tag{4.2.2}$$

Evidently if $B_1 \in \mathscr{B}$, $B_2 \in \mathscr{B}$, and $B_1 \subseteq B_2$, then if (4.2.1) holds for $B = B_2$, it also holds for $B = B_1$. Thus if S is a probability space, then the definition of $\xi_n \Rightarrow \xi$ can be stated as follows:

DEFINITION 4.2.2. *A sequence* $\{\xi_n\}$ $(n = 1, 2, \ldots)$ *of random variables on a probability space* $S = (\Omega, \mathscr{A}, P)$ *is said to converge in probability* to the random variables* ξ *on* S *if for every* $\varepsilon > 0$

$$\lim_{n\to+\infty} P(|\xi_n - \xi| > \varepsilon) = 0 \tag{4.2.3}$$

which implies (4.2.1) *for every* $B \in \mathscr{A}$ *for which* $P(B) > 0$.

Remark: Thus, $\xi_n \Rightarrow \xi$ on the conditional probability space $S = [\Omega, \mathscr{A}, \mathscr{B}, P(A \mid B)]$ if for every $B \in \mathscr{B}$ one has $\xi_n \Rightarrow \xi$ on the probability space $S_B = (\Omega, \mathscr{A}, P_B)$. The stochastic limit, if it exists, is evidently almost surely uniquely determined.

THEOREM 4.2.1. *If* $\xi_n \Rightarrow \xi$, $F_n(x)$ *denotes the distribution function of* ξ_n, *and* $F(x)$ *denotes the distribution function of* ξ, *then we have*

$$\lim_{n\to+\infty} F_n(x) = F(x) \tag{4.2.4}$$

for every point of continuity x *of* $F(x)$.

Proof. Let $A_n(\varepsilon)$ $(\varepsilon > 0)$ denote the event $|\xi_n - \xi| \leqq \varepsilon$; then we have

$$F_n(x) = P(\xi_n < x \mid A_n(\varepsilon))P(A_n(\varepsilon)) + P(\xi_n < x \mid \overline{A_n(\varepsilon)})P(\overline{A_n(\varepsilon)}) \tag{4.2.5}$$

As under condition $A_n(\varepsilon)$ the inequality $\xi_n < x$ implies $\xi < x + \varepsilon$ and is implied by $\xi < x - \varepsilon$, and by supposition $\lim_{n\to+\infty} P(\overline{A_n(\varepsilon)}) = 0$, it follows that

$$F(x - \varepsilon) \leqq \liminf_{n\to+\infty} F_n(x) \leqq \limsup_{n\to+\infty} F_n(x) \leqq F(x + \varepsilon) \tag{4.2.6}$$

As (4.2.6) holds for every $\varepsilon > 0$, Theorem 4.1.1 follows. ∎

* In measure theory this type of convergence is called "convergence in measure." In probability theory "convergence in probability" is also called "stochastic convergence," and if $\xi_n \Rightarrow \xi$, ξ is called the "stochastic limit" of the sequence ξ_n.

Remark: As $\xi_n \Rightarrow \xi$ on $S = (\Omega, \mathscr{A}, P)$ implies $\xi_n \Rightarrow \xi$ on $S_B = (\Omega, \mathscr{A}, P_B)$, it follows that if $\xi_n \Rightarrow \xi$ and $F_n(x \mid B)$ respective of $F(x \mid B)$ denote the conditional distribution function of ξ_n respectively of ξ under condition B, one has (if $P(B) > 0$)

$$\lim_{n \to \infty} F_n(x \mid B) = F(x \mid B) \tag{4.2.7}$$

for every point of continuity of $F(x \mid B)$.

If ξ is equal to a constant c with probability 1, then (4.2.4) can be written in the form

$$\lim_{n \to +\infty} F_n(x) = \begin{cases} 0 & \text{for } x < c \\ 1 & \text{for } x > c \end{cases} \tag{4.2.8}$$

In this special case, clearly (4.2.8) implies $\xi_n \Rightarrow c$. In the general case the validity of (4.2.4) does not imply that ξ_n converges in probability to a limit. (See Example 4.2.1.) However, if the conditional distribution function of ξ_n under condition B tends for $n \to +\infty$ to the conditional distribution function of ξ under condition B (for every point of continuity of the latter) for every $B \in \mathscr{A}$ with $P(B) > 0$, then $\xi_n \Rightarrow \xi$. In other words, the following theorem holds:

THEOREM 4.2.2. *Let ξ_n $(n = 1, 2, \ldots)$ and ξ be random variables on a probability space $S = (\Omega, \mathscr{A}, P)$. For any $B \in \mathscr{A}$ with $P(B) > 0$ let $F_n(x \mid B)$ and $F(x \mid B)$ denote the conditional distribution function (under condition B) of ξ_n and ξ, respectively. If (4.2.7) holds for every $B \in \mathscr{A}$ with $P(B) > 0$ and for every point of continuity of $F(x \mid B)$, then $\xi_n \Rightarrow \xi$.*

Proof. Let $\varepsilon > 0$ be arbitrary and denote by $B_k(\varepsilon)$ the event $k\varepsilon \leqq \xi < (k + 1)\varepsilon$. If $P(B_k(\varepsilon)) > 0$, then by supposition

$$\lim_{n \to +\infty} F_n((k-1)\varepsilon \mid B_k(\varepsilon)) = 0 \tag{4.2.8'}$$

and

$$\lim_{n \to +\infty} F_n((k+2)\varepsilon \mid B_k(\varepsilon)) = 1 \tag{4.2.9}$$

As $|\xi_n - \xi| > 2\varepsilon$ implies, under condition $B_k(\varepsilon)$, that either $\xi_n < (k-1)\varepsilon$ or $\xi_n > (k+2)\varepsilon$, it follows that

$$\lim_{n \to +\infty} P(|\xi_n - \xi| > 2\varepsilon \mid B_k(\varepsilon)) = 0 \tag{4.2.10}$$

Using the theorem of total probability we get, for every $N \geqq 1$,

$$P(|\xi_n - \xi| > 2\varepsilon) \leqq \sum_{k=-N}^{+N} P(|\xi_n - \xi| > 2\varepsilon \mid B_k(\varepsilon))P(B_k(\varepsilon)) + P(|\xi| > N\varepsilon) \tag{4.2.11}$$

and thus,

$$\limsup_{n\to+\infty} P(|\xi_n - \xi| > 2\varepsilon) \leqq P(|\xi| > N\varepsilon) \tag{4.2.12}$$

As N can be chosen arbitrarily large, it follows that

$$\lim_{n\to+\infty} P(|\xi_n - \xi| > 2\varepsilon) = 0 \tag{4.2.13}$$

which proves Theorem 4.2.2. ■

DEFINITION 4.2.3. *Let ξ_n and ξ be random variables on a probability space S having finite second moments. We say that ξ_n converges strongly (in $L_2(S)$) to ξ if*

$$\lim_{n\to+\infty} E((\xi_n - \xi)^2) = 0 \tag{4.2.14}$$

THEOREM 4.2.3. *Strong convergence in $L_2(S)$ implies convergence in probability.*

Proof. By Markov's inequality (2.11.1)

$$P(|\xi_n - \xi| > \varepsilon) \leqq \frac{E((\xi_n - \xi)^2)}{\varepsilon^2} \tag{4.2.15}$$

and thus (4.2.14) implies (4.2.3). ■

DEFINITION 4.2.4. *A sequence ξ_n of random variables on a probability space $S = (\Omega, \mathscr{A}, P)$ is said to converge almost surely to a random variable ξ on S, if denoting by Z the set of those $\omega \in \Omega$ for which $\lim_{n\to+\infty} \xi_n(\omega) = \xi(\omega)$ does not hold, one has $P(Z) = 0$.*

THEOREM 4.2.4. *Almost sure convergence implies convergence in probability.*

Proof. Let $\varepsilon > 0$ be arbitrary and let $A_N(\varepsilon)$ denote the set

$$\{\omega : \sup_{n\geqq N} |\xi_n(\omega) - \xi(\omega)| > \varepsilon\}$$

Clearly, $A_{N+1}(\varepsilon) \subseteq A_N(\varepsilon)$ $(n = 1, 2, \ldots)$ and if Z denotes the set of those ω for which $\lim_{n\to+\infty} \xi_n(\omega) = \xi(\omega)$ does not hold, then $\prod_{N=1}^{\infty} A_N(\varepsilon) \subseteq Z$. As by supposition $P(Z) = 0$, it follows from the continuity of the probability measure that $\lim_{n\to+\infty} P(A_N(\varepsilon)) = 0$. Now evidently,

$$P(|\xi_N - \xi| > \varepsilon) \leqq P(A_N(\varepsilon))$$

which implies the statement of the theorem. ■

Remark: Note that it follows from the proof that the *almost sure convergence of ξ_n to ξ is equivalent to the convergence in probability to* 0 *of* $\zeta_N = \sup_{n\geqq N} |\xi_n - \xi|$.

It is well known that neither does almost sure convergence imply strong convergence, nor conversely. (See Example 4.2.2.) However, if ξ_n ($n = 1, 2, \ldots$) and ξ are random variables on a probability space $S = (\Omega, \mathscr{A}, P)$ such that Ω is denumerable, then the convergence of ξ_n to ξ in probability implies that $\xi_n \to \xi$ almost surely (see Problem P.4.1).

DEFINITION 4.2.5. *Let ξ_n ($n = 1, 2, \ldots$) and ξ be random variables with finite second moments on a probability space S. If for every random variable η on S having finite second moment one has*

$$\lim_{n \to +\infty} E(\xi_n \eta) = E(\xi\eta) \tag{4.2.16}$$

then we call the sequence ξ_n weakly convergent to ξ in L_2 and denote this by $\xi_n \rightharpoonup \xi$.

Remark: Strong convergence evidently implies weak convergence, because by (2.9.19)

$$|E(\xi_n \eta) - E(\xi\eta)|^2 = |E((\xi_n - \xi)\eta)|^2 \leqq E((\xi_n - \xi)^2)E(\eta^2) \tag{4.2.17}$$

Let us now consider some examples.

Example 4.2.1. Let S be the Lebesgue probability space and let $\xi_n = R_n(x)$ be the nth Rademacher function (see Example 3.6.1). As the $R_n(x)$ ($n = 1, 2, \ldots$) form an orthonormal system, it follows by the Bessel inequality (see Lemma 3.6.1) that for every random variable $f \in L_2(S)$, one has

$$\lim_{n \to +\infty} E(f\xi_n) = 0 \tag{4.2.17'}$$

Thus $R_n(x) \rightharpoonup 0$ for $n \to +\infty$. Especially if $f(x)$ is the indicator of an event B with $P(B) > 0$, we obtain from (4.2.17')

$$\lim_{n \to +\infty} \int_B R_n(x)\, dx = 0 \tag{4.2.18}$$

This means that

$$\lim_{n \to +\infty} P(R_n(x) = 1 \mid B) = 1/2 \tag{4.2.19}$$

Thus the conditional probability distribution of $R_n(x)$ with respect to every condition B with $P(B) > 0$ tends to the distribution which attributes the probability $1/2$ to the values ± 1. However, $R_n(x)$ does not tend in probability to a random variable ξ.

As a matter of fact, if $R_n(x)$ would tend in probability to a random variable $\xi = \xi(x)$, then for every measurable subset B of the interval $(0, 1)$ with $P(B) > 0$, ξ would take on the values ± 1 with conditional probability $1/2$ under condition B. Taking for B the set on which $\xi(x) = 1$, we get a contradiction, which shows that $R_n(x)$ cannot converge in probability.

This example shows that the conditions of Theorem 4.2.2 cannot be relaxed so that only the existence of the limit (4.2.7) is assumed for every event B with $P(B) > 0$. It is not sufficient either that the limit $F(x \mid B)$ should be a distribution function for every such B; it is necessary that $F(x \mid B)$ should be for every B the conditional distribution function of the same random variable ξ with respect to the condition B.

Example 4.2.2. Let us define the random variables ξ_n $(n = 1, 2, \ldots)$ on the Lebesgue probability space as follows:

$$\xi_{N^2+k}(x) = \begin{cases} N^\alpha & \text{if} \quad k/(2N+1) \leqq x < (k+1)/(2N+1) \\ 0 & \text{otherwise} \end{cases} \tag{4.2.20}$$

where $\alpha \geqq 0$, $k = 0, 1, \ldots, 2N$; $N = 1, 2, \ldots$. Then $\xi_n \Rightarrow 0$ whatever the value of α. However, $\lim_{n\to+\infty} \xi_n(x)$ does not exist for any x in $(0, 1)$, if $\alpha \geqq 0$, but ξ_n tends strongly to 0 if $\alpha < 1/2$. On the other hand, the subsequence ξ_{N^2} tends almost surely to 0 for every α, but does not tend to 0 strongly if $\alpha \geqq 1/2$; ξ_{N^2} tends weakly to 0 for $\alpha \leqq 1/2$ but not for $\alpha > 1/2$.

Let us introduce the following:

DEFINITION 4.2.6. *Let ξ_n $(n = 1, 2, \ldots)$ be a sequence of random variables on a probability space S and let ξ_n' $(n = 1, 2, \ldots)$ be a sequence of random variables on another probability space S'. Suppose that the joint distribution of the variables $\xi_1, \xi_2, \ldots, \xi_n$ is the same as the joint distribution of the random variables $\xi_1', \xi_2', \ldots, \xi_n'$ for every n. In this case we call the sequences $\{\xi_n\}$ and $\{\xi_n'\}$ isomorphic.*

Probability theory deals mainly with such statements about sequences of random variables which remain valid if the sequence of random variables involved is replaced by another isomorphic sequence. It is easy to see that if $f_n(x_1, x_2, \ldots, x_n)$ is a sequence of Borel-measurable functions, $\eta_n = f_n(\xi_1, \xi_2, \ldots, \xi_n)$ and $\eta_n' = f_n(\xi_1', \ldots, \xi_n')$, then if η_n tends to 0 in probability or almost everywhere or strongly or weakly in S, the same holds for the sequence η_n' in S'. As regards almost sure convergence, this follows from the remark to Theorem 4.2.4, according to which the convergence of η_n to 0 almost everywhere is equivalent to the convergence in probability to 0 of $\sup_{n \geqq N} |\eta_n|$ for $N \to +\infty$, and the latter fact depends evidently only on the joint distributions of the random variables $\xi_1, \ldots, \xi_n$ for $n = 1, 2, \ldots$.

4.3 CONVERGENCE OF PROBABILITY DISTRIBUTIONS

DEFINITION 4.3.1. *Let Ω be a metric space* and $\mathscr{A}$ the σ-algebra of all Borel sets of Ω. Let $P_n(A)$ $(n = 1, 2, \ldots)$ and $P(A)$ be probability measures on $\mathscr{A}$.*

* See Appendix B.

Put for every bounded $\mathscr{A}$-measurable function $f(\omega)$ on Ω

$$E_n(f) = \int_\Omega f(\omega)\, dP_n$$

and

$$E(f) = \int_\Omega f(\omega)\, dP$$

The sequence P_n $(n = 1, 2, \ldots)$ is called weakly convergent to P if for every bounded and continuous function $f(\omega)$ on Ω, one has

$$\lim_{n \to +\infty} E_n(f) = E(f) \tag{4.3.1}$$

If we take $\Omega = R$ where R is the real axis, considered as a metric space with respect to the distance $d(x, y) = |y - x|$, then, as we have seen, every probability distribution is uniquely determined by the corresponding distribution function $F(x)$, where $F(x)$ is the measure of the interval $(-\infty, x)$.

We show now that in this case the weak convergence of measures is equivalent to the convergence of the corresponding distribution functions $F_n(x)$ to the distribution function $F(x)$ in each point of continuity of the latter. In other words, the following theorem is true:

THEOREM 4.3.1. *If on the real line a sequence P_n of probability distributions is given which converges weakly to a probability distribution P, then putting $F_n(x) = P_n((-\infty, x))$ and $F(x) = P((-\infty, x))$, one has*

$$\lim_{n \to +\infty} F_n(x) = F(x) \tag{4.3.2}$$

for each x which is a point of continuity of $F(x)$; conversely, if (4.3.2) *holds at an everywhere dense set of points then one has for every bounded and continuous function $f(x)$ on the real line,*

$$\lim_{n \to +\infty} \int_{-\infty}^{+\infty} f(x)\, dF_n(x) = \int_{-\infty}^{+\infty} f(x)\, dF(x) \tag{4.3.3}$$

i.e., the measure P_n corresponding to $F_n(x)$ converges weakly to the measure P corresponding to $F(x)$.*

Proof. Suppose that (4.3.3) holds for every bounded and continuous function $f(x)$. Let y be a point of continuity of $F(x)$ and put

$$f(\varepsilon, y, x) = \begin{cases} 1 & \text{for } -\infty < x \leqq y \\ \left[1 - \left(\dfrac{x - y}{\varepsilon}\right)^4\right]^4 & \text{for } y \leqq x \leqq y + \varepsilon \\ 0 & \text{for } y + \varepsilon \leqq x \end{cases} \tag{4.3.4}$$

* See (2.9.12).

Applying (4.3.3) for $f(x) = f(\varepsilon, y, x)$ and $f(x) = f(\varepsilon, y - \varepsilon, x)$, we obtain that

$$\limsup_{n\to+\infty} F_n(y) \leqq F(y+\varepsilon) \qquad \text{and} \qquad \liminf_{n\to\infty} F_n(y) \geqq F(y-\varepsilon)$$

which implies that (4.3.2) holds for $x = y$. Conversely, if (4.3.2) holds on an everywhere dense set E of points of continuity of $F(x)$, then clearly (4.3.3) holds for every $f(x)$ which is a step function such that its points of jump belong to the set E; as every continuous function vanishing outside a finite interval can be uniformly approximated by such a step function, (4.3.3) holds if $f(x)$ is continuous and vanishes outside a finite interval. However, if $f(x)$ is bounded, $|f(x)| \leqq k$, then

$$\left|\int_{|x|>A} f(x)\, dF_n(x)\right| \leqq [F_n(-A) + (1 - F_n(A))]-k$$

If $-A$ and A are both points of continuity of $F(x)$, then $\lim_{n\to+\infty} F_n(-A) = F(-A) < \varepsilon$ and $\lim_{n\to+\infty}(1 - F_n(A)) = (1 - F(A)) < \varepsilon$ for any $\varepsilon > 0$, if A is sufficiently large; it follows that (4.3.3) holds for every function $f(x)$ which is bounded and continuous on the real line. ■

Theorem 4.3.1 shows that the weak limit of probability measures on the real line—if it exists—is unique.

Definition 4.3.2. *If $F_n(x)$ $(n = 1, 2, \ldots)$ and $F(x)$ are distribution functions on the real line and* (4.3.2) *holds for every point of continuity x of $F(x)$, we call the sequence $F_n(x)$ weakly convergent to $F(x)$.*

Remark: It should be noted that if one supposes only that $\lim_{n\to+\infty} F_n(x) = F(x)$ for every point of continuity of the nondecreasing function $F(x)$, without supposing that $F(x)$ is a distribution function, it does not follow that the corresponding probability measures converge weakly. The reason for this is that the real line is not compact, and therefore while passing to the limit a positive probability may "escape" to infinity.

This is shown by the following:

Example 4.3.1. Let P_n be the uniform distribution over the sets of points $\{1, 2, \ldots, n\}$. The corresponding distribution function is

$$F_n(x) = \begin{cases} 0 & \text{for } x \leqq 1 \\ k/n & \text{for } k < x \leqq k+1,\ k = 1, 2, \ldots, n-1 \\ 1 & \text{for } n < x \end{cases}$$

In this case, one has for every x

$$\lim_{n\to+\infty} F_n(x) = 0$$

However, the measures P_n do not converge weakly to a measure, because one has for any $f(x)$

$$E_n(f) = \frac{f(1) + f(2) + \cdots + f(n)}{n}$$

and one can choose for $f(x)$ a bounded continuous function so that

$$\lim_{n \to +\infty} E_n(f)$$

should not exist. (For example, one can put $f(x) = +1$ for $2^{2k} \leqq x \leqq 2^{2k+1} - 1$ and $f(x) = -1$ for $2^{2k+1} \leqq x < 2^{2k+2}$ $(k = 0, 1, \ldots)$, and define $f(x)$ to be linear in the intervals $(2^n - 1, 2^n)$ $(n = 0, 1, \ldots)$.)

As mentioned, the phenomenon for Example 4.3.1 is possible because the real line is not compact. If we consider only distributions on a compact subset, e.g., on a finite closed interval of the real line, then if the limit of a sequence of such distributions converges to a function $F(x)$ at all its points of continuity, then $F(x)$ is clearly a distribution function. This is shown by the following:

THEOREM 4.3.2. *If $F_n(x)$ $(n = 1, 2, \ldots)$ is a sequence of distribution functions on the real line, each concentrated in the closed interval $[a, b]$ (i.e., if $F_n(a) = 0$ and $F_n(b) = 1$ where $a < b$), and if the limit*

$$\lim_{n \to +\infty} F_n(x) = F(x)$$

exists for every x in $[a, b]$ (or at least in an everywhere dense set in $[a, b]$), then $F(x)$ is a distribution function and $F_n(x)$ tends weakly to $F(x)$.

The proof is obvious, as $F(x)$ is clearly nondecreasing, $F(a) = 0$, and $F(b) = 1$.

We shall now prove the following:

THEOREM 4.3.3. *If the distribution functions $F_n(x)$ $(n = 1, 2, \ldots)$ converge weakly to a distribution function $F(x)$ which is everywhere continuous, then the convergence is uniform for $-\infty < x < +\infty$.*

Proof. Suppose the statement of the theorem is not true. Then there exists a sequence of points x_k and an increasing sequence of integers n_k such that

$$\lim_{k \to +\infty} |F_{n_k}(x_k) - F(x_k)| = a > 0$$

Now the sequence x_k either contains a subsequence x_{k_j} which converges to a finite x_0, or $|x_k| \to +\infty$. In the first case, we have

$$\lim_{j \to +\infty} |F_{n_{k_j}}(x_{k_j}) - F(x_0)| = a > 0$$

As, however, for sufficiently large values of j, one has $|x_{k_j} - x_0| < \varepsilon$ and consequently $F_{n_{k_j}}(x_0 - \varepsilon) \leqq F_{n_{k_j}}(x_{k_j}) \leqq F_{n_{k_j}}(x_0 + \varepsilon)$, it follows that $F(x_0 + \varepsilon) - F(x_0 - \varepsilon) \geqq a > 0$, which is impossible for a sufficiently small value of ε, because of the continuity of $F(x)$ at x_0. On the other hand, if $|x_k| \to +\infty$, let us choose an A such that $F(-A) < a/4$ and $F(A) > 1 - a/4$, and a value of n_0 such that $|F_n(-A) - F(-A)| < a/2$ and $|F_n(A) - F(A)| < a/2$ for $n \geqq n_0$; choose an n_1 such that $|x_n| > A$ for $n \geqq n_1$. Then

$$|F_n(x_n) - F(x_n)| < 3a/4$$

for $n \geqq \max(n_0, n_1)$, which contradicts our assumption. This proves our theorem. ■

The fact used at the end of the proof can be formulated as follows: If the distribution functions $F_n(x)$ tend to a distribution function $F(x)$, then for every $\varepsilon > 0$ there exists an interval $(-A, +A)$ such that $F_n(A) - F_n(-A) > 1 - \varepsilon$ uniformly in n, i.e., the sequence $F_n(x)$ is *uniformly tight* (see Definition 2.5.4).

We now prove the following:

THEOREM 4.3.4. *If $F_n(x)$ $(n = 1, 2, \ldots)$ is a sequence of distribution functions and $F(x)$ is a distribution function on the real line, and one has*

$$\lim_{n \to +\infty} (F_n(b) - F_n(a)) = F(b) - F(a) \tag{4.3.5}$$

whenever a and $b > a$ are points of continuity of $F(x)$, then (4.3.2) *holds for every x which is a point of continuity of $F(x)$.*

Proof. If x is a point of continuity of $F(x)$, choose $a = x$ and $b > a$ in (4.3.5). It follows that

$$\liminf_{n \to +\infty} (1 - F_n(x)) \geqq F(b) - F(x) \tag{4.3.6}$$

As (4.3.6) holds for arbitrarily large values of b, it follows that

$$\liminf_{n \to +\infty} (1 - F_n(x)) \geqq 1 - F(x)$$

and thus

$$\limsup_{n \to +\infty} F_n(x) \leqq F(x) \tag{4.3.7}$$

Similarly, choosing x for b and $a < x$ in (4.3.4), we obtain

$$\liminf_{n \to +\infty} F_n(x) \geqq F(x) \tag{4.3.8}$$

Now (4.3.7) and (4.3.8) imply (4.3.5). ■

We now introduce the following:

DEFINITION 4.3.3. *Let Ω be a separable metric space and let $\mathscr{A}$ be the σ-algebra of all Borel subsets of Ω. Let $\{P_n\}$ be a sequence of probability measures on $\mathscr{A}$. If for every $\varepsilon > 0$ one can find a compact subset C of Ω such that $P_n(C) > 1 - \varepsilon$, uniformly for all n, then $\{P_n\}$ is called uniformly tight.**

Thus if a sequence of probability measures on the real line converges weakly to a probability measure, then this sequence is uniformly tight. Conversely, from a uniformly tight sequence of probability measures on the real line one can always select a subsequence which is weakly convergent to a probability measure.

This is a special case of the following general theorem, due to Prochorov ([5], see also [6]) which we state here without proof:

THEOREM 4.3.5. *Let Ω be a complete separable metric space and let $\mathscr{A}$ be the family of Borel subsets of Ω. Let $\{P_n\}$ be a sequence of probability measures on $\mathscr{A}$. Then one can select from $\{P_n\}$ a subsequence which converges weakly to a probability measure P on $\mathscr{A}$ if and only if $\{P_n\}$ contains a subsequence that is uniformly tight.*

We shall now add some general remarks on weak convergence of probability distribution on a metric space. Let Ω be a metric space and let $\mathscr{A}$ be the set of Borel subsets of Ω. Let P_n $(n = 1, 2, \ldots)$ and P be probability measures on $\mathscr{A}$. Let $E_n(f)$ denote the expectation of the random variable $f(\omega)$ with respect to the probability space $(\Omega, \mathscr{A}, P_n)$ and let $E(f)$ denote the expectation of $f(\omega)$ with respect to the probability space $(\Omega, \mathscr{A}, P)$. If

$$\lim_{n \to +\infty} E_n(f) = E(f) \tag{4.3.9}$$

holds for $f = f_k$ $(k = 1, 2, \ldots, N)$, it holds by the linearity of the expectation for $f = \sum_{k=1}^{N} c_k f_k$ if the c_k are arbitrary constants. Further, let (4.3.9) hold for $f = g_k$ $(k = 1, 2, \ldots)$ and suppose that the sequence g_k $(k = 1, 2, \ldots)$ tends uniformly on Ω to a function g. Then

$$|E_n(g) - E(g)| \leqq |E_n(g_k) - E(g_k)| + 2 \sup_{\omega} |g_k(\omega) - g(\omega)|$$

and therefore we have, for each k,

$$\limsup_{n \to \infty} |E_n(g) - E(g)| \leqq 2 \sup_{\omega} |g_k(\omega) - g(\omega)|$$

and thus (4.3.9) holds for $f = g$.

Thus we have shown the following:

THEOREM 4.3.6. *If (4.3.9) holds for $f \in \mathscr{F}$ where $\mathscr{F}$ is any family of integrable functions, then it holds for $f \in \overline{\mathscr{F}}$ where $\overline{\mathscr{F}}$ is the closed linear hull of $\mathscr{F}$, i.e.,*

* Compare with Definition 2.5.4. Notice that in a separable metric space the family of compact nonempty sets forms a bunch.

the least family of random variables which contains the family $\mathscr{F}$ and all linear combinations and uniform limits of its elements.

This theorem leads to a series of necessary and sufficient conditions for weak convergence.

COROLLARY TO THEOREM 4.3.6. *If Ω is the interval $[0, 1]$ of the real line, considered as a metric space with the usual metric $d(x, y) = |y - x|$, then a necessary and sufficient condition for the weak convergence of a sequence of probability distributions P_n with distribution function $F_n(x)$ $(n = 1, 2, \ldots)$ to a probability distribution P is the existence of the limits*

$$\lim_{n\to+\infty} E_n(x^k) = \lim_{n\to+\infty} \int_0^1 x^k \, dF_n(x) = M_k \tag{4.3.10}$$

for $k = 1, 2, \ldots$.

Another necessary and sufficient condition is the existence of the limits

$$\lim_{n\to+\infty} E_n(e^{2\pi ikx}) = \lim_{n\to+\infty} \int_0^1 e^{2\pi ikx} \, dF_n(x) = c_k \tag{4.3.11}$$

for $k = 0, \pm 1, \pm 2, \ldots$.

Proof. The proof follows from the remark made above and the theorems of Weierstrass, according to which every real continuous function $f(x)$ in $[0, 1]$ is the uniform limit of a sequence of polynomials,* or trigonometric polynomials. ■

Note that in the Corollary to Theorem 4.3.6 it is not necessary to suppose that $E_n(f)$ tends to the expectation $E(f)$ of f with respect to some measure in $[0, 1]$ because, the interval $[0, 1]$ being compact, this condition is automatically satisfied: if the limit $\lim_{n\to+\infty} F_n(f)$ exists for every continuous function f, in view of Theorem 4.3.2, it is equal to the expectation $E(f)$ of f with respect to a probability distribution in $[0, 1]$. We now state an important theorem.

THEOREM 4.3.7. *Let R be the real line considered as a metric space with the usual metric $d(x, y) = |y - x|$ and let $\mathscr{A}$ be the family of Borel subsets of R. Let $F_n(x)$ be a sequence of probability distribution functions on R and let*

$$\varphi_n(t) = \int_{-\infty}^{+\infty} e^{itx} \, dF_n(x) \qquad (-\infty < t < +\infty) \tag{4.3.12}$$

be the corresponding characteristic functions. If

$$\lim_{n\to+\infty} \varphi_n(t) = \varphi(t) \tag{4.3.13}$$

exists for all real values of t and $\varphi(t)$ is continuous at $t = 0$, then $\varphi(t)$ is the characteristic function of a distribution function $F(x)$, i.e.,

* For a probabilistic proof of Weierstrass' theorems, see Exercise E.4.1.

$$\varphi(t) = \int_{-\infty}^{+\infty} e^{itx}\, dF(x) \tag{4.3.14}$$

and the distribution functions $F_n(x)$ tend weakly to $F(x)$ for $n \to +\infty$.

Remark: This important theorem has already been mentioned in Section 2.9. (For the proof see [55] and [63], further, see [29], [43] and [44].)

We now consider in detail the case when Ω is a denumerably infinite set, $\Omega = \{\omega_1, \omega_2, \ldots, \omega_n, \ldots\}$ and $\mathscr{A} = \mathscr{P}(\Omega)$. In this case each probability measure P on $\mathscr{A}$ is uniquely determined by the sequence $p_k = P(\{\omega_k\})$ where, of course, $p_k \geqq 0$ and $\sum_{k=1}^{\infty} p_k = 1$. We shall prove the following:

THEOREM 4.3.8. *Let $\Omega = \{\omega_1, \omega_2, \ldots, \omega_n, \ldots\}$ be a denumerably infinite set, let $\mathscr{A} = \mathscr{P}(\Omega)$, and let P_n ($n = 1, 2, \ldots$) be a sequence of probability measures on $\mathscr{A}$. Let us put $P_n(\{\omega_k\}) = p_{n,k}$. If the limit*

$$\lim_{n \to +\infty} p_{n,k} = p_k \tag{4.3.15}$$

exists for $k = 1, 2, \ldots$ and

$$\sum_{k=1}^{\infty} p_k = 1 \tag{4.3.16}$$

holds, then putting

$$\Delta_n = \sum_{k=1}^{\infty} |p_{n,k} - p_k| \tag{4.3.17}$$

one has

$$\lim_{n \to +\infty} \Delta_n = 0 \tag{4.3.18}$$

*Proof.** Let $\varepsilon > 0$ be arbitrary. Let us choose N so that

$$\sum_{k=N+1}^{\infty} p_k < \varepsilon/4$$

Fixing the value of N, one can find an n_1 such that for $n \geqq n_1$ one has

$$|p_{n,k} - p_k| < \varepsilon/4N \qquad \text{for } k = 1, 2, \ldots N$$

It follows that

$$\left| \sum_{k=1}^{N} p_{n,k} - \sum_{k=1}^{N} p_k \right| < \varepsilon/4$$

and thus

$$\left| \sum_{k=N+1}^{\infty} p_{n,k} - \sum_{k=N+1}^{\infty} p_k \right| = \left| \sum_{k=1}^{N} p_{n,k} - \sum_{k=1}^{N} p_k \right| < \varepsilon/4$$

* Since $\sum_{k=1}^{\infty} |p_{n,k} - p_k| \leqq \sum_{k=1}^{\infty} (p_{n,k} + p_k) = 2$, the theorem can be obtained as a consequence of Lebesque convergence theorem too.

and therefore

$$\sum_{k=N+1}^{\infty} p_{n,k} < \varepsilon/2$$

for all $n \geqq n_1$. Thus we get for $n \geqq n_1$,

$$\Delta_n \leqq \sum_{k=1}^{N} |p_{n,k} - p_k| + \sum_{k=N+1}^{\infty} p_{n,k} + \sum_{k=N+1}^{\infty} p_k < \varepsilon/4 + \varepsilon/2 + \varepsilon/4 = \varepsilon$$

This proves Theorem 4.3.8. ■

Remark: It follows from Theorem 4.3.8 that if P_n $(n = 1, 2, \ldots)$ is a sequence of probability measures on the family of all subsets of a denumerable space Ω, and $\lim_{n\to+\infty} P_n(A) = P(A)$ for every *finite* subset A of Ω where P is a probability measure on $\mathscr{A}$, then

$$\lim_{n\to+\infty} P_n(A) = P(A) \qquad \text{for every} \quad A \in \mathscr{A}$$

and uniformly in A.

COROLLARY TO THEOREM 4.3.8. *Under the conditions of Theorem* 4.3.8 *one has, putting* $P(A) = \sum_{\omega_k \in A} p_k$,

$$\lim_{n\to+\infty} \sup_{A\in\mathscr{A}} |P_n(A) - P(A)| = 0 \tag{4.3.19}$$

Let $E_n(f)$ *denote the expectation of the random variable* f *with respect to the probability measure* P_n *and let* $E(f)$ *denote the expectation with respect to the probability measure* P. *Then we have, if* f *is bounded,* $|f(\omega_k)| \leqq K$ *for* $k = 1, 2, \ldots,$

$$|E_n(f) - E(f)| \leqq K\Delta_n \tag{4.3.20}$$

*and thus** P_n *converges weakly to* P.

Proof. Evidently $|E_n(f) - E(f)| = |\sum_{k=1}^{\infty} f(\omega_k)(p_{n,k} - p_k)| \leqq K\Delta_n$ and in particular, if f is the indicator of A, we have

$$|P_n(A) - P(A)| = |\sum_{\omega_k\in A} (p_{n,k} - p_k) \leqq \Delta_n$$

for all $A \in \mathscr{A}$. ■

We shall now prove the following:

THEOREM 4.3.9. *Let* $P_n (n = 1, 2, \ldots)$ *be a sequence of probability measures on all subsets of the denumerable space* $\Omega = \{\omega_1, \omega_2, \ldots, \omega_k, \ldots\}$. *Put* $p_{n,k} =$

* If Ω is considered as a metric space, e.g., with respect to the metric $d(\omega_k, \omega_j) = |k - j|$, $k, j = 1, 2, \ldots$.

$P_n(\{\omega_k\})$. *Suppose that for every* $\varepsilon > 0$ *there exists a number* $N = N(\varepsilon)$ *such that*

$$\sum_{k > N(\varepsilon)} p_{n,k} < \varepsilon \qquad \text{for } n = 1, 2, \ldots$$

Then one can select a subsequence P_{n_j} $(j = 1, 2, \ldots)$ *of the sequence* P_n *which converges weakly to a probability measure* P. *In other words, one has*

$$\lim_{j \to +\infty} p_{n_j,k} = p_k \quad (k = 1, 2, \ldots)$$

where $p_k \geqq 0$ *and* $\sum_{k=1}^{\infty} p_k = 1$.

Remark: Clearly one can consider Ω as a metric space by putting $d(\omega_j, \omega_k) = |k - j|$; Ω as a metric space is separable and complete, and a subset of Ω is compact if and only if it is finite. Thus Theorem 4.3.6 is a particular case of Prochorov's Theorem 4.3.9.

Proof. As the numbers $p_{n,1}$ $(n = 1, 2, \ldots)$ are bounded, we can select a sequence of integers $n_1^{(1)} < n_2^{(1)} < \cdots < n_j^{(1)} < \cdots$, such that the sequence $p_{n_j^{(1)},1}$ converges to a limit p_1 for $j \to +\infty$. Similarly, from the sequence $n_j^{(1)}$ we can select a subsequence $n_j^{(2)}$ $(j = 1, 2, \ldots)$ such that $p_{n_j^{(2)},2}$ converges to a limit p_2. Continuing, we get the sequences $n_j^{(l)}$ $(j = 1, 2, \ldots)$ for $l = 1, 2, \ldots$ such that

$$\lim_{j \to +\infty} p_{n_j^{(l)},k} = p_k \qquad \text{for } k \leqq l$$

It follows that

$$\lim_{j \to +\infty} p_{n_j^{(j)},k} = p_k \qquad \text{for } k = 1, 2, \ldots$$

Clearly $\sum_{k=1}^{\infty} p_k \leqq 1$. From the uniform tightness of the measures P_n it follows immediately that for every $\varepsilon > 0$ there exists a number $N(\varepsilon)$ such that $\sum_{k=1}^{N(\varepsilon)} p_k \geqq 1 - \varepsilon$. Thus we have $\sum_{k=1}^{\infty} p_k = 1$, i.e., by Theorem 4.3.4, the measures $P_{n_j^{(j)}}$ converge weakly to the measure P, defined by $P(A) = \sum_{\omega_k \in A} p_k$. ■

THEOREM 4.3.10. *If a sequence* $\{P_n\}$ *of probability measures on the family* $\mathscr{A} = \mathscr{P}(\Omega)$ *of all subsets of the denumerable space* Ω *is uniformly tight (as in Theorem* 4.3.9), *then putting for each* $A \in \mathscr{A}$,

$$\mathscr{Q}(A) = \sup_n P_n(A) \tag{4.3.21}$$

the set function $\mathscr{Q}(A)$ *has the following property: If* $A_j \in \mathscr{A}$ $(j = 1, 2, \ldots)$ *is a sequence of sets such that* $A_{j+1} \subseteq A_j$ $(j = 1, 2, \ldots)$ *and* $\prod_{j=1}^{\infty} A_j = \varnothing$, *then*

$$\lim_{j \to +\infty} \mathscr{Q}(A_j) = 0 \tag{4.3.22}$$

Remark: Equation (4.3.22) can be expressed by saying that the probability measures P_n are *uniformly continuous* (on the empty set).

Proof. By supposition, for every $\varepsilon > 0$ there exists a number N such that putting $B_N = \{\omega_k : k > N\}$ one has

$$P_n(B_N) < \varepsilon \qquad \text{for all } n$$

Let A_j be a sequence of subsets of Ω such that $A_{j+1} \subseteq A_j$ and $\prod_{j=1}^{\infty} A_j = \varnothing$. Then for each k there exists a j_k such that $\omega_k \notin A_j$ if $j \geqq j_k$. Let us put $\max_{k \leq N} j_k = h_N$. Then $A_{h_N} \subseteq B_N$ and thus, $P_n(A_{h_N}) < \varepsilon$ for all $n = 1, 2, \ldots$, i.e., $\mathscr{Q}(A_{h_N}) < \varepsilon$. This proves (4.3.22). ■

Now we are in the position to prove the celebrated theorem of Vitali-Hahn-Saks, which was already used in Section 2.5.

THEOREM 4.3.11. *Let $\mathscr{E} = (\Omega, \mathscr{A})$ be an experiment and let P_n $(n = 1, 2, \ldots)$ be a sequence of probability measures on $\mathscr{E}$. Suppose that the limit*

$$\lim_{n \to +\infty} P_n(A) = P(A) \tag{4.3.23}$$

exists for all $A \in \mathscr{A}$. Then $P(A)$ is a probability measure on $\mathscr{A}$ and putting

$$\mathscr{Q}(A) = \sup_n P_n(A) \tag{4.3.24}$$

$\mathscr{Q}(A)$ is continuous on the empty set, i.e., if $A_j \in \mathscr{A}$, $A_{j+1} \subseteq A_j$ $(j = 1, 2, \ldots)$, and $\prod_{j=1}^{+\infty} A_j = \varnothing$, then one has

$$\lim_{j \to +\infty} \mathscr{Q}(A_j) = 0 \tag{4.3.25}$$

Proof. It can be seen that $P(A)$ is a nonnegative and finitely additive set function and $P(\Omega) = 1$. Thus, in order to show that $P(A)$ is a probability measure it is sufficient to show that it is σ-additive on $\mathscr{A}$, i.e., if the sets $A_j \in \mathscr{A}$ $(j = 1, 2, \ldots)$ are disjoint, then one has

$$P\left(\sum_{j=1}^{+\infty} A_j\right) = \sum_{j=1}^{+\infty} P(A_j) \tag{4.3.26}$$

Let us put $A_0 = \overline{\left(\sum_{j=1}^{\infty} A_j\right)}$. Then (4.3.26) is equivalent to

$$\sum_{j=0}^{+\infty} P(A_j) = 1 \tag{4.3.27}$$

where $(A_0, A_1, \ldots, A_j, \ldots)$ is a partition of Ω.

Evidently,

$$\sum_{j=0}^{N} P(A_j) = \lim_{n \to +\infty} \sum_{j=0}^{N} P_n(A_j) \leq 1$$

Thus, to prove (4.3.27) we have to show only that putting

$$s = \sum_{j=0}^{+\infty} P(A_j) \tag{4.3.28}$$

$s < 1$ is impossible. Suppose that $s < 1$ holds. It follows that putting

$$a_{n,k} = \frac{P_n(A_k) - P(A_k)}{1 - s} \quad \text{and} \quad K = \frac{1+s}{1-s} \tag{4.3.29}$$

the numbers $a_{n,k}$ have the following properties:

(A) $$\sum_{k=0}^{+\infty} a_{n,k} = 1$$

(B) $$\sum_{k=0}^{+\infty} |a_{n,k}| \leqq K$$

(C) $$\lim_{n\to+\infty} a_{n,k} = 0 \qquad \text{for } k = 0, 1, 2, \ldots$$

We shall show that these properties imply the existence of a sequence s_k ($k = 0, 1, \ldots$) of zeros and ones such that putting

$$t_n = \sum_{k=0}^{\infty} a_{n,k} s_k \tag{4.3.30}$$

the limit $\lim_{n\to+\infty} t_n$ does not exist. As a matter of fact, let us choose first an integer k_1 such that $\sum_{k=0}^{k_1} a_{1,k} > 7/8$ and $\sum_{k=k_1+1}^{+\infty} |a_{1,k}| < 1/8$. This is possible by properties (A) and (B) of the numbers $a_{n,k}$. We put $s_k = 1$ for $0 \leqq k \leqq k_1$. Then we have, in whatever way the sequence s_k of zeros and ones is defined for $k > k_1$,

$$t_1 = \sum_{k=0}^{+\infty} a_{1,k} s_k \geqq \sum_{k=0}^{k_1} a_{1,k} - \sum_{k=k_1+1}^{+\infty} |a_{1,k}| \geqq 7/8 - 1/8 = 3/4$$

Now let us choose the value of $n_1 > 1$ so that for $n \geqq n_1$, $\sum_{k=0}^{k_1} |a_{n,k}| < 1/8$. This is possible by virtue of (C). Let us choose further k_2 so that

$$\sum_{k=k_2+1}^{+\infty} |a_{n_1,k}| < 1/8$$

and put $s_k = 0$ for $k_1 < k \leqq k_2$. Then we have, independently of how s_k is defined for $k > k_2$,

$$t_{n_1} \leqq \sum_{k=0}^{k_1} |a_{n_1,k}| + \sum_{k=k_2+1}^{+\infty} |a_{n_1,k}| < 1/8 + 1/8 = 1/4$$

Now let us choose $n_2 > n_1$ so that $\sum_{k=0}^{k_2} |a_{n_2,k}| < 1/8$ and choose k_3 so that $\sum_{k=k_3+1}^{+\infty} |a_{n_2,k}| < 1/16$ should hold, and put $s_k = 1$ for $k_2 < k \leqq k_3$. Then we have, for any choice of the further values of s_k for $k > k_3$,

$$t_{n_2} > \sum_{k=0}^{k_1} a_{n_2,k} + \sum_{k=k_2+1}^{k_3} a_{n_2,k} - \sum_{k=k_3+1}^{+\infty} |a_{n_2,k}| > 3/4$$

Now we choose $n_3 > n_2$ so that we should have for $n \geqq n_3$

$$\sum_{k=0}^{k_3} |a_{n,k}| < 1/8$$

and choose k_4 so that $\sum_{k=k_4+1}^{+\infty} |a_{n_3,k}| < 1/8$ should hold, and put $s_k = 0$ for $k_3 < k \leqq k_4$. It follows that

$$t_{n_3} < \sum_{k=0}^{k_3} |a_{n_3,k}| + \sum_{k=k_4+1}^{+\infty} |a_{n_3,k}| < 1/4$$

Continuing this construction we get a sequence s_k of zeros and ones and a sequence $n_0 = 1 < n_1 < n_2 < \cdots$ of integers such that $t_{n_{2k+1}} < 1/4$ and $t_{n_{2k}} > 3/4$; thus, $\lim_{n\to+\infty} t_n$ does not exist. However, we have, putting

$$A = \sum_{s_j=1} A_j \quad \text{and} \quad P^*(A) = \sum_{s_j=1} p_j$$

that

$$t_n = \frac{P_n(A) - P^*(A)}{1-s}$$

Thus, if $\lim_{n\to+\infty} t_n$ does not exist, then $\lim_{n\to+\infty} P_n(A)$ does not exist either, in contradiction to our assumption. Thus the supposition $s < 1$ leads to a contradiction and therefore we have $s = 1$, which proves that $P(A)$ is σ-additive.

To prove the second statement of our theorem, take a sequence $B_k \in \mathscr{A}$ such that $B_{k+1} \subseteq B_k$ and $\prod_{k=1}^{+\infty} B_k = \varnothing$. We may suppose $B_1 = \Omega$. Let us consider the measures P_n only on the σ-algebra $\mathscr{A}^*$ generated by the sets B_k $(k = 1, 2, \ldots)$. Then $\mathscr{A}^*$ consists of all sets which can be obtained as the union of a subsequence of the sets $A_k = B_k \bar{B}_{k+1}$ $(k = 1, 2, \ldots)$ which form a partition of Ω. As by supposition, $\lim_{n\to+\infty} P_n(A_k) = P(A_k)$ exists and by the already proved part of our theorem, $\sum_{k=1}^{+\infty} P(A_k) = 1$, (4.3.25) follows from Theorem 4.3.10. Thus Theorem 4.3.11 is proved. ■

Remark: The main step of the above proof of the Vitali-Hahn-Saks theorem was the proof of the fact that if the infinite matrix $(a_{n,k})$ satisfies the above conditions (A), (B) and (C), then there exists a sequence s_k of zeros and ones such that, putting

$$t_n = \sum_{k=0}^{+\infty} a_{n,k} s_k \tag{4.3.31}$$

the sequence t_n does not tend to a limit. A matrix $(a_{n,k})$ with the mentioned properties (A), (B) and (C) is called a Toeplitz matrix; the conditions (A), (B) and (C) are known to be sufficient in order that for any convergent sequence s_k with limit A the sequence t_n should also converge to A, i.e., that the Toeplitz summation method (4.3.31) should be permanent (theorem of Toeplitz-Silverman-Steinhaus, see [7] and [8]). The condition (A*): $\lim_{n\to+\infty} \sum_{k=1}^{+\infty} a_{n,k}=1$, further conditions (B) and (C) are necessary and sufficient for the permanence of the summation method (4.3.31). What we proved above means that *to every permanent method of summation there exists a sequence of zeros and ones which is not summable by the given method.*

This theorem is due to Steinhaus [9] (see also [48]). Our proof of the Vitali-Hahn-Saks theorem consisted, therefore, essentially in reducing this theorem to the mentioned theorem of Steinhaus, and proving the latter.

This proof is quite elementary, while the usual proof of the Vitali-Hahn-Saks theorem given in textbooks (see, e.g., Neveu [57], Dunford and Schwarz [10], etc.), which is due to Saks [11], uses much deeper tools, namely the *Baire category theorem.* The idea of the above elementary proof is due essentially to Doubrovsky [13]; his proof has been simplified by Dowker [14].

The Vitali-Hahn-Saks theorem can also be deduced from the following theorem of Schur [12]: *In order that the transformation* (4.3.31) *should transform every bounded sequence s_k into a convergent sequence t_n, the following conditions are necessary and sufficient:*

(A) *The limits* $\lim_{n\to+\infty} a_{n,k} = a_k$ *exist* ($k = 1, 2, \ldots$).

(B) *The series* $\sum_{k=1}^{+\infty} |a_{n,k}|$ *converges uniformly in n.*

Other types of convergence of probability measures, different from weak convergence, are also considered in probability theory. One of them is the uniform convergence, i.e.,

$$\lim_{n\to+\infty} \sup_{A\in\mathscr{A}} |P_n(A) - P(A)| = 0 \tag{4.3.32}$$

If the underlying basic space is denumerable, then as we have seen, uniform convergence is equivalent to weak convergence. In the general case however, this is not true. For instance, if the basic space is the interval [0, 1], and the probability measures P_n and P are all absolutely continuous with respect to the Lebesgue measure with density functions $f_n(x)$ and $f(x)$, respectively, then one can show that (4.3.32) is equivalent to

$$\lim_{n\to+\infty} \int_0^1 |f_n(x) - f(x)|\, dx = 0 \tag{4.3.33}$$

Equation (4.3.33) implies (4.3.32), because

$$|P_n(A) - P(A)| = \left| \int_A (f_n(x) - f(x))\, dx \right| \leq \int_0^1 |f_n(x) - f(x)|\, dx$$

On the other hand, if A_n denotes the set on which $f_n(x) \geq f(x)$, then

$$\int_0^1 |f_n(x) - f(x)|\, dx = 2\, |P_n(A_n) - P(A_n)| \leq 2 \sup_A |P_n(A) - P(A)|$$

and thus (4.3.32) implies (4.3.33).

We shall now prove the following:

THEOREM 4.3.12. *Let for each* $n \geqq 1$ $\mathscr{P}_n = \{p_{n,k}\}$ $(k = 1, 2, \ldots, r)$ *be an r-term probability distribution such that* $p_{n,k} > 0$ *for all n and k. In order that*

$$\lim_{n \to +\infty} p_{n,k} = p_k \quad (k = 1, 2, \ldots, r) \tag{4.3.34}$$

should hold, where $\mathscr{P} = \{p_1, \ldots, p_k)$ *is some r-term probability distribution such that* $p_k > 0$ *for* $k = 1, 2, \ldots, r$, *it is necessary and sufficient that*

$$\lim_{\substack{n \to +\infty \\ m \to +\infty}} D(\mathscr{P}_n, \mathscr{P}_m) = 0 \tag{4.3.35}$$

should hold, where $D(\mathscr{P}_n, \mathscr{P}_m)$ *is the information-theoretical divergence of* the distribution* $\mathscr{P}_n$ *and* $\mathscr{P}_m$.

Proof. The necessity is obvious. To prove the sufficiency let us select an increasing sequence n_j $(j = 1, 2, \ldots)$ of integers such that the limits $\lim_{j \to +\infty} p_{n_j,k} = p_k$ $(k = 1, 2, \ldots, r)$ exist. It is easy to see that $p_k > 0$ for $k = 1, 2, \ldots, r$ because if, for instance, one would have $p_i = 0$ for some i, then for each fixed value of m, $\lim_{n \to +\infty} D(\mathscr{P}_m, \mathscr{P}_n) = +\infty$ would hold, which contradicts (4.3.35).

Thus $\mathscr{P} = \{p_k\}$ is a probability distribution of positive terms, and it follows from (4.3.35) that

$$\lim_{n \to +\infty} D(\mathscr{P}, \mathscr{P}_n) = 0 \tag{4.3.36}$$

Now if $\mathscr{P}_n$ would not converge to $\mathscr{P}$, we could select a subsequence m_j such that $\mathscr{P}_{m_j} \to \mathscr{Q} = \{q_1, \ldots, q_k\}$ where $\mathscr{Q}$ is not identical to $\mathscr{P}$. From (4.3.36) we would have $D(\mathscr{P}, \mathscr{Q}) = 0$, which, however, implies $\mathscr{Q} = \mathscr{P}$ by Lemma 3.8.2. This proves $\mathscr{P}_n \to \mathscr{P}$. ■

Remark: Note that if $\mathscr{P} = \{p_1, \ldots, p_r\}$ and $\mathscr{Q} = (q_1, \ldots, q_r)$, then while $D(\mathscr{P}, \mathscr{Q})$ is a continuous function of the variables $p_1, \ldots, p_r$, it is continuous in the variables $q_1, \ldots, q_r$ only under the restriction that $q_i \geqq \varepsilon > 0$ for $i = 1, 2, \ldots, r$.

Finally, we shall prove the following:

THEOREM 4.3.13. *If* ξ_n *and* η_n $(n = 1, 2, \ldots)$ *are independent random variables and the distributions of* ξ_n *and* η_n *tend for* $n \to +\infty$ *to distributions with distribution functions* $F(x)$ *and* $G(x)$, *respectively, then the distribution of* $\xi_n + \eta_n$ *tends to the distribution having the distribution function* $H(x) = F(x) * G(x)$.

* See Definition 3.8.2.

Proof. Let $f(x)$ be a bounded continuous function and put $H_n(x) = F_n(x) * G_n(x)$, where $F_n(x)$ and $G_n(x)$ denote the distribution functions of ξ_n and η_n, respectively. Then we have

$$\lim_{n\to+\infty}\int_{-\infty}^{+\infty} f(u)\, dH_n(u) = \lim_{n\to+\infty}\int_{-\infty}^{+\infty} f(x+y)\, dF_n(x)\, dG_n(y) = \int_{-\infty}^{+\infty} f(u)\, dH(u)$$

which was to be proved. ■

4.4 THE LAWS OF LARGE NUMBERS

We start by reproducing Bernoulli's (almost forgotten) original proof of his law of large numbers [15], which was historically the first rigorously proved result of this type. This proof is not reproduced in modern textbooks, as it has been replaced by another proof based on Čebyshev's inequality, which is more capable of generalization. Nevertheless, the original proof of Bernoulli is still of interest, not only from an historical but also from a mathematical point of view.

THEOREM 4.4.1. *Let us carry out a sequence of identical independent experiments, in each of which the event A has probability $p = P(A)$ $(0 < p < 1)$. Let ν_n denote the frequency of the occurrence of the event A in the course of the first n experiments. Then one has*

$$\nu_n/n \Rightarrow p \tag{4.4.1}$$

In other words, the relative frequency of the event A tends in probability to the probability of A.

Proof. According to the definition of stochastic convergence, (4.4.1) means that for every $\varepsilon > 0$, one has

$$\lim_{n\to+\infty} P(|\nu_n - np| > n\varepsilon) = 0 \tag{4.4.2}$$

Now, as we have seen,* ν_n has a binomial distribution of order n and parameter p, i.e.,

$$p_{n,k} = P(\nu_n = k) = \binom{n}{k} p^k q^{n-k} \qquad (k = 0, 1, \ldots, n) \tag{4.4.3}$$

where $q = 1 - p$.

Thus we have to prove that

$$\lim_{n\to+\infty} \sum_{|k-np|>n\varepsilon} p_{n,k} = 0 \tag{4.4.4}$$

* See Section 3.4.

Now we have

$$\frac{p_{n,k+1}}{p_{n,k}} = \left(\frac{n-k}{k+1}\right)\frac{p}{q} \tag{4.4.5}$$

This ratio is a decreasing function of k for fixed n and is $\leqq 1$ if $k+1 \geqq (n+1)p$. It follows that $p_{n,k}$, as a function of k, increases for $k < (n+1)p$ and decreases for $k > (n+1)p$; if $(n+1)p$ is an integer, there are two maximal terms (those with indices $(n+1)p$ and $(n+1)p-1$), otherwise there is only one maximal term (with index $[(n+1)p]$). Let us put $k_0 = [(n+1)p]$ and $p_n^* = p_{n,k_0}$. Further, let k_1 denote the greatest integer $\leqq n(p-\varepsilon)$ and k_2 the least integer $\geqq n(p+\varepsilon)$.

It follows that for $k \leqq k_1$, we have

$$p_{n,k} = p_{n,k_1}\prod_{j=k}^{k_1-1}\frac{p_{n,j}}{p_{n,j+1}} = p_{n,k_1}\prod_{j=k}^{k_1-1}\left[\left(\frac{j+1}{n-j}\right)\frac{q}{p}\right]$$

and thus

$$p_{n,k} \leqq p_{n,k_1}\left(\frac{1-\varepsilon/p}{1+\varepsilon/q}\right)^{k_1-k} \qquad \text{for } k \leqq k_1 \tag{4.4.6}$$

Similarly, for $k \geqq k_2$, we have

$$p_{n,k} = p_{n,k_2}\prod_{j=k_2}^{k-1}\frac{p_{n,j+1}}{p_{n,j}} = p_{n,k_2}\prod_{j=k_2}^{k-1}\left[\left(\frac{n-j}{j+1}\right)\frac{p}{q}\right]$$

and thus,

$$p_{n,k} \leqq p_{n,k_2}\left(\frac{1-\varepsilon/q}{1+\varepsilon/p}\right)^{k-k_2} \qquad \text{for } k \geqq k_2 \tag{4.4.7}$$

Thus we obtain

$$\sum_{|k-pn|>n\varepsilon} p_{n,k} \leqq p_{n,k_1}\left(\sum_{j=0}^{\infty}\left(\frac{1-\varepsilon/p}{1+\varepsilon/q}\right)^j\right) + p_{n,k_2}\left(\sum_{j=0}^{\infty}\left(\frac{1-\varepsilon/q}{1+\varepsilon/p}\right)^j\right) \tag{4.4.8}$$

On the other hand,

$$[n\varepsilon]p_{n,k_1} \leqq \sum_{k=0}^{n} p_{n,k} = 1 \tag{4.4.9}$$

and

$$[n\varepsilon]p_{n,k_2} \leqq \sum_{k=0}^{n} p_{n,k} = 1 \tag{4.4.10}$$

Thus we get

$$\sum_{|k-np|>n\varepsilon} p_{n,k} \leqq \frac{2pq+\varepsilon}{[n\varepsilon]\varepsilon} \tag{4.4.11}$$

This proves (4.4.4). ■

Remark 1: Theorem 4.4.1 shows that in a sequence of independent repetitions of an experiment the relative frequency of any event A tends in probability to $P(A)$. This gives an exact meaning to the vague statement, from which we started describing the intuitive background of the notion of probability, that the relative frequency of an event is "in the long run usually near to its probability."

Remark 2: The usual way to prove Theorem 4.4.1 is by Čebyshev's inequality. Recalling that one has $E(\nu_n) = np$ and $D(\nu_n) = \sqrt{npq}$ (see Example 3.5.1), it follows from this inequality (see Theorem 2.11.1) that

$$P(|\nu_n - np| > n\varepsilon) \leqq \frac{pq}{n\varepsilon^2} \tag{4.4.12}$$

Notice that Bernoulli's proof gives essentially the same estimate as Čebyshev's inequality. However, Čebyshev's inequality gives somewhat more, namely that ν_n/n tends to p strongly in $L_2(S)$, and it can be used to prove the following more general result:

THEOREM 4.4.2. *Let $\xi_1, \xi_2, \ldots, \xi_n, \ldots$ be a sequence of random variables having zero expectation and finite variances $D^2(\xi_n) = D_n^2$ and being pairwise uncorrelated. Put $\zeta_n = \xi_1 + \xi_2 + \cdots + \xi_n$ and suppose*

$$\lim_{n \to +\infty} \frac{1}{n^2} \sum_{k=1}^{n} D_k^2 = 0$$

Then ζ_n/n tends to 0 strongly in L_2 and thus

$$\zeta_n/n \Rightarrow 0 \tag{4.4.13}$$

Remark: The statement (4.4.13) is usually expressed as follows: *The weak law of large numbers applies to the random variables ξ_n.*

Proof. According to the linearity of the expectation (see Section 2.9) and Theorem 3.5.6, we have $E(\zeta_n) = 0$ and $D^2(\zeta_n) = \sum_{k=1}^{n} D_k^2$. Thus we get, by Čebyshev's inequality,

$$P\left(\frac{|\zeta_n|}{n} > \varepsilon\right) \leqq \frac{D^2(\zeta_n)}{n^2\varepsilon^2} = \frac{1}{n^2\varepsilon^2} \sum_{k=1}^{n} D_k^2 \tag{4.4.14}$$

which proves Theorem 4.4.2. ■

Remark: Notice that in the special case when ξ_k takes on the values 1 and 0 with probabilities p and q (where $q = 1 - p$) and the ξ_k are not only uncorrelated but mutually independent, the statement of Theorem 4.4.2 reduces to that of Theorem 4.4.1.

We shall now prove the following:

THEOREM 4.4.3. *Let $\xi_1, \xi_2, \ldots, \xi_n, \ldots$ be independent and identically distributed random variables having finite expectation $E = E(\xi_k)$ and put $\zeta_n = \xi_1 + \xi_2 + \cdots + \xi_n$. Then $\zeta_n/n \Rightarrow E$.*

Proof. If the variance of the random variables ξ_k exists, then the statement of Theorem 4.4.3 follows from Theorem 4.4.2. The general case could also be reduced to Theorem 4.4.2 by "truncating" the random variables ξ_k, i.e., by applying Theorem 4.4.2 to the variables ξ_k^* defined as follows: $\xi_k^* = \xi_k$ if $|\xi_k - E| \leqq K$ and $\xi_k^* = 0$, otherwise. We prefer, however, to give another proof, by the method of characteristic functions. Let $\varphi(t)$ be the characteristic function of the common distribution of the random variables ξ_k. Then the characteristic function of ζ_n/n is $\varphi(t/n)^n$. Now it follows from our suppositions that $\varphi'(0)$ exists and $\varphi'(0) = iE$. Thus we get*

$$\lim_{n\to+\infty} \varphi\left(\frac{t}{n}\right)^n = \lim_{n\to+\infty} \left(1 + \frac{itE}{n} + o\left(\frac{1}{n}\right)\right)^n = e^{itE}$$

This means that the distribution of ζ_n/n tends to the (degenerate) distribution of the constant E, i.e., (see the remark to Theorem 4.2.1) that ζ_n/n tends in probability to E. ■

Remark: In the above proof we have deduced $\zeta_n/n \Rightarrow E$ from $\varphi'(0) = iE$. However, $\varphi'(0)$ may also exist in such cases when the expectation does not exist (see Example 4.4.1). The above proof also shows that the condition that $\varphi'(0)$ exists and is equal to iE is necessary for $\zeta_n/n \Rightarrow E$, because if $\varphi(t/n)^n \to e^{itE}$, then $\lim_{n\to\infty} \ln \varphi(t/n)/(t/n) = iE$, i.e., $(\ln \varphi(t))' = \varphi'(t)/\varphi(t)$ exists for $t = 0$ and so does $\varphi'(0)$. Thus we have proved the following result (see Ehrenfeucht and Fisz [46]):

THEOREM 4.4.4. *Let $\xi_n (n = 1, 2, \ldots)$ be a sequence of independent and identically distributed random variables and put $\zeta_n = \xi_1 + \xi_2 + \cdots + \xi_n$. Then we have $\zeta_n/n \Rightarrow E$ where E is a constant if and only if, denoting by $\varphi(t)$ the characteristic function of the random variables ξ_k, $\varphi'(0)$ exists and is equal to iE.*

Example 4.4.1. Let ξ be a random variable having an absolutely continuous distribution with the density function

$$f(x) = \begin{cases} \dfrac{(\log|x| + 1)e}{2x^2 \log^2|x|} & \text{for } |x| > e \\ 0 & \text{for } |x| \leqq e \end{cases}$$

*Here and in what follows we use the usual notation $o(\psi(n))$ (due to Landau) for a term which, when divided by $\psi(n)$, tends to 0 for $n \to +\infty$; especially, $o(1)$ denotes a term tending to zero for $n \to +\infty$.

(It is easy to see that $f(x)$ is a density function because $(\log x + 1)/(x^2 \log^2 x) = (1 - 1/x \log x)'$ and thus $\int_e^{+\infty} (\log x + 1)/(x^2 \log^2 x) = 1/e$).

Then $E(\xi)$ does not exist, however,

$$\frac{1 - \varphi(t)}{t} = \frac{e}{t} \int_e^{+\infty} \frac{(1 - \cos xt)(\log x + 1)}{x^2 \log^2 x} \, dx = o(1) \qquad \text{for } t \to 0$$

and therefore $\varphi'(0) = 0$.

Thus if ξ_n $(n = 1, 2, \ldots)$ are independent random variables, each having the same distribution with the density function $f(x)$ defined above, and $\zeta_n = \xi_1 + \cdots + \xi_n$, then by Theorem 4.4.4 we have $\zeta_n/n \Rightarrow 0$.

Remark: It is easy to see that the same holds if the distribution function $F(x)$ of the independent and identically distributed random variables ξ_k satisfies the following two conditions:

$$\text{(A)} \quad \lim_{A \to +\infty} \int_{-A}^{+A} x \, dF(x) = 0 \qquad \text{(B)} \quad \lim_{A \to +\infty} A(F(-A) + 1 - F(A)) = 0$$

From Theorem 4.4.2 we can deduce a much sharper theorem, under the restriction that ξ_n is bounded, called the *strong law of large numbers*.

THEOREM 4.4.5. *Let $\xi_n(n = 1, 2, \ldots)$ be a sequence of pairwise independent, uniformly bounded random variables such that $E(\xi_n) = 0$. Let us put $\zeta_n = \xi_1 + \xi_2 + \cdots + \xi_n$. Then ζ_n/n tends almost surely to 0.*

Proof. If $|\xi_n| \leq C$, then $D_n^2 = D^2(\xi_n) \leq C^2$. It follows from (4.4.14) that the series $\sum_{n=1}^{+\infty} P(|\zeta_{n^2}|/n^2 > \varepsilon)$ is convergent for every $\varepsilon > 0$. Using the Borel-Cantelli lemma (see Problem P.3.2) it follows that almost surely $|\zeta_{n^2}|/n^2 > \varepsilon$ holds only for a finite number of values of n, i.e., if $B(\varepsilon)$ denotes the set of those $\omega \in \Omega$ for which $|\zeta_{n^2}(\omega)|/n^2 > \varepsilon$ holds for infinitely many values of n, then $P(B(\varepsilon)) = 0$. Thus if $B = \sum_{k=1}^{+\infty} B(1/k)$, we have $P(B) = 0$. Now if $\omega \in \bar{B}$, then for every k we have $|\zeta_{n^2}|/n^2 \leq 1/k$ for $n \geq n_k$; thus we have

$$\lim_{n \to +\infty} \frac{\zeta_{n^2}(\omega)}{n^2} = 0 \qquad \text{for } \omega \in \bar{B} \tag{4.4.15}$$

Now if $n^2 < N < (n+1)^2$, we have $N - n^2 \leq 2n$, and thus if $|\xi_k| \leq C$, we obtain

$$\frac{|\zeta_N|}{N} \leq \frac{|\zeta_{n^2}|}{n^2} + \frac{2C}{n}$$

Thus we get

$$\lim_{N \to +\infty} \frac{\zeta_N(\omega)}{N} = 0 \qquad \text{for } \omega \in \bar{B} \tag{4.4.16}$$

As $P(\bar{B}) = 1$, Theorem 4.4.5 is proved. ■

Remark: If the random variables ξ_n are identically distributed and bounded, the conditions of Theorem 4.4.5 are fulfilled for the variables $\xi_n^* = \xi_n - E$ where $E = E(\xi_n)$. Thus it follows that in this case $(\xi_1 + \cdots + \xi_n)/n$ tends, for $n \to +\infty$, almost surely to E. As a particular case of Theorem 4.4.5 we obtain that *the relative frequency of an event A in a sequence of independent repetitions of an experiment tends almost surely to the probability $P(A)$ of the event.*

The estimate (4.4.12) given by the Čebyshev inequality for the probability $P(|\zeta_n|/n > \varepsilon)$ is rather weak. A much better estimate can be obtained from Bernstein's inequality (Theorem 2.11.3), from which we obtain, for $0 < \varepsilon < \min(p, q)$,

$$P\left(\left|\frac{\nu_n}{n} - p\right| > \varepsilon\right) \leqq [\alpha(p, q, \varepsilon)]^n + [\alpha(q, p, \varepsilon)]^n \tag{4.4.17}$$

where

$$\alpha(p, q, \varepsilon) = \frac{1}{(1 + \varepsilon/p)^{p+\varepsilon}(1 - \varepsilon/q)^{q-\varepsilon}} \leqq e^{-\varepsilon^2/2pq + \varepsilon^3/2p^2}$$

and $\alpha(q, p, \varepsilon)$ is obtained by interchanging p and q in the above formula. Thus one has

$$P\left(\left|\frac{\nu_n}{n} - p\right| > \varepsilon\right) \leqq 2e^{-n\varepsilon^2(1-A\varepsilon)/2pq} \tag{4.4.18}$$

if $\varepsilon < 1/A$, where $A = \max(p/q, q/p)$.

Especially for $\varepsilon < 1/2A$, we have

$$P\left(\left|\frac{\nu_n}{n} - p\right| > \varepsilon\right) \leqq 2e^{-n\varepsilon^2/4pq}$$

Using the Borel-Cantelli lemma (as in the first step of the proof of Theorem 4.4.5) we can immediately deduce from (4.4.18) that ν_n/n tends almost surely to p. The same argument can be applied for arbitrary independent and identically distributed random variables ξ_n, and it follows that the statement of Theorem 4.3.5 remains valid, if the condition that the variables ξ_n should be bounded is replaced by the somewhat weaker condition that the Laplace transform of the variables ξ_n exists in some neighborhood of the point $s = 0$. Still much more is true; the following theorem, called *Kolmogoroff's strong law of large numbers*, is valid:

THEOREM 4.4.6. *Let ξ_n $(n = 1, 2, \ldots)$ be a sequence of independent and identically distributed random variables having finite expectation $E(\xi_n) = E$. Let us put $\zeta_n = \xi_1 + \xi_2 + \cdots + \xi_n$. Then ζ_n/n tends almost surely to E. Conversely, if the ξ_k are independent and identically distributed random variables* *and ζ_n/n tends almost surely to a constant E, then the expectation of the random variables ξ_k exists and is equal to E.*

Proof. The first assertion of Theorem 4.4.6 can be deduced from the following important result, called *Birkhoff's ergodic theorem:*

THEOREM 4.4.7. *Let $S=(\Omega, \mathscr{A}, P)$ be a probability space and let T be a measure-preserving transformation of S. Let $f=f(\omega)$ be a random variable on S with finite expectation. Then the limit*

$$\lim_{N\to+\infty} \frac{1}{N}\sum_{k=0}^{N-1} f(T^k\omega)=f^*(\omega)$$

exists almost surely, and one has $f^*(T\omega)=f^*(\omega)$ a.s. If T is ergodic, then $f^*(\omega)$ is (almost surely) equal to the constant $E(f)$.

The proof of Theorem 4.4.7 can be found in Riesz and Sz. Nagy [16], Halmos [58], Billingsley [59], and Jacobs [60]. For an important recent simplification of the proof see Garsia [17].

To deduce the first assertion of Theorem 4.4.6 from Theorem 4.4.7 it is sufficient to point out that, according to the remark made at the end of Section 4.2, we can replace the sequence ξ_n by another isomorphic sequence. Now according to Theorem 3.4.2, one can construct a sequence ξ_n^* isomorphic to ξ_n in a product space such that the variables ξ_n^* are obtained from ξ_1^* by the shift transformation T and $\xi_1^*(\omega)$, where $\omega=(\omega_1, \omega_2, \ldots)$, depends only on ω_1. As T is evidently mixing (see Section 3.7) and therefore ergodic, it follows that ζ_n/n tends almost surely to E. For another proof of the first assertion of Theorem 4.4.6, based on Kolmogoroff's inequality, see Section 5.3.

The second statement of Theorem 4.4.6 is a consequence of the Borel-Cantelli lemma (Problem P.3.2). As a matter of fact, if $\zeta_n/n\to E$ almost surely, then putting $\xi_k^*=\xi_k-E$, we get that almost surely $\xi_k^*/k\to 0$ for $k\to+\infty$. This means that with probability 1 the inequality $|\xi_k^*|>k$ is satisfied only for a finite number of values of k. Thus, by the Borel-Cantelli lemma, the series $\sum P(|\xi_k^*|>k)$ is convergent. Now let $F(x)$ denote the distribution function of the random variables ξ_k^*; it follows that the series

$$\sum_{n=1}^{+\infty}[F(-n)+(1-F(n))]$$

is convergent. As clearly

$$\int_{-n}^{+n}|x|dF(x)\leqq\sum_{k=1}^{n}[F(-k)+(1-F(k))]$$

it follows that the integral $\int_{-\infty}^{+\infty}|x|dF(x)$ exists. ■

Remark: Theorem 4.4.6 fully justifies the notion of expectation; it shows that the relation between random variables and their expectations is essentially the same as between events and their probabilities.

Example 4.4.2. Let S be the Lebesgue probability space and let the random variable $\varepsilon_n(x)$ be the nth digit in the decimal expansion of x, i.e., we put

$$x = \sum_{n=1}^{+\infty} \frac{\varepsilon_n(x)}{10^n} \qquad (0 \leqq x \leqq 1) \tag{4.4.19}$$

where the possible values of $\varepsilon_n(x)$ are the integers 0, 1, ..., 9. Put for each r

$$\delta_r(j) = \begin{cases} 1 & \text{if } j = r \\ 0 & \text{if } j \neq r \end{cases} \tag{4.4.20}$$

Then for each fixed r $(0 \leqq r \leqq 9)$ the random variables $\delta_r(\varepsilon_n(x))$ $(n = 1, 2, \ldots)$ are independent and identically distributed, and take on the values 1 and 0 with the corresponding probabilities 1/10 and 9/10. It follows from Theorem 4.4.5 (applied to the variables $\delta_r(\varepsilon_n(x)) - 1/10$) that for all x in the interval [0, 1] except if x belongs to a set of Lebesgue measure 0, one has

$$\lim_{N \to +\infty} \frac{\sum_{n=1}^{N} \delta_r(\varepsilon_n(x))}{N} = \frac{1}{10} \qquad (r = 0, 1, \ldots, 9)$$

In other words, the relative frequency of any one of the numbers 0, 1, ..., 9 among the first n digits of the decimal expansion (4.4.19) of the real number x, tends for almost all x to 1/10.

By considering the independent random variables $\delta_{r_1}(\varepsilon_{2n-1}(x))\delta_{r_2}(\varepsilon_{2n}(x))$ $(n = 1, 2, \ldots)$ and also the independent random variables

$$\delta_{r_1}(\varepsilon_{2n}(x))\delta_{r_2}(\varepsilon_{2n+1}(x)) \qquad (n = 1, 2, \ldots)$$

it follows that the relative frequency of the ordered pair of numbers (r_1, r_2) among the pairs of consecutive digits $(\varepsilon_n(x), \varepsilon_{n+1}(x))$ $(n = 1, 2, \ldots, N)$ tends for $N \to +\infty$ and, for almost all x, to $1/10^2$. Similarly, one obtains that for any ordered k-tuple of integers $(r_1, r_2, \ldots, r_k)$ $(0 \leqq r_j \leqq 9; j = 1, 2, \ldots, k)$ the relative frequency of the ordered k-tuple $(r_1, r_2, \ldots, r_k)$ among the k-tuples of consecutive digits $(\varepsilon_{n+1}(x), \ldots, \varepsilon_{n+k}(x))$ $(n = 1, 2, \ldots, N)$ tends (for $N \to +\infty$) for almost all x to $1/10^k$ for $k = 1, 2, \ldots$. As the set of all possible k-tuples $(r_1, \ldots, r_k)$ $(k = 1, 2, \ldots)$ is denumerable, the union of the exceptional sets belonging to these k-tuples is also of measure 0. Thus we get that for almost all x in [0, 1] each k-tuple occurs in the decimal expansion of x with the same limiting relative frequency $1/10^k$ for $k = 1, 2, \ldots$; in Borel's original terminology, *the decimal expansion of almost all real numbers is normal.*

This implies, e.g., that in the decimal expansion of almost all real numbers every one of the ten digits occurs infinitely often; moreover, every possible block of digits of arbitrary length also occurs infinitely often. Of course, the same holds for the q-adic expansion of x, instead of the decimal expansion, for every integer $q \geqq 2$.

Using again that the union of denumerably many sets of measure 0 is also of measure 0, we obtain that *the q-adic expansion of almost all real numbers is normal simultaneously for every $q \geqq 2$.*

Example 4.4.3. Let us define the measure $P^{(p)}$ $(0 < p < 1)$ on the Borel subsets of the interval $[0, 1]$ as follows: If $[a, b)$ denotes the interval $a \leqq x < b$, and r is a binary rational number such that $r = \sum_{k=1}^{n} \varepsilon_k/2^k$, put

$$P^{(p)}\left(\left[r, r + \frac{1}{2^n}\right)\right) = p^{\sum_{k=1}^{n}\varepsilon_k} \cdot q^{n - \sum_{k=1}^{n}\varepsilon_k}$$

where $q = 1 - p$. The measure $P^{(p)}$ can be uniquely extended to the σ-algebra $\mathscr{B}$ of all Borel subsets of $\Omega = [0, 1]$. (Compare with Example 3.4.1.) If

$$x = \sum_{n=1}^{\infty} \frac{\varepsilon_n(x)}{2^n} \qquad (\varepsilon_n(x) = 0 \text{ or } 1)$$

is the binary expansion of x, the random variables $\varepsilon_n(x)$ $(n = 1, 2, \ldots)$ are independent and identically distributed in the probability space $S_p = (\Omega, \mathscr{B}, P^{(p)})$, and one has $P^{(p)}(\varepsilon_n(x) = 1) = p$, and $P^{(p)}(\varepsilon_n(x) = 0) = 1 - p$. Thus by Theorem 4.4.5, if C_p denotes the set of those real numbers x for which

$$\lim_{N \to +\infty} \frac{\sum_{n=1}^{N} \varepsilon_n(x)}{N} = p$$

then $P^{(p)}(C_p) = 1$. Now clearly, if $p_1 \neq p_2$, the sets C_{p_1} and C_{p_2} are disjoint. Thus if $0 < p_1 < 1, 0 < p_2 < 1$, and $p_2 \neq p_1$, the measures $P^{(p_1)}$ and $P^{(p_2)}$ have the property that there exist disjoint sets C_{p_1} and C_{p_2} such that $P^{(p_1)}(C_{p_1}) = 1$ and $P^{(p_2)}(C_{p_2}) = 1$, and thus $P^{(p_1)}(C_{p_2}) = 0$ and $P^{(p_2)}(C_{p_1}) = 0$; such pairs of measures are called *orthogonal*. Thus we have constructed on the Borel subsets of the interval $[0, 1]$, a family of pairwise orthogonal measures having the power of the continuum. Clearly the set $C_{0,5}$ contains all numbers x, the binary expansion of which is normal. Thus the set of numbers, the binary expansion of which is not normal, contains a continuum of disjoint sets C_p $(0 < p < 1, p \neq 1/2)$, each of which has the measure one with respect to the corresponding measure $P^{(p)}$. Notice that each of the sets C_p is everywhere dense in $[0, 1]$, further that the distribution function $F_p(x) = P^{(p)}([0, x))$ if for $p \neq 1/2$ strictly increasing and continuous in $[0, 1]$ and its derivative $F_p'(x)$ is almost everywhere equal to 0; i.e., it is a *singular function*. (Of course, $F_{1/2}(x) = x$ for $0 \leqq x \leqq 1$.)

We shall deal with the generalization of the laws of large numbers to certain sequneces of dependent random variables in Chapter 5.

4.5 APPROXIMATIONS TO THE BINOMIAL AND MULTINOMIAL DISTRIBUTIONS

The method of Bernoulli, used in proving Bernoulli's law of large numbers (Theorem 4.4.1) leads in a straightforward way to the following classical result, called the *Moivre-Laplace theorem* (see [18] and [19]):

THEOREM 4.5.1. *Let us consider the binomial distribution*

$$\left\{\binom{n}{k} p^k q^{n-k};\ 0 \leqq k \leqq n\right\}$$

of order n and parameter p $(0 < p < 1; q = 1 - p)$. *The kth term of the binomial distribution can be approximated as follows:**

$$\binom{n}{k} p^k q^{n-k} \approx \frac{e^{-(k-np)^2/2npq}}{\sqrt{2\pi npq}} \qquad \text{if } |k - np| = o(n^{2/3}) \tag{4.5.1}$$

and if $-\infty < a < b < +\infty$, *we have*

$$\lim_{n\to+\infty} \sum_{a \leqq \frac{k-np}{\sqrt{npq}} \leqq b} \binom{n}{k} p^k q^{n-k} = \frac{1}{\sqrt{2\pi}} \int_a^b e^{-x^2/2}\, dx \tag{4.5.2}$$

Proof. Let us put, as in Section 4.4,

$$p_{n,k} = \binom{n}{k} p^k q^{n-k} \qquad (0 \leqq k \leqq n) \tag{4.5.3}$$

Then we have, putting $k_0 = [(n+1)p]$ (see (4.4.5)) for $k > k_0$,

$$p_{n,k} = p_{n,k_0} \prod_{j=1}^{k-k_0} \left(\frac{1 - j/nq - \vartheta_1/nq}{1 + j/np + \vartheta_2/np}\right) \tag{4.5.4}$$

where $|\vartheta_1| \leqq 1$ and $|\vartheta_2| \leqq 2$.

Similarly, for $k < k_0$, we get

$$p_{n,k} = p_{n,k_0} \prod_{j=1}^{k_0-k} \left(\frac{1 - j/np + \vartheta_2/np}{1 + j/nq + \vartheta_1/nq}\right) \tag{4.5.4'}$$

Thus it follows, using the inequality $e^{x/(1+x)} \leqq 1 + x \leqq e^x$ (valid for $|x| < 1$) and the formula $\sum_{j=1}^{N} j = N(N+1)/2$, that for $|k - np| \leqq n^{2/3}\varepsilon_n$ (where $\varepsilon_n \to 0$), we have†

$$p_{n,k} = p_{n,k_0} \cdot e^{-(k-np)^2/2npq} \cdot (1 + O(\varepsilon_n)) \tag{4.5.5}$$

* The sign $\approx$ in (4.5.1) means that the ratio of the quantities on the left- and right-hand sides of (4.5.1) tends to 1 for $n \to +\infty$, uniformly in k, if $|k - np| \leqq n^{2/3}\varepsilon_n$ where $\lim_{n\to+\infty} \varepsilon_n = 0$.

† Here and in what follows we use the notation $O(\Psi(n))$ (due to Landau) for a term which, when divided by $\Psi(n)$, remains bounded if $n \to +\infty$; especially, $O(1)$ denotes a bounded term.

Thus it follows that for $a < b$,

$$\sum_{a \leqq \frac{k-np}{\sqrt{npq}} \leqq b} p_{n,k} = p_{n,k_0}\left(\sum_{a \leqq \frac{k-np}{\sqrt{npq}} \leqq b} e^{-(k-np)^2/2npq}\right)(1 + O(\varepsilon_n)) \tag{4.5.6}$$

Now we have

$$\lim_{n\to+\infty} \sum_{a \leqq \frac{k-np}{\sqrt{npq}} \leqq b} \frac{1}{\sqrt{npq}} e^{-(k-np)^2/2npq} = \int_a^b e^{-x^2/2}\, dx \tag{4.5.7}$$

as the sum on the left-hand side is a Riemann sum of the integral on the right-hand side. Thus it follows that

$$\sum_{a \leqq \frac{k-np}{\sqrt{npq}} \leqq b} p_{n,k} \approx p_{n,k_0}\sqrt{npq} \int_a^b e^{-x^2/2}\, dx \tag{4.5.8}$$

Now, according to Theorem 4.4.2, we have

$$\sum_{\frac{|k-np|}{\sqrt{npq}} \leqq c} p_{n,k} \geqq 1 - \frac{1}{c^2} \tag{4.5.9}$$

On the other hand, it is well known* from elementary analysis that

$$\int_{-\infty}^{+\infty} e^{-x^2/2} dx = \sqrt{2\pi} \tag{4.5.10}$$

It follows from (4.5.8), (4.5.9) and (4.5.10) that

$$\lim_{n\to+\infty} p_{n,k_0}\cdot\sqrt{npq} = \frac{1}{\sqrt{2\pi}} \tag{4.5.11}$$

Thus we get from (4.5.5)

$$p_{n,k} \approx \frac{e^{-(k-np)^2/2npq}}{\sqrt{2\pi npq}} \tag{4.5.12}$$

for $|k - np| \leqq \varepsilon_n n^{2/3}$, $\varepsilon_n \to 0$, and from (4.5.8)

$$\lim_{n\to+\infty} \sum_{a \leqq \frac{k-np}{\sqrt{npq}} \leqq b} p_{n,k} = \frac{1}{\sqrt{2\pi}} \int_a^b e^{-x^2/2} dx \tag{4.5.13}$$

Thus, both statements of Theorem 4.5.1 are proved. ∎

* A simple proof of (4.5.10) is as follows:

$$\left(\int_{-\infty}^{+\infty} e^{-x^2/2} dx\right)^2 = \int_{-\infty}^{+\infty}\int_{-\infty}^{+\infty} e^{-(x^2+y^2)/2} dxdy = 2\pi \int_{-\infty}^{+\infty} re^{-r^2/2} dr = 2\pi$$

Remark: Formula (4.5.1) is called the *local form* and (4.5.2) the *global form* of the Moivre-Laplace theorem.

DEFINITION 4.5.1. *We call the distribution function*

$$\Phi(x) = \frac{1}{\sqrt{2\pi}} \int_{-\infty}^{x} e^{-u^2/2} du \tag{4.5.14}$$

the standard normal distribution function. We shall say that a random variable ξ *is normally distributed with parameters* m *and* $\sigma(-\infty < m < +\infty, \sigma > 0)$ *(abbreviated as: distributed* $N(m, \sigma)$*) if the distribution function of* ξ *is* $\Phi((x-m)/\sigma)$, *i.e., if the distribution function of* $(\xi - m)/\sigma$ *is equal to* $\Phi(x)$.

Remark: Thus, if ξ is distributed $N(m, \sigma)$, its density function is $1/\sigma\ \varphi((x-m)/\sigma)$ where

$$\varphi(x) = \frac{1}{\sqrt{2\pi}} e^{-x^2/2}$$

The probabilistic meaning of the parameters m and σ is as follows: *If a random variable* ξ *is distributed* $N(m, \sigma)$, *then* $E(\xi) = m$ *and* $D^2(\xi) = \sigma^2$. As a matter of fact, one has

$$E(\xi) = \int_{-\infty}^{+\infty} x \cdot \frac{1}{\sigma} \varphi\left(\frac{x-m}{\sigma}\right) dx = \frac{1}{\sqrt{2\pi}} \int_{-\infty}^{+\infty} (m + \sigma u) e^{-u^2/2} du = m$$

and

$$D^2(\xi) = E((\xi - m)^2) = \int_{-\infty}^{+\infty} (x-m)^2 \frac{1}{\sigma} \varphi\left(\frac{x-m}{\sigma}\right) dx = \frac{\sigma^2}{\sqrt{2\pi}} \int_{-\infty}^{+\infty} u^2 e^{-u^2/2} du$$

and we get, by partial integration,

$$\frac{1}{\sqrt{2\pi}} \int_{-\infty}^{+\infty} u^2 e^{-u^2/2} du = \frac{1}{\sqrt{2\pi}} \int_{-\infty}^{+\infty} e^{-u^2/2} du = 1$$

COROLLARY TO THEOREM 4.5.1. *Let* ν_n *denote the frequency of an event* A *in a sequence of independent repetitions of an experiment in which* A *has the probability* p. *Then the distribution function of* $(\nu_n - np)/\sqrt{npq}$ *tends to the standard normal probability distribution, i.e., for each real value of* x *we have, uniformly in* x,

$$\lim_{n \to +\infty} P\left(\frac{\nu_n - np}{\sqrt{npq}} < x\right) = \Phi(x) \tag{4.5.15}$$

where $\Phi(x)$ *is defined by* (4.5.14).

Proof. Equation (4.5.15) follows from (4.5.2) by Theorem 4.3.4. The uniformity of the convergence follows from Theorem 4.3.3. ■

Remarks: It should be mentioned that (4.5.15) evidently contains Bernoulli's law of large numbers, because for any $\delta > 0$ we can find an A such that $\Phi(-A) = 1 - \Phi(A) < \delta/4$, and thus

$$\lim_{n \to +\infty} P\left(\frac{|\nu_n - np|}{\sqrt{npq}} > A\right) \leq \frac{\delta}{2}$$

Thus, if $n \geqq n_1$,

$$P\left(\frac{|\nu_n - np|}{\sqrt{npq}} > A\right) < \delta \tag{4.5.16}$$

On the other hand, if $\varepsilon > 0$ is arbitrary, we have for $n \geqq n_2$, $n\varepsilon > A\sqrt{npq}$. Thus it follows from (4.5.16) that

$$\overline{\lim_{n \to +\infty}} P(|\nu_n - np| > n\varepsilon) < \delta \tag{4.5.17}$$

As (4.5.17) holds for every $\delta > 0$, it follows that $\lim_{n \to +\infty} P(|\nu_n - np| > n\varepsilon) = 0$ for every $\varepsilon > 0$.

However, the Moivre-Laplace theorem gives much more information on the random fluctuations of the relative frequency ν_n/n of an event A in a sequence of n independent observations around the probability p of A: it shows not only that these fluctuations are small, with probability near to 1 if n is large, but that the order of the deviation $\nu_n/n - p$ is $O(1/\sqrt{n})$ and that $(\nu_n - np)/\sqrt{npq}$ is distributed for large n approximately according to the standard normal distribution.

In Section 4.7 we shall prove a general theorem (Theorem 4.7.1) about the convergence of the distribution of sums of independent random variables to the normal distribution (called the "central limit theorem") which contains the corollary to Theorem 4.5.1 as a particular case. It should be noted that (4.5.1) is usually proved by means of Stirling's formula. We have avoided this by using the identity (4.5.10). Our proof follows essentially that of Abraham de Moivre, who proved Stirling's formula in the form $n! \sim (n/e)^n c\sqrt{n}$ in connection with this proof, without, however, determining the value $\sqrt{2\pi}$ of the constant c. (See David [61].)

Another alternative way to prove (4.5.2) (but not (4.5.1)) is by means of the continuity theorem of characteristic functions (Theorem 4.3.7). As a matter of fact, denoting by $\varphi_n(t)$ the characteristic function of the random variable $(\nu_n - np)/\sqrt{npq}$) we have evidently

$$\varphi_n(t) = \sum_{k=0}^{n} \binom{n}{k} p^k q^{n-k} e^{it(k-np)/\sqrt{npq}} = (pe^{itq/\sqrt{npq}} + qe^{-itp/\sqrt{npq}})^n \tag{4.5.18}$$

and thus,

$$\lim_{n\to+\infty} \varphi_n(t) = e^{-t^2/2} \qquad \text{for } -\infty < t < +\infty \tag{4.5.19}$$

On the other hand, one can show (e.g., by countour integration) that

$$\frac{1}{\sqrt{2\pi}} \int_{-\infty}^{+\infty} e^{itx} e^{-x^2/2}\, dx = e^{-t^2/2} \tag{4.5.20}$$

and thus Theorem 4.3.7 leads to (4.5.2).

On the other hand, as we have proved (4.5.2) directly, this furnishes a probabilistic proof of the purely analytic relation (4.5.20).

We shall now deduce from the Moivre-Laplace theorem the following:

LEMMA 4.5.1. *If η_1 and η_2 are independent random variables having normal distributions $N(m_1, \sigma_1)$, and $N(m_2, \sigma_2)$, respectively,* ($\sigma_1 > 0$, $\sigma_2 > 0$), *then $\eta_1 + \eta_2$ is also normally distributed* with distribution $N(m_1 + m_2, \sqrt{\sigma_1^2 + \sigma_2^2})$.*

Proof. Evidently it is sufficient to prove Lemma 4.5.1 for the case $m_1 = m_2 = 0$, $\sigma_1^2 + \sigma_2^2 = 1$.

Let $\xi_1, \xi_2, \ldots, \xi_n, \ldots$ be independent random variables, each taking on the values 1 and 0 with probabilities p and q ($0 < p < 1$; $q = 1 - p$) and put

$$\eta_1(n) = \frac{\sum_{k<n\sigma_1^2} \xi_k - n\sigma_1^2 p}{\sqrt{npq}}$$

$$\eta_2(n) = \frac{\sum_{n\sigma_1^2 \leq k < n} \xi_k - n\sigma_2^2 p}{\sqrt{npq}}$$

Then $\eta_1(n)$ and $\eta_2(n)$ are independent, and by Theorem 4.5.1 their distribution functions tend to the normal distribution functions $\Phi(x/\sigma_1)$ and $\Phi(x/\sigma_2)$, respectively, if $n \to +\infty$.

On the other hand, the distribution function of $\eta_1(n) + \eta_2(n)$ tends to the normal distribution function $\Phi(x)$. Thus by virtue of Theorem 4.3.13, it follows that $\Phi(x/\sigma_1) * \Phi(x/\sigma_2) = \Phi(x)$, which was to be proved. ∎

We now pass to approximating the terms of the binomial distribution in a wider range. We shall prove the following:

THEOREM 4.5.2. *Putting $k/n = f$, for every $\varepsilon > 0$, uniformly for $0 < \varepsilon < f < 1 - \varepsilon$, if $0 < p < 1$ and $q = 1 - p$, one has*

$$\binom{n}{k} p^k q^{n-k} = \frac{2^{-nd(f,\,p)}}{\sqrt{2\pi n f(1-f)}} \left(1 + O\left(\frac{1}{n}\right)\right) \tag{4.5.21}$$

* Once we know that $\eta_1 + \eta_2$ is normally distributed, it follows that it is $N(m_1 + m_2, \sqrt{\sigma_1^2 + \sigma_2^2})$ distributed, because by adding independent random variables the expectations and variances behave additively.

where

$$d(f, p) = f \log_2 \frac{f}{p} + (1-f)\log_2 \frac{1-f}{q} \tag{4.5.22}$$

Remark: In (4.5.22) $d(f, p)$ is equal to the information-theoretical divergence of the two-term distribution $(f, 1-f)$ from the distribution (p, q) (see Section 3.8).

Notice that Theorem 4.5.2 contains the local form (4.5.1) of the Moivre-Laplace theorem as a special case, in view of the fact that

$$\ln 2 \cdot d(p+\delta, p) = \frac{\delta^2}{2pq} + O(\delta^3) \tag{4.5.23}$$

and thus, if $|k-np| = O(\varepsilon_n n^{2/3})$ where $\varepsilon_n \to 0$, one has

$$2^{-nd(f,p)} = e^{-(k-np)^2/2npq}(1+O(\varepsilon_n{}^3)) \tag{4.5.24}$$

while in this range f can be replaced by p in the denominator on the right-hand side of (4.5.21). Theorem 4.5.2 gives insight also in the asymptotic behavior of the terms of the binomial distribution in the range when $k-np$ is of the order of magnitude n^α with $2/3 \leqq \alpha < 1$; as a matter of fact, one obtains from (4.5.21) for instance, for the case when $k-np = O(n^\alpha)$ with $2/3 \leqq \alpha < 3/4$, that

$$\binom{n}{k} p^k q^{n-k} = \frac{1}{\sqrt{2\pi k\left(1-\frac{k}{n}\right)}} e^{-(k-np)^2/2npq + (k-np)^3(q-p)/6n^2p^2q^2} \cdot \left(1 + O\left(\frac{1}{n^{3-4\alpha}}\right)\right)$$

Taking into account that

$$\sqrt{k\left(1-\frac{k}{n}\right)} = \sqrt{npq}\left(1 + \frac{(k-np)(q-p)}{2npq} + O\left(\frac{(k-np)^2}{n^2}\right)\right)$$

it follows that, putting $x = (k-np)/\sqrt{npq}$,

$$\binom{n}{k} p^k q^{n-k} = \frac{1}{\sqrt{2\pi npq}} \exp\left(-\frac{x^2}{2} + \frac{(x^3-3x)(q-p)}{6\sqrt{npq}}\right) \cdot \left(1 + O\left(\frac{x^4}{n}\right)\right) \tag{4.5.25}$$

(4.5.25) shows that for the special case $p = q = 1/2$ the local form (4.5.1) of the Moivre-Laplace theorem remains valid for $|k-np| = O(n^\alpha)$ with $\alpha < 3/4$, but this is not true if $p \neq q$.

Proof. By using Stirling's formula $n! = (n/e)^n \sqrt{2\pi n}\, e^{\vartheta_n/12n}$ where $0 < \vartheta_n < 1$, we get that if $0 < \varepsilon < k/n < 1-\varepsilon$ where $\varepsilon > 0$, one has, putting $f = k/n$,

$$\binom{n}{k} = \frac{1}{\sqrt{2\pi nf(1-f)}} \cdot \frac{1}{[f^f(1-f)^{1-f}]^n}\left(1 + O\left(\frac{1}{n}\right)\right) \tag{4.5.26}$$

from which (4.5.18) immediately follows. ∎

Remark 1: This proof of (4.5.21) reduces, if $|k-np| = O(n^\alpha)$ with $\alpha < 2/3$, to the traditional proof of (4.5.1).

Remark 2: The asymptotic evaluation of the terms of the binomial distribution "far" from its mean value np (i.e., for such k that $(k-np)^2$ is of much higher order of magnitude than the variance npq of the binomial distribution) is a special instance of the general theory of "big deviations" (see Ibrahimov and Linnik [4]).

By means of (4.5.25), using a slightly more careful analysis, we can prove the following improvement of (4.5.15):

THEOREM 4.5.3. *One has uniformly in x for $-\infty < x < +\infty$*

$$\sum_{k<np+x\sqrt{npq}} \binom{n}{k} p^k q^{n-k} = \Phi(x) + O\left(\frac{1}{\sqrt{n}}\right) \tag{4.5.27}$$

Proof. From (4.5.25), putting $x_k = (k-np)/\sqrt{npq}$ and taking into account that

$$\left|\int_a^b f(x)dx - f(a)(b-a)\right| \leqq {}^1\!/_2 (b-a)^2 \max_{a \leqq x < b} |f'(x)|$$

it follows that

$$\sum_{y \leqq \frac{k-np}{\sqrt{npq}} \leqq x} \binom{n}{k} p^k q^{n-k} = \Phi(x) - \Phi(y) + O\left(\frac{1}{\sqrt{n}}\right) + O\left(\frac{x^4+y^4}{n}\right) \tag{4.5.28}$$

On the other hand, by the inequality of Bernstein (Theorem 2.11.3), we get (see (4.4.18))

$$\sum_{|k-np|>A\sqrt{npq}} \binom{n}{k} p^k q^{n-k} \leqq 2e^{-A^2/4} \qquad \text{if} \qquad A < \frac{1}{2}\sqrt{\frac{n}{pq}} \tag{4.5.29}$$

We get for $x > 0$, by partial integration,

$$1 - \Phi(x) = \frac{1}{\sqrt{2\pi}} \int_x^{+\infty} e^{-u^2/2}\, du = \frac{1}{\sqrt{2\pi}\, x} e^{-x^2/2} - \frac{1}{\sqrt{2\pi}} \int_x^\infty \frac{e^{-u^2/2}}{u^2} du < \frac{e^{-x^2/2}}{\sqrt{2\pi}\, x} \tag{4.5.30}$$

From (4.5.28), (4.5.29) and (4.5.30), we obtain (4.5.27). ∎

Theorem 4.5.2 can be generalized without difficulty for the multinomial distribution (see Exercise E.3.8) instead of the binomial distribution.

THEOREM 4.5.4. *Let us repeat an experiment $\mathscr{E}$ n times independently and let the events $A_1, A_2, \ldots, A_r$ form a partition of the experiment $\mathscr{E}$, and let their probabilities in each experiment be* $P(A_j) = p_j > 0$ $(1 \leqq j \leqq r; \sum_{j=1}^{r} p_j = 1; r \geqq 2)$. *Let ν_j denote the number of those among the experiments which resulted in the occurrence of the event A_j $(1 \leqq j \leqq r)$. Then the joint probability distribution*

$$P(\nu_j = k_j \text{ for } 1 \leqq j \leqq r) = \frac{n!}{k_1!\,k_2!\cdots k_r!} p_1^{k_1} p_2^{k_2} \cdots p_r^{k_r} \tag{4.5.31}$$

(called the multinomial distribution) can be approximated under the condition $f_j = k_j/n > \varepsilon > 0$, *as follows:*

$$\frac{n!}{k_1!k_2!\cdots k_r!} p_1^{k_1} p_2^{k_2} \cdots p_r^{k_r} \approx \frac{1}{(\sqrt{2\pi n})^{r-1}} \frac{2^{-nD(\mathscr{F}_n, \mathscr{P})}}{\sqrt{f_1 f_2 \cdots f_r}} \tag{4.5.32}$$

where $D(\mathscr{F}_n, \mathscr{P})$ is the (information-theoretical) divergence of the r-term probability distribution $\mathscr{F}_n = \{f_1, f_2, \ldots, f_r\}$ from the distribution $\mathscr{P} = \{p_1, p_2, \ldots, p_r\}$, i.e.,

$$D(\mathscr{F}_n, \mathscr{P}) = \sum_{j=1}^{r} f_j \log_2 \frac{f_j}{p_j} \tag{4.5.33}$$

Theorem 4.5.4 is proved by the use of Stirling's formula in the same manner as Theorem 4.5.2.

COROLLARY TO THEOREM 4.5.4. Under the conditions of Theorem 4.5.4 if $k_j - np_j = O(n^{2/3}\varepsilon_n)$ with $\varepsilon_n \to 0$, one has

$$\frac{n!}{k_1!k_2!\cdots k_r!} p_1^{k_1} p_2^{k_2} \cdots p_r^{k_r} \approx \frac{1}{(\sqrt{2\pi n})^{r-1}} \cdot \frac{e^{-1/2n \sum_{j=1}^{r} \frac{(k_j - np_j)^2}{p_j}}}{\sqrt{p_1 p_2 \cdots p_r}} \tag{4.5.34}$$

Proof. To deduce (4.5.34) from (4.5.32) we use the following asymptotic formula: If $\delta_1, \delta_2, \ldots, \delta_r$ are real numbers such that $0 < p_j + \delta_1 < 1$, $\sum_{j=1}^{r} \delta_j = 0$, and $\max_{1 \leqq j \leqq r} |\delta_j| = \delta$, one has

$$\sum_{j=1}^{r} (p_j + \delta_j) \ln \frac{(p_j + \delta_j)}{p_j} = 1/2 \sum_{j=1}^{r} \frac{\delta_j^2}{p_j} + O(\delta^3) \tag{4.5.35}$$

DEFINITION 4.5.1. *If $\mathscr{P} = (p_1, p_2, \ldots, p_r)$ and $\mathscr{Q} = (q_1, q_2, \ldots, q_r)$ are two r-term probability distributions, the χ^2 divergence $\chi^2(\mathscr{Q}, \mathscr{P})$ of $\mathscr{Q}$ and $\mathscr{P}$ is defined by*

$$\chi^2(\mathscr{Q}, \mathscr{P}) = \sum_{j=1}^{r} \frac{(q_j - p_j)^2}{p_j} \tag{4.5.36}$$

Using the terminology of Definition 4.5.1 we may express (4.5.35) roughly as follows: If the distribution $\mathscr{P}$ and $\mathscr{Q}$ are very near to each other, their information-theoretical divergence is asymptotically equal to their χ^2 divergence. The χ^2 divergence plays an important role in mathematical statistics.

Remark: For $r = 2$ (4.5.34) reduces to (4.5.1) because of the identity

$$\frac{(k-np)^2}{p} + \frac{(n-k-nq)^2}{q} = \frac{(k-np)^2}{pq}$$

We shall now introduce the following:

DEFINITION 4.5.2. *The joint distribution of the random variables $\eta_1, \eta_2, \ldots, \eta_r$ is called an r-dimensional normal distribution if its density function is given by*

$$f(x_1, x_2, \ldots, x_r) = \frac{\|A\|^{1/2}}{(\sqrt{2\pi})^{r-1}} e^{-(1/2)A(x)} \tag{4.5.37}$$

where

$$A(x) = \sum_{j=1}^{r} \sum_{k=1}^{r} a_{j,k} x_j x_k \tag{4.5.38}$$

is a positive definite symmetric quadratic form, i.e., $a_{j,k} = a_{k,j}$, $A(x) \geqq 0$ if $x_1, x_2, \ldots, x_r$ are arbitrary real numbers and $A(x) = 0$ only if $x_1 = x_2 = \cdots = x_r = 0$, and $\|A\|$ denotes the determinant of the matrix $A = (a_{j,k})$.

Remark: It is a well known fact (see [20]) that the determinant of a positive definite quadratic form is positive.

From (4.5.30) one can deduce the following corresponding "global" formula:

$$\lim_{n\to+\infty} \sum_{\frac{k_j-np_j}{\sqrt{np_j(1-p_j)}} < x_j (1 \leqq j \leqq r-1)} \frac{n!}{k_1!k_2!\cdots k_r!} p_1^{k_1} p_2^{k_2} \cdots p_r^{k_r}$$
$$= \int_{-\infty}^{x_1} \cdots \int_{-\infty}^{x_{r-1}} f(u_1, \ldots, u_{r-1}) du_1, \ldots du_{r-1} \tag{4.5.39}$$

with

$$f(u_1, \ldots, u_{r-1}) = \frac{\exp(-1/2 \sum_{j=1}^{r-1} \sum_{k=1}^{r-1} a_{j,k} u_j u_k)}{(\sqrt{2\pi})^{r-1} \sqrt{p_1 p_2 \cdots p_r}} \tag{4.5.40}$$

where, putting $\sum_{i=1}^{r-1} p_i = 1 - p_r = s < 1$,

$$a_{jj} = (1-p_j)\left(1 + \frac{p_j}{1-s}\right) \qquad (j = 1, 2, \ldots, r) \tag{4.5.41}$$

and

$$a_{j,k} = \frac{\sqrt{p_j(1-p_j)p_k(1-p_k)}}{(1-s)^2} \qquad \text{for } j \neq k \tag{4.5.42}$$

Thus if ν_j denotes the frequency of the occurrence of the event A_j in a sequence of independent repetitions of an experiment in which the events A_j mutually exclude each other and $P(A_j) = p_j$ $(1 \leqq j \leqq r)$, then the joint distribution of the random variables $(\nu_j - np_j)/\sqrt{np_j(1-p_j)}$ $(1 \leqq j \leqq r-1)$ tends to an $(r-1)$-dimensional normal distribution with density (4.5.40).

4.6 THE POISSON PROCESS

In the previous section we have given approximations for the terms $\binom{n}{k} p^k q^{n-k}$ of the binomial distribution for $n \to +\infty$, while p was kept constant. As regards the asymptotic behavior of the terms of the binomial distribution, if p is not kept constant but is a function of n so that for $n \to +\infty$, we have $p \to 0$ in such a way that np (the expectation of the distribution) tends to a positive constant λ, we get the following result:

THEOREM 4.6.1. *If $p = p(n)$ depends on n so that*

$$\lim_{n \to +\infty} np = \lambda > 0 \tag{4.6.1}$$

and $q = 1 - p$, one has

$$\lim_{n \to +\infty} \binom{n}{k} p^k q^{n-k} = \frac{\lambda^k e^{-\lambda}}{k!} \qquad \text{for } k = 0, 1, 2, \ldots \tag{4.6.2}$$

Proof. To get (4.6.2), we use the evident relation

$$\lim_{n \to +\infty} \frac{\binom{n}{k}}{n^k} = \frac{1}{k!}$$

and the well known relation

$$\lim_{n \to +\infty} (1 + x/n)^n = e^x \quad \blacksquare$$

Remark: Clearly

$$\sum_{k=0}^{+\infty} \frac{\lambda^k e^{-\lambda}}{k!} = e^{-\lambda} e^{\lambda} = 1$$

Thus the limits on the right side of (4.6.2) form a probability distribution.

DEFINITION 4.6.1. *The probability distribution*

$$\frac{\lambda^k e^{-\lambda}}{k!} \qquad (k = 0, 1, 2, \ldots; \lambda > 0) \tag{4.6.3}$$

is called the Poisson distribution with parameter λ.

Thus, in view of Theorem 4.3.8, Theorem 4.6.1 shows that if $n \to +\infty$ and $p \to 0$ so that (4.6.1) holds, the binomial distribution of order n and parameter p tends (weakly) to the Poisson distribution with parameter λ, and putting

$$\Delta_n = \sum_{k=0}^{+\infty} \left| \binom{n}{k} p^k q^{n-k} - \frac{\lambda^k e^{-\lambda}}{k!} \right| \tag{4.6.4}$$

one has

$$\lim_{n \to +\infty} \Delta_n = 0$$

By the same argument used to prove (4.6.2), carried out more carefully, one can show that

$$\Delta_n = O\left(\frac{1}{n}\right) \tag{4.6.5}$$

For every bounded function $f(x)$, (4.6.2) of course implies

$$\lim_{\substack{n \to +\infty \\ np \to \lambda}} \sum_{k=0}^{n} f(k) \binom{n}{k} p^k q^{n-k} = \sum_{k=0}^{+\infty} f(k) \frac{\lambda^k e^{-\lambda}}{k!} \tag{4.6.6}$$

However, (4.6.6) holds for certain unbounded functions too, especially for $f(k) = k$ and $f(k) = k^2$. This can be seen by evaluating the first two moments of the Poisson distribution. If η is a random variable having a Poisson distribution with parameter λ, then the moment-generating function of η can be determined as follows:

$$\psi(s) = E(e^{s\eta}) = \sum_{k=0}^{+\infty} e^{ks} \frac{\lambda^k e^{-\lambda}}{k!} = e^{\lambda(e^s - 1)} \tag{4.6.7}$$

As we have (see Problem P.1.1(g))

$$e^{\lambda(e^s-1)} = 1 + \sum_{k=1}^{+\infty} \lambda^k \sum_{m=k}^{+\infty} \frac{S(m,k)s^m}{m!}$$

we get

$$E(\eta^m) = \sum_{k=1}^{m} S(m,k)\lambda^k \tag{4.6.8}$$

where $S(m,k)$ are the Stirling numbers of the second kind, defined, e.g., by the recursion formula $S(0, 0) = 1$, $S(m, 0) = 0$ for $m \geqq 1$, $S(m + 1, k) = S(m,k - 1) + k \cdot S(m,k)$ for $1 \leqq k \leqq m + 1$, $m = 1, 2, \ldots$.

Thus we have in particular

$$E(\eta) = \lambda \quad \text{and} \quad E(\eta^2) = \lambda + \lambda^2 \tag{4.6.9}$$

and therefore

$$D^2(\eta) = \lambda \tag{4.6.10}$$

Thus *both the expectation and the variance of a Poisson distribution are equal to its parameter.*

On the other hand, the expectation and second moment of the binomial distribution are np and $npq + n^2p^2$ and if (4.6.1) holds, these tend to the expectation λ respective of second moment $\lambda + \lambda^2$.

Note that it follows from (4.6.7) that the characteristic function of the Poisson distribution with parameter λ is given by

$$\varphi(t) = e^{\lambda(e^{it}-1)} \tag{4.6.11}$$

One can interpret Theorem 4.6.1 as follows:* Let us consider a certain event A which occurs repeatedly at certain time instants τ_n $(n = 1, 2, \ldots)$ where $0 < \tau_1 < \tau_2 < \cdots$, depending on chance. To fix our minds let us suppose that the event in question is the disintegration of one of the atoms of a certain radioactive substance. To describe the situation let us set up the following model: We divide the time axis starting with the instant 0 into time intervals having the same length $h > 0$, i.e., into the intervals $[kh, (k + 1)h)$ $(k = 0, 1, 2, \ldots)$. Let us for the moment neglect the probability of more than one radioactive disintegration taking place in a time interval of length h, and suppose that (1) the probability of a disintegration taking place in the time interval $[kh, (k + 1)h)$ is equal to λh, and (2) the events consisting in the occurrence of disintegrations in any number of disjoint time intervals are independent.

This model which we denote by $\mathcal{M}_h$ is only approximately correct but it becomes correct in the limit if $h \to 0$, because if h tends to 0, the effect of excluding the possibility of more than one disintegration taking place in a time interval of length h becomes really negligible.

Let $\zeta(T)$ denote the number of atoms disintegrating in the time interval $(0, T)$. Then if $t > 0$ and $T > 0$, $\zeta(T + t) - \zeta(T)$ denotes the number of atoms disintegrating in the time interval $[T, T + t)$; choosing for h the value $h = t/n$, we have

$$\zeta(T + t) - \zeta(T) = \sum_{k=1}^{n} \left[\zeta\left(\frac{kt}{n} + T\right) - \zeta\left(\frac{(k-1)t}{n} + T\right)\right]$$

where, according to the model $\mathcal{M}_{t/n}$, the random variables $\zeta(kt/n + T) - \zeta((k - 1)t/n + T)$ are independent and each takes on the values 1 and 0 with probability $p = \lambda t/n$ and $q = 1 - \lambda t/n$, respectively. Thus, according to

* The following deduction of the Poisson process is of heuristic character; a precise definition of the Poisson process will be given afterwards (see Definition 4.6.3).

model $\mathscr{M}_{t/n}$, $\zeta(T+t) - \zeta(T)$ possesses a binomial distribution of order n and parameter $p = \lambda t/n$, i.e., denoting probabilities according to model $\mathscr{M}_{t/n}$ by P_n,

$$P_n(\zeta(T+t) - \zeta(T)=k) = \binom{n}{k} p^k q^{n-k} \tag{4.6.12}$$

where $p = \lambda t/n$ and $q = 1 - p$.

According to what has been said, the probability distribution given by (4.6.12) gives only an approximation to the true distribution of $\zeta(T+t) - \zeta(T)$, which is obtained by passing to the limit $n \to +\infty$. By virtue of Theorem 4.6.1, denoting by P the true probability distribution, we get

$$P(\zeta(T+t) - \zeta(T) = k) = \frac{(\lambda t)^k e^{-\lambda t}}{k!} \quad (k = 0, 1, \ldots) \tag{4.6.13}$$

Thus the number of atoms of the radioactive substance considered, disintegrating during the time interval $(T, T+t)$ has a Poisson distribution with parameter (expectation) λt.

Notice that if $t > 0$ and $s > 0$, then the random variables $\zeta(T+t) - \zeta(T)$ and $\zeta(T+1+s) - \zeta(T+t)$ and their sum $\zeta(T+t+s) - \zeta(T)$ all have Poisson distributions with parameters λt, λs and $\lambda(t+s)$. Further, by our assumption, $\zeta(T+t) - \zeta(T)$ and $\zeta(T+t+s) - \zeta(T+t)$ are independent random variables. This is equivalent to the following:

LEMMA 4.6.1. *If ξ and η are independent random variables having Poisson distributions with parameters a and b, respectively, then $\xi + \eta$ also has a Poisson distribution, with parameter $a + b$.*

Lemma 4.6.1 can be verified directly; it also follows from formula (4.6.11) for the characteristic function of the Poisson distribution and Theorem 3.5.10, as

$$e^{a(e^{it}-1)} \cdot e^{b(e^{it}-1)} = e^{(a+b)(e^{it}-1)}$$

Let us denote by τ_k $(k = 1, 2, \ldots)$ the instant in which the kth disintegration (counted from the moment $t = 0$) takes place. According to model $\mathscr{M}_{t/n}$, the probability of the event $m \cdot t/n \leqq \tau_k < (m+1)t/n$ is equal to the probability of the event that in a series of independent repetitions of an experiment the kth occurrence of an event A having the probability $\lambda t/n = p$ occurs at the mth experiment. Thus the negative binomial distribution (see Exercise E.3.7) of order k and parameter p is applicable and we get

$$P_n\left(\frac{mt}{n} \leqq \tau_k < \frac{(m+1)t}{n}\right) = \binom{m-1}{k-1} p^k q^{m-k} \tag{4.6.14}$$

where $p = \lambda t/n$, $q = 1 - p$, and $m = k, k+1, \ldots$.

Now we have evidently, if $m/n \to u$ and $n \to +\infty$,

$$\lim_{\substack{n \to +\infty \\ m/n \to u}} n \cdot \binom{m-1}{k-1} \left(\frac{\lambda t}{n}\right)^k \left(1 - \frac{\lambda t}{n}\right)^{m-k} = \frac{(\lambda t)^k u^{k-1} e^{-\lambda t u}}{(k-1)!} \tag{4.6.15}$$

and the convergence is uniform in u for $\varepsilon \leq u < 1$, for each fixed value of k if $\varepsilon > 0$.

Denoting by P the "true" distribution of τ_k, we have

$$P(\tau_k < t) = \lim_{n \to +\infty} \sum_{m=0}^{n-1} P_n\left(\frac{mt}{n} \leq \tau_k < \frac{(m+1)t}{n}\right) \tag{4.6.16}$$

and thus

$$P(\tau_k < t) = \int_0^t \frac{\lambda^k x^{k-1} e^{-\lambda x}}{(k-1)!} \, dx \tag{4.6.17}$$

Note that according to a well known formula concerning the gamma function,

$$\int_0^{+\infty} \frac{\lambda^k u^{k-1} e^{-\lambda u}}{(k-1)!} \, du = \int_0^{+\infty} \frac{v^{k-1} e^{-v} \, dv}{(k-1)!} = 1$$

Thus the "true" distribution of τ_k is an absolutely continuous distribution with density function

$$f_k(x) = \begin{cases} \dfrac{\lambda^k x^{k-1} e^{-\lambda x}}{(k-1)!} & \text{for } x > 0 \\ 0 & \text{for } x < 0 \end{cases} \tag{4.6.18}$$

DEFINITION 4.6.2. *The absolutely continuous probability distribution with density function $f_k(x)$ given by (4.6.18) is called the Γ distribution of order k and parameter λ. The Γ distribution of order 1, i.e., the distribution having the density function*

$$f_k(x) = \begin{cases} \lambda e^{-\lambda x} & \text{for } x > 0 \\ 0 & \text{for } x < 0 \end{cases} \tag{4.6.19}$$

and thus the distribution function

$$F_1(x) = \begin{cases} 1 - e^{-\lambda x} & \text{for } x \geqq 0 \\ 0 & \text{for } x < 0 \end{cases} \tag{4.6.20}$$

is called the exponential distribution with parameter λ.

If τ_k is distributed according to the Γ distribution of order k and parameter λ, the rth moment of τ_k can be evaluated as follows λ

$$E(\tau_k^r) = \int_0^{+\infty} \frac{\lambda^k x^{k+r-1} e^{-\lambda x} \, dx}{(k-1)!} = \frac{(k+r-1)!}{\lambda^r (k-1)!} \tag{4.6.21}$$

Thus, in particular,

$$E(\tau_k) = k/\lambda \tag{4.6.22}$$

and

$$E(\tau_k^2) = k(k+1)/\lambda^2 \tag{4.6.23}$$

and therefore

$$D^2(\tau_k) = k/\lambda^2 \tag{4.6.24}$$

Especially, *the expectation and variance of the exponential distribution are equal to* $1/\lambda$ *and* $1/\lambda^2$, *respectively.*

An alternative proof of the limit relation (4.6.15) can be obtained as follows: The characteristic function of the negative binomial distribution of order k and parameter p is

$$\varphi_k(u) = \sum_{m=k}^{\infty} \binom{m-1}{k-1} p^k q^{m-k} e^{imu} = \left(\frac{pe^{iu}}{1-qe^{iu}}\right)^k \tag{4.6.25}$$

Thus the characteristic function of the distribution of $t[n\tau_k/t]/n$, according to model $\mathscr{M}_{t/n}$, is

$$\varphi_{n,k}(u) = \left(\frac{(\lambda t/n)e^{iut/n}}{1-(1-\lambda t/n)e^{iut/n}}\right)^k \tag{4.6.26}$$

and we get easily,

$$\lim_{n\to\infty} \varphi_{n,k}(u) = \frac{1}{(1-iu/\lambda)^k} \tag{4.6.27}$$

As

$$\int_0^\infty e^{ixu} \frac{\lambda^k x^{k-1} e^{-\lambda x}}{(k-1)!} dx = \frac{1}{(1-iu/\lambda)^k} \tag{4.6.28}$$

$1/(1-iu/\lambda)^k$ is the characteristic function of the distribution with density function (4.6.18), it follows that τ_k is distributed according to the Γ distribution of order k and parameter λ.

We shall now show the following:

LEMMA 4.6.2. *If* $\xi_1, \xi_2, \ldots, \xi_k$ *are independent random variables, each having the same exponential distribution with the parameter* λ, *then* $\xi_1 + \xi_2 + \cdots + \xi_k$ *has a* Γ *distribution of order* k *and parameter* λ.

Proof. As

$$\int_0^x \frac{\lambda^k (x-y)^{k-1} e^{-\lambda(x-y)}}{(k-1)!} \lambda e^{-\lambda y}\, dy = \frac{\lambda^{k+1} x^k e^{-\lambda x}}{k!}$$

the statement of Lemma 4.6.2 follows by induction on k. ∎

It follows by Exercise 3.7(c) that the random variables $\tau_1, \tau_2 - \tau_1, \ldots, \tau_{k+1} - \tau_k, \ldots$ are independent and identically distributed, i.e., each has an exponential distribution with parameter $\lambda > 0$.

Such a sequence $\{\tau_k\}$ of random variables is called a *Poisson process with density* λ.

DEFINITION 4.6.3. *A sequence* $\{\tau_k\}$ $(k=1, 2, \ldots)$ *of random variables such that* $\tau_1, \tau_2-\tau_1, \ldots, \tau_{k+1}-\tau_k, \ldots$ *are independent and identically distributed, each having an exponential distribution with parameter* $\lambda>0$, *is called a Poisson process with density* λ.

We now prove a theorem, the statement of which is just what one expects on the basis of the preceding heuristic discussion.

THEOREM 4.6.2. *If* $\{\tau_k\}$ $(k=1, 2, \ldots)$ *is a Poisson process with density* $\lambda>0$, I *is an arbitrary finite subinterval of the half line* $x>0$, *and the random variable* $\zeta(I)$ *denotes the number of those of the random variables* τ_k *which fall into the interval* I, *then* $\zeta(I)$ *has a Poisson distribution with parameter* $\lambda|I|$, *where* $|I|$ *denotes the length of the interval* I. *Moreover, if* $I_1, I_2, \ldots, I_r$ *are disjoint intervals, the random variables* $\zeta(I_1), \zeta(I_2), \ldots, \zeta(I_r)$ *are independent.*

Proof. We shall need the formula

$$\sum_{k=1}^{+\infty} f_k(x)=\begin{cases}\lambda & \text{if } x>0\\ 0 & \text{if } x<0\end{cases} \tag{4.6.29}$$

Let us put

$$E(x)=\begin{cases}0 & \text{for } x\leqq 0\\ 1+\lambda x & \text{for } x>0\end{cases} \tag{4.6.30}$$

Then we have

$$\int_0^a e^{\lambda x}\,dE(x)=e^{\lambda a} \qquad \text{for } a>0 \tag{4.6.31}$$

Let $I[a, b)$ denote the interval $[a, b)$ $(0\leqq a<b)$. Then we have, putting $\tau_0=0$,

$$P(\zeta(I[a, b))=0)=\sum_{k=0}^{+\infty} P(\tau_k<a, \tau_{k+1}\geqq b)$$

$$=\int_b^\infty f_1(y)\,dy+\sum_{k=1}^{+\infty}\int_0^a f_k(x)\int_{b-x}^{+\infty} f_1(y)\,dy\,dx=\int_0^a e^{-\lambda(b-x)}\,dE(x)=e^{-\lambda(b-a)} \tag{4.6.32}$$

We have further

$$P(\zeta(I[a, b))=1)=\sum_{k=0}^{+\infty} P(\tau_k<a\leqq\tau_{k+1}<b\leqq\tau_{k+2})$$

$$=\int_a^b f_1(y)\int_{b-y}^{+\infty} f_1(z)\,dz\,dy+\sum_{k=1}^{+\infty}\int_0^a f_k(x)\int_{a-x}^{b-x} f_1(y)\int_{b-y-x}^{+\infty} f_1(z)\,dz\,dy\,dx$$

$$=\lambda(b-a)e^{-\lambda b}\int_0^a e^{\lambda x}\,dE(x)=\lambda(b-a)e^{-\lambda(b-a)} \tag{4.6.33}$$

Finally, for $n \geqq 2$, we get

$$P(\zeta(I[a, b)) = n) = \sum_{k=0}^{+\infty} P(\tau_k < a \leqq \tau_{k+1}, \tau_{k+n} < b \leqq \tau_{k+n+1})$$

$$= \int_0^a \left[\int_{a-x}^{b-x} f_1(y) \int_0^{b-x-y} f_{n-1}(u) \int_{b-x-y-u}^{+\infty} f_1(v)\, dv \right] du\, dy\, dE(x)$$

$$= \frac{[\lambda(b-a)]^n}{n!} e^{-\lambda(b-a)} \tag{4.6.34}$$

It follows from (4.6.32)–(4.6.34) that

$$P(\zeta(I[a, b)) = n) = \frac{[\lambda(b-a)]^n e^{-\lambda(b-a)}}{n!} \tag{4.6.35}$$

holds for $n = 0, 1, 2, \ldots$, i.e., that $\zeta(I)$ has a Poisson distribution with parameter $\lambda|I|$. The independence of $\zeta(I_1), \ldots, \zeta(I_r)$ in case the intervals $I_1, \ldots, I_r$ are disjoint, is proved by the same argument; we leave the obvious details to the reader. ■

It follows from Theorem 4.6.1 and Lemma 4.6.1 that if A is a subset of the half line $0 \leqq x < +\infty$, which is the union of a finite number of disjoint intervals, and $\zeta(A)$ denotes the number of points τ_k of the Poisson process lying in the set A, then $\zeta(A)$ has a Poisson distribution with parameter $\lambda|A|$ where $|A|$ denotes the Lebesgue measure of the set A.

This fact can also be proved directly, without using the independence of the random variables $\zeta(I)$ corresponding to disjoint intervals I, by the same method as was used to prove the statement for the case when I is an interval. On the other hand, it can be shown that the following theorem holds (see Rényi [21]):

THEOREM 4.6.3. *Let $\{\tau_k\}$ $(k = 1, 2, \ldots)$ be any sequence of positive random variables such that $\tau_k < \tau_{k+1}$ $(k \geqq 1)$ and let $\zeta(A)$ denote the number of those τ_k, the value of which lies in the set A; then if $\zeta(A)$ is Poisson distributed with parameter $\lambda|A|$ whenever A is the union of a finite number of disjoint intervals, where $|A|$ denotes the Lebesgue measure of A, then the random variables $\zeta(I_1), \ldots, \zeta(I_r)$ are independent if $I_1, \ldots, I_r$ are disjoint intervals, and thus $\{\tau_k\}$ is a Poisson process.*

Proof. Let $I_1, \ldots, I_r$ be r disjoint intervals. We first show that the events $\zeta(I_j) = 0$ $(j = 1, 2, \ldots, r)$ are independent. This follows simply from the remark that denoting by A the union of the intervals $I_j (1 \leqq j \leqq r)$, by supposition,

$$P(\zeta(I_j) = 0 \text{ for } 1 \leqq j \leqq r) = P(\zeta(A) = 0) = e^{-\lambda|A|} = \prod_{j=1}^{r} e^{-\lambda|I_j|} = \prod_{j=1}^{r} P(\zeta(I_j) = 0) \tag{4.6.36}$$

We have further, by supposition,

$$P(\zeta(I) \geqq 2) = 1 - e^{-\lambda|I|} - \lambda|I|e^{-\lambda|I|} = O(|I|^2) \tag{4.6.37}$$

if $|I| \to 0$. Now let I_1 and I_2 be two disjoint intervals. Then for every natural number N, I_1 is the union of disjoint subintervals $I_{1,k}$ $(1 \leqq k \leqq N)$, each having length $|I_1|/N$, and I_2 is the union of disjoint subintervals $I_{2,k}$ $(1 \leqq k \leqq N)$, each having the length $|I_2|/N$.

Thus we have for every $N \geqq 1$ and every pair n, m of nonnegative integers

$$P(\zeta(I_1) = n, \zeta(I_2) = m) = \sum_{\substack{a_k = 0 \text{ or } 1, \sum_{k=1}^{N} a_k = n \\ b_j = 0 \text{ or } 1, \sum_{j=1}^{N} b_j = m}} P(\zeta(I_{1,k}) = a_k(1 \leqq k \leqq N), \zeta(I_{2,j}) = b_j(1 \leqq j \leqq N) + O(1/N) \tag{4.6.38}$$

As the events $\zeta(I_{1,k}) = 0$, $\zeta(I_{2,j}) = 0$ $(1 \leqq k \leqq N, 1 \leqq j \leqq N)$ are—as we have shown—independent, it follows that if each a_k and each b_j is either 0 or 1, and the conditions of summation in (4.6.38), i.e., the conditions $\sum_{k=1}^{N} a_k = n$ and $\sum_{k=1}^{N} b_j = m$ are fulfilled, we have

$$P(\zeta(I_{1,k}) = a_k, \zeta(I_{2,j}) = b_j, 1 \leqq k \leqq N, 1 \leqq j \leqq N) = \left(\frac{\lambda|I_1|}{N}\right)^n \left(\frac{\lambda|I_2|}{N}\right)^m e^{-\lambda(|I_1|+|I_2|)} \tag{4.6.39}$$

Thus it follows that

$$P(\zeta(I_1) = n, \zeta(I_2) = m) = \binom{N}{n}\binom{N}{m} \frac{(\lambda|I_1|)^n(\lambda|I_2|)^m}{N^{n+m}} e^{-\lambda(|I_1|+|I_2|)} + O\left(\frac{1}{N}\right) \tag{4.6.40}$$

Passing to the limit $N \to +\infty$, it follows that

$$P(\zeta(I_1) = n, \zeta(I_2) = m) = \frac{(\lambda|I_1|)^n e^{-\lambda|I_1|}}{n!} \cdot \frac{(\lambda|I_2|)^m e^{-\lambda|I_2|}}{m!} = P(\zeta(I_1) = n)P(\zeta(I_2) = m) \tag{4.6.41}$$

Thus $\zeta(I_1)$ and $\zeta(I_2)$ are independent. The independence of $\zeta(I_1) \ldots, \zeta(I_r)$ with $r > 2$, if $I_1, \ldots, I_r$ are disjoint intervals, is shown in exactly the same way. Thus Theorem 4.6.3 is proved. ■

Remark: It has been shown by counterexamples by Shepp (see [22]), Moran [23] and Lee [24] that if the supposition in Theorem 4.6.3, that $\zeta(A)$ has a Poisson distribution with parameter $\lambda|A|$, is postulated only for intervals A, then the statement of Theorem 4.6.3 does not remain valid, i.e., the variables $\zeta(I_1), \ldots, \zeta(I_r)$ are not necessarily independent if the intervals $I_1, \ldots, I_r$ are disjoint.

We shall now prove the following characterization of a Poisson process:

THEOREM 4.6.4. *Let τ_k $(k=1, 2, \ldots)$ be a sequence of random variables for which $0 \leqq \tau_1 \leqq \tau_2 \leqq \cdots \leqq \tau_k \leqq \cdots$, such that if, for every interval I of the half line $x \geqq 0$, $\zeta(I)$ denotes the number of values of k for which τ_k lies in I, then* (1) *$\zeta(I)$ has a Poisson distribution with parameter $\lambda|I|$ where $\lambda > 0$ is a constant, and* (2) *if $I_1, \ldots, I_r$ are disjoint intervals, the random variables $\zeta(I_1), \ldots, \zeta(I_r)$ are independent. Then $\{\tau_k\}$ is a Poisson process, i.e., the variables $\tau_1, \tau_2 - \tau_1, \ldots, \tau_{k+1} - \tau_k, \ldots$ are independent and identically distributed, each having an exponential distribution with parameter λ.*

Proof. To prove our theorem it is sufficient to show that the joint distribution of the random variables $\tau_1, \tau_2, \ldots, \tau_n$ can be expressed for every $n \geqq 1$ by means of the joint distribution of some of the random variables $\zeta(I)$ corresponding to disjoint intervals I. This, however, can be seen from the following evident formula: Putting

$$G(x_1, x_2, \ldots, x_n) = P(\tau_1 > x_1, \tau_2 > x_2, \ldots, \tau_n > x_n) \tag{4.6.42}$$

one has

$$G(x_1, x_2, \ldots, x_n) = \sum_{\substack{\varepsilon_k = 0 \text{ or } 1 \\ \sum_{j=0}^{k} \varepsilon_j \leqq k}} P(\zeta(I[x_k, x_{k+1})) = \varepsilon_k \text{ for } 0 \leqq k \leqq n-1) \tag{4.6.43}$$

if $x_0 = 0 \leqq x_1 \leqq x_2 \leqq \cdots \leqq x_n$. (If I is an empty interval, $\zeta(I)$ is supposed to be identically 0.) Clearly, it is sufficient to determine $P(\tau_1 > x_1, \ldots, \tau_n > x_n)$ for the case when $x_1 \leqq x_2 \leqq \cdots \leqq x_n$, because one has for arbitrary nonnegative numbers $x_1, x_2, \ldots, x_n$, putting $x_k^* = \max(x_1, x_2, \ldots, x_k)$ $(k = 1, 2, \ldots, n)$,

$$P(\tau_1 > x_1, \ldots, \tau_n > x_n) = P(\tau_1 > x_1^*, \tau_2 > x_2^*, \ldots, \tau_n > x_n^*)$$

and evidently $x_1^* \leqq x_2^* \leqq \cdots \leqq x_n^*$.

It should be added that the functions $G(x_1, x_2, \ldots, x_n) = G_n$ can be determined by the following formula:

$$\frac{\partial G_n}{\partial x_n} = \lambda^n e^{-\lambda x_n} P_n(x_1, x_2, \ldots, x_n) \tag{4.6.44}$$

where $P_n(x_1, \ldots, x_n)$ is a polynomial of degree $\leqq n-1$ in each of its variables, which can be expressed as follows:

$$P_n(x_1, \ldots, x_n) = \int_{x_{n-1}}^{x_n} \int_{x_{n-2}}^{y_{n-1}} \cdots \int_{x_1}^{y_2} dy\, dy_2 \cdots dy_{n-1} \tag{4.6.45}$$

The polynomial $p_n(x) = P_n(x_1, \ldots, x_{n-1}, x)$ of degree $n-1$ in x can be uniquely defined by the following conditions:

$$p_n^{(k)}(x_{n-1-k}) = 0 \qquad \text{for } k = 0, 1, \ldots, n-2 \tag{4.6.46}$$

and

$$p_n^{(n-1)}(x) \equiv 1 \tag{4.6.47}$$

In analysis, $p_n(x)$ ($n = 1, 2, \ldots$) are called the Abel-Gontcharoff polynomials (see [25]). ■

Remark: The statement of Theorem 4.6.4 is usually taken as the definition of a Poisson process (see, e.g., Doob [26] and Blanc-Lapierre and Fortet [27], where many further properties of the Poisson process are discussed). We have taken a different approach, because this made it possible to construct a Poisson process from a sequence of independent and identically distributed random variables (having an exponential distribution) and thus avoid the difficulties of proving the existence of the process which are encountered in the usual approach.

4.7 THE CENTRAL LIMIT THEOREM

All versions of the central limit theorem state that under fairly general conditions the sum of a large number of independent random variables is approximately normally distributed. We shall restrict ourselves to proving an important theorem of this type, due to Lindeberg [28].

THEOREM 4.7.1. *Let $\xi_1, \xi_2, \ldots, \xi_n, \ldots$ be a sequence of independent random variables. Suppose that ξ_n has expectation 0 and finite variance $D_n^2 = D^2(\xi_n)$ and denote by $F_n(x)$ the distribution function of ξ_n ($n = 1, 2, \ldots$). Let us put*

$$\zeta_n = \xi_1 + \xi_2 + \cdots + \xi_n \tag{4.7.1}$$

and

$$S_n = D(\zeta_n) = \sqrt{D_1^2 + D_2^2 + \cdots + D_n^2} \tag{4.7.2}$$

Let us put for $\varepsilon > 0$

$$L_n(\varepsilon) = \frac{1}{S_n^2} \sum_{k=1}^{n} \int_{|x| > \varepsilon S_n} x^2 dF_k(x) \tag{4.7.3}$$

and suppose that for every $\varepsilon > 0$, one has*

$$\lim_{n \to +\infty} L_n(\varepsilon) = 0 \tag{4.7.4}$$

Then one has, uniformly in x ($-\infty < x < +\infty$),

$$\lim_{n \to +\infty} P\left(\frac{\zeta_n}{S_n} < x\right) = \Phi(x) \tag{4.7.5}$$

* Formula (4.7.4) is called *Lindeberg's condition.*

where

$$\Phi(x) = \frac{1}{\sqrt{2\pi}} \int_{-\infty}^{x} e^{-u^2/2}\, du \tag{4.7.6}$$

denotes the distribution of the standard normal distribution; i.e., the distribution of ζ_n/S_n tends weakly to the standard normal distribution.

Proof. We shall prove the theorem by the use of the calculus of operators. Let $\mathfrak{C}$ denote the set of all bounded and continuous functions $f(x)$ defined on the real axis. Let us put for $f \in \mathfrak{C}$

$$\|f\| = \sup_x |f(x)|$$

A *linear operator* A on $\mathfrak{C}$ is a mapping $f \to Af$ of $\mathfrak{C}$ into itself such that if $f \in \mathfrak{C}$, $g \in \mathfrak{C}$ and a and b are constants, one has $A(af + bg) = a \cdot Af + b \cdot Ag$; further, one has $\|Af\| \leqq k \|f\|$, where $k > 0$ is a constant, not depending on $f \in \mathfrak{C}$. If $k = 1$, the operator A is called a *linear contraction operator*.

The sum of two operators A and B is defined by $(A + B)f = Af + Bf$; clearly, $\|(A + B)f\| \leqq \|Af\| + \|Bf\|$. If c is a constant, the product cA of the constant c and the operator A is defined as the operator $cA \cdot f = c \cdot (Af)$. Evidently, $\|cAf\| = |c| \cdot \|Af\|$. The product AB of the operators A and B is defined by $AB \cdot f = A(Bf)$; i.e., by applying to the function f first the operator B, and then applying on the result Bf obtained the operator A. Clearly if A and B are linear contraction operators, so are AB and BA (which are not necessarily identical).

If $F(x)$ is any probability distribution function, let us define the operator A_F on $\mathfrak{C}$ by the formula

$$A_F f(x) = \int_{-\infty}^{+\infty} f(x + y)\, dF(y) \tag{4.7.7}$$

It is easy to see that A_F is a linear contraction operator. If $G(x)$ is another distribution function, then one can easily show that the two operators A_F and A_G commute and one has

$$A_F \cdot A_G = A_G \cdot A_F = A_{F*G} \tag{4.7.8}$$

As a matter of fact, putting $H = F * G$,

$$\begin{aligned} A_F A_G f(x) &= \int_{-\infty}^{+\infty} \int_{-\infty}^{+\infty} f(x + y + z)\, dF(y)\, dG(z) \\ &= \int_{-\infty}^{+\infty} f(x + u)\, dH(u) = A_H f(x) \end{aligned} \tag{4.7.8'}$$

LEMMA 4.7.1. *Let $A_1, \ldots, A_n, B_1, \ldots, B_n$ be any linear contraction operators. Then we have for every $f \in \mathfrak{C}$*

$$\|A_1 A_2 \cdots A_n f - B_1 B_2 \cdots B_n f\| \leqq \sum_{k=1}^{n} \|A_k f - B_k f\| \tag{4.7.9}$$

Proof. Formula (4.7.9) follows immediately from the obvious identity,

$$A_1A_2\cdots A_n - B_1B_2\cdots B_n = \sum_{k=1}^{n} \left(\prod_{i<k} A_i\right)(A_k - B_k) \prod_{j>k} B_k \tag{4.7.10}$$

(Where an empty product means the identical operator I for which $If = f$ for every $f \in \mathfrak{C}$.) Thus Lemma 4.7.1 is proved. ■

Now let us denote by $G_n(x)$ the distribution function of ζ_n/S_n. Then one evidently has

$$G_n(x) = G_{n,1}(x) * G_{n,2}(x) * \cdots * G_{n,n}(x) \tag{4.7.11}$$

where

$$G_{n,k}(x) = F_k(xS_n) \quad (k = 1, 2, \ldots, n) \tag{4.7.12}$$

Because $G_{n,k}(x)$ is the distribution function of ξ_k/S_n, it follows that

$$\int_{-\infty}^{+\infty} x\, dG_{n,k}(x) = 0 \tag{4.7.13}$$

and

$$\int_{-\infty}^{+\infty} x^2\, dG_{n,k}(x) = \frac{D_k^2}{S_k^2} \tag{4.7.14}$$

In view of Lemma 4.5.1, we have further,

$$\Phi(x) = \Phi_{n,1}(x) * \Phi_{n,2}(x) * \cdots * \Phi_{n,n}(x) \tag{4.7.15}$$

where

$$\Phi_{n,k}(x) = \Phi\left(\frac{xS_n}{D_k}\right) \qquad (k = 1, 2, \ldots, n) \tag{4.7.16}$$

As $\Phi(xS_n/D_k)$ is the distribution function of the normal distribution with expectation 0 and variance D_k^2/S_n^2, it follows that

$$\int_{-\infty}^{+\infty} x\, d\Phi_{n,k}(x) = 0 \tag{4.7.17}$$

and

$$\int_{-\infty}^{+\infty} x^2\, d\Phi_{n,k}(x) = \frac{D_k^2}{S_n^2}. \tag{4.7.18}$$

In other words, $\Phi_{n,k}(x)$ is the normal distribution function which has the same expectation and variance as the distribution function $G_{n,k}(x)$.

Applying Lemma 4.7.1, we get that for every $f \in \mathfrak{C}$

$$\|A_{G_n} f - A_\Phi f\| \leq \sum_{k=1}^{n} \|A_{G_{n,k}} f - A_{\Phi_{n,k}} f\| \tag{4.7.19}$$

Now let us suppose that the function $f(x)$ possesses derivatives of the first three orders, which are uniformly bounded on the real axis. We denote the set of such functions by $\mathfrak{C}_3$. If $f \in \mathfrak{C}_3$, we can expand $f(x+y)$ in a finite Taylor series up to the second and the third term, i.e., we can put

$$f(x+y)=f(x)+yf'(x)+\frac{y^2}{2}f''(x)+\frac{y^3}{6}f'''(x+\vartheta_1 y) \tag{4.7.20}$$

and

$$f(x+y)=f(x)+yf'(x)+\frac{y^2}{2}f''(x+\vartheta_2 y) \tag{4.7.21}$$

where $0 \leqq \vartheta_1 \leqq 1$ and $0 \leqq \vartheta_2 \leqq 1$ (ϑ_1 and ϑ_2 depend, of course, on x and y).

Applying the expansion (4.7.20) on the interval $|y| \leqq \varepsilon$ and the expansion (4.7.21) for $|y| > \varepsilon$, we get, taking (4.7.13) and (4.7.14) into account,

$$A_{G_{n,k}} f = f(x) + \frac{f''(x)D_k^2}{2S_n^2} + \frac{1}{6}\int_{|y|\leqq\varepsilon} y^3 f'''(x+\vartheta_1 y)\, dG_{n,k}(y) \\ + \frac{1}{2}\int_{|y|>\varepsilon} y^2(f''(x+\vartheta_2 y) - f''(x))\, dG_{n,k}(y) \tag{4.7.22}$$

Similarly we get, with respect to (4.7.16) and (4.7.17),

$$A_{\Phi m,k} f = f(x) + \frac{f''(x)D_k^2}{2S_n^2} + \frac{1}{6}\int_{|y|\leqq\varepsilon} y^3 f'''(x+\vartheta_1 y)\, d\Phi_{n,k}(y) \\ + \frac{1}{2}\int_{|y|>\varepsilon} y^2(f''(x+\vartheta_2 y) - f''(x))\, d\Phi_{n,k}(y) \tag{4.7.23}$$

It follows, majorizing for $|y| \leqq \varepsilon$ y^3 by εy^2 and putting $\sup|f''(x)| = k_1$ and $\sup|f'''(x)| = k_2$, that

$$\|A_{G_{n,k}}f - A_{\Phi n,k}f\| \leqq \frac{k_1}{2}\left(\int_{|y|>\varepsilon} y^2\, dG_{n,k}(y) + \int_{|y|>\varepsilon} y^2\, d\Phi_{n,k}(y)\right) \\ + \frac{k_2\,\varepsilon}{6}\left(\int_{|y|\leqq\varepsilon} y^2\, dG_{n,k}(y) + \int_{|y|\leqq\varepsilon} y^2\, d\Phi_{n,k}(y)\right) \tag{4.7.24}$$

The second term on the right of (4.7.24) is clearly $\leqq k_2\,\varepsilon D_k^2/3S_n^2$. Adding the inequalities (4.7.24) and taking into account that

$$\int_{|y|>\varepsilon} y^2\, dG_{n,k}(y) = \frac{1}{S_n^2}\int_{|y|>\varepsilon S_n} y^2\, dF_k(y)$$

and

$$\int_{|y|>\varepsilon} y^2\, d\Phi_{n,k}(y) = \frac{D_k^2}{S_n^2}\int_{|y|>\varepsilon S_n/D_k} \frac{u^2 e^{-u^2/2}\, du}{\sqrt{2\pi}}$$

we get, putting $\max_{1 \leqq k \leqq n} D_k = \Delta_n$,

$$\|A_{G_n} f - A_\Phi f\| \leqq \frac{k_2 \varepsilon}{3} + \frac{k_1}{2}\left(L_n(\varepsilon) + \frac{1}{\sqrt{2\pi}} \int_{|u| > \varepsilon S_n/\Delta_n} u^2 e^{-u^2/2} \, du\right) \tag{4.7.25}$$

Now it is easy to see that for each $k \leqq n$ and every $\varepsilon > 0$

$$\frac{D_k^2}{S_n^2} = \frac{1}{S_n^2} \int_{-\infty}^{+\infty} x^2 \, dF_k(x) \leqq \varepsilon^2 + L_n(\varepsilon)$$

Thus we have

$$\frac{\Delta_n^2}{S_n^2} \leqq \varepsilon^2 + L_n(\varepsilon)$$

which implies

$$\lim_{n \to +\infty} \frac{\Delta_n}{S_n} \leqq \varepsilon^2 \tag{4.7.26}$$

As $\varepsilon > 0$ can be chosen arbitrarily small, (4.7.26) implies

$$\lim_{n \to +\infty} \frac{\Delta_n}{S_n} = 0 \tag{4.7.27}$$

It follows from (4.7.25) and (4.7.27) that for each $f \in \mathfrak{C}_3$ one has, uniformly in x,

$$\lim_{n \to +\infty} A_{G_n} f = A_\Phi f \tag{4.7.28}$$

Thus we have for each $f \in \mathfrak{C}_3$,

$$\lim_{n \to +\infty} \int_{-\infty}^{+\infty} f(y) \, dG_n(y) = \int_{-\infty}^{+\infty} f(y) \, d\Phi(y) \tag{4.7.29}$$

Now note that the functions (4.3.4) belong to $\mathfrak{C}_3$ for every $\varepsilon > 0$. Thus it follows, as in the proof of (4.3.1), that the distribution function $G_n(x)$ tends weakly to the distribution function $\Phi(x)$. Moreover, as $\Phi(x)$ is everywhere continuous, by Theorem 4.3.3, $G_n(x)$ tends uniformly to $\Phi(x)$. Thus Theorem 4.7.1 is proved. ■

Corollary to Theorem 4.7.1. *If the random variables $\xi_1, \ldots, \xi_n, \ldots$ are independent and identically distributed, and their expectation $E(\xi_n) = E$ and variance $D^2(\xi_n) = D^2$ exists, then one has*

$$\lim_{n \to +\infty} P\left(\frac{\xi_1 + \xi_2 + \cdots + \xi_n - nE}{D\sqrt{n}} < x\right) = \Phi(x) \tag{4.7.30}$$

uniformly in x.

Proof. It is easy to see that the conditions of Theorem 4.7.1 are fulfilled for the random variables $\xi_k^* = \xi_k - E$, because in this case one has $S_n^2 = nD^2$ and, denoting by $F(x)$ the distribution function of the variables $\xi_k{}^*$,

$$\lim_{n\to+\infty} L_n(\varepsilon) = \lim_{n\to+\infty} \frac{1}{D^2}\int_{|x|>D\varepsilon\sqrt{n}} x^2\, dF(x) = 0 \quad \blacksquare$$

Remark 1: The above corollary to Theorem 4.7.1 contains the global form of the Moivre-Laplace theorem as a special case.

Remark 2: An alternative way to prove Theorem 4.7.1 is by the method of characteristic functions. We shall show how this method works by giving an alternative proof of the corollary to Theorem 4.7.1. Let $\varphi(t)$ denote the characteristic function of the variables $\xi_k - E$; then, putting $\zeta_n^* = (\xi_1 + \cdots + \xi_n - nE)/D\sqrt{n}$, we get for the characteristic function $\varphi_n(t)$ of ζ_n^*, $\varphi_n(t) = [\varphi(t/D\sqrt{n})]^n$. Now by supposition,

$$\varphi'(0) = iE(\xi_k^*) = 0$$

and

$$\varphi''(0) = -D^2$$

Thus one has

$$\varphi\left(\frac{t}{D\sqrt{n}}\right) = 1 - \frac{t^2}{2n} + o\left(\frac{1}{n}\right)$$

and therefore

$$\lim_{n\to+\infty} \varphi_n(t) = e^{-t^2/2}$$

As we have seen (see (4.5.20)) that $e^{-t^2/2}$ is the characteristic function of the standard normal distribution, it follows by Theorem 4.3.7 that the distribution of $\zeta_n{}^*$ tends weakly to the standard normal distribution.

Example 4.7.1. Let $\xi_1, \xi_2, \ldots, \xi_n, \ldots$ be independent random variables, each having the same Poisson distribution with parameter c. Then the corollary to Theorem 4.7.1 can be applied and thus we get that the distribution of $(\xi_1 + \xi_2 + \cdots + \xi_n - nc)/\sqrt{nc}$ tends for $n \to +\infty$ to the standard normal distribution. However, according to Lemma 4.6.1, $\xi_1 + \xi_2 + \cdots + \xi_n$ has a Poisson distribution with parameter nc. Thus, our result can be written in the form

$$\lim_{\lambda\to+\infty} \sum_{k\leq\lambda+x\sqrt{\lambda}} \frac{\lambda^k e^{-\lambda}}{k!} = \Phi(x) \tag{4.7.31}$$

(see Exercise E.4.2).

Example 4.7.2. Let $\xi_1, \xi_2, \ldots, \xi_n, \ldots$ be independent random variables, each having the same exponential distribution with parameter λ. Then the corollary to Theorem 4.7.1 can be applied and thus we get that the distribution of $(\xi_1 + \xi_2 + \cdots + \xi_n - n \cdot \lambda^{-1})/\lambda^{-1} \cdot \sqrt{n}$ tends for $n \to +\infty$ to the standard normal distribution. As by Lemma 4.6.2, the distribution of $\xi_1 + \xi_2 + \cdots + \xi_n$ is a Γ distribution of order n with parameter λ, our result can be written in the following form:

$$\lim_{n \to +\infty} \int_0^{(n+x\sqrt{n})/\lambda} \frac{\lambda^n t^{n-1} e^{-\lambda t}\, dt}{(n-1)!} = \Phi(x) \tag{4.7.32}$$

The relation (4.7.32) can also be deduced from (4.7.31). First, one can verify by partial integration the identity

$$\sum_{k=0}^{n-1} \frac{\mu^k e^{-\mu}}{k!} = 1 - \frac{\lambda^n}{(n-1)!} \int_0^{\mu/\lambda} t^{n-1} e^{-\lambda t}\, dt \tag{4.7.33}$$

valid for all $\lambda > 0$ and $\mu > 0$. Putting $\mu = n + x\sqrt{n}$, we have $n = \mu - x\sqrt{\mu} + O(1)$. Thus it follows from (4.7.31) that (4.7.32) holds. Equations (4.7.31) and (4.7.32) can also be expressed as follows:

Let $\{\tau_n\}$ be a Poisson process with parameter λ. Let $\zeta(t)$ denote the number of values of n for which $\tau_n < t$. Then we have

$$\lim_{t \to +\infty} P\left(\frac{\zeta(t) - \lambda t}{\sqrt{\lambda t}} < x\right) = \Phi(x) \tag{4.7.34}$$

and

$$\lim_{n \to +\infty} P\left(\frac{\tau_n - n/\lambda}{\sqrt{n}/\lambda} < x\right) = \Phi(x) \tag{4.7.35}$$

As clearly, $\zeta(\tau_n) = n - 1$, we have

$$P\left(\tau_n \geqq \frac{\mu}{\lambda}\right) = P\left(\zeta\left(\frac{\mu}{\lambda}\right) \leqq n - 1\right) \tag{4.7.36}$$

Notice that (4.7.36) is identical to (4.7.33). Thus we obtained a probabilistic proof of the purely analytic identity (4.7.33).

4.8 LAWS OF FLUCTUATION

In this section we shall present certain further laws of chance concerning the "fine structure" of randomness and the laws of fluctuation, through the study of the most simple special case. We consider a sequence $\eta_1, \eta_2, \ldots, \eta_n, \ldots$ of independent random variables, each η_n taking on the values 1 and

0, with probability 1/2. We may interpret the $\{\eta_n\}$ as the results of a series of throws with a fair coin: $\eta_n = 1$ if we get a "head" and $\eta_n = 0$ if we get a "tail" at the nth throw. Thus,

$$S_n = \eta_1 + \eta_2 + \cdots + \eta_n \tag{4.8.1}$$

is the number of throws among the first n which are favorable for the player betting on heads. It follows from the strong law of large numbers, that almost surely $S_n \sim n/2$; moreover, it follows from the central limit theorem that the difference $S_n - n/2$ will be, in general, of order of magnitude $\sqrt{n}$; as a matter of fact, the distribution of $(2S_n - n)/\sqrt{n}$ tends to the standard normal distribution. It is more convenient to introduce the notation $\xi_n = 2\eta_n - 1$ $(n = 1, 2, \ldots)$, i.e., $\xi_n = +1$ or $\xi_n = -1$, according to whether the nth throw resulted in a head or a tail, and we put

$$\zeta_n = \xi_1 + \xi_2 + \cdots + \xi_n = 2S_n - n \tag{4.8.2}$$

Thus ζ_n is the *net gain* of the player betting on heads after n throws. Throughout this section, ξ_n and ζ_n will have this meaning. Thus we know that almost surely $\zeta_n/n \to 0$ and that the distribution of $\zeta_n/\sqrt{n}$ tends to the standard normal distribution.

In this section we shall study exceptionally large values of ζ_n and, more generally, exceptionally long periods of luck, i.e., such pairs of values of n and k for which k is large and $\xi_{n+1} + \xi_{n+2} + \cdots + \xi_{n+k}$ is a large positive (or negative) number. We shall show that even the "exceptional" long runs of luck are subject to certain laws. We want to emphasize that these laws are valid for general sequences of independent random variables under fairly weak restrictions (see, e.g., [3] and [29]); we shall restrict ourselves to the special case mentioned only for the sake of simplicity—to be able to present these laws with a minimum of technical difficulty. It will be seen that even in the simple case under consideration rather powerful estimates and delicate arguments are necessary; this is, however, unavoidable as we are dealing with the "fine structure" of random fluctuations.

The results presented in this section can be formulated in a rather picturesque way in terms of "*random walk.*" Imagine a point moving on the lattice points of the (x, y) plane as follows: it is at time $t = n$ situated at the lattice point (n, ζ_n). Thus, during the time interval $(n, n+1)$ the point moves one unit to the right and one unit up or down according to whether $\xi_{n+1} = +1$ or $\xi_{n+1} = -1$. Connecting the points (n, ζ_n) and $(n+1, \zeta_{n+1})$ by a straight segment (forming an angle of $+45°$ or $-45°$ with the direction of the positive x axis), we get a zigzag line which contains all information about the sequence ξ_n, respectively about ζ_n. The reader is encouraged to formulate all the results presented in this section in terms of the mentioned random-walk diagram.

We start with the celebrated *law of the iterated logarithm*, discovered by Khintchine [30], for the special case presented here. Its validity for general sequences of independent random variables under fairly mild restrictions has been proved by Kolmogoroff [31], Feller [32], Erdös [33] and others; recently, a far-reaching generalization has been achieved by Strassen [34], which gives deep insight into the real meaning of this theorem. However, this generalization, based on the study of the Brownian movement, lies outside the scope of the present book.

For the special case which we consider, the law of the iterated logarithm can be formulated as follows:

THEOREM 4.8.1. *With probability* 1, *one has*

$$\limsup_{n\to+\infty} \frac{\zeta_n}{\sqrt{2n \log\log n}} = +1 \tag{4.8.3}$$

Proof. To prove the theorem we have to prove two statements: (a) For every $\varepsilon > 0$ with probability 1, the inequality

$$\zeta_n > (1+\varepsilon)\sqrt{2n \log\log n} \tag{4.8.4}$$

is satisfied only for a finite number of values of n. (b) For every ε, with $0 < \varepsilon < 1$, the inequality

$$\zeta_n > (1-\varepsilon)\sqrt{2n \log\log n} \tag{4.8.5}$$

is satisfied with probability 1 for an infinity of values of n.

First we shall prove statement (b). According to Theorem 4.5.3,

$$P(\zeta_n > \sqrt{2(1-\varepsilon)n \log\log n}) > \frac{1}{\sqrt{2\pi}} \int_{\sqrt{2(1-\varepsilon)n \log\log n}} e^{-u^2/2}\, du - \frac{A_1}{\sqrt{n}}$$

where A_1 is a constant.

Now we have (see (4.5.30))

$$\frac{1}{\sqrt{2\pi}} \int_x^{+\infty} e^{-u^2/2}\, du = \frac{e^{-x^2/2}}{\sqrt{2\pi}\, x} - \frac{1}{\sqrt{2\pi}} \int_x^{+\infty} \frac{e^{-u^2/2}}{u^2}\, du \geqq \frac{e^{-x^2/2}}{\sqrt{2\pi}\, x}\left(1 - \frac{1}{x^2}\right)$$

Thus

$$P(\zeta_n > \sqrt{2(1-\varepsilon)n \log\log n}) \geqq \frac{A_2}{(\log n)^{1-\varepsilon}(\log\log n)^{1/2}} \tag{4.8.6}$$

where $A_2 > 0$ is a constant.

On the other hand, we have, in view of (4.4.18),

$$P(|\zeta_n| > \sqrt{2(1+\varepsilon)n \log\log n}) < \frac{A_3}{(\log n)^{1+\varepsilon}} \tag{4.8.7}$$

where $A_3 > 0$ is a constant. Let us now put

$$n_k = k^k \tag{4.8.8}$$

Then we have for sufficiently large values of k, if $\varepsilon < 1/2$,

$$\sqrt{2(1-\varepsilon)(n_{k+1}-n_k)\log\log(n_{k+1}-n_k)} > \left(1-\frac{3\varepsilon}{4}\right)\sqrt{2n_{k+1}\log\log n_{k+1}}$$

Applying (4.8.6) for $n = n_{k+1} - n_k$ to $\zeta_{n_{k+1}} - \zeta_{n_k}$, we get if k is sufficiently large, that

$$P\left(\zeta_{n_{k+1}} - \zeta_{n_k} > \left(1-\frac{3\varepsilon}{4}\right)\sqrt{2n_{k+1}\log\log n_{k+1}}\right) > \frac{1}{k^{1-\varepsilon/2}} \tag{4.8.9}$$

The random variables $\zeta_{n_{k+1}} - \zeta_{n_k}$ $(k = 1, 2, \dots)$ being independent and the series $\sum 1/(k^{1-\varepsilon/2})$ being divergent, it follows by the Borel-Cantelli lemma, that with probability 1 for an infinity of values of k, one has

$$\zeta_{n_{k+1}} - \zeta_{n_k} > \left(1-\frac{3\varepsilon}{4}\right)\sqrt{2n_{k+1}\log\log n_{k+1}}$$

It follows by the same argument from (4.8.7) that with probability 1 the inequality

$$\zeta_{n_k} < -(1+\varepsilon)\sqrt{2n_k \log\log n_k}$$

holds only for a finite number of values of k. Thus, with probability 1 for an infinity of values of k,

$$\zeta_{n_{k+1}} > \left(1-\frac{3\varepsilon}{4}\right)\sqrt{2n_{k+1}\log\log n_{k+1}} - (1+\varepsilon)\sqrt{2n_k \log\log n_k}$$
$$> (1-\varepsilon)\sqrt{2n_{k+1}\log\log n_{k+1}}$$

i.e., with probability 1 the inequality (4.8.5) holds for an infinity of values of n. This proves part (b) of Theorem 4.8.1.

Now we turn to the proof of statement (a). As $\sqrt{n \log\log n}$ is an increasing function of n if $n \geqq 3$, if we put $\zeta_0 = 0$,

$$\vartheta_n = \max_{0 \leq k \leq n} \zeta_k$$

and

$$m_k = \left[2^{\frac{k}{\log k}}\right]$$

then if (4.8.4) is satisfied for at least one n in the interval $m_{k-1} \leqq n < m_k$, then we have, if k is sufficiently large, in view of $m_k/m_{k-1} \to 1$,

$$\vartheta_{m_k} > \left(1+\frac{\varepsilon}{2}\right)\sqrt{2m_k \log\log m_k} \tag{4.8.10}$$

Thus, to prove statement (a) it is sufficient to show that with probability 1, (4.8.10) holds for a finite number of values of k only. In order to prove this we compute the distribution of ϑ_n.

LEMMA 4.8.1. *One has*

$$P(\vartheta_n = k) = \frac{1}{2^n}\binom{n}{\left[\frac{n-k}{2}\right]} \qquad (k = 0, 1, \ldots, n) \tag{4.8.11}$$

Proof. Since ϑ_n may take on the values $0, 1, \ldots, n$, let us put

$$p_{n,k} = P(\vartheta_n = k) \qquad (0 \leqq k \leqq n)$$

Then we have, putting $\vartheta_n^* = \max_{2 \leqq k \leqq n+1} \sum_{j=2}^k \xi_j$,

$$p_{n+1,k} = P(\xi_1 = 1, \vartheta_n^* = k-1) + P(\xi_1 = -1, \vartheta_n^* = k+1) \qquad \text{for } k \geqq 1$$

As, however, ξ_1 and ϑ_n^* are independent, and ϑ_n^* has the same distribution as ϑ_n, we get

$$p_{n+1,k} = \tfrac{1}{2}(p_{n,k-1} + p_{n,k+1}) \qquad \text{for } k \geqq 1 \tag{4.8.12}$$

We get similarly,

$$p_{n+1,0} = P(\xi_1 = -1, \vartheta_n^* \leqq 1) = \tfrac{1}{2}(p_{n,1} + p_{n,0}) \tag{4.8.13}$$

Now we prove (4.8.11) by induction. As $p_{1,0} = p_{1,1} = 1/2$, (4.8.12) holds for $n = 1$. It is easy to see that the numbers $p_{n,k} = 1/2^n\binom{n}{[(n-k)/2]}$ satisfy the recursion formulas (4.8.12) and (4.8.13). Thus, (4.8.11) holds for every n and Lemma 4.8.1 is proved.

Let us now consider the moment generating function of ϑ_n, i.e.,

$$\psi_n(s) = E(e^{\vartheta_n s}) = \frac{1}{2^n}\sum_{k=0}^n \binom{n}{\left[\frac{n-k}{2}\right]} e^{ks} \tag{4.8.14}$$

We prove the following:

LEMMA 4.8.2. *One has for any real value of s,*

$$\psi_{2n}(s) \leq (1 + e^s)\left(\frac{e^s + e^{-s}}{2}\right)^{2n} \tag{4.8.15}$$

Proof. Multiplying out the right-hand side, we get all terms on the left-hand side, and some further nonnegative terms; thus (4.8.15) holds and Lemma 4.8.2 is proved. ∎

Let us now apply Bernstein's inequality. We obtain that for $x > 0$

$$P(\vartheta_{2n} \geqq x) \leqq \min_s \left[e^{-sx}(1+e^s)\left(\frac{e^s+e^{-s}}{2}\right)^{2n}\right]$$

Using the inequality

$$\frac{e^s+e^{-s}}{2} \leqq e^{s^2/2} \qquad (-\infty < s < +\infty) \tag{4.8.16}$$

(which follows by comparing the coefficients of the two power series) and choosing $s = x/2n$, we get

$$P(\vartheta_{2n} \geqq x) \leqq 2e^{-x^2/4n} \tag{4.8.17}$$

Thus we have

$$P\left(\vartheta_{2n} \geqq \left(1+\frac{\varepsilon}{2}\right)\sqrt{4n \operatorname{loglog} 2n}\right) \leqq \frac{2}{(\log 2n)^{1+\varepsilon/2}} \tag{4.8.18}$$

It follows that if $m_k = 2^{[k/\log k]}$, the series

$$\sum_{k=1}^{+\infty} P\left(\vartheta_{m_k} > \left(1+\frac{\varepsilon}{2}\right)\sqrt{2m_k \operatorname{loglog} m_k}\right) \tag{4.8.19}$$

is convergent for every $\varepsilon > 0$. Thus, by the Borel-Cantelli lemma, the inequality (4.8.10) is fulfilled with probability 1 for a finite number of values of k only.

As was pointed out, this proves statement (a). Thus Theorem 4.8.1 is proved. ■

It can be seen that ϑ_n is, with probability 1, for infinitely many values of n approximately equal to $\sqrt{2n \operatorname{loglog} n}$. However, for a fixed n such a large value of ϑ_n is highly improbable, and the usual order of magnitude of ϑ_n is $\sqrt{n}$; as a matter of fact, $\vartheta_n/\sqrt{n}$ has a limiting distribution. We shall now determine this distribution.

THEOREM 4.8.2. *One has uniformly for* $x \geqq 0$

$$\lim_{n\to+\infty} P\left(\frac{\vartheta_n}{\sqrt{n}} < x\right) = 2\Phi(x) - 1 = \sqrt{\frac{2}{\pi}}\int_0^x e^{-u^2/2}\,du \tag{4.8.20}$$

and

$$P(\vartheta_n = k) \approx \frac{2}{\sqrt{2\pi n}} e^{-k^2/2n} \qquad \text{for } k = o(n^{2/3}) \tag{4.8.21}$$

Proof. Both (4.8.20) and (4.8.21) follow immediately from the global and local forms, respectively, of the Moivre-Laplace theorem; as a matter of fact, we get (4.8.21) from Lemma 4.8.1, and we also get

$$P\left(\frac{\vartheta_n}{\sqrt{n}} < x\right) = \frac{1}{2^n} \sum_{k < x\sqrt{n}} \binom{n}{\left[\frac{n-k}{2}\right]}$$

$$= \frac{2}{2^n} \sum_{\frac{n-x\sqrt{n}}{2} < j < \frac{u}{2}} \binom{n}{j} - \frac{\varepsilon_n}{2^n} \binom{n}{\left[\frac{n}{2}\right]}$$

where $\varepsilon_n = 1$ if n is even and 0 if n is odd. The factor of ε_n is of order $1/\sqrt{n}$. From the Moivre-Laplace theorem we have

$$\lim_{n\to\infty} \frac{1}{2^n} \sum_{\frac{n-x\sqrt{n}}{2} < j < \frac{u}{2}} \binom{n}{j} = \frac{1}{\sqrt{2\pi}} \int_x^0 e^{-u^2/2}\, du = \frac{1}{\sqrt{2\pi}} \int_0^x e^{-u^2/2}\, du$$

Thus we get (4.8.20). ■

Remark: Formula (4.8.20) means that the limit distribution of $\vartheta_n/\sqrt{n}$ is that of the absolute value $|\xi|$ of a random variable ξ having the standard normal distribution.

The expectation of $|\xi|$, if ξ is $N(0, 1)$-distributed, is

$$E(|\xi|) = \sqrt{\frac{2}{\pi}} \int_0^{+\infty} x e^{-x^2/2}\, dx = \sqrt{\frac{2}{\pi}} \tag{4.8.22}$$

Thus we may guess that

$$\lim_{n\to+\infty} E\left(\frac{\vartheta_n}{\sqrt{n}}\right) = \sqrt{\frac{2}{\pi}} \tag{4.8.23}$$

(Note that this does *not* follow directly from Theorem 4.8.2.) In fact, we have

$$E\left(\frac{\vartheta_n}{\sqrt{n}}\right) = \frac{1}{\sqrt{n}} \sum_{k=0}^{n} \frac{k\binom{n}{\left[\frac{n-k}{2}\right]}}{2^n} = \frac{\sqrt{n}}{2^{n-1}} \binom{n-1}{\left[\frac{n-1}{2}\right]} - \delta_n \tag{4.8.24}$$

where

$$\delta_{2k} = \frac{1 - \binom{2k}{k} \cdot \frac{1}{2^{2k}}}{2\sqrt{2k}}$$

and

$$\delta_{2k+1} = \frac{1}{2\sqrt{2k+1}}$$

and therefore $\lim_{n\to+\infty} \delta_n = 0$.

As, by the Moivre-Laplace theorem,

$$\binom{n-1}{\left[\frac{n-1}{2}\right]} \approx \frac{2^n}{\sqrt{2\pi n}}$$

it follows that (4.8.23) holds, as expected.

Let us consider now the infinite sequence ζ_n $(n=1, 2, \ldots)$; let ν_k denote the least positive integer for which $\zeta_{\nu_k}=k$ $(k=0, 1, \ldots)$. Clearly, ν_k is a random variable and we have

$$P(\nu_k > n) = P(\vartheta_n < k) \tag{4.8.25}$$

Thus we can reformulate Lemma 4.8.1 in terms of ν_k as follows:

LEMMA 4.8.2. *One has for $k \geqq 1$ and $n \geqq 1$*

$$P(\nu_k > n) = \frac{1}{2^n} \sum_{j=0}^{k-1} \binom{n}{\left[\frac{n-j}{2}\right]} \tag{4.8.26}$$

Remark: Notice that as ζ_n has the same parity as n, ν_{2k} takes on only even values and ν_{2k+1} only odd values. Let us mention further that the random variables $\nu_1, \nu_2-\nu_1, \ldots, \nu_{k+1}-\nu_k, \ldots$ are clearly independent and identically distributed. As a matter of fact, $\nu_{k+1}-\nu_k$ is the least value of n such that $\zeta_{\nu_{k+n}}-\zeta_{\nu_k}=\sum_{i=\nu_k+1}^{\nu_k+n} \xi_i=1$ and thus has the same distribution as ν_1, and is independent from the random variables $\xi_1, \ldots, \xi_{\nu_k}$ and thus also from $\nu_1, \nu_2-\nu_1, \ldots, \nu_k-\nu_{k-1}$. It follows from (4.8.26) that

$$P(\nu_1 > n) = \frac{1}{2^n} \binom{n}{\left[\frac{n}{2}\right]}$$

and thus we get

$$P(\nu_1 = 2m+1) = \frac{\binom{2m}{m}}{(m+1)2^{2m+1}} \quad (m=0, 1, \ldots) \tag{4.8.27}$$

We now prove the following limit theorem:

THEOREM 4.8.3. *For $x \geqq 0$ we have*

$$\lim_{k\to+\infty} P\left(\frac{\nu_k}{k^2} < x\right) = \frac{1}{\sqrt{2\pi}} \int_0^x \frac{e^{-1/2v}}{v^{3/2}}\, dv \tag{4.8.28}$$

Proof. From (4.8.25) we get for any $x > 0$

$$P\left(\frac{\nu_k}{k^2} > x\right) = P(\vartheta_{[xk^2]} < k) = P\left(\frac{\vartheta_{[xk^2]}}{\sqrt{[xk^2]}} < \frac{k}{\sqrt{[xk^2]}}\right)$$

As $k/\sqrt{[xk^2]} \to 1/\sqrt{x}$ for $k \to +\infty$, we obtain

$$\lim_{k\to+\infty} P\left(\frac{\nu_k}{k^2} > x\right) = \sqrt{\frac{2}{\pi}} \int_0^{1/\sqrt{x}} e^{-u^2/2}\, du = \frac{1}{\sqrt{2\pi}} \int_x^{+\infty} \frac{e^{-1/2v}}{v^{3/2}}\, dv$$

As $1/\sqrt{2\pi} \int_0^{+\infty} e^{-1/2v}/v^{3/2}\, dv = 1$, (4.8.28) follows. ■

Remark: As pointed out, we can represent ν_k in the form $\nu_k = \sum_{j=1}^{k} (\nu_j - \nu_{j-1})$ as the sum of k independent and identically distributed random variables ($\nu_0 \equiv 0$). However, the central limit theorem is not applicable for this sum, because the terms have infinite expectation. As a matter of fact, it follows from (4.8.27) that $P(\nu_1 = 2m+1) \approx 1/2\sqrt{\pi}\, m^{3/2}$ and thus $E(\nu_1) = +\infty$.

The limit distribution occurring in Theorem 4.8.3 having the density $f(x) = (1/\sqrt{2\pi})(e^{-1/2x}/x^{3/2})$ for $x > 0$ is a so-called *stable* distribution with parameter $\alpha = 1/2$. Its characteristic function is $e^{-\sqrt{-2it}}$, from which it can be seen that the family of distribution with density $(1/a)f(x/a)$ $(a > 0)$ is closed with respect to convolution, because

$$\frac{1}{a} f\left(\frac{x}{a}\right) * \frac{1}{b} f\left(\frac{x}{b}\right) = \frac{1}{c} f\left(\frac{x}{c}\right) \qquad \text{where} \quad c = (\sqrt{a} + \sqrt{b})^2$$

Now let us consider those values of $n \geqq 1$ for which $\zeta_n = 0$. Let these values in increasing order be $\rho_1 < \rho_2 < \cdots < \rho_k < \cdots$. Clearly, ρ_k is even for $k = 1, 2, \ldots$. Now if $\xi_1 = -1$, then ρ_1 is the least value of n for which $-1 + \sum_{k=2}^{n} \xi_k = 0$, i.e., the least value of n for which $\sum_{k=2}^{n} \xi_k = 1$. As ξ_1 is independent from the sequence $\sum_{k=2}^{n} \xi_k$ $(n = 2, 3, \ldots)$, it follows that

$$P(\rho_1 = 2l \mid \xi_1 = -1) = P(\nu_1 = 2l - 1) \qquad (l = 1, 2, \ldots)$$

On the other hand, if $\xi_1 = +1$, then ρ_1 is the least value of n for which $\sum_{k=2}^{n} \xi_k = -1$. As the distribution of the ξ_k is symmetric, it follows that

$$P(\rho_1 = 2l \mid \xi_1 = +1) = P(\nu_1 = 2l - 1)$$

Thus we get

$$P(\rho_1 = 2l) = P(\nu_1 = 2l - 1) = \frac{\binom{2l-2}{l-1}}{l \cdot 2^{2l-1}} \qquad (l = 1, 2, \ldots) \tag{4.8.29}$$

Let us compute the generating function of ρ_1 and thus of $\nu_1 + 1$. We have

$$\sum_{j=0}^{+\infty} \binom{2j}{j} \frac{u^j}{2^{2j}} = \sum_{j=0}^{+\infty} \binom{-\frac{1}{2}}{j} (-u)^j = \frac{1}{\sqrt{1-u}} \tag{4.8.30}$$

and thus

$$G(z) = E(z_{\rho_1}) = \frac{1}{2}\sum_{j=0}^{+\infty}\binom{2j}{j}\frac{z^{2j+2}}{(j+1)2^{2j}} = \int_0^{z^2}\frac{du}{\sqrt{1-u}} = 1 - \sqrt{1-z^2} \tag{4.8.31}$$

It follows that

$$\sum_{l=1}^{+\infty} P(\rho_1 = 2l) = G(1) = 1$$

Notice that this means that with probability 1 there is a value of n such that $\zeta_n = 0$; as the properties of the sequence ζ_{ρ_1+n} $(n = 1, 2, \ldots)$ are the same as those of the sequence ζ_n, it follows that *with probability* 1 *there is an infinity of zeros in the sequence* ζ_n.

The random variables $\rho_1, \rho_2 - \rho_1, \ldots, \rho_k - \rho_{k-1}, \ldots$ are independent and identically distributed. As the distribution of ρ_1 is identical with that of $\nu_1 + 1$, it follows that the distribution of $\rho_k - k$ is identical to that of ν_k, i.e.,

$$P(\rho_k - k > n) = P(\nu_k > n) = P(\vartheta_n < k) \tag{4.8.32}$$

From (4.8.32) and Theorem 4.8.3 we easily get the following:

THEOREM 4.8.4. *For* $x \geqq 0$

$$\lim_{k\to+\infty} P\left(\frac{\rho_k}{k^2} < x\right) = \frac{1}{\sqrt{2\pi}}\int_0^x \frac{e^{-1/2v}}{v^{3/2}}\,dv \tag{4.8.33}$$

Another way to express our result is as follows. Let Z_{2n} denote the number of zeros in the sequence $\zeta_1, \zeta_2, \ldots, \zeta_{2n}$. It follows that for $0 \leqq k \leqq n$

$$P(Z_{2n} = k) = P(\vartheta_{2n-k} = k) = \frac{1}{2^{2n-k}}\binom{2n-k}{n} \tag{4.8.34}$$

Of course, (4.8.34) implies the identity

$$\sum_{k=0}^{n}\frac{\binom{2n-k}{n}}{2^{2n-k}} = 1 \tag{4.8.35}$$

Taking into account that $Z_{2n+1} = Z_{2n}$, we get the following:

THEOREM 4.8.5. *For* $x \geqq 0$,

$$\lim_{n\to+\infty} P\left(\frac{Z_n}{\sqrt{n}} < x\right) = \sqrt{\frac{2}{\pi}}\int_0^x e^{-u^2/2}\,du \tag{4.8.36}$$

Now we consider the random variables π_{2n} defined as follows: π_{2n} is the number of those terms of the sequence $\zeta_1, \zeta_2, \ldots, \zeta_{2n}$ which are positive or which are equal to 0 but the preceding term of which is positive.

We shall now prove the following:

LEMMA 4.8.3. *For* $k = 0, 1, \ldots, n$,

$$P(\pi_{2n} = 2k) = \frac{\binom{2k}{k}\binom{2n-2k}{n-k}}{2^{2n}} \tag{4.8.37}$$

Proof. Clearly $\pi_{2n} = 0$ means that $\zeta_k \leqq 0$ for $k = 1, 2, \ldots, 2n$, i.e., that $\vartheta_{2n} = 0$. Thus we have

$$P(\pi_{2n} = 0) = P(\vartheta_{2n} = 0) = \frac{1}{2^{2n}}\binom{2n}{n}$$

i.e., (4.8.37) holds for $k = 0$. Similarly it also holds for $k = n$.

Now we use induction on n. Suppose that (4.8.37) is true for $n \leqq N - 1$ and consider $P(\pi_{2N} = 2k)$ for $1 \leq k \leq N - 1$. If $\pi_{2N} = 2k$ and $1 \leqq k \leqq N - 1$, then the sequence $\zeta_1, \ldots, \zeta_{2N}$ has to contain both positive and negative terms, and thus it contains at least one term equal to 0. Let $\rho_1 = 2l$ be the least value of n for which $\zeta_n = 0$. Then either $\zeta_n > 0$ for $n < 2l$ and $\zeta_{2l} = 0$, or $\zeta_n < 0$ for $n < 2l$ and $\zeta_{2l} = 0$; by symmetry both possibilities have the probability (see (4.8.29))

$$\frac{1}{2}P(\rho_1 = 2l) = \frac{\binom{2l-2}{l-1}}{l \cdot 2^{2l}}$$

Now evidently if $\zeta_n > 0$ for $n < 2l$ and $\zeta_{2l} = 0$, further if $\pi_{2N} = 2k$, then among the numbers $\zeta_{2l+1}, \ldots, \zeta_{2N}$ there are $2k - 2l$ positive ones or zeros preceded by a positive term, while in case $\zeta_n < 0$ for $n < 2l$, $\zeta_{2l} = 0$, and $\pi_{2N} = 2k$, the number of such terms is $2k$. Thus we get

$$P(\pi_{2N} = 2k) = \frac{1}{2}\sum_{l=1}^{k} P(\rho_1 = 2l)P(\pi_{2N-2l} = 2k - 2l) + \frac{1}{2}\sum_{l=1}^{N-k} P(\rho_1 = 2l)P(\pi_{2N-2l} = 2k) \tag{4.8.38}$$

From (4.8.38) it follows that if (4.8.37) holds for $n < N$, it also holds for $n = N$.

In order to show this, it is sufficient to prove the identity

$$\sum_{l=1}^{k}\binom{2l-2}{l-1}\frac{1}{l \cdot 2^{2l}}\binom{2k-2l}{k-l}\frac{1}{2^{2k-2l}} = \frac{\binom{2k}{k}}{2^{2k+1}} \quad \text{for } k \geqq 1 \tag{4.8.39}$$

Formula (4.8.39) follows from the remark that, with respect to (4.8.30) and (4.8.31), the left-hand side of (4.8.39) is the coefficient of x^k in the product of the power series $\sum_{l=1}^{+\infty}\binom{2l-2}{l-1} x^l/l\cdot 2^{2l}=(1-\sqrt{1-x})/2$ and the power series $\sum_{j=0}^{+\infty}\binom{2j}{j} x^j/2^{2j}=1/\sqrt{1-x}$, i.e., of the power series of the function

$$\frac{1-\sqrt{1-x}}{2\sqrt{1-x}}=\frac{1}{2\sqrt{1-x}}-\frac{1}{2}=\sum_{j=1}^{+\infty}\binom{2j}{j}\frac{x^j}{2^{2j+1}}$$

which proves (4.8.39). Thus the proof of Lemma 4.8.3 is complete. ■

Remark: It follows from Lemma 4.8.3 that

$$\frac{1}{2^{2n}}\sum_{k=0}^{n}\binom{2k}{k}\binom{2n-2k}{n-k}=1 \qquad (n=1,2,\ldots) \tag{4.8.40}$$

Another way to verify this interesting identity is to compare coefficients in the identity

$$\left(\sum_{j=0}^{+\infty}\binom{2j}{j}\frac{x^j}{2^{2j}}\right)^2=\left(\frac{1}{\sqrt{1-x}}\right)^2=\frac{1}{1-x}=\sum_{n=0}^{+\infty}x^n$$

From Lemma 4.8.3 we deduce the celebrated "arc sine law":

Theorem 4.8.6. *For* $0\leqq x\leqq 1$

$$\lim_{n\to+\infty}P\left(\frac{\pi_n}{n}<x\right)=\frac{2}{\pi}\arcsin\sqrt{x} \tag{4.8.41}$$

Proof. From the Moivre-Laplace theorem we get for $0<\varepsilon<k/n<1-\varepsilon$,

$$\frac{\binom{2k}{k}\binom{2n-2k}{n-k}}{2^{2n}}\approx\frac{1}{\pi n\sqrt{k/n(1-k/n)}}$$

Thus for $0<y<x<1$ we get

$$\lim_{n\to+\infty}P\left(y<\frac{\pi_{2n}}{2n}<x\right)=\frac{1}{\pi}\int_y^x\frac{du}{\sqrt{u(1-u)}}$$

In view of $(\arcsin\sqrt{x})'=1/2\sqrt{x(1-x)}$, we get

$$\frac{1}{\pi}\int_0^1\frac{du}{\sqrt{u(1-u)}}=\frac{2}{\pi}\arcsin\sqrt{1}=1$$

Thus, applying Theorem 4.3.4, it follows that

$$\lim_{n \to +\infty} P\left(\frac{\pi_{2n}}{2n} < x\right) = \frac{2}{\pi} \text{ arc sin } \sqrt{x}$$

As $0 \leqq \pi_{2n+1} - \pi_{2n} \leqq 1$, (4.8.41) follows. ■

Remark 1: It is interesting to remark that the characteristic function of $\pi_{2n} - n$, i.e., the function

$$E(e^{it(\pi_{2n}-n)}) = \frac{1}{2^{2n}} \sum_{k=0}^{n} \binom{2k}{k} \binom{2n-2k}{n-k} e^{it(2k-n)}$$

is equal to $P_n(\cos t)$, where $P_n(x)$ is the nth *Legendre polynomial*

$$P_n(x) = \frac{1}{2^n \cdot n!} \frac{d^n}{dx^n} (x^2 - 1)^n \tag{4.8.42}$$

The polynomials $\sqrt{(2n+1)/2}\, P_n(x)$ are, as is well known, orthonormal in the interval $[-1, +1]$. The characteristic function

$$\mathscr{J}_0(t) = \frac{1}{\pi} \int_0^1 \frac{e^{itx}\, dx}{\sqrt{x(1-x)}} \tag{4.8.43}$$

of the arc sine law is equal to the Bessel function of order 0,

$$\mathscr{J}_0(t) = \sum_{k=0}^{+\infty} \frac{(-1)^k}{(k!)^2} \left(\frac{t}{2}\right)^{2k} \tag{4.8.44}$$

Thus Theorem 4.8.6 can be deduced also from the relation

$$\lim_{n \to +\infty} P_n\left(\cos \frac{t}{n}\right) = \mathscr{J}_0(t) \tag{4.8.45}$$

(see Szegő [35]). Of course, conversely, (4.8.45) follows from Theorem 4.8.6.

Remark 2: Notice that the distribution (4.8.37) is symmetric about $k = n/2$, and is u-shaped. Thus, the most improbable value of π_{2n} is the value n if n is even, and the values $n - 1$ and $n + 1$ is n is odd; the most probable values are the values 0 and $2n$. Accordingly, the density function $1/\pi\sqrt{x(1-x)}$ of the limiting distribution of π_n/n is u-shaped and symmetric about $x = 1/2$, at which point it takes on its minimum. This shows that in questions concerning random fluctuation the naive intuition may be very misleading.

Theorems 4.8.2–4.8.6 are examples showing that by a deeper analysis one can detect in random fluctuations quite a number of hidden regularities, which we call "laws of chance." It should be emphasized again that while we presented these laws only for the special case of coin tossing, all these laws possess generalizations for sequences of independent random variables subject only to rather mild restrictions. Here we will not deal with these generalizations.

We shall present one more group of "laws of chance" which were discovered recently. We consider

$$\vartheta(N, k) = \max_{n \leqq N-k} \frac{\xi_{n+1} + \xi_{n+2} + \cdots + \xi_{n+k}}{k} \tag{4.8.46}$$

The quantity $\vartheta(N, k)$ can be interpreted as the maximal average gain of the player, betting on heads, over k consecutive throws, during a game consisting of N throws. We shall prove the following result (see Erdős and Rényi [36]):

THEOREM 4.8.7. *Let α be any number such that $0 < \alpha \leqq 1$. Put*

$$c(\alpha) = \left[\left(\frac{1+\alpha}{2}\right) \log_2 (1+\alpha) + \left(\frac{1-\alpha}{2}\right) \log_2 (1-\alpha)\right]^{-1} \tag{4.8.47}$$

Then one has, with probability 1,

$$\lim_{N \to +\infty} \vartheta(N, [c(\alpha)\log_2 N]) = \alpha \tag{4.8.48}$$

and, with probability 1,

$$\limsup_{n \to +\infty} \frac{\xi_{n+1} + \cdots + \xi_{n+[c(\alpha)\log_2 n]}}{c(\alpha)\log n} = \alpha \tag{4.8.49}$$

Remark: Let $\alpha = \alpha(c)$ for $c \geqq 1$, denote the unique solution, in the interval $0 < \alpha \leqq 1$, of the equation

$$\frac{1}{c} = \left(\frac{1+\alpha}{2}\right)\log_2(1+\alpha) + \left(\frac{1-\alpha}{2}\right)\log_2(1-\alpha) \tag{4.8.50}$$

Then (4.8.48) can be rewritten in the form: For every $c \geqq 1$ one has, with probability 1,

$$\lim_{N \to +\infty} \vartheta(N, [c \log_2 N]) = \alpha(c) \tag{4.8.51}$$

Proof. According to (4.5.21) and (4.4.5), if $k = [c(\alpha)\log_2 N/2]$ and $\varepsilon > 0$,

$$P\left(\frac{\xi_{n+1} + \cdots + \xi_{n+k}}{k} > \alpha + \varepsilon\right) = O(2^{-d((1+\alpha+\varepsilon)/2, 1/2)[c \log_2 N]})$$

where $d(f, p)$ is defined by (4.5.22); thus

$$d\left(\frac{1+\alpha+\varepsilon}{2}, \frac{1}{2}\right) = \frac{1}{c} + \delta_1$$

where $\delta_1 = \delta_1(\varepsilon) > 0$, because $x \log_2 2x + (1-x)\log_2 2(1-x)$ is strictly increasing in the interval $1/2 \leqq x \leqq 1$. Thus

$$P\left(\frac{\xi_{n+1} + \cdots + \xi_{n+k}}{k} > \alpha + \varepsilon\right) = O\left(\frac{1}{N^{1+c\delta_1}}\right) \tag{4.8.52}$$

It follows that

$$P\left(\vartheta\left(N, \left[c(\alpha)\log_2 \frac{N}{2}\right]\right) > \alpha + \varepsilon\right) = O\left(\frac{1}{N^{c\delta_1}}\right) \tag{4.8.53}$$

Similarly, we obtain from (4.5.21) that for $\varepsilon > 0$

$$P\left(\frac{\xi_{n+1} + \cdots + \xi_{n+k}}{k} > \alpha - \varepsilon\right) \geqq \frac{A}{N^{1-c\delta_2}} \tag{4.8.54}$$

where $\delta_2 = \delta_2(\varepsilon) > 0$ and $A > 0$. Thus we get

$$P\left(\vartheta\left(N, \left[c(\alpha) \log_2 \frac{N}{2}\right]\right) < \alpha - \varepsilon\right) \leqq \prod_{r=1}^{[N/k]-1} P\left(\frac{\xi_{kr+1} + \cdots + \xi_{kr+k}}{k} \leqq \alpha - \varepsilon\right)$$

$$\leqq \left(1 - \frac{A}{N^{1-c\delta_2}}\right)^{N/k-2} \leqq e^{-AN^{c\delta_2}/2 \log_2 N} \tag{4.8.55}$$

if $N \geqq N_0$; thus, the series

$$\sum_{N=2}^{+\infty} P\left(\vartheta\left(N, \left[c(\alpha)\log_2 \frac{N}{2}\right]\right) < \alpha - \varepsilon\right) \tag{4.8.56}$$

converges for every $\varepsilon > 0$. It follows from (4.8.53) that

$$P(\vartheta(2^{k+1/c(\alpha)}, k) > \alpha + \varepsilon) = O\left(\frac{1}{2^{k\delta_1}}\right)$$

where $\delta_1 > 0$, which implies that the series

$$\sum_{k=1}^{\infty} P(\vartheta(2^{k+1/c(\alpha)}, k) > \alpha + \varepsilon) \tag{4.8.57}$$

is convergent for every $\varepsilon > 0$. Now $[c(\alpha)\log_2 N] = k$ if $2^{k/c(\alpha)} \leqq N < 2^{(k+1)/c(\alpha)}$ and thus

$$\vartheta(N, [c(\alpha)\log_2 N]) \leqq \vartheta(2^{k+1/c(\alpha)}, k) \tag{4.8.58}$$

for $2^{k/c(\alpha)} \leqq N < 2^{(k+1)/c(\alpha)}$. Now it follows from (4.8.57), by the Borel-Cantelli lemma, that with probability 1, one has

$$\vartheta(2^{k+1/c(\alpha)}, k) < \alpha + \varepsilon$$

for all but a finite number of values of k. In view of (4.8.58) this implies that with probability 1

$$\vartheta(N, [c(\alpha)\log_2 N]) < \alpha + \varepsilon$$

for all but a finite number of values of N. On the other hand, it follows from the convergence of the series (4.8.56) that with probability 1

$$\vartheta(N, [c(\alpha)\log_2 N]) > \alpha - \varepsilon$$

for all but a finite number of values of N; this proves (4.8.48). One has further, from (4.8.52),

$$P\left(\frac{\xi_{n+1} + \cdots + \xi_{n+[c \log_2 n]}}{[c \log_2 n]} > \alpha + \varepsilon\right) = O\left(\frac{1}{n^{1+c\delta_1}}\right)$$

and thus the series

$$\sum_{n=2}^{+\infty} P\left(\frac{\xi_{n+1} + \cdots + \xi_{n+[c \log_2 n]}}{[c \log_2 n]} > \alpha + \varepsilon\right)$$

is convergent for every $\varepsilon > 0$; therefore it follows also from the Borel-Cantelli lemma, that with probability 1

$$\limsup_{n \to +\infty} \frac{\xi_{n+1} + \cdots + \xi_{n+[c \log_2 n]}}{c \log_2 n} \leqq \alpha \tag{4.8.59}$$

Similarly, it follows from (4.8.54) that, putting $n_k = [ck \log k]$, the series

$$\sum_{k=1}^{\infty} P\left(\frac{\xi_{n_k+1} + \cdots + \xi_{n_k+[c \log_2 n_k]}}{[c \log_2 n_k]} < \alpha - \varepsilon\right)$$

is divergent; as $n_{k+1} > n_k + [c \log_2 n_k]$, it follows from the divergence part of the Borel-Cantelli lemma that with probability 1

$$\limsup_{n \to +\infty} \frac{\xi_{n+1} + \cdots + \xi_{n+[c \log_2 n]}}{c \log_2 n} \geqq \alpha \tag{4.8.60}$$

Inequalities (4.8.59) and (4.8.60) imply that (4.8.50) holds with probability 1. Thus the proof of Theorem 4.8.7 is complete. ■

EXERCISES

E.4.1. Let $f(x)$ be a real, continuous function in the interval [0, 1]. Consider the polynomials

$$B_n(x) = \sum_{k=0}^{n} f\left(\frac{k}{n}\right)\binom{n}{k} x^k (1-x)^{n-k}$$

(called Bernstein's polynomials) and prove that one has

$$\lim_{n \to +\infty} B_n(x) = f(x) \tag{E.4.1.1}$$

uniformly in x.

Hint: If ν_n has a binomial distribution of order n and parameter x $(0 < x < 1)$, then $B_n(x) = E(f(\nu_n/n))$. Thus (E.4.1.1) follows immediately from Bernoulli's law of large numbers (Theorem 4.4.1) and from the remark to Theorem 4.2.1, according to which if $\xi_n \Rightarrow c$, where c is a constant, then the distribution of ξ_n tends weakly to the degenerate distribution of the constant c, and therefore if $f(x)$ is a bounded and continuous function, $E(f(\xi_n)) \to f(c)$. The uniformity of the convergence follows from the fact that ν_n/n tends to x strongly in L_2.

Remark: Thus we obtained a probabilistic proof of the theorem of Weierstrass, according to which every continuous function on a finite interval can be uniformly approximated arbitrarily closely by a polynomial. It should be noted that while this proof is perhaps the simplest, the convergence of $B_n(x)$ to $f(x)$ is rather slow: much better approximation can be obtained by other methods.

E.4.2. Let ρ_k $(k = 1, 2, \ldots)$ be an increasing sequence of positive numbers, such that $\sum_{k=1}^{+\infty} 1/\rho_k < +\infty$. Let us consider the entire function

$$f(z) = \prod_{k=1}^{+\infty} \left(1 + \frac{z}{\rho_k}\right)$$

Let the power series of $f(z)$ be

$$f(z) = \sum_{n=0}^{+\infty} a_n z^n$$

(Clearly $a_n > 0$ for $n = 0, 1, \ldots$.) Show that, putting for $r > 0$,

$$A(r) = \sum_{k=1}^{+\infty} \frac{r}{r + \rho_k} \quad \text{and} \quad B(r) = \left[\sum_{k=1}^{+\infty} \frac{r\rho_k}{(r+\rho_k)^2}\right]^{1/2}$$

one has, for every fixed value of x $(-\infty < x < +\infty)$,

$$\lim_{r \to +\infty} \frac{\sum_{n < A(r) + xB(r)} a_n r^n}{f(r)} = \Phi(x) = \frac{1}{\sqrt{2\pi}} \int_{-\infty}^{x} e^{-u^2/2} du \qquad \text{(E.4.2.1)}$$

and the convergence is uniform in x.

Hint: Notice that if $\xi_1(r), \ldots, \xi_k(r), \ldots$ are independent random variables taking on the values 0 and 1 with the corresponding probabilities

$$P(\xi_k(r) = 0) = \frac{\rho_k}{r + \rho_k} \quad \text{and} \quad P(\xi_k(r) = 1) = \frac{r}{r + \rho_k}$$

then $\{a_n r^n / f(r)\}$ is the probability distribution of the sum $\sum_{k=1}^{+\infty} \xi_k(r)$. Thus the method of proof of Theorem 4.7.1 can be applied.

Remark: Sharper, more general results of this type have been obtained by Hayman [37] and Rényi [38]. Notice that if $\rho_k = k^2$ $(k = 1, 2, \ldots)$, we have $f(z) = \sinh \pi\sqrt{z}/\pi\sqrt{z} = \sum_{n=0}^{+\infty} z^n/(2n+1)!$ In this case, $A(r) \sim B^2(r) \sim \pi\sqrt{r}/2$ and (E.4.2.1) can also be deduced from (4.7.31)

E.4.3. (a) Show that the exponential distribution can be uniquely characterized by its following property: If ξ is a positive random variable, it has an exponential distribution if and only if for every $x \geqq 0$ and $y \geqq 0$, one has

$$P(\xi \geqq x + y \mid \xi \geqq x) = P(\xi \geqq y) \tag{E.4.3.1}$$

Hint: If $G(x) = P(\xi \geqq x)$, (E.4.3.1) means that

$$G(x + y) = G(x)G(y) \tag{E.4.3.2}$$

and it is easy to show that the only monotonically decreasing solutions of this functional equation are $G(x) = e^{-\lambda x}$ with $\lambda > 0$.

(b) Show that the negative binomial distribution of order 1 (the Pascal distribution) can be characterized by its following property: If ξ is a nonnegative integral-valued random variable, it has negative binomial distribution of order 1 if and only if for all integers $n \geqq 0$ and $k \geqq 0$,

$$P(\xi \geqq n + k \mid \xi \geqq k) = P(\xi \geqq n)$$

Remark: Parts (a) and (b) imply that if ξ is exponentially distributed, then $[\xi]$ has a negative binomial distribution; of course, this can be verified directly.

E.4.4. (a) Let ξ_n $(n = 1, 2, \ldots)$ be independent random variables, taking on the values ± 1 with probability 1. Put $\zeta_0 = 0$, $\zeta_n = \xi_1 + \xi_2 + \cdots + \xi_n$ for $n \geqq 1$. Let L_{2n} denote the largest integer $k \leqq 2n$ such that $\zeta_k = 0$. Prove that

$$P(L_{2n} = 2k) = P(L_{2n} = 2n - 2k)$$

(b) Denote by $P(m, n)$ the probability that there is at least one k with $m \leqq k < m + n$ such that $\xi_{2k} = 0$. Deduce from (a) that

$$P(m, n) + P(n, m) = 1$$

(c) Show that

$$P(L_{2n} = 2k) = \frac{\binom{2k}{k}\binom{2n-2k}{n-k}}{2^{2n}}$$

(d) Show that if $c > 0$,

$$\lim_{n \to +\infty} P(n, cn) = 1 - \frac{2}{\pi} \text{arc sin} \frac{1}{\sqrt{1 + c}}$$

Hint: Let ρ_k denote, as in Section 4.8, the kth value of n for which $\zeta_n = 0$. Then we have, taking into account that $\rho_1, \rho_2 - \rho_1, \ldots, \rho_{j+1} - \rho_j, \ldots$ are independent and identically distributed,

$$\begin{aligned}
P(L_{2n} = 2k) &= \sum_{j=1}^{k} P(\rho_j = 2k, \rho_{j+1} > 2n) \\
&= \sum_{j=1}^{k} P(\rho_j = 2k)\, P(\rho_{j+1} - \rho_j > 2n - 2k) \\
&= P(\rho_1 > 2n - 2k) \sum_{j=1}^{k} P(\rho_j = 2k) = P(\rho_1 > 2n - 2k)\, P(\zeta_{2k} = 0) \\
&= \frac{\binom{2k}{k}\binom{2n-2k}{n-k}}{2^{2k} \cdot 2^{2n-2k}}
\end{aligned}$$

This proves (c) and (a). To deduce (b) from (a) one has to notice that

$$P(m, n) = P(L_{2m+2n-2} \geqq 2m) \tag{E.4.4.1}$$

and $1 - P(n, m) = P(L_{2m+2n-2} \leqq 2n - 2)$, and these two quantities are equal according to (a). Finally, (d) follows from (c) and the arc sine law.

Remark: Parts (a) and (b) are due to Blackwell, Deuel and Freedman (see [50]).

E.4.5. Give an example of a sequence of absolutely continuous probability distribution functions $F_n(x)$ $(n = 1, 2, \ldots)$ which converge weakly to an absolutely continuous probability distribution function $F(x)$, but the corresponding density functions $f_n(x) = F'(x)_n$ do not converge to $f(x) = F'(x)$ for any value of x such that $f(x) > 0$.

Hint: Consider the distribution functions

$$F_n(x) = x + \frac{\sin 2\pi nx}{2\pi n} \qquad 0 \leqq x \leqq 1$$

Then $F_n(x) \to F(x) = x$ for $0 \leqq x \leqq 1$, but $f_n(x) = F_n'(x) = 1 + \cos 2\pi nx$ does not converge to $f(x) = 1$ for any x in $[0, 1]$.

Moreover, we have

$$\int_0^1 |f_n(x) - f(x)|\, dx = \int_0^1 |\cos 2\pi nx|\, dx = \frac{2}{\pi}$$

E.4.6. Let us consider the Cantor series of the real number $x (0 < x < 1)$ with respect to an arbitrary sequence $\{q_n\}$ of positive integers such that $q_n \geqq 2$ $(n = 1, 2, \ldots)$, i.e., the representation of x in the form

$$x = \sum_{n=1}^{+\infty} \frac{c_n(x)}{q_1 q_2 \cdots q_n} \tag{E.4.6.1}$$

where $c_n(x)$ may take on the values $0, 1, \ldots, q_n - 1$.

Show that for almost all values of x

$$\lim_{N \to +\infty} \frac{\sum_{n=1}^{N} (c_n(x) + 1/2)/q_n}{N} = 1/2 \tag{E.4.6.2}$$

Hint: Notice that the $c_n(x)$, considered as random variables on the Lebesgue probability space, are independent and $P(c_n(x) = k) = 1/q_n$ for $k = 0, 1, \ldots, q_n - 1$. To obtain the statement under (a), apply Theorem 4.4.5 to the sequence $c_n(x)/q_n - (q_n - 1)/2q_n$ of uniformly bounded and independent random variables, each having expectation 0.

Remark: In the special case when $q_n = 2$ for all n, the statement of Exercise E.4.6 reduces to the following: that the relative frequency of the digit 1 in the binary expansion of x tends for almost all x to 1/2 (see Example 4.4.2).

E.4.7. Let us consider the expansion

$$x = \frac{1}{q_1} + \sum_{k=1}^{+\infty} \frac{1}{q_1(q_1 - 1)q_2(q_2 - 1) \cdots q_k(q_k - 1)q_{k+1}} \tag{E.4.7.1}$$

of a real number $x (0 < x < 1)$ defined as follows: Put

$$\xi_1(x) = x, q_1 = [1/\xi_1(x)], \xi_{k+1}(x) = q_k(q_k - 1)(\xi_k(x) - 1/q_k)$$

and $q_{k+1} = [1/\xi_{k+1}(x)]$ for $k = 1, 2, \ldots$.

(a) Prove that if $f(x)$ is any bounded measurable function, then for almost all x

$$\lim_{n \to +\infty} \frac{1}{n} \sum_{k=1}^{n} f(\xi_k(x)) = \int_0^1 f(x)\, dx \tag{E.4.7.2}$$

(b) Prove that if $g(k)$ is any function defined on the positive integers such that series the $\sum_{k=2}^{+\infty} g(k)/k(k-1)$ is convergent, then we have for almost all x

$$\lim_{n \to +\infty} \frac{1}{n} \sum_{k=1}^{n} g(q_k) = \sum_{k=2}^{+\infty} \frac{g(k)}{k(k-1)} \tag{E.4.7.3}$$

Hint: Define the transformation Tx of the interval $(0, 1)$ as follows: $Tx = [1/x]\,([1/x]-1)\,(x-[1/x]^{-1})$. It can be seen that Tx is a measure-preserving and ergodic transformation of the Lebesgue probability space. As further, $\xi_k(x) = T^k x$ follows from Birkhoff's ergodic theorem (Theorem 4.4.7), (E.4.7.3) is a special case of (E.4.7.2). Notice that the $\xi_k(x)$ $(k \geqq 1)$ considered as random variables on the Lebesgue probability space, are independent and identically distributed (all being uniformly distributed in $(0, 1)$); thus, Theorem 4.4.6 can also be applied.

E.4.8. The Γ distribution of order α ($\alpha > 0$ is real, but not necessarily an integer) and parameter $\lambda > 0$ is defined as the probability distribution on the positive half-line having the density function

$$f(x, \alpha, \lambda) = \frac{\lambda^\alpha x^{\alpha-1} e^{-\lambda x}}{\Gamma(\alpha)} \text{ for } x > 0, \text{ where } \Gamma(\alpha) = \int_0^{+\infty} t^{\alpha-1} e^{-t}\, dt$$

is the gamma function. Prove that $f(x, \alpha, \lambda) * f(x, \beta, \lambda) = f(x, \alpha + \beta, \lambda)$.

Hint:

$$f(x, \alpha, \lambda) * f(x, \beta, \lambda) = \frac{\lambda^{\alpha+\beta} x^{\alpha+\beta-1} e^{-\lambda x}}{\Gamma(\alpha)\Gamma(\beta)} \cdot \int_0^1 t^{\alpha-1}(1-t)^{\beta-1}\, dt$$

As $f(x, \alpha, \lambda) * f(x, \beta, \lambda)$ is a density function and the formula obtained for it differs from that for $f(x, \alpha + \beta, \lambda)$ only in the expression of the constant factor, these two expressions have to be equal; thus we get a probabilistic proof for the purely analytical identity

$$\int_0^1 t^{\alpha-1}(1-t)^{\beta-1}\, dt = \frac{\Gamma(\alpha)\Gamma(\beta)}{\Gamma(\alpha+\beta)} \tag{E.4.8.1}$$

due to L. Euler. The integral in (E.4.8.1) is called Euler's complete beta-function.

E.4.9. Let ξ have the density function

$$g(x, \alpha, \beta) = \frac{\Gamma(\alpha+\beta)}{\Gamma(\alpha)\Gamma(\beta)} x^{\alpha-1}(1-x)^{\beta-1} \quad (0 < x < 1)$$

The distribution function of ξ,

$$G(x, \alpha, \beta) = \frac{\Gamma(\alpha+\beta)}{\Gamma(\alpha)\Gamma(\beta)} \int_0^x t^{\alpha-1}(1-t)^{\beta-1}\,dt \quad (0 \leq x \leq 1)$$

is called the *incomplete beta integral* and ξ is said to have the beta distribution with parameters α and β.

(a) Prove that if η has a binomial distribution of order n and parameter p, and $q = 1 - p$, then the distribution function of η can be written in the form

$$P(\eta < r) = \frac{n!}{(r-1)!(n-r)!} \int_0^q x^{n-r}(1-x)^{r-1}\,dx \qquad (r = 1, 2, \ldots n) \quad \text{(E.4.9.1)}$$

i.e., can be expressed as the value of the incomplete beta integral with parameters $n - r + 1$, r at q.

(b) Using the result of (a), prove that if $\xi(\alpha, \beta)$ denotes a random variable having the beta distribution with integer parameters α and β, then for every real x we have

$$\lim_{\substack{\alpha\to+\infty,\beta\to+\infty,\\ \alpha/\beta\to c>0}} P\left(\xi(\alpha, \beta) < \frac{\alpha}{\alpha+\beta} + x\sqrt{\frac{\alpha\beta}{(\alpha+\beta)^3}}\right) = \Phi(x)$$

Hint: The result follows from the Moivre-Laplace theorem.

E.4.10. Deduce (4.7.33) from the identity (E.4.9.1) by substituting $p = \lambda/n$ and passing to the limit $n \to +\infty$.

Hint: Transform the integral on the right-hand side of (E.4.9.1) as follows:

$$\frac{n!}{(r-1)!(n-r)!} \int_0^q x^{n-r}(1-x)^{r-1}\,dx = \frac{\prod_{j=1}^{r-1}(1-j/n)}{(r-1)!} \int_{np}^n u^{r-1}(1-u/n)^{n-r}\,du$$

PROBLEMS

P.4.1. Let $\xi_n = \xi_n(\omega)$ $(n = 1, 2, \ldots)$ be a sequence of random variables on the probability space $S = (\Omega, \mathcal{A}, P)$, where Ω is a denumerable set. Prove that if the sequence ξ_n converges for $n \to +\infty$ in probability to a random variable ξ, then it converges to ξ almost surely; moreover, if S does not contain atoms having probability 0, then ξ_n converges to ξ everywhere.

P.4.2. (a) Find the necessary and sufficient condition that a probability distribution $\{w_n\}$ $(n = 0, 1, 2, \ldots)$ on the nonnegative integers should be the mixture of binomial distributions of different order with the same parameter p $(0 < p < 1)$.

Hint: Put

$$w(z) = \sum_{n=0}^{+\infty} w_n z^n$$

If $\{w_n\}$ is the mixture of binomial distributions with parameter p and $q = 1 - p$, we must have

$$w(z) = \sum_{N=0}^{+\infty} A_N (pz+q)^N \quad \text{where } A_N \geq 0$$

and $\sum_{N=0}^{+\infty} A_N = 1$. Thus $w(z)$ is an analytic function of z regular for $|z + q/p| < 1/p$ and its Taylor series around the point $-q/p$ should be convergent in the closed circle $|z + q/p| \leqq 1/p$. As by Taylor's formula,

$$w(z) = \sum_{N=0}^{+\infty} \frac{w^{(N)}(-q/p)}{N!\,p^N} (pz+q)^N$$

we must have $w^{(N)}(-q/p) \geqq 0$ for $N = 0, 1, \ldots$ and $\sum_{N=0}^{+\infty} w^{(N)}(-q/p)/N!p^N = 1$, and these conditions are evidently sufficient.

(b) A distribution $\{w_n\}$ $(n = 0, 1, 2, \ldots)$ is called a *compound Poisson distribution* if the probabilities w_n are of the form

$$w_n = \int_0^{+\infty} \frac{\lambda^n e^{-\lambda}}{n!} \, dF(\lambda)$$

where $F(\lambda)$ is a distribution function which is continuous for $\lambda = 0$ and $F(0) = 0$. Show that a distribution $\{w_n\}$ is the mixture of binomial distributions with parameter p for *all* values of p with $0 < p < 1$, if and only if $\{w_n\}$ is a compound Poisson distribution.

Hint: Use the following theorem of Bernstein [39]: If the function $f(x)$ is defined for $x \geqq 0$, if all its derivatives $f^{(n)}(x)$ exist $(n = 1, 2, \ldots)$, and if $(-1)^n f^{(n)}(x) \geqq 0$ for all x and for $n = 0, 1, 2, \ldots$, then there exists a distribution function $G(t)$ on $[0, +\infty)$ with $G(0) = 0$, and a constant $c > 0$ such that

$$f(x) = c \int_0^{+\infty} e^{-xt} \, dG(t)$$

(Such functions $f(x)$ are called *completely monotonic*.)

P.4.3. (a) Let $\xi_1, \xi_2, \ldots, \xi_n, \ldots$ be a Bernoulli sequence of random variables, that is, $P(\xi_n = 1) = p$, $P(\xi_n = 0) = 1 - p$ $(n = 1, 2, \ldots)$ where $0 < p < 1$, and let the variables ξ_n be independent. Put $\zeta_n = \xi_1 + \xi_2 + \cdots + \xi_n - [np]$ where $[x]$ denotes the integral part of x. Show that

$$\lim_{n \to +\infty} \frac{P(\zeta_n = k)}{P(\zeta_n = l)} = 1 \quad \text{for } k, l = 0, \pm 1, \pm 2, \ldots.$$

Remark: The result also can be stated as follows: The probability space generated by ζ_n on the set of integers tends for $n \to +\infty$ in the sense of Section 2.5 to the uniform conditional probability space on the set of integers.

(b) Generalize the statement under (a) for the case when $\xi_1, \xi_2, \ldots, \xi_n, \ldots$ are independent integer-valued random variables with the same distribution and with finite expectation p, and such that the greatest common divisor of those values which are taken on with positive probability by the variables ξ_n is equal to 1. (See Erdös and Chung [40].)

P.4.4. Let us consider an infinite sequence of independent repetitions of an experiment, the possible, mutually exclusive, outcomes of which are the events $A_1, A_2, \ldots, A_r$, having the probabilities $P(A_j) = p_j > 0$ $(j = 1, 2, \ldots, r)$. Put $\xi_n = j$ if the result of the nth experiment is the event A_j. Consider the vector $(\xi_1, \xi_2, \ldots, \xi_n) = \zeta_n$; its possible values are the sequences $(j_1, j_2, \ldots, j_n)$ having the probabilities $p_{j_1}p_{j_2}\cdots p_{j_n}$. Let π_n denote the probability of that value of ζ_n which actually was observed, i.e., put $\pi_n = p_{\xi_1}p_{\xi_2}\cdots p_{\xi_n}$. Prove that with probability 1,

$$\lim_{n\to+\infty}\frac{1}{n}\log_2\frac{1}{\pi_n} = H(\mathscr{P}) = \sum_{j=1}^{r} p_j \log_2 \frac{1}{p_j} \tag{P.4.4.1}$$

Hint: Let $\zeta_n(j)$ denote the frequency of the event A_j in the first n experiments. Then by the strong law of large numbers, one has almost surely

$$\frac{\zeta_n(j)}{n} \to p_j \qquad (j = 1, 2, \ldots, r) \tag{P.4.4.2}$$

As

$$\frac{1}{n}\log_2\frac{1}{\pi_n} = \sum_{j=1}^{r}\frac{\zeta_n(j)}{n}\log_2\frac{1}{p_j}$$

relation (P.4.4.1) follows from (P.4.4.2).

Remark: The statement of this problem is a special case of a general theorem of McMillan (see [41], further, [59] and [62]).

P.4.5. Let $\xi_1, \xi_2, \ldots, \xi_n, \ldots$ be a sequence of independent, identically distributed random variables having an everywhere continuous distribution function. Let us call a term ξ_n of the sequence a "record," if $\xi_n > \xi_k$ for $k = 1, 2, \ldots n-1$. Let A_n denote the event that ξ_n is a record and let ν_k $(k \geqq 1)$ denote the index of the kth record in the sequence ξ_n; i.e., suppose that the event A_n occurs if and only if n belongs to the increasing sequence ν_k (evidently, $\nu_1 = 1$).

Let $R(N)$ denote the number of records among $\xi_1, \xi_2, \ldots, \xi_N$

(a) Prove that the events A_n ar e in dependent and

$$P(A_n) = \frac{1}{n} \qquad (n = 2, 3, \ldots)$$

(b) Prove that for $n \geqq k$, we have

$$P(\nu_k = n) = \frac{1}{n(n-1)}\sum_{2\leq l_2<l_3<\cdots<l_{k-1}<n}\frac{1}{(l_2-1)(l_3-1)\cdots(l_{k-1}-1)} \tag{P.4.5.1}$$

(c) Prove that $R(N)/\log N \Rightarrow 1$.

Hints: (1) We may suppose that the numbers $\xi_1, \xi_2, \ldots, \xi_n$ are all different, because this holds with probability 1. Let us arrange the numbers $\xi_1, \ldots, \xi_n$ according to their order of magnitude. Let this arrangement be $\xi_{\rho_{n,1}} < \xi_{\rho_{n,2}} < \cdots < \xi_{\rho n,n}$ where $(\rho_{n,1}, \rho_{n,2}, \ldots, \rho_{n,n})$ is a (random) permutation of the integers $1, 2, \ldots, n$. Evidently, all $n!$ permutations have the same probability $1/n!$ as there are $(n-1)!$ permutations for which $\rho_{n,n} = n$, i.e., for which the event A_n occurs, we get $P(A_n) = (n-1)!/n! = 1/n$. Now let $B_n(r)$ denote the event that the permutation $(\rho_{n,1}, \rho_{n,2}, \ldots, \rho_{n,n})$ is equal to any fixed permutation $r = (r_1, r_2, \ldots, r_n)$ of the integers $1, 2, \ldots, n$; then under the condition $B_n(r)$ the permutation $(\rho_{n+1,1}, \ldots, \rho_{n+1,n+1})$ is obtained by inserting the number $n+1$

into the permutation $r = (r_1, r_2, \ldots, r_n)$ at any one of the possible $n + 1$ places, each such possibility having the same conditional probability $1/(n+1)$. Thus, $P(A_{n+1} \mid B_n(r)) = 1/(n+1)$ for each of the $n!$ permutations r of the numbers 1, 2, ..., n. Thus, it follows that A_{n+1} is independent from each $B_n(r)$ and thus from every event of the form $A_1^{\varepsilon_1} \ldots, A_n^{\varepsilon_n}$ where $\varepsilon_k = \pm 1$, $1 \leqq k \leqq n$, i.e., the events $A_1, A_2, \ldots, A_{n+1}$ are independent.

(2) It follows from (1) that if $2 \leqq l_2 < l_3 < \cdots < l_{k-1} < n$ and the l_j are integers, then

$$P(\nu_2 = l_2, \ldots, \nu_{k-1} = l_{k-1}, \nu_k = n) = \frac{1}{(l_2 - 1)(l_3 - 1)(l_{k-1} - 1)n(n-1)}$$

By summation over $l_2, \ldots, l_{k-1}$ we get (P.4.5.1).

(3) Let α_k denote the indicator of the event A_k. We have $R(N) = \sum_{k=1}^{N} \alpha_k$, and thus $E(R(N)) = 1 + 1/2 + \cdots + 1/N$ and

$$D^2(R(N)) = \sum_{k=1}^{N} \frac{1}{k} - \sum_{k=1}^{N} \frac{1}{k^2}$$

As $\sum_{k=1}^{N} 1/k \approx \log N$, it follows that $R(N)/\log N$ tends strongly in L_2 and thus in probability to 1.

Remark: We shall later prove (see Problem P.5.3) that $R(N)/\log N$ tends almost surely to 1, which implies that almost surely $\sqrt[k]{\nu_k} \to e$.

P.4.6. Show that if a sequence $\{\xi_n\}$ of random variables tends in probability to a random variable ξ, then one can find a subsequence $\{\xi_{n_k}\}$ $(n_1 < n_2 < \cdots)$ such that ξ_{n_k} tends almost surely to ξ for $k \to +\infty$.

Hint: We may suppose $\xi \equiv 0$. By definition for every $\varepsilon > 0$, we have

$$\lim_{n \to +\infty} P(|\xi_n| > \varepsilon) = 0$$

Let us choose n_k so that $n_k > n_{k-1}$ and $P(|\xi_{n_k}| > 1/2^k) \leqq 1/2^k$ $(k = 1, 2, \ldots)$. Then evidently, the series $\sum_{k=1}^{+\infty} P(|\xi_{n_k}| > \varepsilon)$ is convergent for every $\varepsilon > 0$, and thus by the Borel-Cantelli lemma $\xi_{n_k} \to 0$ almost surely for $k \to +\infty$.

P.4.7. Let $\xi_1, \xi_2, \ldots, \xi_n, \ldots$ be independent random variables, each taking on the values ± 1 with probability 1/2. Put $\zeta_0 = 0$ and $\zeta_n = \xi_1 + \xi_2 + \cdots + \xi_n$. Denote by R_n the number of different integers in the sequence $\zeta_0, \zeta_1, \ldots, \zeta_n$. Compute $E(R_n)$.

Hint: Put $\delta_0 = 1$, and for $k \geqq 1$ define δ_k as follows:

$$\delta_k = \begin{cases} 1 \text{ if } \zeta_k \text{ is different from each of the numbers } \zeta_0, \zeta_1, \ldots, \zeta_{k-1} \\ 0 \text{ otherwise} \end{cases}$$

Clearly, $R_n = \sum_{k=0}^{n} \delta_k$. Now $\delta_k = 1$ if and only if $\zeta_k - \zeta_j \neq 0$ for $j = 0, 1, \ldots, k-1$. Thus we have

$$\begin{aligned} P(\delta_k = 1) &= P\Big(\sum_{h=j+1}^{k} \xi_h \neq 0 \quad \text{for} \quad 0 \leqq j \leqq k-1\Big) \\ &= P(\zeta_j \neq 0 \quad \text{for} \quad j = 1, 2, \ldots, k) \\ &= P(\nu_1 > k - 1) \end{aligned}$$

Thus from (4.8.26) we get

$$P(\delta_k = 1) = \frac{1}{2^{k-1}} \binom{k-1}{[(k-1)/2]} \quad \text{for } k \geqq 1$$

It follows that

$$E(R_n) = \sum_{k=0}^{n} P(\delta_k = 1) = 1 + \sum_{k=1}^{n} \frac{1}{2^{k-1}} \binom{k-1}{[(k-1)/2]} \tag{P.4.7.1}$$

Remark: As $1/2^{k-1}\binom{k-1}{[(k-1)/2]} \approx \sqrt{2/\pi k}$, it follows that

$$E(R_n) \approx 2\sqrt{\frac{2n}{\pi}} \tag{P.4.7.2}$$

An alternative proof for this is as follows: Evidently,

$$R_n = \max_{0 \leqq k \leqq n} \zeta_k - \min_{0 \leqq k \leqq n} \zeta_k + 1$$

We have seen (see (4.8.23)) that

$$E(\max_{0 \leqq k \leqq n} \zeta_k) \sim \sqrt{\frac{2n}{\pi}}$$

By symmetry, we have

$$E(\min_{0 \leqq k \leqq n} \zeta_k) \sim -\sqrt{\frac{2n}{\pi}}$$

Thus we again get (P.4.7.2).

P.4.8. Let us define the "Lévy distance" $L(F, G)$ of two distribution functions F and G on the real line as the greatest lower bound of those positive numbers ε for which the inequality

$$F(x - \varepsilon) - \varepsilon \leqq G(x) \leqq F(x + \varepsilon) + \varepsilon$$

holds for all x.

(a) Show that $L(F, G)$ is a metric, i.e., $L(G, F) = L(F, G) \geqq 0$; $L(F, G) = 0$ if and only if $F = G$; for any three distribution functions F, G and H, we have $L(F, G) \leqq L(F, H) + L(H, G)$.

(b) Show that a sequence F_n of distribution functions tends weakly to a distribution function F if and only if $L(F_n, F) \to 0$.

P.4.9. The Hausdorff dimension of a subset Z of the interval $(0, 1)$ is defined as the greatest lower bound of those numbers β $(0 < \beta)$ for which for every $\varepsilon > 0$, the set Z can be covered with a finite system $I_1, I_2, \ldots, I_n$ of intervals such that $\sum_{k=1}^{n} |I_k|^\beta < \varepsilon$, where $|I_k|$ denotes the length of the interval I_k.*

* The notion of Hausdorff dimension is used to distinguish between "smaller" and "larger" sets of measure zero. Of course, the Hausdorff dimension of a set of positive measure is equal to 1. See [742].

(a) Let Z_c $(0 < c < 1/2)$ denote the set of those numbers x in the interval $(0, 1)$ which have the following property: If $x = \sum_{n=1}^{+\infty} \varepsilon_n(x)/2^n$ is the binary expansion of x, then

$$\limsup_{n\to+\infty} \frac{\sum_{k=1}^{n} \varepsilon_k(x)}{n} \leqq c < \frac{1}{2}$$

Show that the Hausdorff dimension of Z_c is $c \log_2 1/c + (1-c)\log_2 1/(1-c)$.

Hint: If $x \in Z_c$, then for every c' for which $c < c' < 1/2$ and for every sufficiently large n,

$$\frac{\sum_{k=1}^{n} \varepsilon_k(x)}{n} < c' \tag{P.4.9.1}$$

Dividing the interval $(0, 1)$ into 2^n subintervals $(j/2^n, (j+1)/2^n)$ $(j = 0, 1, \ldots, 2^n - 1)$ of length $1/2^n$, denoting by N_c the number of those such subintervals for which (P.4.9.1) holds, we have $N_c = 2^n P(\sum_{k=1}^{n} \varepsilon_k(x) < c'n)$. According to Theorem 4.5.2,

$$N_c \approx \frac{2^{nh(c')}}{\sqrt{2\pi nc'(1-c')}}$$

where $h(x) = x \log_2 1/x + (1-x)\log_2 1/(1-x)$. The sum of the βth powers of the lengths of these intervals is therefore $2^{n(h(c')-\beta)}/\sqrt{2\pi nc'(1-c')}$ and this can be made arbitrarily small if $\beta > h(c')$. It follows that the Hausdorff dimension of Z_c is $\leqq h(c)$, and by a slightly more detailed analysis one gets that the Hausdorff dimension of Z_c is in fact equal to $h(c)$.

P.4.9. Let $\xi_1, \xi_2, \ldots, \xi_n, \ldots$ be independent and identically distributed random variables having the same distribution function $F(x)$. Put $\eta_n = \max(\xi_1, \xi_2, \ldots, \xi_n)$.

(a) Show that if the ξ_n are uniformly distributed in the interval $(0, 1)$, then

$$\lim_{n\to+\infty} P(n(1-\eta_n) < x) = 1 - e^{-x} \qquad \text{for } x \geqq 0$$

i.e., $n(1-\eta_n)$ is in the limit exponentially distributed.

(b) Show that if the ξ_n have a standard normal distribution, then for every real x

$$\lim_{n\to+\infty} P\left(\eta_n < \sqrt{2\log n} - \frac{\log(2\sqrt{\pi}\log n)}{2\sqrt{2\log n}} + \frac{x}{2\sqrt{2\log n}}\right) = e^{-e^{-x}}$$

(c) Show that if the variables ξ_n have the density function $f(x) = 1/\pi(1+x^2)$ (i.e., the ξ_n have Cauchy distribution), then

$$\lim_{n\to+\infty} P\left(\eta_n < \frac{nx}{\pi}\right) = e^{-1/x} \qquad \text{for } x > 0$$

Hint: Use $P(\eta_n < y) = F^n(y)$ and $\lim_{n\to+\infty}(1 + a/n)^n = e^a$.

Remark: The question of the possible types of limit distributions of η_n for arbitrary $F(x)$ has been solved by Gnedenko [43].

P.4.10. Let us consider a Poisson process $\{\tau_n\}$ with density λ.

(a) For every n let us make a random choice: let us keep τ_n with probability p $(0 < p < 1)$ and remove it from the process with probability $q = 1 - p$. Suppose that these random choices are independent from each other as well as from the whole process $\{\tau_n\}$. Show that the remaining variables τ_{ν_k} again form a Poisson process with density $p\lambda$.

(b) Let us consider those τ_n which lie in the interval (A, B). Show that under the condition that the number of these τ_n is equal to $k \geqq 1$, these k points are distributed in the interval (A, B) as if we would have chosen k points in this interval independently with uniform distribution; i.e., the number of points in any subinterval (a, b) of (A, B) has a binomial distribution of order k and parameter $p = (b - a)/(B - A)$.

Hints: (a) Let τ_{ν_k} $(k = 1, 2, \ldots; \nu_1 < \nu_2 < \cdots)$ be the remaining elements of the sequence $\{\tau_n\}$. Clearly, $P(\nu_1 = k) = pq^{k-1}$ $(k = 1, 2, \ldots)$. Thus the distribution of τ_{ν_k} is, by the theorem of total probability,

$$P(\tau_{\nu_1} < x) = \sum_{k=1}^{+\infty} P(\tau_k < x) pq^{k-1} = 1 - e^{-\lambda p x}$$

We get similarly, that

$$P(\tau_{\nu_{k+1}} - \tau_{\nu_k} < x) = 1 - e^{-\lambda p x}$$

As the variables $\tau_{\nu_{k+1}} - \tau_{\nu_k}$ are independent, it follows that $\{\tau_{\nu_k}\}$ is a Poisson process with parameter $p\lambda$.

(b) Using Theorem 4.6.2, it follows that if $A < a < b < B$ and $p = (b - a)/(B - A)$ and $q = 1 - p$, then

$$P(\zeta(I[a, b)) = l \mid \zeta(I[A, B)) = k) = \binom{k}{l} p^l q^{k-l} \qquad (0 \leqq l \leqq k)$$

Remark: It can be shown that property (a) is characteristic for the Poisson process (see Rényi [51]). Concerning limit theorems connected with the transformation, described under (a), of a general process $\{\tau_n\}$, see Rényi [51] and [54], Nawrotzki [52], Belaev [53].

REFERENCES

[1] B. V. Gnedenko and A. N. Kolmogoroff, *Limit Distributions for Sums of Independent Random Variables*, trans. and annoted by K. L. Chung with an appendix by J. L. Doob, Addison-Wesley, Cambridge, Mass., 1954.

[2] P. Révész, *The Laws of Large Numbers*, Academic Press, New York, 1967.

[3] F. Spitzer, *Principles of Random Walk*, Van Nostrand, Princeton, 1964.

[4] I. A. Ibrahimov and Yu. V. Linnik, *Independent and Stationarily Dependent Random Variables* (in Russian), Nauka, Moscow, 1965.

[5] Yu. V. Prochorov, "Convergence of Random Processes and Limit Theorems in Probability Theory," *Theor. Prob. Appl.*, 1: 157–214, 1956.

[6] K. R. Parthasarathy, *Probability Measures on Metric Spaces*, Academic Press, New York, 1967.

[7] H. G. Hardy, *Divergent Series*, Oxford University Press, New York, 1949.
[8] W. Zeller, *Allgemeine Limitierungsverfahren*, Springer, Berlin, 1953.
[9] H. Steinhaus, "Some Remarks on the Generalization of Limit" (in Polish), *Prace Matematyczno Fizyczne*, **22**: 121–134, 1911.
[10] N. Dunford and J. T. Schwarz, *Linear Operators I*, Interscience, New York, 1958.
[11] S. Saks, "On Some Functionals," *Trans. Am. Math. Soc.*, **35**: 549–556 and 965–970, 1933.
[12] I. Schur, "Über lineare Transformationen in der Theorie der unendlichen Reihen," *J. Reine Angew. Math.*, **151**: 79–111, 1921.
[13] V. Doubrovsky, "On Some Properties of Completely Additive Set Functions," *Isvestia Akad. Nauk. SSSR Ser. Math.*, **9**: 311–320, 1945.
[14] Y. N. Dowker, "Finite and σ-Finite Invariant Measures," *Annals Math.*, **54**: 595–608, 1951.
[15] J. Bernoulli, *Ars Coniectandi*, Basel, 1713.
[16] F. Riesz, B. Sz. Nagy, *Functional Analysis*, Blackie, London, 1956.
[17] A. M. Garsia, "A Simple Proof of Hopf's Maximal Ergodic Theorem," *J. Math. Mech.* **14**: 381–382, 1965.
[18] A. de Moivre, *The Doctrine of Chances*, 3rd edition, London, 1756. (A part is reprinted in F. N. David, *Games, Gods and Gambling*, Griffin, London, 1962.)
[19] P. S. Laplace, *Théorie Analytique de Probabilités* (1791), *Oeuvres Complètes de Laplace*, Vol. 7, Gauthier-Villars, Paris, 1886.
[20] A. C. Aitken, *Determinants and Matrices*, Oliver and Boyd, Edinburgh, 1948.
[21] A. Rényi, "Remarks on the Poisson Process," *Studia Sci. Math. Hung.*, **2**: 119–123, 1967.
[22] J. Goldman, "Stochastic Point Processes: Limit Theorems," *Ann. Math. Stat.*, **38**: 771–779, 1967.
[23] P. A. P. Moran, "A Non-Markovian Quasi-Poisson Process," *Studia Sci. Math. Hung.*, **2**: 425–429, 1967.
[24] P. M. Lee, "Some Examples of Infinitely Divisible Point Processes," *Studia Sci. Math. Hung.*, **3**: 219–224, 1968.
[25] M. A. Evgrafov, *The Abel-Gontcharov Interpolation Problem* (in Russian), Gostechizdat, Moscow, 1954.
[26] J. L. Doob, *Stochastic Processes*, Wiley, New York, 1953.
[27] A. Blanc-Lapierre and R. Fortet, *Théorie des Fonctions Aléatoires*, Masson, Paris, 1953.
[28] W. Lindeberg, "Eine neue Herleitung des Exponentialgesetzes in der Wahrscheinlichkeitsrechnung," *Math. Zeitschrift*, **15**: 211–225, 1922.
[29] W. Feller, *An Introduction to Probability Theory and Its Applications*, Vol. 2, Wiley, New York, 1965.
[30] A. Khintchine, "Über dyadische Brüche," *Math. Zeitschrift*, **18**: 109–116, 1923.
[31] A. N. Kolmogoroff, "Über das Gesetz des iterierten Logarithmus," *Math. Ann.* **101**: 126–135, 1929.
[32] W. Feller, "The General Form of the So-Called Law of the Iterated Logarithm," *Trans. Am. Math. Soc.*, **54**: 373–402, 1943.

[33] P. Erdős, "On the Law of the Iterated Logarithm," *Annals. Math.*, **43**: 419–436, 1942.

[34] V. Strassen, "An Invariance Principle for the Law of Iterated Logarithm," *Zeitschrift Wahrscheinlichkeitstheorie*, **3**: 221–226, 1964.

[35] G. Szegő, "Orthogonal Polynomials," *Am. Math. Soc. Coll. Publ.* 23, Am. Math. Soc., New York, 1959.

[36] P. Erdös and A. Rényi, "Some New Limit Theorems in Probability Theory," (in print in *Journal d'Analyse*).

[37] W. K. Hayman, "A Generalization of Stirling's Formula," *J. Reine Angew. Math.*, **196**: 67–95, 1956.

[38] A. Rényi, "Probabilistic Methods in Analysis" (in Hungarian), *Matematika Lapok*, **18**: 5–35, 1967.

[39] S. Bernstein, "Demonstration du Théorème de Weierstrass Fondée sur la Calcul des Probabilités," *Soob. Charkov Mat. Obs.*, **13**: 1–2, 1912.

[40] P. Erdös, K. L. Chung, "Probability Limit Theorems Assuming only the First Moments," *Memoirs Am. Math. Soc.*, **6**: 1–19, 1950.

[41] B. McMillan, "The Basic Theorems of Information Theory," *Annals Math. Stat.*, **24**: 169–219., 1953.

[42] F. Hausdorff, *Set Theory*, Chelsea, New York, 1957.

[43] B. V. Gnedenko, *The Theory of Probability*, Chelsea, New York, 1962.

[44] M. Loève, *Probability Theory*, 2nd edition, Van Nostrand, New York, 1966.

[45] U. Grenander, *Probabilities on Algebraic Structures*, Wiley, New York, 1963.

[46] A. Ehrenfeucht and M. Fisz, "A Necessary and Sufficient Condition for the Validity of the Weak Law of Large Numbers," *Bull. Polon. Acad. Sci.*, **8**: 583–585, 1960.

[47] D. Blackwell and J. L. Hodges, Jr., "Elementary Path Counts," *Am. Math. Monthly*, **74**: 801–804, 1967.

[48] R. G. Cooke, *Infinite Matrices and Sequence Spaces*, MacMillan, London, 1950.

[49] Lucretius, *On the Nature of the Universe*, trans. by R. E. Latham, pp. 63–64, Penguin Books, London, 1951.

[50] D. Blackwell, P. Deuel, and D. Freedman, "The Last Return to Equilibrium in a Coin-Tossing Game," *Annals, Math. Stat.*, **35**: 1344, 1964.

[51] A. Rényi, "A Characterization of the Poisson Process" (in Hungarian), *Publ. Math. Inst. Hung. Acad. Sci.*, **1**: 519–527, 1956.

[52] K. Nawrotzki, "Ein Grenzwertsatz für homogene zufällige Punktfolgen (Verallgemeinerung eines Satzes von A. Rényi)," *Math. Nachrichten*, **24**: 202–218, 1962.

[53] J. K. Belaev, "Limit Theorems for Rearing Random Flows" (in Russian), *Teor. Veroyat. Primen.*, **8**: 175–184, 1963.

[54] A. Rényi, "On Two Mathematical Models of the Traffic on a Divided Highway," *J. Appl. Prob.*, **1**: 311–320, 1964.

[55] A. Rényi, *Wahrscheinlichkeitsrechnung, mit einem Anhang über Informationstheorie*, VEB Deutscher Verlag der Wissenschaften, Berlin, 1962.

[56] W. Blaschke, *Vorlesungen über Integralgeometrie*, 3rd edition, VEB Deutscher Verlag der Wissenschaften, Berlin, 1955.

[57] J. Neveu, *Mathematical Foundations of the Calculus of Probability*, Holden-Day, Inc., 1965.

[58] P. Halmos, *Lectures on Ergodic Theory*, The Mathematical Society of Japan, Tokyo, 1956.

[59] P. Billingsley, *Ergodic Theory and Information*, Wiley, New York, 1965.

[60] K. Jacobs, *Neuere Methoden und Ergebnisse der Ergodentheorie*, *Ergebnisse der Mathematik und ihrer Grenzgebiete*, *Nr*. 29, Springer.

[61] F. N. David, *Games, Gods and Gambling*, Griffin, London, 1962.

[62] A. Feinstein, *Foundations of Information Theory*, McGraw-Hill, New York, 1958.

[63] E. Lukacs, *Characteristic Functions*, Griffin, London, 1960.

CHAPTER 5

DEPENDENCE

5.1 CONDITIONAL EXPECTATIONS WITH RESPECT TO A σ-ALGEBRA

DEFINITION 5.1.1. *Let η be a random variable on the probability space $S = (\Omega, \mathscr{A}, P)$ such that $E(\eta)$ exists. Let $\mathscr{B} \subseteq \mathscr{A}$ be a purely atomic σ-algebra, with atoms B_k having the probabilities $P(B_k) = p_k \geqq 0$ $(k = 1, 2, \ldots)$. The conditional expectation of η with respect to the σ-algebra $\mathscr{B}$, denoted by $E(\eta \mid \mathscr{B})$, is a random variable defined as follows:*

$$E(\eta \mid \mathscr{B}) = E(\eta \mid B_k) \qquad \text{if } \omega \in B_k \text{ and } P(B_k) > 0 \tag{5.1.1}$$

Remark 1: The value of the conditional expectation $E(\eta \mid \mathscr{B})$ is not specified if $\omega \in B_k$ where $P(B_k) = 0$. Thus, $E(\eta \mid \mathscr{B})$ is not necessarily uniquely defined. However, as by definition $\sum_{P(B_k)>0} P(B_k) = 1$, it is uniquely defined almost everywhere; in other words, any two versions of $E(\eta \mid \mathscr{B})$ coincide almost everywhere. Let us recall, however, our convention that two random variables which are almost surely equal, are considered as identical. Thus the slight ambiguity about the definition of $E(\eta \mid \mathscr{B})$ is irrelevant.

Clearly, $E(\eta \mid \mathscr{B})$ is additive, i.e., if η_1 and η_2 are any random variables with finite expectations and c_1, c_2 are constants, we have

$$E(c_1\eta_1 + c_2\eta_2 \mid \mathscr{B}) = c_1 E(\eta_1 \mid \mathscr{B}) + c_2 E(\eta_2 \mid \mathscr{B}) \tag{5.1.2}$$

If $\mathscr{A}_\eta$ and $\mathscr{B}$ are independent σ-algebras, then

$$E(\eta \mid \mathscr{B}) = E(\eta) \tag{5.1.3}$$

Remark 2: Let ξ be an arbitrary discrete random variable such that the σ-algebra $\mathscr{A}_\xi$ generated by ξ coincides with $\mathscr{B}$; in other words, suppose that $\xi(\omega) = x_k$ for $\omega \in B_k$ $(k = 1, 2, \ldots)$ where the real numbers $x_1, x_2, \ldots$ are all different. Then the value of $E(\eta \mid \mathscr{B}) = E(\eta \mid \mathscr{A}_\xi)$ depends (almost surely) only on the value of ξ; thus, $E(\eta \mid \mathscr{A}_\xi)$ can be interpreted as *the conditional expectation of η given the value of ξ*. (For this reason the notation $E(\eta \mid \xi)$ for $E(\eta \mid \mathscr{A}_\xi)$ will also be used.) Note that $E(\eta \mid \mathscr{A}_\xi)$ does not depend on the actual values of ξ. If, for instance, $f(x)$ is any function such that $f(x_k) \neq f(x_l)$ if $x_k \neq x_l$ and $P(\xi = x_k) > 0$, $P(\xi = x_l) > 0$, then $E(\eta \mid \mathscr{A}_\xi) = E(\eta \mid \mathscr{A}_{f(\xi)})$.

We now prove the following:

Theorem 5.1.1. *Let η be a random variable on the probability space $S = (\Omega, \mathscr{A}, P)$ such that $E(\eta)$ exists. Let $\mathscr{B} \subseteq \mathscr{A}$ be a purely atomic sub-σ-algebra of $\mathscr{A}$. Then the conditional expectation $E(\eta \mid \mathscr{B})$ is a random variable on the probability space $S' = (\Omega, \mathscr{B}, P)$, having the following property: The expectation of $E(\eta \mid \mathscr{B})$ exists and if ζ is any random variable on S' such that $E(\eta\zeta)$ exists, then*

$$E(E(\eta \mid \mathscr{B})\zeta) = E(\eta\zeta) \tag{5.1.4}$$

This property characterizes $E(\eta \mid \mathscr{B})$, that is, if η^* is a random variable on S' such that for any random variable ζ on S' such that $E(\eta\zeta)$ exists, one has*

$$E(\eta^*\zeta) = E(\eta\zeta) \tag{5.1.5}$$

then $\eta^* = E(\eta \mid \mathscr{B})$ almost surely.

Proof. Let B_k $(k = 1, 2, \ldots)$ be all atoms with positive probability of $\mathscr{B}$. Then any random variable ζ on S' is necessarily constant on each B_k and thus if $\zeta = z_k$ on B_k,

$$E(\eta\zeta) = \sum_k P(B_k) z_k E(\eta \mid B_k) = E(E(\eta \mid \mathscr{B})\zeta)$$

Conversely, if (5.1.5) holds for every random variable ζ on S' for which $E(\eta\zeta)$ exists, take $\zeta = 1/P(B_k)$ for $\omega \in B_k$ and $\zeta = 0$ otherwise; then, as η^* has to be constant on B_k,

$$\eta^* = E(\eta \mid B_k) \qquad \text{for } \omega \in B_k$$

for $k = 1, 2, \ldots$, i.e., $\eta^* = E(\eta \mid \mathscr{B})$ almost surely. ∎

Taking for ζ the indicator of an event $B \in \mathscr{B}$ with $P(B) > 0$, we get the following:

Corollary to Theorem 5.1.1. *If $B \in \mathscr{B}$ and $P(B) > 0$, then*

$$E(E(\eta \mid \mathscr{B}) \mid B) = E(\eta \mid B) \tag{5.1.6}$$

In particular, choosing $B = \Omega$, we get

$$E(E(\eta \mid \mathscr{B})) = E(\eta) \tag{5.1.7}$$

Notice that (5.1.7) is equivalent to the theorem on total expectation (see Theorem 2.10.1).

Our aim in this section is to extend—following Kolmogoroff [1]—the definition of $E(\eta \mid \mathscr{B})$ for an arbitrary σ-algebra $\mathscr{B} \subseteq \mathscr{A}$, so that the statement of Theorem 5.1.1 should remain valid. This can be attained by means of the Radon-Nikodym theorem (see Appendix A), according to which if the

* It is sufficient even to suppose that (5.1.5) holds for every random variable ζ on S' taking on a finite number of values only.

bounded signed measure μ on the σ-algebra $\mathscr{B}$ is absolutely continuous with respect to the probability measure P on $\mathscr{B}$, then there exists a $\mathscr{B}$-measurable function $d\mu/dP$ called the Radon-Nikodym derivative of μ with respect to P, such that for every $B \in \mathscr{B}$,

$$\mu(B) = \int_B \left(\frac{d\mu}{dP}\right) dP$$

This theorem can be applied to the signed measure

$$\mu(B) = \int_B \eta dP \tag{5.1.8}$$

which is clearly absolutely continuous with respect to P. Thus we arrive at the following:

DEFINITION 5.1.2. *Let η be a random variable having finite expectation, on the probability space $S = (\Omega, \mathscr{A}, P)$. Let $\mathscr{B}$ be an arbitrary σ-algebra of subsets of Ω such that $\mathscr{B} \subseteq \mathscr{A}$. The conditional expectation $E(\eta \mid \mathscr{B})$ of η with respect to $\mathscr{B}$ is defined as the Radon-Nikodym derivative of the signed measure* (5.1.8) *on $\mathscr{B}$, with respect to P.*

Remark: It follows from Theorem 5.1.1 that in case $\mathscr{B}$ is purely atomic, the two definitions of $E(\eta \mid \mathscr{B})$, as given in Definitions 5.1.1 and 5.1.2, coincide. In other words, if $\mathscr{B}$ is purely atomic, and its atoms having positive probabilities are the sets B_k $(k = 1, 2, \ldots)$, and $\mu(B)$ is defined by (5.1.8) on $\mathscr{B}$, then

$$\frac{d\mu}{dP} = E(\eta \mid B_k) \qquad \text{for } \omega \in B_k \ (k = 1, 2, \ldots) \tag{5.1.9}$$

Of course, for $\mathscr{B} = \mathscr{A}$ we have

$$E(\eta \mid \mathscr{A}) = \eta \tag{5.1.10}$$

According to the definition, $E(\eta \mid \mathscr{B})$ is a random variable on the probability space $S' = (\Omega, \mathscr{B}, P)$, and in view of (5.1.8), the first statement of Theorem 5.1.1 also holds in the general case. The second statement also remains valid, as the Radon-Nikodym derivative $d\mu/dP$ is unique in the sense that if for two $\mathscr{B}$-measurable functions f and g one has $\int_B f dP = \int_B g dP$ for every $B \in \mathscr{B}$, then $f = g$ almost surely.*

Thus we have proved the following:

THEOREM 5.1.2. *The statement of Theorem* 5.1.1 *remains valid for an arbitrary σ-algebra $\mathscr{B} \subseteq \mathscr{A}$ by defining $E(\eta \mid \mathscr{B})$ according to Definition* 5.1.2.

* As a matter of fact, if B_+ is the set on which $f > g$ and B_- is the set on which $f < g$, then $B_+ \in \mathscr{B}$ and $B_- \in \mathscr{B}$, and therefore $\int_{B_+} (f - g)\, dP = \int_{B_-} (g - f)\, dP = 0$, which implies $P(B_+) = P(B_-) = 0$.

COROLLARY TO THEOREM 5.1.2. *For an arbitrary σ-algebra* $\mathscr{B} \subseteq \mathscr{A}$, (5.1.6) *and* (5.1.7) *hold.*

Remark: Notice that thus we obtained a generalization of the theorem of total expectation.

The following theorems, which follow immediately from Definition 5.1.2, state certain properties of conditional expectation which will often be used:

THEOREM 5.1.3. *The conditional expectation* $E(\eta \mid \mathscr{B})$ *is almost surely additive, i.e., if* η_1 *and* η_2 *are random variables with finite expectations and* c_1 *and* c_2 *are constants, then one has, almost surely,*

$$E(c_1\eta_1 + c_2\eta_2 \mid \mathscr{B}) = c_1E(\eta_1 \mid \mathscr{B}) + c_2E(\eta_2 \mid \mathscr{B}) \tag{5.1.11}$$

THEOREM 5.1.4. *If* $\mathscr{A}_\eta$ *and* $\mathscr{B}$ *are independent, then*

$$E(\eta \mid \mathscr{B}) = E(\eta) \tag{5.1.12}$$

Conversely, if $\mathscr{A}^* \subset \mathscr{A}$ *is a σ-algebra such that for every random variable* η *on* $(\Omega, \mathscr{A}^*, P)$ *having finite expectation* (5.1.12) *holds almost surely, then* $\mathscr{A}^*$ *and* $\mathscr{B}$ *are independent.*

Remark: In particular, (5.1.12) holds if $\mathscr{B}$ is the trivial σ-algebra $\mathscr{B} = \{\Omega, \varnothing\}$.

THEOREM 5.1.5. *If* ζ *is a random variable on* $S' = (\Omega, \mathscr{B}, P)$ *and* η *on* $(\Omega, \mathscr{A}, P)$ *where* $\mathscr{B} \subseteq \mathscr{A}$, *and the expectations of* η *and of* $\eta\zeta$ *(on* S*) exist; then we have*

$$E(\eta\zeta \mid \mathscr{B}) = \zeta E(\eta \mid \mathscr{B}) \tag{5.1.13}$$

Proof. Equation (5.1.13) follows immediately from (5.1.4), which holds by Theorem 5.1.2. ■

Let us consider now the case when $E(\eta^2)$ exists. In this case there is an alternative way to define $E(\eta \mid \mathscr{B})$, as can be seen from the following:

THEOREM 5.1.6. *Let* η *be a random variable on the probability space* $S = (\Omega, \mathscr{A}, P)$ *such that* $E(\eta^2) < +\infty$. *Let* $\mathscr{B}$ *be a σ-algebra of subsets of* Ω *such that* $\mathscr{B} \subseteq \mathscr{A}$ *and put* $S' = (\Omega, \mathscr{B}, P)$. *Let* $L_2(S)$ *and* $L_2(S')$ *denote the Hilbert space of all random variables on* S *respective on* S' *having finite second moment. Then* $E(\eta \mid \mathscr{B})$ *is the projection of* $\eta \in L_2(S)$ *onto the subspace* $L_2(S')$ *of* $L_2(S)$, *and thus* $E(E^2(\eta \mid \mathscr{B}))$ *exists and*

$$E(E^2(\eta \mid \mathscr{B})) \leqq E(\eta^2) \tag{5.1.14}$$

Proof. Let η^* be the projection of η onto $L_2(S')$. Then by definition $\eta - \eta^*$ is orthogonal to every $\zeta \in L_2(S')$. Thus we have for each $\zeta \in L_2(S')$

$$E(\eta\zeta) = E(\eta^*\zeta) + E((\eta - \eta^*)\zeta) = E(\eta^*\zeta) \tag{5.1.15}$$

Thus η^* satisfies (5.1.5) for every $\zeta \in L_2(S')$. Thus—in view of the footnote to Theorem 5.1.1—the first statement of Theorem 5.1.6 follows from Theorem 5.1.1. It follows from (5.1.15) that

$$E(\eta^2) = E(\eta^{*2}) + E((\eta - \eta^*)^2)$$

Thus (5.1.14) holds. ■

The following result is an immediate consequence of Definition 5.1.2:

THEOREM 5.1.7. *Let η be a random variable on the probability space $S = (\Omega, \mathscr{A}, P)$ such that $E(\eta)$ exists. Let $\mathscr{B}_1$ and $\mathscr{B}_2$ be two σ-algebras of subsets of Ω such that $\mathscr{B}_1 \subseteq \mathscr{B}_2 \subseteq \mathscr{A}$. Then we have*

$$E(\eta \mid \mathscr{B}_1) = E(E(\eta \mid \mathscr{B}_2) \mid \mathscr{B}_1) \tag{5.1.16}$$

The general notion of conditional expectation also leads to a natural extension of the notion of conditional probability.

DEFINITION 5.1.3. *If A is any event in the probability space $S = (\Omega, \mathscr{A}, P)$ and $\mathscr{B}$ is a σ-algebra of subsets of Ω such that $\mathscr{B} \subseteq \mathscr{A}$, the conditional probability $P(A \mid \mathscr{B})$ of A given $\mathscr{B}$ is defined as the conditional expectation $E(\alpha \mid \mathscr{B})$, where α stands for the indicator of the event A.*

It follows from Definition 5.1.3 and Theorem 5.1.1 that almost surely $0 \leqq P(A \mid \mathscr{B}) \leqq 1$ and for every $B \in \mathscr{B}$ such that $P(B) > 0$, one has

$$P(A \mid B) = E(P(A \mid \mathscr{B}) \mid B) \tag{5.1.17}$$

and especially, choosing $B = \Omega$, we get

$$P(A) = E(P(A \mid \mathscr{B})) \tag{5.1.17'}$$

Notice that (5.1.17) can be regarded as a generalization of the theorem of total probability.

We may regard $P(A \mid \mathscr{B})$ as a function of both $\omega \in \Omega$ and of the set A. Considered as a function of the set A, $P(A \mid \mathscr{B})$ is σ-additive in the restricted sense that for any fixed sequence $\{A_n\}$ of disjoint sets ($A_n \in \mathscr{A}$),

$$P\left(\sum_{n=1}^{\infty} A_n \mid \mathscr{B}\right) = \sum_{n=1}^{\infty} P(A_n \mid \mathscr{B}) \tag{5.1.18}$$

holds almost surely. However, the reader should be warned that this does not necessarily mean that $P(A \mid \mathscr{B})$ is for almost all $\omega \in \Omega$, as a function of the set A, a measure on $\mathscr{A}$. As a matter of fact, as (5.1.18) holds not everywhere but except for a set of measure zero only, the union of these exceptional sets corresponding to different sequences $\{A_n\}$ of disjoint events may contain a set of positive measure—or even cover the full space Ω—as the set of such sequences $\{A_n\}$ is in general not denumerable.

The example below shows how one can overcome this difficulty. (For a more thorough discussion of this problem see Doob [2].)

Example 5.1.1. Let ξ and η be random variables on the probability space S such that their joint distribution is absolutely continuous with the density function $h(x, y)$. Then, as we have seen, the distributions of ξ and η are also absolutely continuous, with the density functions

$$f(x)=\int_{-\infty}^{+\infty} h(x, y)dy \tag{5.1.19}$$

and

$$g(y)=\int_{-\infty}^{+\infty} h(x, y)dx \tag{5.1.19'}$$

respectively. Let $\mathscr{A}_\xi$ denote the σ-algebra generated by ξ and let $U(y)$ be a Borel function such that the expectation of $U(\eta)$ exists, i.e., the integral

$$E(U(\eta))=\int_{-\infty}^{+\infty} U(y)g(y)dy$$

is convergent. Then the conditional expectation of $U(\eta)$ with respect to A_ξ can be expressed as follows:

$$E(U(\eta)\,|\,\mathscr{A}_\xi)=\int_{-\infty}^{+\infty} U(y)g(y\,|\,x)dy \quad \text{if } \xi(\omega)=x \text{ and } f(x)>0$$

where

$$g(y\,|\,x)=\frac{h(x, y)}{f(x)} \quad \text{if } f(x)>0 \tag{5.1.20}$$

Notice that the set, on which $E(U(\eta)\,|\,\mathscr{A}_\xi)$ is undefined, is the set on which $f(\xi)=0$, and this has, by definition, the probability 0.

Choosing for $U(y)$ the function

$$U_Y(y)=\begin{cases}1 & \text{for } y<Y\\ 0 & \text{for } y\geqq Y\end{cases}$$

$E(U_Y(\eta)\,|\,\mathscr{A}_\xi)$ is equal to the conditional probability of the event $\eta<Y$ with respect to $\mathscr{A}_\xi$.

Thus we get

$$P(\eta<Y\,|\,\mathscr{A}_\xi)=G(Y\,|\,\xi)$$

where

$$G(Y\,|\,x)=\int_{-\infty}^{Y} g(y\,|\,x)dy \tag{5.1.21}$$

Thus we may interpret $g(y \mid x)$ defined by (5.1.20) as *the conditional density function of η under condition $\xi = x$*, and accordingly, $G(Y \mid x)$ defined by (5.1.21) as the conditional distribution function of η under the condition $\xi = x$.

Notice that $G(Y \mid x)$ is for every x, such that $f(x) > 0$, in fact, a probability distribution function and $g(y \mid x)$ is a density function, because $g(y \mid x) \geqq 0$ and

$$\int_{-\infty}^{+\infty} g(y \mid x)dy = \frac{\int_{-\infty}^{+\infty} h(x, y)dy}{f(x)} = 1$$

It should be noted also that if $P(x - h \leqq \xi < x + h) > 0$, denoting by $g(y, x, h)$ the conditional density function of η under the condition $x - h \leqq \xi < x + h$ (having positive probability), we have

$$g(y, x, h) = \frac{\int_{x-h}^{x+h} h(u, y)du}{\int_{x-h}^{x+h} f(u)du}$$

As one has, for almost all x, such that $f(x) > 0$,

$$\lim_{h \to 0} \frac{1}{2h} \int_{x-h}^{x+h} f(v)dv = f(x)$$

and

$$\lim_{h \to 0} \frac{1}{2h} \int_{x-h}^{x+h} h(u, y)du = h(x, y)$$

it follows that for almost all x such that $f(x) > 0$, one has

$$g(y \mid x) = \lim_{h \to 0} g(y, x, h) \tag{5.1.22}$$

Notice that

$$\int_{-\infty}^{+\infty} f(t)g(y \mid t)dt = \int_{-\infty}^{+\infty} h(t, y)dt = g(y) \tag{5.1.23}$$

Interchanging the role of ξ and η, putting

$$f(x \mid y) = \frac{h(x, y)}{g(y)} \quad \text{if } g(y) > 0 \tag{5.1.24}$$

comparing (5.1.24) with (5.1.20), and using (5.1.23), we get the following relation:

$$f(x \mid y) = \frac{f(x)g(y \mid x)}{g(y)} = \frac{f(x)g(y \mid x)}{\int_{-\infty}^{+\infty} f(t)g(y \mid t)dt} \tag{5.1.25}$$

Equation (5.1.25) can be regarded as a variant of the formula of Bayes (compare with Exercise E.2.7).

If ξ and η are independent, then $h(x, y) = f(x)g(y)$, and thus

$$f(x \mid y) = \frac{f(x)g(y)}{g(y)} = f(x)$$

and

$$g(y \mid x) = \frac{f(x)g(y)}{f(x)} = g(y)$$

in accordance with our intuition.

Example 5.1.2. Let ξ be a random variable on the probability space $S = (\Omega, \mathscr{A}, P)$. Let $B \in \mathscr{A}$ be an event such that $0 < P(B) < 1$. Suppose that the conditional density function of ξ with respect to both B and $\bar{B}$ exists; let us denote the density functions by $f(x \mid B)$ and $f(x \mid \bar{B})$ and let $f(x) = P(B)f(x \mid B) + P(\bar{B})f(x \mid \bar{B})$ denote the unconditional density function of ξ. Let $P(B \mid x)$ denote the conditional probability of the event B under condition $\xi = x$; in other words, we put

$$P(B \mid x) = P(B \mid \mathscr{A}_\xi) \qquad \text{if } \xi(\omega) = x$$

Then we have

$$P(B \mid x) = \frac{f(x \mid B)P(B)}{f(x)} \tag{5.1.26}$$

for all those x for which $f(x) > 0$. As a matter of fact, if A is an event in A_ξ, then $A = \xi^{-1}(C)$, where C is a Borel subset of the real line. Now clearly, we have, if $P(B \mid x)$ is defined by (5.1.26),

$$\int_A P(B \mid \mathscr{A}_\xi)dP = \int_C P(B \mid x)f(x)dx = P(B)\int_C f(x \mid B)dx = P(BA) \tag{5.1.27}$$

That is, the defining property of the conditional probability is satisfied.

Especially, we get from (5.1.27),

$$P(B) = \int_{-\infty}^{+\infty} P(B \mid x)f(x)dx \tag{5.1.28}$$

In view of (5.1.28), we can write (5.1.26) in the form

$$f(x \mid B) = \frac{P(B \mid x)f(x)}{\int_{-\infty}^{+\infty} P(B \mid y)f(y)dy} \tag{5.1.29}$$

Evidently (5.1.29) can be regarded as another extension of the formula of Bayes.

Using the notion of generalized conditional probability introduced in this section, we can extend the notion of a Markov chain to sequences of arbitrary random variables. Generalizing the definition given in Section 3.10, we introduce the following:

DEFINITION 5.1.4. *A sequence $\{\xi_n\}$ of random variables on a probability space S is called a Markov chain if for every real x and for $n=2, 3, \ldots$, one has almost surely*

$$P(\xi_{n+1} < x \mid \mathscr{A}_n) = P(\xi_{n+1} < x \mid \mathscr{A}_{\xi_n})$$

where $\mathscr{A}_{\xi_n}$ denotes the σ-algebra generated by ξ_n, and $\mathscr{A}_n$ the σ-algebra generated by $\xi_1, \xi_2, \ldots, \xi_n$. (For a general theory of Markov chains see Chung [3].)

We introduce some further definitions which will be needed in the sequel.

DEFINITION 5.1.5. *Let η be a random variable on a probability space $S=(\Omega, \mathscr{A}, P)$ which belongs to $L_2(S)$. Let $\mathscr{B} \subseteq \mathscr{A}$ be a σ-algebra of subsets of Ω. Then the variance of the random variable $E(\eta \mid \mathscr{B})$ will be called the conditional variance of η with respect to $\mathscr{B}$, and it will be denoted by $D^2(\eta \mid \mathscr{B})$.*

By Theorem 5.1.6 $D^2(\eta \mid \mathscr{B})$ exists. Thus we have

$$D^2(\eta) = D^2(\eta \mid \mathscr{B}) + E([\eta - E(\eta \mid \mathscr{B})]^2) + 2E([E(\eta \mid \mathscr{B}) - E(\eta)][\eta - E(\eta \mid \mathscr{B})]) \tag{5.1.30}$$

As, however, by the definition of the conditional expectation,

$$E([E(\eta \mid \mathscr{B}) - E(\eta)]E(\eta \mid \mathscr{B})) = E([E(\eta \mid \mathscr{B}) - E(\eta)]\eta)$$

it follows that the third term on the right of (5.1.30) vanishes; thus we get

$$D^2(\eta) = D^2(\eta \mid \mathscr{B}) + E([\eta - E(\eta \mid \mathscr{B})]^2) \tag{5.1.31}$$

It follows that the ratio

$$K^2_{\mathscr{B}}(\eta) = \frac{D^2(\eta \mid \mathscr{B})}{D^2(\eta)} \tag{5.1.32}$$

always lies between 0 and 1.

The positive square root of the quantity defined by (5.1.32) is called the *correlation ratio* of η with respect to $\mathscr{B}$.

It is easy to prove the following:

THEOREM 5.1.8. *One has*

$$K^2_{\mathscr{B}}(\eta) = \sup_{\mathscr{A}_\xi \subseteq \mathscr{B}} R^2(\xi, \eta) \tag{5.1.33}$$

where $R(\xi, \eta)$ denotes the correlation coefficient of ξ and η and the supremum is taken over all random variables ξ for which $A_\xi \subseteq \mathscr{B}$ and for which $E(\xi^2)$ exists, i.e., for all random variables $\xi \in L_2(S')$ where $S' = (\Omega, \mathscr{B}, P)$.

Proof. According to the definition of conditional expectation, we have for every $\xi \in L_2(S')$,

$$E^2([\xi - E(\xi)][\eta - E(\eta)]) = E^2([\xi - E(\xi)][E(\eta \mid \mathscr{B}) - E(\eta)]) \leqq D^2(\xi)D^2_{\mathscr{B}}(\eta) \tag{5.1.34}$$

Dividing (5.1.34) by $D^2(\xi)D^2(\eta)$, we get (5.1.33). ■

Remark: Evidently we have equality in (5.1.34) if $\xi = E(\eta \mid \mathscr{B})$.

5.2 MARTINGALES

Let $S = (\Omega, \mathscr{A}, P)$ be a probability space, and let $\mathscr{B}_n$ $(n = 1, 2, \ldots)$ be a sequence of σ-algebras of subsets of Ω such that $\mathscr{B}_n \subseteq \mathscr{B}_{n+1} \subseteq \mathscr{A}$ for $n = 1, 2, \ldots$. Let η be a random variable on S having a finite expectation and put

$$\eta_n = E(\eta \mid \mathscr{B}_n) \qquad (n = 1, 2, \ldots) \tag{5.2.1}$$

Let $\mathscr{A}_n$ denote the σ-algebra generated by the random variables $n_1, \eta_2, \ldots, \eta_n$. As each η_k with $k \leqq n$ is measurable with respect to $\mathscr{B}_k$ and thus with respect to $\mathscr{B}_n$, we have $\mathscr{A}_n \subseteq \mathscr{B}_n \subseteq \mathscr{B}_{n+1}$. Thus it follows from Theorem 5.1.7 that

$$E(\eta_{n+1} \mid \mathscr{A}_n) = E(E(\eta \mid \mathscr{B}_{n+1}) \mid \mathscr{A}_n) = E(\eta \mid \mathscr{A}_n) \tag{5.2.2}$$

Applying Theorem 5.1.7 once more, we get

$$E(\eta \mid \mathscr{A}_n) = E(E(\eta \mid \mathscr{B}_n) \mid \mathscr{A}_n) = E(\eta_n \mid \mathscr{A}_n) = \eta_n \tag{5.2.3}$$

From (5.2.2) and (5.2.3), we get

$$E(\eta_{n+1} \mid \mathscr{A}_n) = \eta_n \tag{5.2.4}$$

which can be written in the form

$$E(\eta_{n+1} \mid \eta_1, \eta_2, \ldots, \eta_n) = \eta_n \qquad (n = 1, 2, \ldots) \tag{5.2.5}$$

Expressed in words: The sequence η_n defined by (5.2.1) has the property that the conditional expectation of η_{n+1}, given the values of $\eta_1, \eta_2, \ldots, \eta_n$, is equal to η_n.

Sequences $\{\eta_n\}$ of random variables, having the property (5.2.5), play an important role in probability theory, therefore they deserve a special name: they are called "martingales." Thus we introduce the following:

DEFINITION 5.2.1. *A sequence η_n $(n = 1, 2, \ldots)$ of random variables on a probability space S such that $E(\eta_1)$ exists and* (5.2.5) *holds, is a martingale.*

The notion of a martingale is due to Ville [4] and Lévy [5]. The theory of martingales owes very much to Doob [2]. See also Krickeberg [6], Neveu [7] and Loève [33].

We now shall prove some immediate consequences of the definition of a martingale.

THEOREM 5.2.1. *If η_n is a martingale, then*

$$E(\eta_n \mid \eta_1, \eta_2, \ldots, \eta_k) = \eta_k \qquad \text{for } 1 \leqq k \leqq n \text{ and } n = 2, 3, \ldots \tag{5.2.6}$$

further, $E(\eta_n)$ exists for $n \geqq 2$ and

$$E(\eta_n) = E(\eta_1) \tag{5.2.7}$$

Putting $\eta_0 = 0$ and

$$\xi_n = \eta_n - \eta_{n-1} \qquad (n = 1, 2, \ldots) \tag{5.2.8}$$

the random variables ξ_n satisfy

$$E(\xi_n \mid \xi_1, \xi_2, \ldots, \xi_k) = 0 \qquad \text{for } 1 \leqq k < n \tag{5.2.9}$$

Moreover, if $E(\eta_n^2)$ exists for $n = 1, 2, \ldots$, so does $E(\xi_n^2)$ $(n = 1, 2, \ldots)$, and the random variables ξ_n $(n = 1, 2, \ldots)$ are orthogonal.

Proof. Equation (5.2.6) follows from Theorem 5.1.7. The existence of $E(\eta_n)$ follows by Theorem 5.1.2 and (5.2.7) from (5.1.7). Formula (5.2.9) follows by the remark that the σ-algebra generated by $\xi_1, \ldots, \xi_n$ is identical to that generated by $\eta_1, \ldots, \eta_n$.

Finally, the orthogonality of the random variables ξ_n is proved as follows: If $E(\eta_n^2)$ exists, so does $E(\xi_n^2)$ and for $1 \leqq k < n$ we have, by (5.1.7), Theorem 5.1.5, and (5.2.9),

$$E(\xi_k \xi_n) = E(E(\xi_k \xi_n \mid \xi_1, \xi_2, \ldots, \xi_k)) = E(\xi_k E(\xi_n \mid \xi_1, \xi_2, \ldots, \xi_k)) = 0 \qquad \blacksquare$$

Example 5.2.1. Let $\delta_1, \delta_2, \ldots, \delta_n, \ldots$ be independent random variables, each having 0 expectation. Let $f_k(x_1, \ldots, x_k)$ be any bounded Borel-measurable function of its variables. Put

$$\xi_1 = \delta_1 \quad \text{and} \quad \xi_{k+1} = \delta_{k+1} f_k(\delta_1, \delta_2, \ldots, \delta_k) \qquad (k = 1, 2, \ldots)$$

and

$$\eta_n = \xi_1 + \xi_2 + \cdots + \xi_n \qquad (n = 1, 2, \ldots)$$

Then $\{\eta_n\}$ is a martingale. As a matter of fact,

$$\begin{aligned} E(\eta_{n+1} \mid \eta_1, \ldots, \eta_n) &= E(\eta_{n+1} \mid \delta_1, \ldots, \delta_n) \\ &= E(\xi_{n+1} \mid \delta_1, \ldots, \delta_n) + E(\eta_n \mid \delta_1, \ldots, \delta_n) \end{aligned}$$

As η_n is a function of $\delta_1, \ldots, \delta_n$, the second term on the right is equal to η_n; as regards the first term, we have by Theorem 5.1.5,

$$E(\xi_{n+1} \mid \delta_1, \ldots, \delta_n) = f_n(\delta_1, \ldots, \delta_n) E(\delta_{n+1} \mid \delta_1, \ldots, \delta_n) = 0$$

The word "martingale" is borrowed from the field of gambling where it is the name of a certain gambling system.

Let us consider a fair game of chance, in which in every round a player can put any sum at stake and if his stake is $a > 0$, he gains* a random sum ξ_a such that $E(\xi_a) = 0$.

If the player starts with a random capital η_1, and if η_{n+1} $(n = 1, 2, \ldots)$ denotes his capital after n rounds, then whatever the values of $\eta_1, \ldots, \eta_n$ and of the stake in the nth round, the conditional expectation of η_{n+1} is equal to η_n, i.e., the variables η_n form a martingale. If the gain of the player in the nth round is denoted by ξ_n, then of course, $\eta_{n+1} = \eta_1 + \xi_1 + \cdots + \xi_n$. If the stake of the player is constant, the random variables ξ_k are independent. However, the player's choice of his stake at the nth round may depend on his capital and on the results of the previous rounds; in this case the variables ξ_n are not necessarily independent. However, in the mentioned case the martingale $\{\eta_n\}$ is always of the type described in Example 5.2.1.

Let us consider two examples.

Example 5.2.2. Let the game consist in throwing a fair coin, and suppose that in every round the player stakes on heads the qth part of his whole capital. Putting $\delta_n = +1$ or $\delta_n = -1$, according to whether the result of the nth throw is a head or tail, if η_{n+1} denotes the player's capital after n rounds, and he starts with the capital $\eta_1 = 1$, then clearly

$$\eta_{n+1} = \prod_{k=1}^{n} (1 + \delta_k q)$$

Thus,

$$\xi_{n+1} = \eta_{n+1} - \eta_n = \delta_n \cdot q \cdot \prod_{k=1}^{n-1} (1 + q\delta_k)$$

which shows that $\{\eta_n\}$ is a martingale of the type described in Example 5.2.1.

Example 5.2.3. Another system, popular among gamblers, is as follows: Let the game consist again in throwing a fair coin. Suppose that the player stakes in the first round a unit sum, doubles the stake each time he looses and returns to the unit stake each time he wins. Supposing that the player has unlimited funds (or credit), i.e., is never forced to abandon his system for lack of funds, we get for the net gain η_{n+1} of the gambler after the nth round,

* A loss is counted as a negative gain.

the formula

$$\eta_{n+1} = \sum_{k=1}^{n} \delta_k \left[\frac{1+\delta_{k-1}}{2} + \sum_{j=1}^{k-2} \left(\frac{1+\delta_{k-j-1}}{2} \right) \prod_{i=1}^{j} (1-\delta_{k-i}) + \prod_{i=1}^{k-1} (1-\delta_{k-i}) \right]$$

Thus $\{\eta_n\}$ is again a martingale of the type described in Example 5.2.1.

Let ξ_n $(n = 1, 2, \ldots)$ be a sequence of random variables having zero expectations and finite second moment, and put $\eta_n = \xi_1 + \xi_2 + \cdots + \xi_n$. Then the three statements

(A) the $\{\xi_n\}$ are orthogonal
(B) the η_n form a martingale such that $E(\eta_n^2)$ exists for $n \geqq 1$
(C) the ξ_n are independent

are in the following relation:

$$(C) \to (B) \to (A)$$

Thus the supposition of $\{\eta_n\}$ being a martingale is intermediate between the suppositions that the η_n are partial sums of a sequence of orthogonal, respectively, independent random variables. Thus we may expect that the laws of large numbers can be extended to martingales and we get results which are intermediate between the laws valid for orthogonal, respectively, independent random variables. This is in fact true; moreover, we shall prove that essentially the same laws of large numbers hold for martingales as for sums of independent random variables. To show this we prove in the next section certain inequalities.

However, we need still one more result about martingales.

THEOREM 5.2.2. *Let η_n $(n = 1, 2, \ldots)$ be a sequence of random variables on the probability space $S = (\Omega, \mathscr{A}, P)$ having finite expectations, and let $\mathscr{B}_n$ $(n = 1, 2, \ldots)$ be a sequence of σ-algebras of subsets of Ω such that $\mathscr{B}_n \subseteq \mathscr{B}_{n+1} \subseteq \mathscr{A}$ $(n = 1, 2, \ldots)$. Suppose that*

$$E(\eta_{n+1} \mid \mathscr{B}_n) = \eta_n \qquad (n = 1, 2, 3, \ldots) \tag{5.2.10}$$

Then $\{\eta_n\}$ is a martingale.

Proof. It follows from (5.2.10) that η_n is $\mathscr{B}_n$-measurable. As $\mathscr{B}_k \subseteq \mathscr{B}_n$ for $k < n$, it follows that denoting by $\mathscr{A}_n$ the σ-algebra spanned by the random variables $\eta_1, \ldots, \eta_n$, we have $\mathscr{A}_n \subseteq \mathscr{B}_n$. Thus it follows from Theorem 5.1.6 that

$$E(\eta_{n+1} \mid \mathscr{A}_n) = E(E(\eta_{n+1} \mid \mathscr{B}_n) \mid \mathscr{A}_n) = E(\eta_n \mid \mathscr{A}_n) = \eta_n \tag{5.2.11}$$

which was to be proved. ■

COROLLARY TO THEOREM 5.2.2. *If $\{\eta_n\}$ is a martingale, then for each $m \geqq 1$, $\eta_n - \eta_m$ $(n = m + 1, m + 2, \ldots)$ is also a martingale.*

Proof. Let $\mathscr{A}_n$ denote the σ-algebra spanned by the random variables $\eta_1, \ldots, \eta_n$. Then we have, by the definition of a martingale,

$$E(\eta_{n+1} - \eta_m \mid \mathscr{A}_n) = \eta_n - \eta_m \qquad \text{for } n \geqq m+1$$

As $\mathscr{A}_n \subseteq \mathscr{A}_{n+1} \subseteq \mathscr{A}$, Theorem 5.2.2 can be applied and it follows that $\{\eta_n - \eta_m\}$ is a martingale.

Example 5.2.4. Let us consider Haar's orthonormal system $\{H_n(x)\}$ $(0 \leqq x \leqq 1)$ defined as follows (see Haar [8]):

$$H_0(x) = 1 \qquad \text{for } 0 \leqq x < 1$$

$$H_{2^k}(x) = \begin{cases} 2^{k/2} & \text{if } 0 \leqq x < 1/2^{k+1} \\ -2^{k/2} & \text{if } 1/2^{k+1} \leqq x < 1/2^k \\ 0 & \text{if } 1/2^k \leqq x < 1 \end{cases}$$

if $k = 0, 1, \ldots$, further, for $k = 1, 2, \ldots$ and $1 \leqq j \leqq 2^k - 1$,

$$H_{2^k+j}(x) = \begin{cases} H_{2^k}(x - j/2^k) & \text{if } j/2^k \leqq x < (j+1)/2^k \\ 0 & \text{otherwise} \end{cases}$$

It is easy to see that the functions $\{H_n(x)\}$ are orthonormal on the Lebesgue probability space S. Let $f = f(x)$ be any random variable belonging to $L_2(S)$ (i.e., let $f(x)$ be a measurable function on $[0, 1)$ such that $\int_0^1 f^2(x)dx < +\infty$). Let us consider the expansion of $f(x)$ according to the Haar functions, i.e., the series

$$\sum_{n=0}^{+\infty} c_n H_n(x)$$

where

$$c_n = \int_0^1 f(t)H_n(t)dt \qquad (n = 0, 1, \ldots)$$

and let us investigate the partial sums

$$S_N(x) = \sum_{n=0}^{N} c_n H_n(x)$$

of this series. We prove that $\{S_N(x)\}$ $(N = 0, 1, \ldots)$ is a martingale. Let $\mathscr{B}_N$ denote the algebra of subsets of $[0, 1)$ generated by $H_1(x), \ldots, H_N(x)$. If $N = 2^k + j$ $(0 \leqq j \leqq 2^k - 1)$, then the atoms of $\mathscr{B}_N$ are the intervals $(l/2^{k+1}, (l+1)/2^{k+1})$ $(l = 0, 1, \ldots, 2j+1)$ and $(m/2^k, (m+1)/2^k)$ $(m = j+1, \ldots, 2^k - 1)$. Thus $\mathscr{B}_N$ has $N+1$ atoms, and these are obtained from the N atoms of $\mathscr{B}_{N-1}$ so that one of these is halved and the others are left unchanged, and $H_N(x)$ has values $\pm 2^{k/2}$ on the two halves of the halved atom of $\mathscr{B}_{N-1}$, and is 0 on all other atoms of $\mathscr{B}_{N-1}$. Thus, it follows that the integral of $H_N(x)$ is 0 on every atom of $\mathscr{B}_{N-1}$, i.e., $E(H_N(x) \mid \mathscr{B}_{N-1}) = 0$. This implies that $\{S_N(x)\}$ is a martingale. It is easy to see that

$$S_N(x) = E(f(x) \mid \mathscr{B}_N) \qquad (N = 0, 1, \ldots)$$

5.3 INEQUALITIES FOR MARTINGALES

Let $\{\eta_n\}$ be a martingale such that $E(\eta_n^2)$ exists. Then we have, putting $\eta_n - \eta_{n-1} = \xi_n$,

$$E(\eta_n^2 \mid \eta_1, \ldots, \eta_{n-1}) = E(\eta_{n-1}^2 + \xi_n^2 + 2\xi_n \eta_{n-1} \mid \eta_1, \ldots, \eta_{n-1}) \tag{5.3.1}$$

As, by Theorems 5.1.5 and 5.2.1,

$$E(\xi_n \eta_{n-1} \mid \eta_1, \ldots, \eta_{n-1}) = \eta_{n-1} E(\xi_n \mid \eta_1, \ldots, \eta_{n-1}) = 0$$

we obtain

$$E(\eta_n^2 \mid \eta_1, \ldots, \eta_{n-1}) \geqq \eta_{n-1}^2 \tag{5.3.2}$$

Let us put $\zeta_n = \eta_n^2$. As the σ-algebra generated by $\zeta_1, \ldots, \zeta_n$ is a sub-algebra of the σ-algebra generated by $\eta_1, \ldots, \eta_n$, we get, by Theorem 5.1.7,

$$\begin{aligned} E(\zeta_{n+1} \mid \zeta_1, \ldots, \zeta_n) &= E(E(\zeta_{n+1} \mid \eta_1, \ldots, \eta_n) \mid \zeta_1, \ldots, \zeta_n) \\ &\geqq E(\zeta_n \mid \zeta_1, \ldots, \zeta_n) = \zeta_n \end{aligned}$$

Thus the sequence $\{\zeta_n\}$ has the property

$$E(\zeta_{n+1} \mid \zeta_1, \ldots, \zeta_n) \geqq \zeta_n \qquad (n = 1, 2, \ldots) \tag{5.3.3}$$

We introduce the following:

DEFINITION 5.3.1. *A sequence of random variables ζ_n such that $E(\zeta_n)$ exists for $n \geqq 1$ and (5.3.3) holds is called a submartingale.*

Thus, if η_n is a martingale such that $E(\eta_n^2)$ exists for every $n \geqq 1$, $\{\eta_n^2\}$ is a submartingale. Clearly, if $\{\zeta_n\}$ is a submartingale, then

$$\begin{aligned} E(\zeta_{n+2} \mid \zeta_1, \ldots, \zeta_n) &= E(E(\zeta_{n+2} \mid \zeta_1, \ldots, \zeta_{n+1}) \mid \zeta_1, \ldots, \zeta_n) \\ &\geqq E(\zeta_{n+1} \mid \zeta_1, \ldots, \zeta_n) \geqq \zeta_n \end{aligned}$$

and similarly, we get by induction

$$E(\zeta_{n+j} \mid \zeta_1, \ldots, \zeta_n) \geqq \zeta_n \qquad \text{for all } n \geqq 1 \text{ and } j \geqq 1$$

Now we shall prove the following:

THEOREM 5.3.1. *Let ζ_n $(n = 1, 2, \ldots)$ be a sequence of nonnegative random variables which is a submartingale. Let b_n $(n = 1, 2, \ldots)$ be a nonincreasing sequence of nonnegative constants, such that $\lim_{n\to+\infty} b_n = 0$ and the series $\sum_{n=1}^{\infty}(b_n - b_{n+1})E(\zeta_n)$ is convergent. Then we have for every $x > 0$*

$$P(\sup_{n\geqq 1} b_n \zeta_n > x) \leqq \frac{1}{x} \sum_{n=1}^{\infty} (b_n - b_{n+1}) E(\zeta_n) \tag{5.3.4}$$

Proof. Let $\mathscr{A}_n$ denote the σ-algebra generated by the random variables $\zeta_1, \ldots, \zeta_n$ and let A_k $(k = 1, 2, \ldots)$ denote the event that k is the least value of the integer n such that $b_n \zeta_n > x$, i.e., A_k is the event consisting in the joint occurrence of the events $b_k \zeta_k > x$ and $b_j \zeta_j \leqq x$ for $1 \leqq j \leqq k-1$. We have $A_k \in \mathscr{A}_k$ $(k = 1, 2, \ldots)$. Let us put

$$\eta = \sum_{n=1}^{\infty} (b_n - b_{n+1}) \zeta_n \tag{5.3.5}$$

As the terms of the series on the right-hand side of (5.3.5) are nonnegative and their expectations form by supposition a convergent series, the series (5.3.5) is, by the Beppo Levi theorem (see Appendix A), almost surely convergent, and

$$E(\eta) = \sum_{n=1}^{\infty} (b_n - b_{n+1}) E(\zeta_n) \tag{5.3.6}$$

Now we have

$$\int_{A_k} \eta \, dP = \int_{A_k} E(\eta \mid \mathscr{A}_k) dP \geqq \int_{A_k} \sum_{n=k}^{\infty} (b_n - b_{n+1}) E(\zeta_n \mid \mathscr{A}_k) dP$$

As, by supposition, $\{\zeta_n\}$ is a submartingale, it follows that

$$E(\zeta_n \mid \mathscr{A}_k) \geqq \zeta_k \qquad \text{for } n \geqq k$$

Thus,

$$\int_{A_k} \eta \, dP \geqq b_k \int_{A_k} \zeta_k \, dP \geqq x P(A_k)$$

It follows that

$$E(\eta) \geqq \sum_{k=1}^{\infty} \int_{A_k} \eta \, dP \geqq x \sum_{k=1}^{\infty} P(A_k) = x P(\sup_n b_n \zeta_n > x)$$

Thus Theorem 5.3.1 is proved. ∎

Let us note some special cases. If $b_n = 1$ for $n \leqq N$ and $b_n = 0$ for $n > N$, we get the following:

Corollary 1 to Theorem 5.3.1. *If ζ_n is a positive submartingale, we have for every $x > 0$,*

$$P(\max_{n \leqq N} \zeta_n > x) \leqq \frac{1}{x} E(\zeta_N) \tag{5.3.7}$$

In particular, if $\xi_1, \xi_2, \ldots, \xi_k, \ldots$ are independent random variables having 0 expectation and finite variances $D^2(\xi_k) = D_k^2$ $(k \geqq 1)$, then, putting $\eta_n = \xi_1 + \cdots + \xi_n$ and $\zeta_n = \eta_n^2$, all our suppositions are fulfilled and thus we get from Corollary 1 the following important inequality due to Kolmogoroff:

Corollary 2 to Theorem 5.3.1. *If $\xi_1, \xi_2, \ldots, \xi_k, \ldots$ are independent random variables with 0 expectation and finite variances $D^2(\xi_k) = D_k^2$, then, putting $\eta_n = \xi_1 + \xi_2 + \cdots + \xi_n$ $(n \geqq 1)$, one has for every $x > 0$ and $N \geqq 1$,*

$$P(\max_{n \leqq N} |\eta_n| > x) \leqq \frac{1}{x^2} \sum_{k=1}^{N} D_k^2 \tag{5.3.8}$$

Remark: Note that Kolmogoroff's inequality (5.3.8) gives exactly the same upper bound for $P(\max_{n \leqq N} |\eta_n| > x)$ as Čebyshev's inequality (2.11.4) for $P(|\eta_N| > x)$.

If $\xi_1, \xi_2, \ldots, \xi_k, \ldots$ are independent random variables with 0 expectation and finite variances $D^2(\xi_k) = D_k^2$ such that the series $\sum_{k=1}^{\infty} D_k^2/k^2$ is convergent, putting $\eta_n = \xi_1 + \cdots + \xi_n$ and considering the submartingale $\zeta_k = \eta_{N+k-1}^2$ $(k = 1, 2, \ldots)$, and choosing $b_k = 1/(N + k - 1)^2$, we get the following inequality, due to Hájek (see [9]):

Corollary 3 to Theorem 5.3.1. *If $\xi_1, \ldots, \xi_k, \ldots$ are independent random variables with expectation 0 and finite variance $D_k^2 = D^2(\xi_k)$ $(k = 1, 2, \ldots)$ and if the series $\sum_{k=1}^{\infty} D_k^2/k^2$ converges, then, putting $\eta_n = \xi_1 + \cdots + \xi_n$ $(n \geqq 1)$, we have*

$$P\left(\sup_{n \geqq N} \frac{|\eta_n|}{n} > x\right) \leqq \frac{1}{x^2}\left(\frac{1}{N^2} \sum_{k=1}^{N} D_k^2 + \sum_{k=N+1}^{\infty} \frac{D_k^2}{k^2}\right) \tag{5.3.9}$$

For sums of uniformly bounded independent random variables we now prove an inequality which is a counterpart in the opposite direction of Kolmogoroff's inequality (5.3.8).

Theorem 5.3.2. *Let $\xi_1, \xi_2, \ldots, \xi_n, \ldots$ be a sequence of independent random variables. Suppose that the ξ_n are uniformly bounded, i.e.,*

$$|\xi_n| \leqq C \qquad \text{for } n = 1, 2, \ldots \tag{5.3.10}$$

and that $E(\xi_n) = 0$ for $n = 1, 2, \ldots$. Put $\eta_n = \xi_1 + \cdots + \xi_n$ and $D_k^2 = D^2(\xi_k)$ $(k = 1, 2, \ldots)$. Then we have for every ε with $0 < \varepsilon < 1$,

$$P\left(\max_{1 \leqq k \leqq N} |\eta_n| > \sqrt{(1 - \varepsilon) \sum_{k=1}^{N} D_k^2} - C\right) \geqq \varepsilon \tag{5.3.11}$$

Proof. Let $x > 0$ be arbitrary and let A_n $(n \geqq 1)$ denote the event $|\eta_k| \leqq x$ for $k < n$ and $|\eta_n| > x$. Then $\sum_{n=1}^{N} A_n = A$ is just the event $\max_{1 \leqq n \leqq N} |\eta_n| > x$ for which we want to prove a lower estimate. As $A_1, \ldots, A_N$ and $\bar{A}$ form a partition of the basic space, we have, by the theorem of total expectation,

$$E(\eta_N^2) = \sum_{n=1}^{N} E(\eta_N^2 \mid A_n) P(A_n) + E(\eta_N^2 \mid \bar{A}) P(\bar{A}) \tag{5.3.12}$$

Now if α_n denotes the indicator of A_n, then $\eta_n \alpha_n$ and $\eta_N - \eta_n$ are independent and $E(\eta_N - \eta_n) = 0$; thus,

$$E((\eta_N - \eta_n)\eta_n \mid A_n)P(A_n) = E((\eta_N - \eta_n)\eta_n \alpha_n) = 0$$

It follows that

$$E(\eta_N^2 \mid A_n) = E(\eta_n^2 \mid A_n) + E((\eta_N - \eta_n)^2 \mid A_n) \tag{5.3.13}$$

Again, by the independence of $\eta_N - \eta_n$ and α_n, we get

$$E((\eta_N - \eta_n)^2 \mid A_n) = E((\eta_N - \eta_n)^2) \leqq E(\eta_N^2) = \sum_{n=1}^{N} D_n^2 \tag{5.3.14}$$

On the other hand,

$$E(\eta_n^2 \mid A_n) = E(\eta_{n-1}^2 + 2\eta_{n-1}\xi_n + \xi_n^2 \mid A_n) \leqq x^2 + 2Cx + C^2 = (x + C)^2 \tag{5.3.15}$$

Finally, as

$$E(\eta_N^2 \mid \bar{A}) \leqq x^2 \tag{5.3.16}$$

we get from (5.3.12)–(5.3.16),

$$\sum_{1}^{N} D_n^2 = E(\eta_N^2) \leqq \left[\sum_{n=1}^{N} D_n^2 + (x + C)^2\right] P(A) + x^2(1 - P(A))$$

and thus

$$P(A) \geqq \frac{\sum_{n=1}^{N} D_n^2 - x^2}{\sum_{n=1}^{N} D_n^2 - x^2 + (x + C)^2} \geqq 1 - \frac{(x + C)^2}{\sum_{n=1}^{N} D_n^2} \tag{5.3.17}$$

Choosing for x, the value

$$x = \sqrt{(1 - \varepsilon)\sum_{n=1}^{N} D_n^2} - C$$

if $x > 0$, we get the inequality (5.3.11) from (5.3.17). If $x \leqq 0$, then (5.3.11) is trivially valid. ■

5.4 A MARTINGALE-CONVERGENCE THEOREM AND ITS APPLICATIONS

We first prove the following convergence theorem for martingales:

THEOREM 5.4.1. *Let $\{\eta_n\}$ be a martingale, and suppose that η_n tends strongly in L^2-norm to a limit η. Then $\lim_{n\to+\infty} \eta_n = \eta$ holds almost surely.*

Proof. According to Theorem 5.2.2, $\eta_n - \eta_m$ $(n \geqq m)$ is a martingale. By our supposition, $E((\eta_n - \eta_m)^2)$ exists and

$$\lim_{\substack{n\to+\infty \\ m\to+\infty}} E((\eta_n - \eta_m)^2) = 0 \tag{5.4.1}$$

We also have, for each $m \geqq 1$ and $N > m$, $E((\eta_{N+1} - \eta_m)^2) \geqq E((\eta_N - \eta_m)^2)$ and

$$\lim_{N \to +\infty} E((\eta_N - \eta_m)^2) = E((\eta - \eta_m)^2) \tag{5.4.2}$$

further,

$$\lim_{m \to +\infty} E((\eta - \eta_m)^2) = 0 \tag{5.4.3}$$

Now, from Corollary 1 of Theorem 5.3.1, we get that for any $\varepsilon > 0$

$$P(\max_{m+1 \leqq n \leqq N} (\eta_n - \eta_m)^2 > \varepsilon^2) \leqq \frac{E((\eta_N - \eta_m)^2)}{\varepsilon^2} \tag{5.4.4}$$

It follows that

$$P(\max_{n \geqq m+1} (\eta_n - \eta_m)^2 > \varepsilon^2) \leqq \frac{E((\eta - \eta_m)^2)}{\varepsilon^2} \tag{5.4.5}$$

Now let us choose an increasing sequence $\{n_j\}$ of positive integers, such that

$$\sum_{j=1}^{\infty} E((\eta - \eta_{n_j})^2) < +\infty \tag{5.4.6}$$

It follows from (5.4.5) that

$$\sum_{j=1}^{\infty} P(\max_{n \geqq n_j+1} (\eta_n - \eta_{n_j})^2 > \varepsilon^2) < +\infty \tag{5.4.7}$$

Thus by the Borel-Cantelli lemma, we have almost surely, for all sufficiently large values of j,

$$(\eta_n - \eta_{n_j})^2 \leqq \varepsilon^2 \qquad \text{if } n > n_j$$

and thus

$$(\eta_n - \eta_m)^2 \leqq 2_\varepsilon^2 \qquad \text{if both } n \text{ and } m \text{ are sufficiently large} \tag{5.4.8}$$

Choosing for ε the values $1/k$ $(k = 1, 2, \ldots)$, as the union of a denumerable infinity of sets of measure 0 is also of measure 0, it follows that the limit

$$\lim_{n \to +\infty} \eta_n = \eta^* \tag{5.4.9}$$

exists almost surely. By the Lebesgue convergence theorem, it follows that

$$\lim_{n \to +\infty} E((\eta_n - \eta_m)^2) = E((\eta^* - \eta_m)^2)$$

and thus

$$\lim_{m \to +\infty} E((\eta^* - \eta_m)^2) = 0$$

On the other hand,

$$E((\eta - \eta^*)^2) \leqq 2[E((\eta - \eta_m)^2) + E((\eta^* - \eta_m)^2)]$$

Thus we get

$$E((\eta - \eta^*)^2) = 0$$

That is, $\eta^* = \eta$ almost surely. Thus Theorem 5.4.1 is proved. ∎

We shall deduce from Theorem 5.4.1 a corollary which contains an apparently stronger result, namely, it deduces the same conclusion from a seemingly weaker supposition.

COROLLARY TO THEOREM 5.4.1. *Let $\{\eta_n\}$ be a martingale such that $E(\eta_n^2)$ exists for all $n \geqq 1$ and is bounded, i.e.,*

$$E(\eta_n^2) \leqq K \qquad \text{for } n = 1, 2, \ldots$$

Then η_n tends almost surely to a random variable η, such that $E(\eta^2)$ exists and

$$\lim_{n \to +\infty} E(\eta_n^2) = E(\eta^2)$$

and η_n tends to η strongly in $L_2(S)$, too.

Proof. It follows from (5.3.2) that the sequence $E(\eta_n^2)$ is nondecreasing; thus the boundedness of $E(\eta_n^2)$ implies that $\lim_{n\to+\infty} E(\eta_n^2) = c \leqq K$ exists. On the other hand, if $m < n$, we have

$$E(\eta_n \eta_m) = E(E(\eta_n \eta_m \mid \eta_1, \ldots, \eta_m)) = E(\eta_m^2)$$

Thus,

$$E((\eta_n - \eta_m)^2) = E(\eta_n^2) - E(\eta_m^2) \qquad \text{if } m < n$$

and therefore,

$$\lim_{\substack{n \to +\infty \\ m \to +\infty}} E((\eta_n - \eta_m)^2) = 0$$

That is, η_n converges strongly in $L_2(s)$ to a limit η and Theorem 5.4.1 yields $\lim_{n\to+\infty} \eta_n = \eta$ almost surely. ∎

Remark: If $\{\eta_n\}$ is a bounded martingale, $|\eta_n| \leqq K$, then the conditions of the corollary are satisfied. Thus we get that *every bounded martingale converges almost surely*.

Example 5.4.1. Let us consider the martingale $\{S_N(x)\}$ of Example 5.2.4 formed by the partial sum of the expansion of a square integrable function f according to Haar's orthonormal system. As $E(S_N{}^2(x))$ is bounded by $\int_0^1 f^2(x)dx$, it follows from the corollary of Theorem 5.4.1 that $\lim_{N\to+\infty} S_N(x) = f^*(x)$ exists almost surely. Now, as every binary interval is contained in $\mathscr{B}_N$ if N is sufficiently large, it follows that the Haar system is complete; thus, $S_N(x)$ tends strongly in $L_2(S)$ to $f(x)$, and therefore, $f^*(x) = f(x)$ almost surely. Thus we have proved that the Haar expansion of a square integrable function converges to the function almost everywhere.

Remark: This proof by martingales of Haar's theorem (see [8]) is due to Krickeberg [6].

From Theorem 5.4.1 we now deduce a series of important results on independent random variables.

THEOREM 5.4.2.* *Let ξ_n $(n = 1, 2, \ldots)$ be a sequence of independent random variables such that $E(\xi_n) = 0$; $D_n^2 = D^2(\xi_n)$ exists and the series*

$$\sum_{n=1}^{+\infty} D_n^2 \tag{5.4.10}$$

is convergent. Then the series

$$\sum_{n=1}^{+\infty} \xi_n \tag{5.4.11}$$

converges with probability 1 *to a random variable η such that $E(\eta) = 0$ and*

$$E(\eta^2) = \sum_{n=1}^{+\infty} D_n^2 \tag{5.4.12}$$

Proof. Let us put

$$\eta_n = \sum_{k=1}^{n} \xi_k \tag{5.4.13}$$

Then

$$E((\eta_n - \eta_m)^2) = \sum_{k=m+1}^{n} D_k^2 \qquad \text{if } m < n \tag{5.4.14}$$

Thus, η_n converges strongly to a random variable η such that (5.4.12) holds, and applying Theorem 5.4.1 to the martingale $\{\eta_n\}$ we obtain that η_n tends almost surely to η. ■

Example 5.4.2. Let $R_n(x)$ $(n = 1, 2, \ldots)$ denote the Rademacher functions, and let $\{c_n\}$ denote a sequence of real numbers such that the series

$$\sum_{n=1}^{\infty} c_n^2 \tag{5.4.15}$$

is convergent. It follows from Theorem 5.4.2 that the series

$$\sum_{n=1}^{\infty} c_n R_n(x) \tag{5.4.16}$$

* This theorem is a special case of Kolmogoroff's "three series theorem" (see [1]).

is convergent for almost all x. As to every infinite sequence of signs $\varepsilon_k = \pm 1$ ($k = 1, 2, \ldots$) in which infinitely many terms are equal to -1 there corresponds a real number x such that $R_k(x) = \varepsilon_k$ ($k = 1, 2, \ldots$), one can state the result obtained somewhat vaguely as follows: *If the series* (5.4.15) *is convergent, then, giving random signs to the terms* c_n, *the series* $\sum \pm c_n$ *obtained is almost surely convergent.*

Using Theorem 5.3.2 it can be shown that the convergence of the series (5.4.15) is not only sufficient but also necessary for the almost sure convergence of the series (5.4.16). (See Problem P.5.6.)

Thus the probability of the convergence of the series $\sum_{n=1}^{\infty} R_n(x)c_n$ is either 1 or 0 (according to whether $\sum_1^{\infty} c_n^2 < +\infty$ or $\sum_1^{\infty} c_n^2 = +\infty$), no other value is possible. The background of this surprising fact is the celebrated "zero or one law" (see Kolmogoroff [1]).

THEOREM 5.4.3. *Let* $\xi_1, \xi_2, \ldots, \xi_n, \ldots$ *be independent random variables on a probability space* $S = (\Omega, \mathscr{A}, P)$. *Let* $\mathscr{B}_N$ *denote the* σ*-algebra of subsets of* Ω *spanned by the random variables* $\xi_{N+1}, \xi_{N+2}, \ldots$ ($N \geqq 0$). *Let* $\mathscr{B}_\infty = \prod_{N=1}^{\infty} \mathscr{B}_N$ *denote the family of those sets* B *which belong to every* $\mathscr{B}_N$. *Then for any* $B \in \mathscr{B}_\infty$, *one has either* $P(B) = 1$ *or* $P(B) = 0$.

Proof. Let $\mathscr{A}_n$ denote the σ-algebra spanned by the random variables $\xi_1, \xi_2, \ldots, \xi_n$. Clearly, an event $A \in \mathscr{A}_n$ is independent from every $B \in \mathscr{B}_N$ if $N \geqq n$. Thus, if $B \in \mathscr{B}_\infty$, B is independent from every event $A \in \sum_{n=1}^{\infty} \mathscr{A}_n$, from every $A \in \sigma(\sum_{n=1}^{\infty} \mathscr{A}_n) = \mathscr{B}_0$ (see Theorem 3.1.1), and from every $A \in \mathscr{B}_\infty$. But this implies that B is independent from itself, i.e., $P(B) = P(B \cdot B) = P(B)P(B) = P^2(B)$, and thus either $P(B) = 1$ or $P(B) = 0$. ■

THEOREM 5.4.4. *Let* ξ_n ($n = 1, 2, \ldots$) *be independent random variables such that* $E(\xi_n) = 0$, $D^2(\xi_n) = D_n^2 < +\infty$, *and the series*

$$\sum_{n=1}^{+\infty} \frac{D_n^2}{n^2} \tag{5.4.17}$$

is convergent. Then the strong law of large numbers is valid for the sequence $\{\xi_n\}$, *i.e., putting* $\eta_n = \xi_1 + \xi_2 + \cdots + \xi_n$, *one has almost surely*

$$\lim_{n \to +\infty} \frac{\eta_n}{n} = 0 \tag{5.4.18}$$

Proof. For the proof we need the following lemma, due to Kronecker:

LEMMA 5.4.1. *Let* $\sum_{k=1}^{+\infty} a_k$ *be a convergent series and let* c_k ($k = 1, 2, \ldots$) *be a monotonically increasing sequence of positive numbers tending to* $+\infty$. *Then we have*

$$\lim_{n\to+\infty} \frac{\sum_{k=1}^{n} c_k a_k}{c_n} = 0 \tag{5.4.19}$$

For the convenience of the reader we give here the very simple and elementary proof.

Proof. Let us put

$$r_n = \sum_{k=n}^{\infty} a_k \quad (n \geqq 1) \tag{5.4.20}$$

Then by supposition,

$$\lim_{n\to+\infty} r_n = 0$$

Thus, r_n is bounded, say $|r_n| \leqq R$ for $n \geqq 1$. As

$$\frac{1}{c_n} \sum_{k=1}^{n} c_k a_k = \frac{1}{c_n} \sum_{k=1}^{n} r_k (c_k - c_{k-1}) - r_{n+1}$$

it follows that for every fixed m and for $n > m$,

$$\frac{1}{c_n} \left| \sum_{k=1}^{n} c_k a_k \right| \leqq \frac{R c_m}{c_n} + 2 \max_{k \geqq m+1} |r_k|$$

Now let $\varepsilon > 0$ be given. Choosing first m so large that $\max_{k \geqq m+1} |r_k| < \varepsilon/2$, and then $n > m$ so large that $c_m R/c_n < \varepsilon/2$, we get

$$\frac{1}{c_n} \left| \sum_{k=1}^{n} c_k a_k \right| < \varepsilon$$

This proves Lemma 5.4.1. ■

Now Theorem 5.4.2 can be applied to the series

$$\sum_{n=1}^{+\infty} \frac{\xi_n}{n} \tag{5.4.21}$$

and it follows that it is almost surely convergent. Using Lemma 5.4.1 with $c_n = n$, we obtain that

$$\frac{\eta_n}{n} = \frac{1}{n} \sum_{k=1}^{n} k \cdot \frac{\xi_k}{k}$$

tends almost surely to 0. ■

Remark: An alternative way to prove Theorem 5.4.4 is the following: Applying Corollary 3 of Theorem 5.3.1 to the martingale η_n, we get for every $\varepsilon > 0$,

$$P\left(\sup_{n \geqq N} \frac{|\eta_n|}{n} > \varepsilon \right) \leqq \frac{1}{\varepsilon^2} \left(\frac{\sum_1^N D_k^2}{N^2} + \sum_{k=N+1}^{\infty} \frac{D_k^2}{k^2} \right) \tag{5.4.22}$$

As, by Kronecker's lemma (with $c_k = k^2$), it follows from the convergence of the series (5.4.17) that

$$\lim_{n \to +\infty} \frac{1}{n^2} \sum_{k=1}^{n} D_k^2 = 0 \tag{5.4.23}$$

it follows from (5.4.22) that

$$\lim_{N \to +\infty} P\left(\sup_{n \geqq N} \frac{|\eta_n|}{n} > \varepsilon\right) = 0 \tag{5.4.24}$$

which implies that $\eta_n/n \to 0$ almost surely.

Remark: Kolmogoroff's Theorem 4.4.5 can be deduced from Theorem 5.4.4 (see Exercise E.5.4).

We now deduce, from Theorem 5.3.3, the following:

THEOREM 5.4.5. *Let ξ_n $(n = 1, 2, \ldots)$ be a sequence of independent random variables with expectations $E(\xi_n) = M_n > 0$ and variances $D^2(\xi_n) = D_n^2 < +\infty$ $(n = 1, 2, \ldots)$. Suppose that the series $\sum_{n=1}^{\infty} M_n$ is divergent and the series*

$$\sum_{n=1}^{\infty} \frac{D_n^2}{(M_1 + M_2 + \cdots + M_n)^2} \tag{5.4.25}$$

is convergent. Then one has almost surely

$$\lim_{n \to +\infty} \frac{\sum_{k=1}^{n} \xi_k}{\sum_{k=1}^{n} M_k} = 1 \tag{5.4.26}$$

Proof. We can apply Theorem 5.4.2 to the series

$$\sum_{k=1}^{\infty} \frac{(\xi_k - M_k)}{M_1 + M_2 + \cdots + M_k} \tag{5.4.27}$$

It follows that the series (5.4.27) is almost surely convergent; thus, applying Lemma 5.4.1 with $c_n = M_1 + \cdots + M_n$, we get

$$\lim_{n \to +\infty} \frac{\sum_{k=1}^{n} (\xi_k - M_k)}{\sum_{k=1}^{n} M_k} = 0 \tag{5.4.28}$$

which implies (5.4.26). ∎

Remark: Theorem 5.4.5 is a generalization of Theorem 5.4.2. As a matter of fact, if the ξ_k are random variables subject to the conditions of Theorem 5.4.2, we can apply Theorem 5.4.5 to the random variables $\xi_k + 1$ (for which $M_k = 1$), and we get the statement of Theorem 5.4.2.

Example 5.4.3. Let $\{\xi_n\}$ be a sequence of independent and identically distributed random variables with a continuous distribution. Let us consider the random variables α_n (introduced in Problem P.4.5) defined as follows: $\alpha_n = 1$ if $\xi_n > \xi_k$ for $k = 1, 2, \ldots, n-1$, and $\alpha_n = 0$, otherwise. We have shown that $P(\alpha_n = 1) = 1/n$ and that the random variables α_n are independent. As $M(\alpha_n) = 1/n$ and $\sum_{n=1}^{N} 1/n \approx \log N$, it follows from Theorem 5.4.5 that, putting $R(N) = \sum_{n=1}^{N} \alpha_n$, we have

$$\lim_{N \to +\infty} \frac{R(N)}{\log N} = 1$$

which holds almost surely. Denoting by $\nu_1 < \nu_2 < \cdots < \nu_k < \cdots$ those indices n for which $\alpha_n = 1$ (i.e., ξ_{ν_k} is the kth "record" in the sequence $\xi_1, \xi_2, \ldots$) we have $R(\nu_k) = k$. Thus, $\lim_{n \to +\infty} R(N)/\log N = 1$ implies that almost surely we have

$$\lim_{k \to +\infty} \frac{k}{\log \nu_k} = 1$$

i.e., almost surely we have

$$\lim_{k \to +\infty} \sqrt[k]{\nu_k} = e$$

Now we turn to some applications of Theorem 5.4.1 to other questions.

THEOREM 5.4.6. *Let $S = (\Omega, \mathscr{A}, P)$ be a probability space and $\mathscr{B}_n$ $(n = 1, 2, \ldots)$ an increasing sequence of sub-σ-algebras of $\mathscr{A}$, i.e., $\mathscr{B}_n \subseteq \mathscr{B}_{n+1} \subseteq \mathscr{A}$. Let $\mathscr{B}$ denote the least σ-algebra containing all the σ-algebras $\mathscr{B}_n$. Then we have, for every random variable $\xi \in L_2(S)$,*

$$\lim_{n \to +\infty} E(\xi \mid \mathscr{B}_n) = E(\xi \mid \mathscr{B}) \tag{5.4.29}$$

almost surely.

Proof. It is no restriction to suppose that $\mathscr{B} = \mathscr{A}$, because all random variables occurring in the theorem are $\mathscr{B}$-measurable, therefore we can consider them on the probability space $S' = (\Omega, \mathscr{B}, P)$. In case $\mathscr{B} = \mathscr{A}$, (5.4.29) can be written in the form

$$\lim_{n \to +\infty} E(\xi \mid \mathscr{B}_n) = \xi \text{ almost surely} \tag{5.4.30}$$

As

$$\eta_n = E(\xi \mid \mathscr{B}_n) \tag{5.4.31}$$

is a martingale, according to the corollary to Theorem 5.4.1, it is sufficient to show that the sequence $E(\eta_n^2)$ is bounded. Now we have, by Theorem 5.1.6,

$$E(\eta_n^2) \leqq E(\xi^2) \tag{5.4.32}$$

Thus the corollary of Theorem 5.4.1 can be applied, and it follows that η_n tends strongly and also almost surely to an $\eta \in L_2(S)$. It remains to prove that $\eta = \xi$ almost surely. Now, by definition, putting $S_k = (\Omega, \mathscr{B}_k, P)$, for every $\zeta \in L_2(S_k)$ we have, for $n \geqq k$,

$$E(\zeta\eta_n) = E(\zeta\xi) \tag{5.4.33}$$

Thus, as strong convergence implies weak convergence, we get for each $\zeta \in L_2(S_k)$,

$$E(\zeta\eta) = E(\zeta\xi) \tag{5.4.34}$$

Thus, $\eta - \xi$ is orthogonal to every $\zeta \in L_2(S_k)$ for $k = 1, 2, \ldots$. It remains to prove that this implies that $\eta - \xi$ is orthogonal to every $\zeta \in L_2(S)$. This can be shown as follows: If $\zeta \in L_2(S)$, by the definition of the integral, one can for every $\varepsilon > 0$ find a random variable ζ_1 which takes on a finite number of values only and

$$E((\zeta - \zeta_1)^2) < \varepsilon/4 \tag{5.4.35}$$

Let $x_1, \ldots, x_N$ be the possible different values of ζ_1 and put $A_k = \zeta_1^{-1}(\{x_k\})$ $(k = 1, 2, \ldots, N)$. The events A_k belong to $\mathscr{A}$, which is by supposition the least σ-algebra containing the union $\mathscr{B}_\infty$ of the σ-algebras $\mathscr{B}_n$ $(n = 1, 2, \ldots)$. Now $\mathscr{B}_\infty$ is an algebra and $\mathscr{A}$ is the least σ-algebra containing $\mathscr{B}_\infty$. Therefore, it follows from the construction of the extension of a measure (see Appendix A) that each A_k can be approximated arbitrarily closely by a set A_k^* in $\mathscr{B}_\infty$ (and thus in $\mathscr{B}_n$ for some n) so that $P(A_k \circ A_k^*)$ is arbitrarily small. Thus one can find a random variable ζ_2 (taking on the values $x_1, \ldots, x_N$ only) which is $\mathscr{B}_\infty$-measurable, and thus $\mathscr{B}_n$-measurable for some n so that

$$E((\zeta_1 - \zeta_2)^2) < \varepsilon/4 \tag{5.4.36}$$

It follows that

$$E((\zeta - \zeta_2)^2) < \varepsilon$$

and therefore,

$$E^2(\zeta(\eta - \xi)) = E^2((\zeta - \zeta_2)(\eta - \xi)) \leqq \varepsilon E((\eta - \xi)^2)$$

As $\varepsilon > 0$ can be chosen arbitrarily small, it follows that $\eta - \xi$ is orthogonal to every $\zeta \in L_2(S)$ and to itself, and thus, $\eta = \xi$ almost surely. This proves* Theorem 5.4.6. ■

* Equation (5.4.33) can also be deduced, in view of Theorem 5.1.6, from the following well known fact in Hilbert-space theory: If $\prod_n$ is the projection of the Hilbert space H to a subspace H_n (where $H_n \subseteq H_{n+1}$ $(n = 1, 2, \ldots)$ and $\prod$ denotes the projection to the least subspace H^* containing the subspaces H_n $(n = 1, 2, \ldots)$, then for every element $x \in H$, $\prod_n x$ tends strongly to $\prod x$. This is a particular case of a more general theorem of Vigier (see Riesz and Sz. Nagy [10], pp. 261 and 266).

COROLLARY TO THEOREM 5.4.6. *If $S = (\Omega, \mathscr{A}, P)$ is a probability space, $A \in \mathscr{A}$ is an arbitrary event in S, α is the indicator of A, and $\mathscr{B}_n$ is a sequence of σ-algebras of subsets of Ω such that $\mathscr{B}_n \subseteq \mathscr{B}_{n+1} \subseteq \mathscr{A}$ and $\mathscr{A}$ is the least σ-algebra containing all the σ-algebras $\mathscr{B}_n$, then we have almost surely*

$$\lim_{n \to +\infty} P(A \mid \mathscr{B}_n) = \alpha \tag{5.4.37}$$

Proof. We obtain (5.4.37) from (5.4.31) by choosing for ξ the indicator α of the event A, and taking into account that by Definition 5.3.1,

$$E(\alpha \mid \mathscr{B}_n) = P(A \mid \mathscr{B}_n) \quad \blacksquare$$

Remark: Equation (5.4.29) is also valid under the weaker supposition that $E(\xi)$ exists (see Problem P.5.7). This follows from the following important theorem of Doob [2], which we state here without proof: *If η_n is a submartingale such that $E(|\eta_n|)$ $(n \geqq 1)$ is bounded, then $\lim_{n\to+\infty} \eta_n = \eta$ exists almost surely, and $E(\eta)$ exists.*

Example 5.4.4. Let $f(x)$ be a square integrable function on the interval $(0, 1)$. Let $W_n(x)$ $(n = 0, 1, \ldots)$ denote the Walsh functions (see Section 3.6). Let

$$f(x) \sim \sum_{n=1}^{\infty} a_n W_n(x) \tag{5.4.38}$$

be the Walsh expansion of $f(x)$, i.e., put

$$a_n = \int_0^1 f(t) W_n(t) dt \qquad (n = 0, 1, \ldots) \tag{5.4.39}$$

and put

$$S_N(x) = \sum_{n=0}^{N} a_n W_n(x) \tag{5.4.40}$$

We shall prove that

$$\lim_{n \to +\infty} S_{2^n - 1}(x) = f(x) \tag{5.4.41}$$

holds for almost all x.

To prove (5.4.41) let us consider the σ-algebra $\mathscr{B}_n$ spanned by the first n Rademacher functions, i.e., the finite algebra of subsets of the interval (0, 1) which has as atoms the 2^n binary rational intervals $(k/2^n, (k+1)/2^n)$ $(k = 0, 1, \ldots, 2^n - 1)$. Then by Theorem 5.1.6, $E(f \mid \mathscr{B}_N)$ is the projection of $f(x)$ to the subspace spanned by the function $R_k(x)$ $(k \leqq N)$, and therefore,

$$E(f \mid \mathscr{B}_n) = S_{2^n - 1}(x) \tag{5.4.42}$$

Thus (5.4.41) follows from Theorem 5.4.6.

Finally, as an application of Theorem 5.4.5, we give a proof for Theorem 3.6.4, which was stated without proof in Section 3.6.

Let ξ_n be a sequence of bounded random variables on the probability space $S=(\Omega, \mathscr{A}, P)$ where Ω is a separable metric space and suppose that the variables ξ_n form a spanning system. Let $\mathscr{B}_n$ denote the σ-algebra of subsets of Ω spanned by the variables $(\xi_1, \ldots, \xi_n)$. Then $\mathscr{B}_n \subseteq \mathscr{B}_{n+1} \subseteq \mathscr{A}$. Without restricting the generality we may suppose that $\mathscr{A}$ is the least σ-algebra containing all the σ-algebras $\mathscr{B}_n$. Suppose that $\eta \in L_2(S)$ is a random variable which is orthogonal to all power products of the variables ξ_k. Then η is orthogonal to any polynomial of the variables $\xi_1, \ldots, \xi_n$ for every n, and thus if $f(x_1, x_2, \ldots, x_n)$ is a continuous function of its variables, then η is also orthogonal to $f(\xi_1, \ldots, \xi_n)$. It follows that η is orthogonal to the whole space $L_2(S_n)$ where $S_n=(\Omega, \mathscr{B}_n, P)$ and thus, by Theorem 5.1.6, $E(\eta \mid \mathscr{B}_n)=0$ for $n=1, 2, \ldots$, which implies, by Theorem 5.4.5, that $\eta=0$ almost surely. Thus we have proved that if $\{\xi_n\}$ is a spanning system, the power products of the ξ_n form a complete set. To prove the converse statement, let $\mathscr{B}$ denote the σ-algebra spanned by the random variables ξ_n. Let us orthogonalize the system of power products of the random variables ξ_n; we get by supposition a complete orthonormal system X. Thus for every $\eta \in L_2(S)$ there exists a sequence of random variables η_n (namely the partial sums of the orthogonal expansion of η in the system X) such that each η_n belongs to some $L_2(S_N)$ where $S_N=(\Omega, \mathscr{B}_n, P)$ and

$$\lim_{n \to +\infty} E((\eta-\eta_n)^2)=0$$

This implies that $E(\eta \mid \mathscr{B}_N)$ tends for $N \to +\infty$ strongly to η, because

$$E([\eta-E(\eta \mid \mathscr{B}_N)]^2) \leqq E((\eta-\eta^*)^2)$$

for every $\eta^* \in L_2(S_N)$. Thus, by Theorem 5.4.1, $E(\eta \mid \mathscr{B}_N)$ tends to η almost surely.

As, by Theorem 5.4.6, $E(\eta \mid \mathscr{B}_n)$ tends to $E(\eta \mid \mathscr{B})$ almost surely, it follows that

$$E(\eta \mid \mathscr{B})=\eta \tag{5.4.43}$$

almost surely. Thus, in particular, if $A \in \mathscr{A}$ is any event and α is its indicator, we have

$$P(A \mid \mathscr{B})=\alpha$$

almost surely. Let A^* denote the set on which $P(A \mid \mathscr{B})=1$. Then $A^* \in \mathscr{B}$ and $P(A \circ A^*)=0$. This implies that $\{\xi_n\}$ is a spanning system.

In a recent paper [50] Waterman has given the following simpler proof for the second statement of Theorem 3.6.4 for the case when $S=(\Omega, \mathscr{A}, P)$ is the Lebesgue probability space: If the power products of the random variables

ξ_k $(k=1, 2, \ldots)$ form a complete set, then there exists a sequence $P_n = P_n(\xi_1, \xi_2, \ldots, \xi_n)$ of polynomials which tends to the random variable $\eta(\omega) = \omega$ $(0 \leq \omega < 1)$ in L_2-norm. It follows that an appropriate subsequence P_{n_k} of these polynomials tends to $\eta(\omega) = \omega$ almost everywhere, i.e., except for a set Z of measure 0. Thus if $\omega_1 \notin Z$, $\omega_2 \notin Z$ and $\omega_1 \neq \omega_2$, then there exists a value of n for which $\xi_n(\omega_1) \neq \xi_n(\omega_2)$. Thus $\{\xi_k\}$ is a separating system, and therefore by Theorem 3.6.3 it is a spanning system.

Example 5.4.5. Let us consider the following question: Let $\xi_1, \xi_2, \ldots, \xi_n, \ldots$ be a sequence of independent random variables, each taking on the values 1 and 0, with the corresponding probabilities $P(\xi_n = 1) = p_n$, $P(\xi_n = 0) = 1 - p_n$ $(0 < p_n < 1; n = 1, 2, \ldots)$. Let us construct these random variables on the probability space $S = (I, \mathscr{B}, P)$, where I is the interval $[0, 1)$ and $\mathscr{B}$ is the family of Borel subsets of I; put $\xi_n = (1 + R_n(x))/2$, where $R_n(x)$ $(n = 1, 2, \ldots)$ are the Rademacher functions, and the probability measure P is defined by the condition that the random variables ξ_n are independent and $P(\xi_n = 1) = p_n$, $P(\xi_n = 0) = 1 - p_n$ $(n \geqq 1)$. By this condition, P is uniquely defined for all binary intervals $(k/2^n, (k+1)/2^n)$ and thus for every Borel subset B of I. The question arises, under what conditions the measure P will be absolutely continuous with respect to the Lebesgue measure. We have seen in Chapter 4 (see Example 4.4.3) that if $p_n = p$ for $n = 1, 2, \ldots$, where $0 < p < 1$, $p \neq 1/2$, then the measure P is orthogonal to the Lebesgue measure. From this we may guess that for the absolute continuity of the measure P with respect to the Lebesgue measure it is *necessary* that $p_n \to 1/2$. This is, in fact, true; we shall show that *the necessary and sufficient condition for the absolute continuity of the measure* P is that the series

$$\sum_{n=1}^{+\infty} (2p_n - 1)^2 \tag{5.4.44}$$

should be convergent.

This can be proved as follows: Suppose first that the series (5.4.44) is convergent. Let us denote by $I_n(x)$ the interval $([2^n x]/2^n, ([2^n x] + 1)/2^n)$ $(0 < x < 1, n = 1, 2, \ldots)$. Then we have, putting $a_n = 2p_n - 1$,

$$P(I_n(x)) = \frac{1}{2^n} \prod_{k=1}^{n} (1 + a_k R_k(x)) \tag{5.4.45}$$

Thus, $2^n P(I_n(x))$ is the $(2^n - 1)$st partial sum of the Walsh series,

$$1 + \sum_{k_1 < k_2 < \cdots < k_r} a_{k_1} a_{k_2} \cdots a_{k_r} R_{k_1}(x) \cdots R_{k_r}(x) \tag{5.4.46}$$

As

$$1 + \sum (a_{k_1} a_{k_2} \cdots a_{k_r})^2 = \prod_{k=1}^{+\infty} (1 + a_k^2) < +\infty \tag{5.4.47}$$

it follows from the result of Example 5.4.2 that

$$\lim_{n \to +\infty} 2^n P(I_n(x)) = g(x) \tag{5.4.48}$$

for almost all x, where $g(x) \in L_2(S)$ is the function having the Walsh–Fourier expansion (5.4.46); thus, for this function, we have, in particular,

$$\int_0^1 g(x)\, dx = 1 \tag{5.4.49}$$

Now let $F(x)$ denote the P measure of the interval $[0, x)$ $(0 < x < 1)$. Then $F(x)$ is a distribution function and the measure P is absolutely continuous if and only if the function $F(x)$ is equal to the indefinite integral of its derivative, (which exists almost everywhere). To prove this, as by Fatou's lemma $\int_0^x F'(x)\, dx \leq F(x)$, it is sufficient to show that

$$\int_0^1 F'(t)\, dt = 1 \tag{5.4.50}$$

Now we have

$$\lim_{n \to +\infty} 2^n P(I_n(x)) = F'(x) \tag{5.4.51}$$

in every point where $F'(x)$ exists. Comparing (5.4.51) and (5.4.48), it follows that $F'(x) = g(x)$ and $F'(x) \in L_2(S)$; from (5.4.49) it follows that (5.4.50) holds.

Thus we have shown that if the series (5.4.44) is convergent, P is absolutely continuous with respect to the Lebesgue measure λ; moreover, $dP/d\lambda = F'(x) \in L_2(S)$.

On the other hand, let us suppose now that the series (5.4.44) is divergent. In this case the series $\sum_{n=1}^{\infty} a_n R_n(x)$ is almost surely (with respect to the Lebesgue measure) divergent; moreover (see Problem P.5.6), we have almost surely

$$\liminf_{N \to +\infty} \sum_{n=1}^{N} a_n R_n(x) = -\infty$$

As by (5.4.45) and (5.4.51),

$$F'(x) = \lim_{N \to \infty} \prod_{n=1}^{N} (1 + a_n R_n(x)) \leq \exp\left(\liminf_{N \to +\infty} \sum_{n=1}^{N} a_n R_n(x)\right)$$

it follows that $F'(x) = 0$ almost surely. Thus, $\int_0^1 F'(x)\, dx = 0$, i.e., the measure P is even orthogonal to the Lebesgue measure.

Remark: The result of this example is a special case of a general theorem of Kakutani [48].

5.5 EXISTENCE THEOREMS

In dealing with sequences of independent random variables there was no need to bother about questions of existence; the existence of a probability space on which a sequence of independent random variables with the required properties can be given follows from Theorem 3.4.2 about the product of probability spaces. As, however, in this chapter we deal with more general sequences of random variables, we need a corresponding existence theorem. For this reason, we now prove the following theorem, called the fundamental theorem of Kolmogoroff* (see [1]):

THEOREM 5.5.1. *Let $F_n(x_1, x_2, \ldots, x_n)$ be for each $n \geqq 1$ an n-dimensional probability distribution function. Let us suppose that the following compatibility condition is satisfied:*

$$\lim_{x_n \to +\infty} F_n(x_1, \ldots, x_{n-1}, x_n) = F_{n-1}(x_1, \ldots, x_{n-1}) \qquad (n = 2, 3, \ldots) \quad (5.5.1)$$

Then there exists a probability space $S = (\Omega, \mathscr{A}, P)$, and on this space a sequence ξ_n of random variables such that for each n, $F_n(x_1, x_2, \ldots, x_n)$ is the joint distribution function of $\xi_1, \xi_2, \ldots, \xi_n$, i.e.,

$$P(\xi_1 < x_1, \ldots, \xi_n < x_n) = F_n(x_1, x_2, \ldots, x_n) \qquad (n = 1, 2, \ldots) \quad (5.5.2)$$

Proof. Let Ω be the space of all infinite sequences $\omega = (\omega_1, \omega_2, \ldots, \omega_n, \ldots)$ of real numbers. We call ω_n the nth coordinate of the point ω, and put

$$\omega_n = c_n(\omega) \qquad (n = 1, 2, \ldots) \quad (5.5.3)$$

We call a subset A of Ω a cylinder set with base B if B is for some n a Borel set in the n-dimensional Euclidean space and $\omega \in A$ if and only if the n-tuple of the first n coordinates $[c_1(\omega), \ldots, c_n(\omega)]$ of ω belong to the set B.

Clearly, if A is a cylinder set with base $B \in R^n$, then A can also be regarded as a cylinder set with base $B' \in R^{n+k}$, where B' is the set of all those points of R^{n+k} whose first n coordinates are the coordinates of a point in B. To each cylinder set A there corresponds a least positive integer n such that A is a cylinder set with base B in R^n; we call this number n the *order* of the cylinder set A.

The distribution function $F_n(x_1, \ldots, x_n)$ defines a (Lebesgue-Stieltjes) measure on the Borel subsets of the n-dimensional Euclidean space. We

* We shall state and prove Kolmogoroff's theorem only for denumerable sequences of random variables, because in this book we do not deal systematically with non-denumerable families of random variables; if we need such families we construct them by means of a denumerable sequence of random variables (as in the discussion of the Poisson process in Section 4.6, or in Problem P.5.10).

denote this measure by $P_n(B)$. Now we define a set function $P(A)$ on all cylinder sets $A \subseteq \Omega$ as follows: If A is a cylinder set with base B, where B is a Borel set in R^n, we put

$$P(A) = P_n(B) \tag{5.5.4}$$

It follows easily from (5.5.1) that if A is a cylinder set of order n with base B in R^n and B' is the base of A in R^{n+k}, then $P_{n+k}(B') = P_n(B)$, i.e., the definition of $P(A)$ by (5.5.4) does not depend on which base of A we choose. It is easy to see that, denoting by $\mathscr{A}_0$ the family of all cylinder sets $A \subseteq \Omega$ (of any order), $\mathscr{A}_0$ is an algebra (though not a σ-algebra) of sets and $P(A)$ is a nonnegative and finitely additive set function on $\mathscr{A}_0$. Let $\mathscr{A} = \sigma(\mathscr{A}_0)$ denote the least σ-algebra of subsets of Ω containing $\mathscr{A}_0$. In order to extend $P(A)$ to a measure on $\mathscr{A}$ it is, as we know, sufficient to show that P has the property that if A_n is a sequence of sets such that $A_n \in \mathscr{A}_0$, $A_{n+1} \subseteq A_n$ $(n = 1, 2, \ldots)$ and $\prod_{n=1}^{\infty} A_n = \emptyset$, then

$$\lim_{n \to +\infty} P(A_n) = 0 \tag{5.5.5}$$

We shall prove (5.5.5) indirectly. Suppose that there exists a sequence $\{A_n\}$ with the required properties for which (5.5.5) does not hold. As $P(A_n)$ is evidently nonincreasing, this means that there exists a positive number c such that

$$P(A_n) \geqq c > 0 \tag{5.5.6}$$

If k_n is the order of A_n, then $k_n \leqq k_{n+1}$. Clearly, if (5.5.6) holds, then $k_n \to +\infty$, because $P(A)$ is a measure on the family of all cylinder sets of order $\leqq N$ (being equal there to the Lebesgue-Stieltjes measure P_N).

Without restricting the generality we may suppose, therefore, that $k_n = n$, i.e., that A_n is a cylinder set of order n with base $B_n \in R^n$.

Now we shall use the fact that each Lebesgue-Stieltjes measure $P_n(A)$ is *perfect*, i.e., for each Borel subset A of R^n, $P_n(A)$ is equal to the least upper bound of $P_n(Z)$ for all compact (closed and bounded) sets Z contained in A. It follows that we can find for each n a compact set Z_n in R^n such that $Z_n \subseteq B_n$ and

$$P_n(Z_n) \geqq P_n(B_n) - \frac{c}{2^{n+1}} \qquad (n = 1, 2, \ldots)$$

Let us denote by A_n' the cylinder set of order n with base Z_n, and put $D_n = A_1'A_2' \cdots A_n'$. It follows that $D_n \subseteq A_n' \subseteq A_n$ and

$$P(A_n \bar{D}_n) \leqq \sum_{k=1}^{n} P(A_n \bar{A}_k') \leqq \sum_{k=1}^{n} P(A_k \bar{A}_k') = \sum_{k=1}^{n} P_k(B_k \bar{Z}_k)$$

Thus,

$$P(A_n \bar{D}_n) \leqq \sum_{k=1}^{n} [P_k(B_k) - P_k(Z_k)] \leqq \sum_{k=1}^{n} \frac{c}{2^{k+1}} \leqq \frac{c}{2}$$

Therefore,

$$P(D_n) = P(A_n D_n) = P(A_n) - P(A_n \bar{D}_n) \geqq \frac{c}{2}$$

This means that we have obtained a sequence D_n of cylinder sets having the following properties:

(1) D_n is a cylinder set of order n with a compact base Z_n.

(2) $D_{n+1} \subseteq D_n$.

(3) $\prod_{n=1}^{\infty} D_n = \varnothing$.

(4) D_n is not empty.

However, these properties are contradictory: (1), (2) and (4) imply that (3) cannot hold. As a matter of fact, choose in each D_n a point $\omega^{(n)} = (\omega_1^{(n)}, \ldots, \omega_k^{(n)}, \ldots)$. Select (by the diagonal process) a sequence n_j of integers such that the finite limits

$$\lim_{j \to +\infty} \omega_k^{(n_j)} = \omega_k^*$$

exist for $k = 1, 2, \ldots$. As all sets Z_k are closed and $(\omega_1^{(n)}, \ldots, \omega_k^{(n)}) \in Z_k$ for all $n \geqq k$, we have

$$(\omega_1^*, \ldots, \omega_k^*) \in Z_k$$

It follows that the point $\omega^* = (\omega_1^*, \ldots, \omega_k^*, \ldots)$ belongs to all sets D_n and thus also to $\prod_{n=1}^{\infty} D_n$, which therefore cannot be empty.

Thus we arrived at a contradiction which shows that P is a measure on $\mathscr{A}_0$, and therefore it can be extended to a probability measure on $\mathscr{A}$.

Now consider on the probability space $S = (\Omega, \mathscr{A}, P)$ thus obtained, the functions $c_n(\omega)$. It is easy to see that $\xi_n = c_n(\omega)$ is a random variable on S and that these random variables $\xi_1, \ldots, \xi_n, \ldots$ satisfy (5.5.2) for every n. This proves Theorem 5.5.1. ■

From Theorem 5.5.1 we can deduce the following special case which is all that is needed for sequences of discrete random variables:

THEOREM 5.5.2. *Let $X = \{x_1, x_2, \ldots, x_k, \ldots\}$ be a denumerable set. Let for each $n \geqq 1$ be given a discrete probability distribution $p_n(k_1, k_2, \ldots, k_n)$ on the set of all n-tuples of positive integers $(k_1, k_2, \ldots, k_n)$. Suppose that the following compatibility condition is fulfilled: For all $n \geqq 1$ and all n-tuples $(k_1, k_2, \ldots, k_n)$ of positive integers,*

$$\sum_{j=1}^{\infty} p_{n+1}(k_1, \ldots, k_n, j) = p_n(k_1, \ldots, k_n) \tag{5.5.7}$$

Then there exists a probability space $S=(\Omega, \mathscr{A}, P)$ and on S a sequence $\{\xi_n\}$ $(n=1, 2, \ldots)$ of discrete random variables such that for each $n \geqq 1$ and every n-tuple $(k_1, k_2, \ldots, k_n)$ of positive integers,

$$P(\xi_1 = x_{k_1}, \ldots, \xi_n = x_{k_n}) = p_n(k_1, \ldots, k_n) \tag{5.5.8}$$

Proof. Put

$$F_n(x_1, x_2, \ldots, x_n) = \sum_{\substack{k_j < x_j \\ (j=1,2,\ldots,n)}} p_n(k_1, k_2, \ldots, k_n) \tag{5.5.9}$$

Here, $F_n(x_1, x_2, \ldots, x_n)$ is for each n an n-dimensional distribution function; it follows from (5.5.7) that $F_n(x_1, x_2, \ldots, x_n)$ satisfies the condition (5.5.1). Thus, Theorem 5.5.1 can be applied and Theorem 5.5.2 follows. ■

Remark: The probability measure of which (5.5.9) is the distribution function is clearly concentrated on the n-tuples of positive integers. Thus we may modify the construction of the probability space S, by including into Ω only those points $\omega = (\omega_1, \ldots, \omega_n, \ldots)$ where all ω_n are positive integers. Thus we obtain the following:

THEOREM 5.5.3. *Let Ω be the set of all infinite sequences $\omega = (\omega_1, \omega_2, \ldots, \omega_n, \ldots)$ of positive integers. Let $\mathscr{A}_0$ be the family of cylinder sets of Ω, i.e., the family of all such subsets A of Ω which are defined by a condition of the following form: $\omega \in A$ of its first n coordinates $(\omega_1, \omega_2, \ldots, \omega_n)$ belong to a certain set B of n-tuples of positive integers. Let $\mathscr{A}$ be the least σ-algebra of subsets Ω containing $\mathscr{A}_0$. Then under the conditions of Theorem 5.5.2 there exists a probability measure P on $\mathscr{A}$ such that if $A(k_1, \ldots, k_n)$ denotes the set of all $\omega = (\omega_1, \ldots, \omega_n, \ldots)$ such that $c_j(\omega) = k_j$ for $j \leqq n$, then we have*

$$P(A(k_1, \ldots, k_n)) = p_n(k_1, \ldots, k_n) \tag{5.5.10}$$

Example 5.5.1. As an example, let us consider the construction of a homogeneous Markov chain. We want to construct a Markov chain $\{\xi_n\}$ such that each ξ_n is capable of the values $1, 2, \ldots, r$ $(r \geqq 2)$, the transition probabilities

$$P(\xi_{n+1} = k \mid \xi_n = j) = p_{j,k} \qquad (1 \leqq j, k \leqq r)$$

do not depend on n (where $(p_{j,k})$ is an arbitrary stochastic matrix), and the initial distribution of ξ_1 is

$$P(\xi_1 = k) = w_k \qquad (1 \leqq k \leqq r)$$

Let us put for every n-tuple of positive integers $(k_1, \ldots, k_n)$, such that $1 \leqq k_j \leqq r$ for $j = 1, 2, \ldots, n$,

$$p_n(k_1, \ldots, k_n) = w_{k_1} \prod_{i=1}^{n-1} p_{k_i, k_{i+1}} \tag{5.5.11}$$

In this case the compatibility condition (5.5.7) is satisfied. Thus the existence of a Markov chain $\{\xi_n\}$ with the prescribed properties follows from Theorem 5.5.2 (or from Theorem 5.5.3). In the special case considered, Ω may be taken as the set of all sequences $\omega = \{\omega_1, \omega_2, \ldots, \omega_n, \ldots\}$, where for every n ω_n is equal to one of the numbers $1, 2, \ldots, r$.

5.6 LIMIT THEOREMS FOR MARKOV CHAINS

Let $\{\xi_n\}$ be a homogeneous Markov chain such that ξ_n is capable of the values $1, 2, \ldots, r$ only ($r \geqq 2$).

Markov chains are used, in the first place, to describe the behavior of *systems* which are capable of a certain number of distinct *states* and which change their state from time to time at random. For instance, the system may consist of a certain number of elevators in a building and the state of the system in a certain instant may be characterized by the position of the elevators at that instant. For this reason the values taken on by the variables of a Markov chain are usually called the "states" of the chain.

Let the transition probabilities of the chain be

$$p_{j,k} = P(\xi_{n+1} = k \mid \xi_n = j) \qquad (j,k = 1, 2, \ldots, r) \tag{5.6.1}$$

and let the initial distribution be

$$w_k = P(\xi_1 = k) \qquad (k = 1, 2, \ldots, r) \tag{5.6.2}$$

where $\{w_1, \ldots, w_r\}$ is a probability distribution and $(p_{j,k})$ is an r-by-r stochastic matrix, i.e., $p_{j,k} \geqq 0$ and

$$\sum_{k=1}^{r} p_{j,k} = 1 \text{ for } \qquad 1 \leqq j \leqq r \tag{5.6.3}$$

Let $p_{j,k}^{(n)}$ denote the n-step transition probabilities, i.e., $p_{j,k}^{(1)} = p_{j,k}$, and

$$p_{j,k}^{(n)} = P(\xi_{n+1} = k \mid \xi_1 = j) \qquad (n = 2, 3, \ldots) \tag{5.6.4}$$

Then, as we have seen, the matrix $p_{j,k}^{(n)} = \prod_n$ is equal to the nth power Π_1^n of the matrix $\Pi_1 = (p_{j,k})$.

We shall first prove the following:

THEOREM 5.6.1. *If there exists a positive integer s such that all elements of the matrix Π_s are strictly positive, then the limit*

$$\lim_{n \to +\infty} p_{j,k}^{(n)} = p_k \tag{5.6.5}$$

exists, for $j,k = 1, 2, \ldots, r$, the limit p_k does not depend on j, the numbers p_k are all positive and they satisfy the system of equations

$$\sum_{j=1}^{r} p_j p_{j,k} = p_k \qquad (k = 1, 2, \ldots, r) \tag{5.6.6}$$

and

$$\sum_{k=1}^{r} p_k = 1 \tag{5.6.7}$$

Remark: It is usual to express the conclusion of Theorem 5.6.1 by saying that the Markov chains in question are *ergodic*.

Proof. The homogeneous system of linear equations (5.6.6) always has nontrivial solutions, because the matrix $(p_{j,k})$ has the property that its row sums are all equal to 1, and therefore its determinant is 0. Thus, the system (5.6.6)–(5.6.7) has a nontrivial solution $(x_1, \ldots, x_r)$ for which we have $\sum_{j=1}^{r} x_j p_{j,k} = x_k$ $(1 \leqq k \leqq r)$ and $\sum_{k=1}^{r} x_k = 1$. It follows that

$$|x_k| \leqq \sum_{j=1}^{r} |x_j| \, p_{j,k} \qquad \text{for } k = 1, 2, \ldots, r \tag{5.6.8}$$

Summing (5.6.8) for $1 \leqq k \leqq r$, we get

$$\sum_{k=1}^{r} |x_k| \leqq \sum_{j=1}^{r} |x_j| \tag{5.6.9}$$

As in (5.6.9) there stands equality, we must have equality in (5.6.8) for every k, and therefore $p_k = |x_k|$ is a solution of the system (5.6.6). Thus the system (5.6.6)–(5.6.7) possesses a nontrivial solution $(p_1, \ldots, p_k)$ consisting of nonnegative numbers, i.e., there exists a probability distribution $\mathscr{P} = (p_1, \ldots, p_r)$ such that (5.6.6) holds. Now let us consider the information-theoretical divergence $D(\mathscr{P}, \mathscr{P}_j^{(n)})$ of the distribution $\mathscr{P}$ and the distribution $\mathscr{P}_j^{(n)} = (p_{j,1}^{(n)}, \ldots, p_{j,r}^{(n)})$. As by supposition all elements of Π_1^s are positive, the same holds for Π_n in view of

$$p_{j,k}^{(n)} = \sum_{l=1}^{r} p_{j,l}^{(n-s)} p_{l,k}^{(s)} \tag{5.6.10}$$

Thus $D(\mathscr{P}, \mathscr{P}_j^{(n)})$ is well defined for $n \geqq s$. Mulitplying both sides of (5.6.6) by $p_{k,l}$ and adding for k, we obtain that the numbers $(p_1, \ldots, p_k)$ also satisfy the system

$$\sum_{j=1}^{r} p_j p_{j,k}^{(2)} = p_k \qquad (k = 1, 2, \ldots, r) \tag{5.6.11}$$

Repeating the same procedure, we get by induction that the numbers $p_1, \ldots, p_r$ satisfy the system of equations

$$\sum_{j=1}^{r} p_j p_{j,k}^{(n)} = p_k \qquad (k = 1, 2, \ldots, r) \tag{5.6.12}$$

for every $n \geqq 1$. Thus (5.6.12) holds for $n = s$ too, and as $p_{j,k}^{(s)} > 0$ for all j and k, it follows that $p_k > 0$ for $k = 1, 2, \ldots, r$.

We shall show now that

$$D(\mathscr{P}, \mathscr{P}_j^{(n+1)}) \leqq D(\mathscr{P}, \mathscr{P}_j^{(n)}) \; (n = s, s+1, \ldots) \tag{5.6.13}$$

To prove (5.6.13) we need Jensen's well known inequality,* according to which if $f(x)$ is a convex function in a real interval I with $x_1, x_2, \ldots, x_r$ arbitrary real numbers in I and $c_1, c_2, \ldots, c_r$ positive numbers (called "weights") such that $\sum_{l=1}^{r} c_l = 1$, then

$$f\left(\sum_{l=1}^{r} c_l x_l\right) \leqq \sum_{k=1}^{r} c_l f(x_l) \tag{5.6.14}$$

and equality holds in (5.6.16) if and only if all x_l are equal.

Applying (5.6.16) to the convex function $f(x) = \log_2 1/x$, the numbers $x_l = p_{j,l}^{(n)}/p_l$ and the "weights" $c_l = p_l p_{l,k}/p_k$ which have, according to (5.6.6) the sum 1, we get

$$\log_2 \frac{p_k}{p_{j,k}^{(n+1)}} = \log_2 \left[\frac{1}{\sum_{l=1}^{r} p_l p_{l,k}/p_k \cdot p_{j,l}^{(n)}/p_l}\right] \leqq \sum_{l=1}^{r} \frac{p_l p_{l,k}}{p_k} \log_2 \frac{p_l}{p_{j,l}^{(n)}} \tag{5.6.15}$$

Multiplying both sides of (5.6.15) by p_k and adding the inequalities thus obtained for $k = 1, 2, \ldots, r$, we get (5.6.13).

As $D(\mathscr{P}, \mathscr{P}_j^{(n)}) \geqq 0$, it follows that

$$\lim_{n \to +\infty} D(\mathscr{P}, \mathscr{P}_j^{(n)}) = d_j \geqq 0 \tag{5.6.16}$$

exists. As the numbers $p_{j,k}^{(n)}$ are all between 0 and 1, we can select an increasing sequence $N_1 < N_2 < \cdots$ of integers such that the limits

$$\lim_{i \to +\infty} p_{j,k}^{(N_i)} = q_{j,k} \tag{5.6.17}$$

exist for all j and k; clearly, for each fixed value of j, the $q_{j,k}$ form a probability distribution $\mathscr{Q}_j = (q_{j,1}, \ldots, q_{j,r})$. It follows that

$$D(\mathscr{P}, \mathscr{Q}_j) = d_j \tag{5.6.18}$$

which implies that none of the numbers $q_{j,k}$ is 0, because otherwise $D(\mathscr{P}, \mathscr{Q}_j)$ would not be finite.

Let us now put

$$q'_{j,k} = \sum_{l=1}^{k} q_{j,l} p_{l,k}^{(s)} \tag{5.6.19}$$

* The Jensen inequality expresses the intuitively evident fact that if we assign positive masses to certain points on a convex curve, then the center of gravity of this system of mass points lies in the convex hull of the points and thus above the curve, if the points do not all coincide.

Clearly, $\mathscr{Q}'_j = (q'_{j,1}, \ldots, q'_{j,r})$ is a probability distribution too, and

$$\lim_{i \to +\infty} p_{j,k}^{(N_i+s)} = q'_{j,k} \qquad (j, k = 1, 2, \ldots, r) \tag{5.6.20}$$

It follows that

$$D(\mathscr{P}, \mathscr{Q}'_j) = d_j \qquad (j = 1, 2, \ldots, r) \tag{5.6.21}$$

i.e.,

$$D(\mathscr{P}, \mathscr{Q}_j) = D(\mathscr{P}, \mathscr{Q}'_j) \tag{5.6.22}$$

On the other hand, by the same method with which (5.6.15) was proved, we obtain from Jensen's inequality applied to $f(x) = \log_2 1/x$, $x_l = q_{j,l}/p_l$, and $c_l = p_l p_{l,k}^{(s)}/p_k$, that

$$p_k \log_2 \frac{p_k}{q_{j,k}} \leqq \sum_{l=1}^{r} \frac{p_l p_{l,k}^{(s)}}{p_k} \log_2 \frac{p_l}{q_{j,l}} \tag{5.6.23}$$

Adding the inequalities (5.6.23) we get $D(\mathscr{P}, \mathscr{Q}'_j)$ on the left-hand side and $D(\mathscr{P}, \mathscr{Q}_j)$ on the right-hand side, and these quantities are equal according to (5.6.22); thus there has to be equality in (5.6.23) for all k, but in Jensen's inequality equality is possible only if all the x_l coincide. Thus we get that the numbers $q_{j,l}/p_l$ are all equal to each other, and since $\sum_{l=1}^{r} p_l(q_{j,l}/p_l) = 1$, we have

$$q_{j,l} = p_l \qquad \text{for } l = 1, 2, \ldots, k \tag{5.6.24}$$

This implies that $\mathscr{Q}_j = \mathscr{P}$, and therefore,

$$d_j = D(\mathscr{P}, \mathscr{P}) = 0 \tag{5.6.25}$$

i.e.,

$$\lim_{n \to +\infty} D(\mathscr{P}, \mathscr{P}_j^{(n)}) = 0 \tag{5.6.26}$$

This is, however, possible (see the proof of Theorem 4.3.12) only if

$$\lim_{n \to +\infty} p_{j,k}^{(n)} = p_k \tag{5.6.27}$$

for all k and j. Thus, (5.6.27) holds for all j and k and Theorem 5.6.1 is proved. ■

Remark 1: It follows from Theorem 5.6.1 that the system of equations (5.6.6)–(5.6.7) has a unique solution* $(p_1, \ldots, p_r)$. It follows also that whatever the initial distribution $w_1, \ldots, w_r$ is, we have

$$\lim_{n \to +\infty} P(\xi_n = k) = p_k \qquad \text{for } k = 1, 2, \ldots, r \tag{5.6.28}$$

* This can be proved, of course, by methods of matrix theory (see, e.g., Gantmacher [12]) by showing that the matrix $\prod_1$ has rank $r - 1$.

As a matter of fact, we have

$$P(\xi_n = k) = \sum_{j=1}^{r} w_j p_{j,k}^{(n)} \tag{5.6.29}$$

From (5.6.27) and (5.6.29) we get (5.6.28). Notice that in the special case when $w_j = p_j$ $(1 \leqq j \leqq r)$ it follows from (5.6.29) and (5.6.13) that

$$P(\xi_n = k) = p_k \qquad \text{for } n = 1, 2, \ldots \tag{5.6.30}$$

Thus, if the distribution of ξ_1 is equal to the limiting distribution $\mathscr{P}$, then the variables ξ_n are identically distributed, all having the distribution $\mathscr{P}$.

Such a distribution which, once established, remains valid forever, is called a *stationary distribution*. As any stationary distribution has to satisfy the system of equations (5.6.6)–(5.6.7), the stationary distribution is necessarily unique. If the initial distribution is identical to the stationary distribution, the Markov chain itself is called *stationary*.

Remark 2: It can be shown that under the conditions of Theorem 5.6.1 $p_{j,k}^{(n)}$ tends exponentially fast to p_k, i.e., there exists a constant λ such that $0 < \lambda < 1$, and a constant $C > 0$ so that

$$|p_{j,k}^{(n)} - p_k| \leqq C\lambda^n \qquad \text{for } n \geqq 1, j, k = 1, 2, \ldots, r$$

Following Kendall, this is expressed by saying that the Markov chains in question are *geometrically ergodic*. This follows from the usual method of proof of Theorem 5.6.1. We have chosen another method of proof, which is interesting because it shows that limit distribution theorems can be proved by considering the information-theoretical divergence $D(\mathscr{P}, \mathscr{Q})$. This has been discovered by Linnik [11], who has proved Lindeberg's theorem by this method. The above method has been extended by Kendall [36] to Markov chains with countably infinitely many states.*

Now we prove the strong law of large numbers for homogeneous Markov chains with a finite number of states.

THEOREM 5.6.2. *Let $\{\xi_n\}$ be a homogeneous Markov chain such that ξ_n takes on the values* $1, 2, \ldots, r$ $(r \geqq 2)$. *Let the transition probabilities of the chain be*

$$p_{j,k} = P(\xi_{n+1} = k \mid \xi_n = j) \qquad (j, k = 1, 2, \ldots, r)$$

Suppose that putting $\Pi = (p_{j,k})$, there exists a positive integer s such that all elements of Π^s are positive. Let $(p_1, \ldots, p_r)$ be the (positive) solution of the system of equations (5.6.6)–(5.6.7), *i.e., $\mathscr{P} = \{p_k\}$ is the stationary distribution of the chain. Let $N_n(j)$ denote the frequency of the number j $(1 \leqq j \leqq r)$ in the*

* For another version of the proof for Markov chains with a finite number of states, see Csiszár [37].

sequence $\xi_1, \xi_2, \ldots, \xi_n$. *Then we have, almost surely,*

$$\lim_{n\to+\infty} \frac{N_n(j)}{n} = p_j \qquad \text{for } j = 1, 2, \ldots, r \tag{5.6.31}$$

Proof. We suppose first that the initial distribution is the stationary distribution $\mathscr{P}$. Let us realize the random variables $\{\xi_n\}$ in the space Ω of all sequences $\omega = (\omega_1, \ldots, \omega_n, \ldots)$ such that each ω_n is equal to one of the numbers $1, 2, \ldots, r$. (See the remark at the end of Example 5.5.1.) Let $\mathscr{A}_0$ denote the set of all cylinder sets in Ω, i.e., the sets defined by imposing some condition on a finite number of coordinates of ω only. Let $\mathscr{A}$ denote the least σ-algebra containing $\mathscr{A}_0$. Let P be the measure constructed as in Example 5.5.1, according to which the random variables $\xi_n(\omega) = c_n(\omega)$ form a Markov chain satisfying the suppositions of Theorem 5.6.2. Let T denote the shift transformation $T\omega = \omega'$ where $\omega' = (\omega_2, \omega_3, \ldots, \omega_{n+1}, \ldots)$ if $\omega = (\omega_1, \omega_2, \ldots, \omega_n, \ldots)$. It is easy to see that the measure P is invariant with respect to T.

We shall prove now that T is mixing. Let A be the set of those $\omega = (\omega_1, \ldots, \omega_n, \ldots)$ for which $\omega_1 = j_1, \ldots, \omega_k = j_k$, where $j_1, \ldots, j_k$ are fixed integers, $1 \leqq j_i \leqq r$ $(i = 1, 2, \ldots, k)$. Then $T^{-n}A$ is the set of those $\omega = (\omega_1, \omega_2, \ldots)$ for which $\omega_{n+1} = j_1, \ldots, \omega_{n+k} = j_k$. Similarly, let B be the set of those $\omega = (\omega_1, \omega_2, \ldots)$ for which $\omega_1 = i_1, \ldots, \omega_h = i_h$ $(1 \leqq i_l \leqq r$, $l = 1, 2, \ldots, h)$, then we have, if $n \geqq h$,

$$P(T^{-n}A \mid B) = P(\xi_{n+1} = j_1, \ldots, \xi_{n+k} = j_k \mid \xi_h = i_h)$$

and thus,

$$P(T^{-n}A \mid B) = p_{i_h, j_1}^{(n+1-h)} \cdot p_{j_1, j_2} p_{j_2, j_3} \cdots p_{j_{k-1}, j_k}$$

Thus we get

$$\lim_{n\to+\infty} P(T^{-n}A \mid B) = p_{j_1} p_{j_1, j_2} \cdots p_{j_{k-1}, j_k} = P(A)$$

i.e.,

$$\lim_{n\to+\infty} P(T^{-n}A \cdot B) = P(A)P(B) \tag{5.6.32}$$

It follows that (5.6.32) holds for all cylinder sets A and B, and thus also for all sets $A \in \mathscr{A}$ and $B \in \mathscr{A}$, i.e., T is mixing. Therefore (see Section 3.8), T is ergodic* and we can apply Birkhoff's ergodic theorem to the function

$$f_j(x) = \begin{cases} 1 & \text{if } x = j \\ 0 & \text{otherwise} \end{cases}$$

* This explains the terminology introduced in the remark to Theorem 5.6.1.

and we get for $j = 1, 2, \ldots, r$ that almost surely

$$\lim_{n \to +\infty} \frac{N_j(n)}{n} = \lim_{n \to +\infty} \frac{1}{n} \sum_{k=0}^{n-1} f_j(T^k\omega) = E(f_j) = p_j \tag{5.6.33}$$

Thus we have proved (5.6.31) for the special case when the Markov chain is stationary. The general case can be reduced to this special case as follows: As (5.6.33) holds almost surely, it holds almost surely on every subset A of Ω, thus it holds especially on each of the sets E_k defined by $\xi_1 = k$. Thus we get

$$P\left(\lim_{n \to +\infty} \frac{N_j(n)}{n} = p_j \quad \text{for} \quad 1 \leq j \leq r \mid E_k \right) = 1 \tag{5.6.34}$$

for $k = 1, 2, \ldots, r$. Now, if the initial distribution of the chain is $P(\xi_1 = k) = w_k$, the probability P^* of any event A can be expressed in the form

$$P^*(A) = \sum_{k=1}^{r} w_k P(A \mid E_k) \tag{5.6.35}$$

where $P(A \mid E_k)$ does not depend on the initial distribution. Relations (5.6.34) and (5.6.35) imply that (5.6.31) holds whatever the initial distribution $\{w_k\}$ is. Thus Theorem 5.6.2 is proved. ■

COROLLARY TO THEOREM 5.6.2. *Under the conditions of the theorem,*

$$\lim_{n \to +\infty} \frac{f(\xi_1) + \cdots + f(\xi_n)}{n} = \sum_{k=1}^{r} p_k f(k)$$

holds almost surely for any function $f(x)$.

Remarks: If the rows of the matrix $(p_{j,k})$ are identical, i.e., if $p_{j,k}$ does not depend on j, then the variables $\{\xi_n\}$ are independent and identically distributed. Thus in this case Theorem 5.6.2 reduces to the strong law of large numbers for independent, identically distributed random variables, which take on only a finite number of values.

If the column sums of the matrix $\Pi = (p_{j,k})$ are also equal to 1 (i.e., if the transposed matrix $\Pi' = (p_{k,j})$ is a stochastic matrix too), then Π is called a *doubly stochastic matrix*. In this case evidently, $p_k = 1/r$ $(k = 1, 2, \ldots, r)$ satisfies equations (5.6.6) and (5.6.7). As the positive solution of this system of equations is, as we have seen, unique, it follows that $p_k = 1/r$ $(1 \leq k \leq r)$. Thus if Π is doubly stochastic, the stationary distribution of the Markov chain is uniform on the set $\{1, 2, \ldots, r\}$.

Example 5.6.1. Let $\delta_1, \ldots, \delta_n, \ldots$ be independent and identically distributed random variables, each taking on positive integral values $\leq r$ with the corresponding probabilities

$$P(\delta_n = k) = q_k > 0 \qquad (k = 1, 2, \ldots, r; n = 1, 2, \ldots)$$

where $\sum_{k=1}^{r} q_k = 1$. Let us define q_n for every integer n by putting $q_n = q_m$, if $n \equiv m \bmod r$. Thus we put $q_{r+k} = q_k$ and $q_{-k+1} = q_{r-k+1}$ $(k = 1, 2, \ldots, r)$. Let us define the random variables ξ_n as follows:

$$\xi_n \equiv \delta_1 + \delta_2 + \cdots + \delta_n \bmod r$$

and $1 \leqq \xi_n \leqq r$. Thus the value of ξ_n is obtained by reducing the sum $\delta_1 + \delta_2 + \cdots + \delta_n \bmod r$. Then $\{\xi_n\}$ is a homogeneous Markov chain, with transition probabilities

$$p_{j,k} = P(\xi_{n+1} = k \mid \xi_n = j) = q_{k-j}$$

The matrix $(p_{j,k})$ is doubly stochastic and all its elements are positive. Thus, the distribution of ξ_n tends for $n \to +\infty$ to the uniform limit distribution.

This example is relevant for the construction of "random numbers." It shows, e.g., that if we start from a sequence of independent digits, each digit 0, 1, ..., g having a positive probability, then adding a large number of these digits mod 10, we get digits which take on all ten possible values with approximately the same probability.

Example 5.6.2. Let us consider a regular $(2r+1)$-gon with vertices $V_1, V_2, \ldots, V_{2r+1}$, and suppose that a unit mass is distributed among the points $V_1, V_2, \ldots, V_{2r+1}$ so that the mass w_k is placed in the point V_k $\left(\sum_{k=1}^{2r+1} w_k = 1\right)$. Let us carry out a transformation $\mathscr{T}$ in which the mass in each point is replaced by the arithmetic mean of the masses in the two adjacent points. Thus, after the transformation there will be at V_k the mass $w_k^{(2)} = (w_{k-1} + w_{k+1})/2$ $(k = 1, 2, \ldots, 2r+1)$, where by definition $w_0 = w_{2r+1}$ and $w_{2r+2} = w_1$. After carrying out this transformation n times, let $w_k^{(n)}$ denote the mass at V_k. Then, whatever the initial distribution $\{w_k\}$ is, one has

$$\lim_{n \to +\infty} w_k^{(n)} = \frac{1}{2r+1} \qquad \text{for } k = 1, 2, \ldots, 2r+1$$

As a matter of fact, if $\{\xi_n\}$ is a Markov chain with initial distribution $\{w_k\}$ and transition probabilities $p_{j,k} = 1/2$ when $k - j = \pm 1 \bmod (2r+1)$, then $w_k^{(n)}$ denotes the distribution of ξ_n. The matrix $(p_{j,k})$ is doubly stochastic. If we extend the definition of w_n for every integer n by putting $w_n = w_m$ if $n \equiv m \bmod (2r+1)$, we have

$$w_k^{(2n+1)} = \frac{1}{2^{2n}} \sum_{j=-n}^{+n} \binom{2n}{n+j} w_{k+2j}$$

and

$$w_k^{(2n)} = \frac{1}{2^{2n-1}} \sum_{j=-n}^{n-1} \binom{2n-1}{n+j} w_{k+2j+1}$$

which shows that all elements of Π^s are positive if $s > r$. Thus Theorem 5.6.1 can be applied. Of course, the result also follows directly from the above explicit formulas. Notice that the corresponding result is not valid for a $2r$-gon. For example, if we put the weights 0, 1/2, 0, 1/2 in the four vertices of a square, carrying out the transformation $\mathscr{T}$ any number of times, we get alternatingly the initial distribution rotated by 90° and the initial distribution itself, and thus no limit distribution exists.

The central limit theorem can also be extended for Markov chains, but we do not deal with this problem here (see Problem P.5.4).

We finish this section with a remark about inverse Markov chains. Let $\{\xi_n\}$ be a homogeneous stationary Markov chain satisfying the conditions of Theorem 5.6.1. Then we have

$$P(\xi_m = l \mid \xi_{m+1} = k_1, \ldots, \xi_{m+n} = k_n) = \frac{p_l p_{l,k_1}}{p_{k_1}} = P(\xi_m = l \mid \xi_{m+1} = k_1)$$

Putting $q_{k,l} = p_l p_{l,k}/p_k$, we get from (5.6.11),

$$\sum_{l=1}^{r} q_{k,l} = 1 \qquad (k = 1, 2, \ldots, r)$$

Thus $(q_{k,l})$ is a stochastic matrix, and for every $N \geqq 1$ the sequence ξ_N, $\xi_{N-1}, \ldots, \xi_1$ is a (finite) homogeneous and stationary Markov chain with the transition probabilities $q_{k,l}$; an infinite homogeneous Markov chain $\{\xi_n^*\}$ with the transition probabilites $q_{k,l}$ is called the *inverse chain* of the chain $\{\xi_n\}$. Clearly the inverse chain has the same stationary distribution. Notice that the weights used in the step when we applied Jensen's inequality in the proof of Theorem 5.6.1 were just the transition probabilities $q_{k,l}$ of the inverse chain.

5.7 STABLE SEQUENCES OF EVENTS

We now introduce the useful notion of "*stable*" sequences of events (see Rényi [13]).

DEFINITION 5.7.1. *Let $S = (\Omega, \mathscr{A}, P)$ be a probability space. We call a sequence $\{A_n\}$ of events ($A_n \in \mathscr{A}$, $n = 1, 2, \ldots$) of the space S stable (of order 1) if the limit*

$$\lim_{n \to +\infty} P(A_n B) = Q(B) \tag{5.7.1}$$

exists for every $B \in \mathscr{A}$ and

$$0 < Q(\Omega) < 1$$

THEOREM 5.7.1. *If $\{A_n\}$ is a stable sequence of events in the probability space $S = (\Omega, \mathscr{A}, P)$ and $Q(B)$ is defined by (5.7.1), then $Q(B)/Q(\Omega)$ is a probability measure on $\mathscr{A}$, which is absolutely continuous with respect to P, and $Q(B)$ can be represented in the form*

$$Q(B) = \int_B \alpha \, dP \tag{5.7.2}$$

where $\alpha = \alpha(\omega)$ is a random variable on S, such that $0 \leqq \alpha(\omega) \leqq 1$ almost surely, and α is not almost surely 0, nor almost surely 1.

Proof. By supposition,

$$\lim_{n \to +\infty} P(A_n) = Q(\Omega) > 0 \tag{5.7.3}$$

Thus, $P(A_n) > 0$ for $n \geqq n_1$. It follows that

$$\lim_{n \to +\infty} P(B \mid A_n) = \frac{Q(B)}{Q(\Omega)} \tag{5.7.4}$$

Thus the limit of the probability measures $P(B \mid A_n)$ exists for every $B \in \mathscr{A}$. Therefore, by the Vitali-Hahn-Saks theorem, $Q(B)/Q(\Omega)$ is a probability measure. As $P(B) = 0$ implies $Q(B) = 0$, the measure $Q(B)$ is absolutely continuous with respect to the measure P; as, further, $0 \leqq P(BA_n) \leqq P(B)$, it follows that $0 \leqq Q(B) \leqq P(B)$ and thus $0 \leqq \alpha = dQ/dP \leqq 1$ a.s. ■

DEFINITION 5.7.2. *We call the random variable α figuring in (5.7.2) the density* of the stable sequence of events $\{A_n\}$.*

Remark: If $\{A_n\}$ is a stable sequence with density α, then evidently $\{\bar{A}_n\}$ is also a stable sequence with density $1 - \alpha$.

Example 5.7.1. Let S be the Lebesgue probability space and let $\alpha(x)$ be a continuous function in the interval $0 \leqq x \leqq 1$ such that $0 \leqq \alpha(x) \leqq 1$ and $\alpha(x)$ is not identically equal to 0 or to 1. Define A_n as the union of the intervals $(k/n, (k + \alpha(k/n))/n)$ $(k = 0, 1, \ldots, n-1)$. Then $\{A_n\}$ is a stable sequence of events with density α. This can be shown as follows: If B is an interval, $B = [c, d]$, then evidently

$$P(A_n B) = \sum_{c < \frac{k}{n} < d} \alpha\left(\frac{k}{n}\right) \cdot \frac{1}{n} + O\left(\frac{1}{n}\right) \to \int_c^d \alpha(x)dx$$

* Evidently the density α of a stable sequence of events is uniquely determined up to a set of measure 0.

Thus, putting

$$Q(B) = \int_B \alpha(x)dx$$

for every Borel subset B of the interval $[0, 1]$, (5.7.1) holds for every interval B and therefore it holds for every Borel subset B of the interval $[0, 1]$.

We shall now prove the following criterion (see [13]) which is useful to show that a sequence of events is stable:

THEOREM 5.7.2. *If $\{A_n\}$ $(n = 1, 2, \ldots)$ is a sequence of events in the probability space $S = (\Omega, \mathscr{A}, P)$, such that, putting $A_0 = \Omega$, the limit*

$$\lim_{n \to +\infty} P(A_n A_k) = \lambda_k \tag{5.7.5}$$

exists for $k = 0, 1, \ldots$ and $0 < \lambda_0 < 1$, then the sequence $\{A_n\}$ is stable and its density α is uniquely determined by the conditions

$$\int_{A_k} \alpha \, dP = \lambda_k \qquad (k = 0, 1, \ldots) \tag{5.7.6}$$

Proof. Let α_n denote the indicator of the event A_n $(n = 1, 2, \ldots)$. By supposition, the limits

$$\lim_{n \to +\infty} E(\alpha_n \alpha_k) = \lambda_k \qquad (k = 0, 1, \ldots) \tag{5.7.7}$$

exist. From this we shall now deduce that the limit

$$\lim_{n \to +\infty} E(\alpha_n \xi) = A(\xi) \tag{5.7.8}$$

exists for every $\xi \in L_2(S)$. Let us denote by L_2^* the least (closed) linear subspace of $L_2(S)$ which contains all the α_k $(k = 0, 1, 2, \ldots)$. We prove first that the limit (5.7.8) exists if $\xi \in L_2^*$. If ξ_0 is a linear combination of a finite number of the α_k, i.e.,

$$\xi_0 = \sum_{k=0}^{N} c_k \alpha_k \tag{5.7.9}$$

where $c_0, c_1, \ldots, c_N$ are constants, then clearly,

$$\lim_{n \to +\infty} E(\alpha_n \xi_0) = \sum_{k=0}^{N} c_k \lambda_k = A(\xi_0) \tag{5.7.10}$$

exists. Now, by definition every $\xi_1 \in L_2^*$ can be approximated in L_2-norm by a ξ_0 of the form (5.7.9) so that

$$E((\xi_1 - \xi_0)^2) < \varepsilon^2$$

where $\varepsilon > 0$ can be chosen arbitrarily small. It follows that

$$|E(\xi_1 \alpha_n) - E(\xi_0 \alpha_n)| \leq \varepsilon$$

and thus

$$\left|\limsup_{n\to+\infty} E(\xi_1\alpha_n) - A(\xi_0)\right| \leqq \varepsilon$$

and

$$\left|\liminf_{n\to+\infty} E(\xi_1\alpha_n) - A(\xi_0)\right| \leqq \varepsilon$$

i.e.,

$$\left|\limsup_{n\to+\infty} E(\xi_1\alpha_n) - \liminf_{n\to+\infty} E(\xi_1\alpha_n)\right| \leqq 2\varepsilon$$

As $\varepsilon > 0$ can be chosen arbitrarily small, it follows that the limit (5.7.8) exists for every $\xi_1 \in L_2^*$. Now take any $\xi \in L_2(S)$. Denoting by ξ_1 the projection of ξ to the subspace L_2^*, we can write

$$\xi = \xi_1 + \xi_2 \tag{5.7.11}$$

where $\xi_1 \in L_2^*$ and ξ_2 is orthogonal to every element of the subspace L_2^*; in particular, we have

$$E(\xi_2\alpha_n) = 0 \qquad \text{for } n = 0, 1, \ldots \tag{5.7.12}$$

Thus,

$$E(\alpha_n\xi) = E(\alpha_n\xi_1)$$

As $\xi_1 \in L_2^*$ and we already know that $\lim_{n\to+\infty} E(\alpha_n\xi_1)$ exists for every $\xi_1 \in L_2^*$, it follows that the limit (5.7.8) exists for all $\xi \in L_2(S)$. In other words, the sequence α_n, as a sequence of elements of the Hilbert space $L_2(S)$, is weakly convergent. Denoting by α the weak limit of the sequence $\{\alpha_n\}$, we have $0 \leqq \alpha \leqq 1$ and

$$\lim_{n\to+\infty} E(\alpha_n\xi) = A(\xi) = E(\alpha\xi)$$

Thus, if β is the indicator of an event $B \in \mathscr{A}$, we have

$$\lim_{n\to+\infty} P(BA_n) = \lim_{n\to+\infty} E(\beta\alpha_n) = A(\beta) = E(\alpha\beta) = \int_B \alpha\, dP$$

Thus we have shown that the sequence $\{A_n\}$ is stable with density α. As $\lambda_0 = \int_\Omega \alpha\, dP$, and by supposition $0 < \lambda_0 < 1$, it follows that α is not almost surely 0 nor almost surely 1. ■

Remark 1: Definitions 5.7.1 and 5.7.2 can be brought to the following equivalent form. *A sequence $\{A_n\}$ of events is stable with density α if, denoting by α_n the indicator of the event A_n, α_n tends for $n \to +\infty$ weakly (in $L_2(S)$) to α.* Theorem 5.7.2 is essentially equivalent to the following theorem due to

E. Schmidt [14]: *For the weak convergence of a bounded sequence* α_n $(n=0, 1, \dots)$ *of elements of a Hilbert space it is sufficient that the limits* $\lim_{n\to+\infty}(\alpha_n, \alpha_k)$ *should exist for* $k=0, 1, \dots$.

Remark 2: It follows from the weak convergence of α_n to α that if ξ is any random variable such that $E(\xi)$ exists, one has

$$\lim_{n\to+\infty} E(\alpha_n \xi) = E(\alpha\xi) \tag{5.7.13}$$

As a matter of fact, put for $N=1, 2, \dots$

$$\xi_N = \begin{cases} \xi & \text{if } |\xi| < N \\ 0 & \text{otherwise} \end{cases}$$

Then for every $\varepsilon > 0$, if N is sufficiently large,

$$\int_\Omega |\xi - \xi_N|\, dP = \int_{|\xi|\geqq N} |\xi|\, dP < \varepsilon$$

Thus

$$|E(\alpha_n \xi_N) - E(\alpha_n \xi)| < \varepsilon$$

Similarly,

$$|E(\alpha\xi_N) - E(\alpha\xi)| < \varepsilon$$

As $\xi_N \in L_2(S)$, and therefore

$$\lim_{n\to+\infty} E(\alpha_n \xi_N) = E(\alpha\xi_N) \qquad \text{for } N=1, 2, \dots$$

it follows that (5.7.13) holds.

As every bounded sequence of elements of a Hilbert space contains a weakly convergent subsequence, it follows that from every sequence $\{A_n\}$ of events such that $0 < a \leqq P(A_n) \leqq b < 1$ $(n=1, 2, \dots)$ one can select a stable subsequence. This remark shows that stability is a rather weak supposition.

Therefore we now introduce a somewhat stronger notion.

DEFINITION 5.7.3. *A sequence* $\{A_n\}$ *of events in a probability space* $S=(\Omega, \mathscr{A}, P)$ *is called doubly stable if for every* $B \in \mathscr{A}$, *the limit*

$$\lim_{\substack{n\to+\infty \\ m\to+\infty}} P(BA_n A_{n+m}) = Q_2(B) \tag{5.7.14}$$

exists, where n and m tend independently to $+\infty$, *and* $0 < Q_2(\Omega) < 1$.

We prove first that doubly stable sequences of events are stable. More exactly, we prove the following:

THEOREM 5.7.3. *If $\{A_n\}$ is a doubly stable sequence of events, then $\{A_n\}$ is stable; denoting by α the density of the sequence $\{A_n\}$, the set function $Q_2(B)$ in (5.7.14) is a measure and can be written in the form*

$$Q_2(B) = \int_B \alpha^2 \, dP \tag{5.7.15}$$

Proof. Let α_n denote the indicator of the event A_n. Let us select from the sequence α_n a subsequence α_{n_k} ($k = 1, 2, \ldots, n_1 < n_2 < \cdots$) which converges weakly in $L_2(S)$ to a limit α. Then if B is any event and β is the indicator of B, we can, for every $\varepsilon > 0$, find a k such that

$$|E(\alpha_{n_k} \alpha\beta) - E(\alpha^2\beta)| < \varepsilon \tag{5.7.16}$$

We can choose k so that n_k is larger than any given number N. Fixing the value of n_k, we can find a number l such that

$$|E(\alpha_{n_l} \alpha_{n_k} \beta) - E(\alpha_{n_k} \alpha\beta)| < \varepsilon \tag{5.7.17}$$

Clearly, we can choose l so that $n_l - n_k$ is larger than any given number M. Now if N and M are appropriately chosen, then by definition we have

$$|E(\alpha_{n_l} \alpha_{n_k} \beta) - Q_2(B)| < \varepsilon \tag{5.7.18}$$

It follows from (5.7.16)–(5.7.18) that

$$|E(\alpha^2\beta) - Q_2(B)| < 3\varepsilon \tag{5.7.19}$$

As $\varepsilon > 0$ was arbitrary, it follows that

$$Q_2(B) = E(\alpha^2\beta) = \int_B \alpha^2 \, dP \tag{5.7.20}$$

for every $B \in \mathscr{A}$. Thus, α^2 (and thus as $0 \leqq \alpha \leqq 1$, α itself) is uniquely determined by $Q_2(B)$, as $\alpha^2 = dQ_2(B)/dP$ a.s.

This means that every weakly convergent subsequence of the sequence $\{\alpha_n\}$ has the same weak limit α, and thus the sequence $\{\alpha_n\}$ tends weakly to α, i.e., $\{A_n\}$ is stable with density α. Thus Theorem 5.7.3 is proved. ■

The notion of double stability can be generalized still further, by introducing the notion of stability of order k as follows:

DEFINITION 5.7.4. *A sequence $\{A_n\}$ of events is called stable of order $k \geqq 3$ (for the sake of brevity: k-stable) if the limit*

$$\lim_{\substack{n_i \to +\infty \\ (i=1,2,\ldots,k)}} P(BA_{n_1} A_{n_1+n_2} \cdots A_{n_1+n_2+\cdots+n_k}) = Q_k(B) \tag{5.7.21}$$

exists for all $B \in \mathscr{A}$.

By the same argument as was used to prove Theorem 5.7.3, one can show that *if a sequence $\{A_n\}$ is k-stable, then it is l-stable for $1 \leqq l < k$ and if its density is α, then $Q_k(B) = \int_B \alpha^k \, dP$.* (See Problem P.5.9.)

We now prove the following:

THEOREM 5.7.4. *If $\{A_n\}$ is a 2-stable sequence of events, and α_n is the indicator of the event A_n, then $(\alpha_1+\alpha_2+\cdots+\alpha_n)/n$ tends strongly (in $L_2(S)$) to the density α of the sequence $\{A_n\}$.*

Proof. Let $\varepsilon>0$ be arbitrary. Let α_n denote the indicator of A_n and let α be the density of the sequence $\{A_n\}$. By definition we can find numbers N and M such that if $k>N$ and $l>M$, then

$$|E(\alpha_k\alpha)-E(\alpha^2)|<\varepsilon$$

and

$$|E(\alpha_k\alpha_{k+l})-E(\alpha^2)|<\varepsilon$$

and thus

$$|E((\alpha_k-\alpha)(\alpha_{k+l}-\alpha))|<3\varepsilon$$

It follows that

$$E\left(\left(\frac{\sum_{j=1}^{n}\alpha_j}{n}-\alpha\right)^2\right)\leqq 3\varepsilon+O\left(\frac{N+M}{n}\right)$$

Thus it follows that

$$\lim_{n\to+\infty}E\left(\left(\frac{\sum_{j=1}^{n}\alpha_j}{n}-\alpha\right)^2\right)=0 \tag{5.7.22}$$

which was to be proved.* ■

Remark: Theorem 5.7.4 can be deduced also from the following theorem of Parzen (see [44]): *If $\{\xi_n\}$ is a sequence of random variables such that $E(\xi_n)=0$ and $D^2(\xi_n)\leqq K$ $(n=1,2,\ldots)$ and $\zeta_n=\xi_1+\xi_2+\cdots+\xi_n$, then a necessary and sufficient condition for $E(\zeta_n^2)\to 0$ (i.e., for ζ_n tending strongly to 0) is that*

$$\lim_{n\to+\infty}E(\xi_n\zeta_n)=0$$

should hold. On the other hand, Parzen's theorem can be proved by an argument similar to the proof of Theorem 5.7.4.

We now prove a criterion for stability of order k, analogous to Theorem 5.7.2.

THEOREM 5.7.5. *A sequence $\{A_n\}$ of events is stable of order k, if for every k-tuple $(l_1,l_2,\ldots,l_k)$ of positive integers the limit*

$$\lim_{\substack{n_i\to+\infty\\(1\leqq i\leqq k)}}P(A_{n_1}A_{n_1+n_2}\cdots A_{n_1+n_2+\cdots+n_k}A_{l_1}A_{l_2}\cdots A_{l_k})=\lambda(l_1,l_2,\ldots,l_k) \tag{5.7.22'}$$

* An essentially equivalent result has been proved by Fischler (see [40] and [41]).

exists. Denoting by α *the density of the sequence* $\{A_n\}$, *one has*

$$\int_{A_{l_1}A_{l_2}\cdots A_{l_k}} \alpha^k \, dP = \lambda(l_1, l_2, \ldots, l_k) \; (l_i \geqq 1, 1 \leqq i \leqq k) \tag{5.7.23}$$

Proof. The proof follows that of Theorem 5.7.2 step-by-step. It follows from (5.7.22′) that denoting by α_n the indicator of the event A_n, the limit

$$\lim_{\substack{n_i \to +\infty \\ (1 \leqq i \leqq k)}} E(\xi \prod_{i=1}^{k} \alpha_{n_1+n_2+\cdots+n_i}) = A(\xi) \tag{5.7.24}$$

exists for every ξ which belongs to the least closed linear subspace L_2^* of $L_2(S)$ containing all random variables $\alpha_{l_1}, \alpha_{l_2}, \cdots, \alpha_{l_k}$ $(l_i \geqq 1, 1 \leqq i \leqq k)$. As the limit (5.7.24) is 0 if ξ belongs to the subspace of elements of $L_2(S)$, orthogonal to L_2^*, it follows that (5.7.24) exists for every $\xi \in L_2(S)$, thus also for the indicator β of an event $B \in \mathscr{A}$, i.e., the limit (5.7.21) exists for every $B \in \mathscr{A}$, which was to be proved. ■

5.8 MIXING SEQUENCES OF EVENTS

In this section we shall deal with an important special class of stable sequences of events, called "mixing" (see Rényi [15]).

DEFINITION 5.8.1. *A stable sequence of events* $\{A_n\}$ *is called mixing if its density* α *is almost surely constant. If* $\alpha(\omega) = c$ *a.s., where c is a constant,* $0 < c < 1$, *the sequence* $\{A_n\}$ *is called mixing with density c. It is called k-mixing if it is mixing and k-stable* $(k \geqq 2)$.

Remark: Thus a sequence $\{A_n\}$ of events is mixing with density c if for every event B, one has

$$\lim_{n \to +\infty} P(A_n B) = cP(B) \tag{5.8.1}$$

If $\{A_n\}$ is mixing with density c, then evidently $\{\overline{A}_n\}$ is mixing with density $1 - c$. If for every event $B \in \mathscr{A}$, one has

$$\lim_{\substack{n \to +\infty \\ m \to +\infty}} P(A_n A_{n+m} B) = c^2 P(B) \tag{5.8.2}$$

then $\{A_n\}$ is 2-mixing, with density c. It follows from Theorem 5.7.3 that (5.8.2) implies (5.8.1.). From Theorem 5.7.2 we get the following criterion:

THEOREM 5.8.1. *If for a sequence* $\{A_n\}$ *of events one has*

$$\lim_{n \to +\infty} P(A_n) = c \qquad (0 < c < 1) \tag{5.8.3}$$

and

$$\lim_{n\to+\infty} P(A_n A_k) = cP(A_k) \qquad \textit{for } k = 1, 2, \ldots \tag{5.8.4}$$

then $\{A_n\}$ *is mixing with density* c.

Proof. It follows from (5.8.4) and Theorem 5.7.2 that $\{A_n\}$ is stable with density α such that

$$\int_{A_k} \alpha\, dP = cP(A_k) \qquad (k = 1, 2, \ldots) \tag{5.8.5}$$

and

$$E(\alpha) = c \tag{5.8.6}$$

It remains to be proved that $\alpha = c$ almost surely. To show this let us evaluate $\lim_{n\to+\infty} E(\alpha_n \alpha)$ in two different ways. As α_n tends weakly to α, we have

$$\lim_{n\to+\infty} E(\alpha_n \alpha) = E(\alpha^2) \tag{5.8.7}$$

On the other hand, by (5.8.5),

$$E(\alpha_n \alpha) = \int_{A_n} \alpha\, dP = cP(A_n) \tag{5.8.8}$$

In view of (5.8.3), it follows that

$$\lim_{n\to+\infty} E(\alpha_n \alpha) = c^2 \tag{5.8.9}$$

Comparing (5.8.7) and (5.8.9), we get

$$E(\alpha^2) = c^2 \tag{5.8.10}$$

i.e., with respect to (5.8.6),

$$D^2(\alpha) = 0 \tag{5.8.11}$$

which shows that $\alpha = c$ almost surely. ■

Remark: One can characterize a mixing sequence $\{A_n\}$ of events somewhat vaguely as a sequence such that A_n is, in the limit, independent from any event $B \in \mathscr{A}$.

With this terminology the content of Theorem 5.8.1 can be described as follows: *If the sequence of events* $\{A_n\}$ *has the property that the event* A_n *is for* $n \to +\infty$ *independent in the limit from every* A_k *with* k *fixed, then* A_n *is in the limit for* $n \to +\infty$ *independent from every event* $B \in \mathscr{A}$, *provided that* (5.8.3) *holds*.

Let us consider some examples.

Example 5.8.1. If T is a measure-preserving mixing transformation of the probability space $S = (\Omega, \mathscr{A}, P)$ and $A_1 \in \mathscr{A}$ is any event such that $P(A_1) = c$

$(0 < c < 1)$, then, putting $A_{n+1} = T^{-n}A_1$ $(n = 1, 2, \ldots)$, the sequence $\{A_n\}$ is mixing with density c.

Example 5.8.2. Let the events $\{A_n\}$ be pairwise independent and let all have the same probability c $(0 < c < 1)$. Then the sequence $\{A_n\}$ is mixing with density c. As a matter of fact, we have $P(A_n A_k) = c^2 = cP(A_k)$ if $n > k$, and thus the conditions of Theorem 5.8.1 are satisfied. If the events A_n are independent by three, then the sequence $\{A_n\}$ is evidently 2-mixing.

Remark: What we have proved can be stated somewhat vaguely as follows: If the events A_n are pairwise independent and each has probability c $(0 < c < 1)$, then for large values of n the set A_n must be so "well spread" over the whole space Ω that with every set $B \in \mathscr{A}$ with $P(B) > 0$ it has an intersection BA_n, the measure of which is nearly proportional to the measure of B. The puzzling thing about this result is that clearly we may rearrange the events A_n in an arbitrary way, and the sequence thus obtained still satisfies our suppositions. On the other hand, the set A_1 (and thus every single element of the sequence) may be a quite arbitrary set with probability c. Of course, the explanation of this puzzle is that in an infinite sequence A_n of pairwise independent events, each having the probability c, all but a finite number of these sets are rather "well spread" (with density nearly everywhere near to c) over the whole basic space. This vague statement will be made precise by Theorem 5.8.2 below.

Example 5.8.3. Let the sequence $\{A_n\}$ be 2-mixing with density c and define the sequence A_n^* as follows: Let $1 = n_1 < n_2 < \cdots$ be an increasing sequence of positive integers such that $n_{k+1} - n_k \to +\infty$, and put

$$A_n^* = A_{k^2} \qquad \text{for } n_k \leqq n < n_{(k+1)^2} \qquad (k = 1, 2, \ldots)$$

Clearly, $\{A_n^*\}$ is also a mixing sequence of events with density c, but it is not 2-mixing, because

$$\lim_{k \to \infty} P(A_{n_k}^* A_{n_{k+1}}^* B) = \lim_{k \to +\infty} P(A_{k^2} A_{(k+1)^2} B) = c^2 P(B)$$

since $(k+1)^2 - k^2 = 2k + 1 \to +\infty$, while

$$\lim_{k \to \infty} P(A_{n_k}^* A_{n_{k+1}-1}^* B) = \lim_{k \to +\infty} P(A_{k^2} B) = cP(B)$$

If n_k grows very fast, e.g., if $n_k = 2^{2^k}$, then denoting by α_n and α_n^* the indicators of A_n and A_n^*, respectively, it is easy to see that $\gamma_n = (\alpha_1^* + \cdots + \alpha_n^*)/n$ does not tend strongly to c. As a matter of fact,

$$\gamma_{2^{2^{k+1}}-1} = \frac{\sum_{j=1}^{k} (2^{2^{j+1}} - 2^{2^j}) \alpha_{j^2}}{2^{2^{k+1}} - 1} = \alpha_{k^2} + O(2^{-2^k})$$

and

$$\lim_{k \to +\infty} E((\alpha_k - c)^2) = c(1-c) > 0$$

Example 5.8.4. Let $\{\xi_n\}$ be a homogeneous Markov chain for which the conditions of Theorem 5.6.1 are fulfilled. Using the same notation, let $A_n^{(j)}$ denote the event that $\xi_n = j$. Then we have for $k < l < n$,

$$P(A_n^{(j)} A_{n+m}^{(j)} A_k^{(j)} A_l^{(j)}) = P(A_k^{(j)} A_l^{(j)}) P(A_n^{(j)} \mid A_l^{(j)}) P(A_{n+m}^{(j)} \mid A_n^{(j)})$$

and thus

$$\lim_{\substack{n \to +\infty \\ m \to +\infty}} P(A_n^{(j)} A_{n+m}^{(j)} A_k^{(j)} A_l^{(j)}) = p_j^2 P(A_k^{(j)} A_l^{(j)})$$

Thus it follows from Theorem 5.7.5 that $\{A_n^{(j)}\}$ is 2-mixing with density p_j. This implies by Theorem 5.7.4 that $N_j(n)/n$ tends strongly to p_j. Of course this follows also from Theorem 5.6.2, as $N_j(n)/n$ is bounded.

Definition 5.8.2. *We call a sequence $\{A_n\}$ of events k-mixing (or mixing of order k) ($k \geqq 3$), if it is k-stable and its density is constant.*

It follows from Theorem 5.7.5 that the sequence $\{A_n\}$ of events in Example 5.8.3 is mixing of any order.

Now we shall prove the following result which throws light on the puzzling character of Example 5.8.2. We first introduce the following:

Definition 5.8.3. *Let $\{A_n\}$ be a sequence of events in a probability space $S = (\Omega, \mathscr{A}, P)$ such that $0 < P(A_n) < 1$ and, putting**

$$\rho_{n,m} = \frac{P(A_n A_m) - P(A_n)P(A_m)}{\sqrt{P(A_n)(1-P(A_n))P(A_m)(1-P(A_m))}} \tag{5.8.12}$$

the quadratic form $\sum \rho_{n,m} x_n x_m$ is bounded, i.e., there exists a constant $K > 0$ such that for every sequence $\{x_n\}$ of real numbers for which $\sum_{n=1}^{\infty} x_n^2 < +\infty$, one has

$$\left| \sum_{n=1}^{\infty} \sum_{m=1}^{\infty} \rho_{n,m} x_n x_m \right| \leqq K \sum_{n=1}^{+\infty} x_n^2 \tag{5.8.13}$$

Then $\{A_n\}$ is called a (pairwise) quasi-independent sequence of events with modulus K.

Theorem 5.8.2. *If $\{A_n\}$ is a (pairwise) quasi-independent sequence of events with modulus K, then for every $B \in \mathscr{A}$, one has*

$$\sum_{n=1}^{\infty} \frac{[P(A_n B) - P(A_n)P(B)]^2}{P(A_n)(1-P(A_n))} \leqq KP(B)(1-P(B)) \tag{5.8.14}$$

* Note that denoting by α_n the indicator of A_n, $\rho_{n,m}$ is the coefficient of correlation of α_n and α_m.

Proof. Put $\xi_n = (\alpha_n - P(A_n))/\sqrt{P(A_n)(1-P(A_n)}$, where α_n is the indicator of the event A_n. Then the random variables ξ_n are *quasi-orthogonal* in the sense of Problem P.3.5, and thus for every random variable $\eta \in L_2(S)$, putting $c_n = E(\eta\xi_n)$, one has

$$\sum_{n=1}^{+\infty} c_n^2 \leqq KE(\eta^2) \tag{5.8.15}$$

Applying (5.8.15) for $\eta = \beta - P(B)$ where β is the indicator of B, we get from (5.8.15) the inequality (5.8.14). ■

COROLLARY TO THEOREM 5.8.2. *If the events $\{A_n\}$ are pairwise independent and each has probability c, then for every $B \in \mathscr{B}$, we have*

$$\sum_{n=1}^{+\infty} (P(A_n B) - cP(B))^2 \leqq c(1-c)P(B)(1-P(B)) \tag{5.8.16}$$

Proof. If the events $\{A_n\}$ are independent and $P(A_n) = c$ for $n = 1, 2, \ldots,$ we have

$$\rho_{n,m} = \begin{cases} 0 & \text{if } n \neq m \\ 1 & \text{if } n = m \end{cases}$$

Thus the A_n are quasi-independent with modulus $K = 1$, and therefore (5.8.16) follows from (5.8.14). ■

Remark: It follows from Theorem 5.8.2 that if $\{A_n\}$ is any pairwise quasi-independent sequence of events and

$$\lim_{n\to+\infty} P(A_n) = c \qquad (0 < c < 1) \tag{5.8.17}$$

then the sequence $\{A_n\}$ is mixing with density c.

The supposition of pairwise quasi-independence together with the requirement (5.8.17) is stronger than that of mixing, because if $\{A_n\}$ is mixing but $P(A_n B)$ tends too slowly to $cP(B)$, then the A_n cannot be qausi-independent. On the other hand, quasi-independence does not imply that (5.8.17) holds. For instance, if the events $\{A_n\}$ are pairwise independent, $P(A_{2n}) = c_1$ and $P(A_{2n+1}) = c_2$ where $n = 1, 2, \ldots$ and $0 < c_1 < c_2 < 1$, then $\{A_n\}$ is quasi-independent, but not mixing.

We have seen that if $\{A_n\}$ is a mixing sequence of events with density c in the probability space $S = (\Omega, \mathscr{A}, P)$, then denoting by α_n the indicator of A_n, α_n tends weakly (in $L_2(S)$) to c. This implies (see Remark 2 to Theorem 5.7.2) that for every random variable η for which $E(\eta)$ exists, one has

$$\lim_{n\to+\infty} E(\eta\alpha_n) = cE(\eta) \tag{5.8.18}$$

Let us denote by $\mathscr{A}_n$ the four-element algebra with elements Ω, $\varnothing$, A_n and $\bar{A}_n$. Then (5.8.18) can be written in the form below. One has almost surely,

$$\lim_{n\to+\infty} E(\eta \mid \mathscr{A}_n) = E(\eta) \tag{5.8.19}$$

Notice that the density c does not appear in (5.8.19). Thus we are led to the following:

DEFINITION 5.8.4. *Let $S = (\Omega, \mathscr{A}, P)$ be a probability space. A sequence $\mathscr{A}_n$ of σ-algebras of events of S ($\mathscr{A}_n \subseteq \mathscr{A}$ for $n \geqq 1$) is called quasi-mixing if* (5.8.19) *holds for every random variable η on S for which $E(\eta)$ exists. A sequence $\{A_n\}$ of events in S is called quasi-mixing if the sequence of four-element σ-algebras $\mathscr{A}_n = (\Omega, \varnothing, A_n, \bar{A}_n)$ is quasi-mixing. A sequence $\{\xi_n\}$ of random variables is called quasi-mixing if the corresponding σ-algebras $\mathscr{A}_{\xi_n}$ are quasi-mixing.*

Remark: It follows from Definition 5.8.4 that a sequence $\{A_n\}$ of events is quasi-mixing if and only if for every $B \in \mathscr{A}$, we have

$$\lim_{n\to+\infty} \frac{P(BA_n)}{P(A_n)} = P(B) \tag{5.8.20}$$

Thus every mixing sequence $\{A_n\}$ is quasi-mixing, further, if for each fixed k ($k = 1, 2, \ldots$) the sequence $A_n^{(k)}$ of events is mixing with some density c_k ($0 < c_k < 1$) and the sequence $\{A_n^*\}$ is obtained by arranging all sets $A_n^{(k)}$ ($n \geqq 1$, $k \geqq 1$) into a single sequence, then $\{A_n^*\}$ is quasi-mixing. Conversely, if $\{A_n\}$ is a quasi-mixing sequence of events and $\{A_{n_k}\}$ is a subsequence of the sequence $\{A_n\}$ such that $\lim_{k\to+\infty} P(A_{n_k}) = c$ ($0 < c < 1$), then $\{A_{n_k}\}$ is mixing with density c. It is easy to see that if $\{A_n\}$ is a quasi-mixing sequence of events such that $0 < P(A_n) < 1$, then

$$\liminf_{n\to+\infty} P(A_n) > 0 \tag{5.8.21}$$

and

$$\limsup_{n\to+\infty} P(A_n) < 1 \tag{5.8.22}$$

It is sufficient to prove (5.8.21). Suppose that $\{A_n\}$ is a quasi-mixing sequence of sets and $\liminf_{n\to+\infty} P(A_n) = 0$. Then we may select a subsequence A_{n_k} ($k = 1, 2, \ldots$) such that $P(A_{n_k}) > 0$ and $\sum_{k=1}^{+\infty} P(A_{n_k}) < 1$. Putting $B = \sum_{k=1}^{+\infty} A_{n_k}$, we have $P(\bar{B}) > 0$ and

$$\lim_{k\to+\infty} \frac{P(\bar{B}A_{n_k})}{P(A_{n_k})} = 0 \neq P(\bar{B})$$

which contradicts the supposition that $\{A_n\}$ is quasi-mixing.

It can be shown that any sequence ξ_n of pairwise independent random variables $\{\xi_n\}$ is quasi-mixing (see Problem P.5.8).

Let us add that a still weaker definition of mixing has been introduced by Sucheston [39]. He calls a sequence $\{A_n\}$ of events of a probability space $S = (\Omega, \mathscr{A}, P)$ mixing if

$$\lim_{n \to +\infty} [P(BA_n) - P(B)P(A_n)] = 0 \qquad \text{for every } B \in \mathscr{A} \tag{5.8.23}$$

To avoid misunderstandings, let us call a sequence $\{A_n\}$ which satisfies the condition (5.8.23) "mixing in the generalized sense." Evidently, if $\{A_n\}$ is quasi-mixing, then it is also mixing in the generalized sense, but not conversely. For instance, if $\{A_n\}$ is any sequence of events such that $P(A_n) \to 0$, then $\{A_n\}$ is mixing in the generalized sense, but—according to what has been proved above—it is not quasi-mixing. However, if $\{A_n\}$ is a sequence of events such that $\lim_{n\to+\infty} P(A_n) = c > 0$ where $0 < c < 1$, then if $\{A_n\}$ is mixing in the generalized sense (and thus, of course, also if it is quasi-mixing), then it is mixing with density c.

Finally we mention some other—stronger—notions of mixing.

Let $\{A_n\}$ $(n \geqq 1)$ be a sequence of events in a probability space $S = (\Omega, \mathscr{A}, P)$. Let $\mathscr{A}_n$ denote the algebra generated by the sets $A_1, \ldots, A_n$ and let $\mathscr{B}_n$ denote the algebra generated by the sets $A_{n+1}, A_{n+2}, \ldots$. In what follows we always suppose that $\lim_{n\to+\infty} P(A_n) = c$ where $0 < c < 1$.

The sequence $\{A_n\}$ is called *$*$-mixing* (see Blum, Hanson, and Koopmans [47]) if there exists a sequence ε_n $(n = 1, 2, \ldots)$ of positive numbers such that $\lim_{n\to+\infty} \varepsilon_n = 0$ and one has

$$|P(AB) - P(A)P(B)| < \varepsilon_n P(A)P(B) \qquad \text{if } A \in \mathscr{A}_k \text{ and } B \in \mathscr{B}_{n+k}$$

We call the sequence *R-mixing* if, putting,

$$\sup_{A \in \mathscr{A}_k, B \in \mathscr{B}_{n+k}} |P(AB) - P(A)P(B)| = \rho(k, n)$$

one has

$$\lim_{n \to +\infty} \rho(k, n) = 0 \qquad \text{for } k = 1, 2, \ldots$$

This notion is due to Rosenblatt (see [45]).

We call the sequence $\{A_n\}$ *S-mixing*, if every sequence $\{A_n^*\}$ of events, such that $A_n^* \in \mathscr{B}_n$ $(n = 1, 2, \ldots)$ and for which $\lim_{n\to+\infty} P(A_n^*) = p$, is mixing with density p. This notion is due to Sucheston (see [39]).

Evidently $*$-mixing implies R-mixing, because $\rho(k, n) \leqq \varepsilon_n$. It can be shown (see Nemetz and Varga [46]) that R-mixing implies S-mixing. Evidently, S-mixing implies mixing of order k for every $k \geqq 1$.

We mention without proof some results concerning the above mentioned notions of mixing.

If $\{A_n\}$ is $*$-mixing and $P(A_n) = c$ $(n = 1, 2, \ldots)$, then denoting by α_n the indicator of A_n, we have (see [47]) almost surely

$$\lim_{n \to +\infty} \frac{\alpha_1 + \cdots + \alpha_n}{n} = c$$

Let us put $\prod_{n=1}^{+\infty} \mathscr{B}_n = \mathscr{B}_\infty$. Then every event $B \in \mathscr{B}_\infty$ has probability 0 or 1 (i.e., the zero-one law holds for the sequence $\{A_n\}$) if and only if $\{A_n\}$ is S-mixing (see [39]).

5.9 EXCHANGEABLE EVENTS

DEFINITION 5.9.1. *The events $\{A_n\}$ $(n = 1, 2, \ldots)$ in a probability space $S = (\Omega, \mathscr{A}, P)$ are called exchangeable if for any $k \geqq 1$, the probability*

$$P(A_{n_1} A_{n_2} \cdots A_{n_k}) = w_k \qquad (k = 1, 2, \ldots) \tag{5.9.1}$$

does not depend on the choice of the different integers $n_1 < n_2 < \cdots < n_k$, but only on k. The numbers w_k $(k = 1, 2, \ldots)$ are called the de Finetti constants of the sequence $\{A_n\}$ of exchangeable events.

According to Theorem 5.7.5, if $\{A_n\}$ is a sequence of exchangeable events, then $\{A_n\}$ is stable of order k for every $k \geqq 1$. Let α denote the density of the sequence $\{A_n\}$ of exchangeable events, considered as a stable sequence. Then we get, by passing to the limit $+\infty$, one after the other with $n_k, n_{k-1}, \ldots, n_1$, that

$$w_k = E(\alpha^k) \qquad (k = 1, 2, \ldots) \tag{5.9.2}$$

Let $F(x)$ denote the distribution function of the random variable α; then $F(x) = 0$ for $x < 0$ and $F(x) = 1$ for $x > 1$ and we get from (5.9.2) that

$$w_k = \int_0^{1+0} x^k \, dF(x) \quad (k = 1, 2, \ldots) \tag{5.9.3}$$

Thus we proved the following theorem, due to de Finetti [18]:

THEOREM 5.9.1. *The de Finetti constants w_k $(k = 1, 2, \ldots)$ of a sequence of exchangeable events can be represented as the moments of the distribution function of a probability distribution on the interval* [0, 1].

Remark: If $\{A_n\}$ is a sequence of exchangeable events, then it follows easily from (5.9.1) that if the integers $n_1, n_2, \ldots, n_k$ and $m_1, m_2, \ldots, m_l$ are all different $(k \geqq 1, l \geqq 1)$, we have

$$P(A_{n_1} A_{n_2} \cdots A_{n_k} \bar{A}_{m_1} \bar{A}_{m_2} \cdots \bar{A}_{m_l}) = \sum_{j=0}^{l} \binom{l}{j} (-1)^j w_{k+j} \tag{5.9.4}$$

and thus, in view of 5.9.2,

$$P(A_{n_1} A_{n_2} \cdots A_{n_k} \bar{A}_{m_1} \bar{A}_{m_2} \cdots \bar{A}_{m_l}) = E(\alpha^k (1 - \alpha)^l) \tag{5.9.5}$$

Example 5.9.1. If the events A_n are independent and each has probability p $(0 < p < 1)$, then they are exchangeable, and their de Finetti numbers are $W_k = p^k$ $(k = 1, 2, \ldots)$. In other words, the events A_n are not only stable (of any order) but also mixing of any order with density p, i.e., $\alpha = p$ almost surely, and $F(x)$ figuring in (5.9.3) is the (degenerate) distribution function of the constant p.

Example 5.9.2. Let $\Omega_1, \Omega_2, \ldots, \Omega_r$ $(r \geqq 2)$ be a partition of the basic space Ω of the probability space $S = (\Omega, \mathscr{A}, P)$. Suppose that the events A_n $(n = 1, 2, \ldots)$ are conditionally independent under condition Ω_j for each $j = 1, 2, \ldots, r$ and $P(A_n \mid \Omega_j) = p_j$ for $n = 1, 2, \ldots$ and $j = 1, 2, \ldots, r$ where the numbers $p_1, \ldots, p_r$ are all different and lie in the interior of the interval (0, 1). Put $P(\Omega_j) = q_j$ $(j = 1, 2, \ldots, r)$. Then we have, for $n_1 < n_2 < \cdots < n_k$,

$$P(A_{n_1}A_{n_2} \cdots A_{n_k}) = \sum_{j=1}^{r} P(A_{n_1}A_{n_2} \cdots A_{n_k} \mid \Omega_j)q_j = \sum_{j=1}^{r} p_j^k q_j = w_k$$

Thus the events A_n are exchangeable, and their density $\alpha = \alpha(\omega)$ equals p_j for $\omega \in \Omega_j$ $(j = 1, 2, \ldots, r)$; the distribution function $F(x)$ is that of the discrete distribution attributing the probability q_j to the point $x = p_j$. In Example 5.9.2 the events A_n are conditionally independent and have the same conditional probability α under the condition that the value of α is fixed.

The following theorem (Révész and Rényi [17]) shows that this is also true in the general case:

THEOREM 5.9.2. *Let $\{A_n\}$ be an arbitrary sequence of exchangeable events in the probability space $S = (\Omega, \mathscr{A}, P)$; let $\alpha = \alpha(\omega)$ denote the density of the sequence $\{A_n\}$ considered as a stable sequence. Then one has for any $k \geqq 2$ and any integers $1 \leqq n_1 < n_2 < \cdots < n_k$,*

$$P(A_{n_1}A_{n_2} \cdots A_{n_k} \mid \alpha) = \prod_{j=1}^{k} P(A_{n_j} \mid \alpha) = \alpha^k \tag{5.9.6}$$

almost surely, i.e., the events A_n $(n = 1, 2, \ldots)$ are conditionally independent and have all the conditional probability α under the condition that the value of α is fixed.

Proof. We shall prove that if $k \geqq 1$ and $1 \leqq n_1 < n_2 < \cdots < n_k$, then almost surely

$$P(A_{n_1}A_{n_2} \cdots A_{n_k} \mid \alpha) = \alpha^k \tag{5.9.7}$$

As by definition, denoting by α_n the indicator of A_n,

$$P(A_{n_1}A_{n_2} \cdots A_{n_k} \mid \alpha) = E(\alpha_{n_1}\alpha_{n_2} \cdots \alpha_{n_k} \mid \alpha)$$

we have to prove that

$$E(\alpha_{n_1}\alpha_{n_2}\cdots\alpha_{n_k} \mid \alpha) = \alpha^k \tag{5.9.8}$$

In view of the definition of conditional expectation, to prove (5.9.8) we have to show that for any Borel-measurable function $f(x)$ taking on a finite number of values, we have for $k \geqq 1$

$$E(\alpha_{n_1}\alpha_{n_2}\cdots\alpha_{n_k} f(\alpha)) = E(\alpha^k f(\alpha)) \tag{5.9.9}$$

We prove first that (5.9.9) holds for $f(\alpha) = \alpha^l$ ($l = 0, 1, \ldots.$). This can be shown as follows: We start from the formula

$$E(\alpha_{n_1}\alpha_{n_2}\cdots\alpha_{n_k}\alpha_{m_1}\alpha_{m_2}\cdots\alpha_{m_l}) = w_{k+l} = E(\alpha^{k+l}) \tag{5.9.10}$$

which is valid for $1 \leqq n_1 < n_2 < \cdots < n_k < m_1 < m_2 < \cdots < m_l$. By passing to the limit $+\infty$, one after the other with $m_l, m_{l-1}, \ldots, m_1$, we get from (5.9.10), using that α_n is a stable sequence with density α,

$$E(\alpha_{n_1}\alpha_{n_2}\cdots\alpha_{n_k}\alpha^l) = E(\alpha^{k+l}) \tag{5.9.11}$$

Thus (5.9.9) holds if $f(\alpha) = \alpha^l$; therefore, it holds if $f(\alpha)$ is any polynomial of α. By approximating any function $f(x)$ which is continuous in the interval [0, 1] by polynomials it follows that (5.9.9) holds if $f(\alpha)$ is an arbitrary continuous function; thus, again by approximation, we get that (5.9.9) holds if $f(x)$ is any bounded Borel-measurable function. This proves (5.9.8); as mentioned above, (5.9.8) implies (5.9.7), and (5.9.7) implies (5.9.6). Thus Theorem 5.9.2 is proved.* ■

We now prove the following:

THEOREM 5.9.3. *If $\{A_n\}$ is a sequence of exchangeable events on a probability space $S = (\Omega, \mathscr{A}, P)$, α_n denotes the indicator of the event A_n ($n = 1, 2, \ldots$) and α is the density of the sequence $\{A_n\}$ considered as a stable sequence, then one has almost surely, putting $s_n = \alpha_1 + \alpha_2 + \cdots + \alpha_n$,*

$$\lim_{n\to+\infty} \frac{s_n}{n} = \alpha \tag{5.9.12}$$

Proof. The random variables $(\alpha_k - \alpha)/\sqrt{w_1 - w_2}$ ($k = 1, 2, \ldots$) form an orthonormal system because, in view of (5.9.11), if $k \neq l$,

$$E((\alpha_k - \alpha)(\alpha_l - \alpha)) = w_2 - 2w_2 + w_2 = 0$$

and

$$E((\alpha_k - \alpha)^2) = w_1 - 2w_2 + w_2 = w_1 - w_2 \tag{5.9.13}$$

* For a different proof see Kendall [21].

Expressed otherwise, the variables $(\alpha_k - \alpha)$ are pairwise uncorrelated, have expectation 0 and variance $w_1 - w_2$ and are bounded.

Thus we can apply Theorem 4.3.3, which shows that the statement of Theorem 5.9.3 holds. ■

Remark 1: From the above proof it follows that

$$E\left(\left(\frac{s_n}{n} - \alpha\right)^2\right) = \frac{w_1 - w_2}{n} \to 0 \tag{5.9.14}$$

Of course, the strong convergence of s_n/n to α follows already from Theorem 5.7.4, the sequence $\{A_n\}$ being doubly stable. Also, Theorem 5.9.3 implies that s_n/n tends in probability to α and thus its distribution tends to that of α, i.e.,

$$\lim_{n\to+\infty} P\left(\frac{s_n}{n} < x\right) = F(x) \qquad (0 \leqq x \leqq 1) \tag{5.9.15}$$

Remark 2: From Theorem 5.9.2 we can get the exact distribution of $s_n = \alpha_1 + \cdots + \alpha_n$, as follows:

$$P(s_n = k) = \binom{n}{k} \int_0^1 x^k (1-x)^{n-k}\, dF(x) \qquad (0 \leqq k \leqq n) \tag{5.9.16}$$

Thus the distribution of s_n is a mixed binomial distribution. Formula (5.9.15) can be deduced directly from (5.9.16) too. Theorem 5.9.2 also implies that

$$P\left(\frac{s_n - n\alpha}{\sqrt{n\alpha(1-\alpha)}} < y\right) = \int_0^1 \left[\sum_{\frac{k-nx}{\sqrt{nx(1-x)}} < y} \binom{n}{k} x^k (1-x)^{n-k}\right] dF(x) \tag{5.9.17}$$

It follows from the Moivre-Laplace theorem that

$$\lim_{n\to+\infty} P\left(\frac{s_n - n\alpha}{\sqrt{n\alpha(1-\alpha)}} < y\right) = \Phi(y) \tag{5.9.18}$$

It follows also from (5.9.17) that

$$\lim_{n\to+\infty} P\left(\frac{s_n - n\alpha}{\sqrt{n}} < z\right) = \int_0^1 \Phi\left(\frac{z}{\sqrt{x(1-x)}}\right) dF(x) \tag{5.9.19}$$

Thus the limit distribution of $(s_n - n\alpha)/\sqrt{n}$ is a mixture of normal distributions $N(0, \sigma)$. If $F(x)$ is absolutely continuous, $F'(x) = f(x)$, one can write the right-hand side of (5.9.19) in the form

$$\int_0^{1/2} \Phi\left(\frac{z}{\sigma}\right) \frac{2\sigma}{\sqrt{1-4\sigma^2}} \left[f\left(\frac{1+\sqrt{1-4\sigma^2}}{2}\right) + f\left(\frac{1-\sqrt{1-4\sigma^2}}{2}\right)\right] d\sigma$$

Example 5.9.3. Let us suppose that an urn contains one red and one black ball. Let us draw at random a ball from the urn, replace it and add one additional ball of the same color as the ball drawn. Let us repeat this process indefinitely. (This is called Pólya's urn model, see [19].) Let A_n denote the event that at the nth drawing we have drawn a red ball. Let η_n denote the proportion of red balls in the urn after the nth drawing. Then, denoting by α_n the indicator of the event A_n,

$$\eta_n = \frac{1 + \alpha_1 + \alpha_2 + \cdots + \alpha_n}{n+2}$$

We prove the following statements:

(A) The events A_n $(n = 1, 2, \ldots)$ are exchangeable with the de Finetti constants

$$w_k = \frac{1}{k+1} \qquad (k = 1, 2, \ldots)$$

Thus the density α of the sequence $\{A_n\}$ has the distribution function $F(x) = x$ $(0 \leq x \leq 1)$, i.e., α is uniformly distributed in the interval $(0, 1)$.

(B) The variables η_n form a martingale.

To prove (A) we evaluate the joint distribution $P(\alpha_1 = \varepsilon_1, \ldots, \alpha_n = \varepsilon_n)$ of the variables $\alpha_1, \alpha_2, \ldots, \alpha_n$, where $\varepsilon_i = 1$ or 0 $(i = 1, 2, \ldots, n)$.

Putting $s_k = \varepsilon_1 + \varepsilon_2 + \cdots + \varepsilon_k$, we have

$$P(\alpha_1 = \varepsilon_1, \ldots, \alpha_n = \varepsilon_n) = \frac{1}{n!} \prod_{k=1}^{n-1} [\varepsilon_{k+1}(1 + s_k) + (1 - \varepsilon_{k+1})(1 + k - \varepsilon_k)] \tag{5.9.20}$$

Thus we get for each sequence $\varepsilon_1, \varepsilon_2, \ldots, \varepsilon_n$,

$$P(\alpha_1 = \varepsilon_1, \ldots, \alpha_n = \varepsilon_n) = \frac{s_n!(n - s_n)!}{(n+1)!} \tag{5.9.21}$$

It follows for $n_1 < n_2 < \cdots < n_k$ that

$$P(A_{n_1} A_{n_2} \cdots A_{n_k}) = \frac{\sum_{m=k}^{n_k} \binom{m}{k}}{\binom{n_k}{k}(n_k + 1)}$$

As, however,

$$\sum_{m=k}^{n_k} \binom{m}{k} = \binom{n_k + 1}{k + 1}$$

we get

$$P(A_{n_1} A_{n_2} \cdots A_{n_k}) = \frac{1}{k+1} \quad (k = 1, 2, \ldots) \tag{5.9.22}$$

This proves our statement (A). Note that from (5.9.21) the surprising result

$$P(s_n = r) = \frac{1}{n+1} \qquad \text{for } r = 0, 1, \ldots, n \tag{5.9.23}$$

follows, i.e., s_n takes on all its $n+1$ possible values with the same probability! Of course it follows from the general theory that the limit distribution of s_n/n is uniform in (0, 1) because by (5.9.15) it is the same as that of the density α of the sequence $\{A_n\}$.

To prove (B) it is sufficient to observe that

$$E(\alpha_n \mid \alpha_1, \alpha_2, \ldots, \alpha_{n-1}) = \frac{1 + \alpha_1 + \cdots + \alpha_{n-1}}{n+1} = \eta_{n-1}$$

and therefore

$$E(\eta_n \mid \eta_1, \ldots, \eta_{n-1}) = \frac{n+1}{n+2}\eta_{n-1} + E\left(\frac{\alpha_n}{n+2}\middle| \alpha_1, \ldots, \alpha_{n-1}\right)$$

$$= \frac{n+1}{n+2}\eta_{n-1} + \frac{1}{n+2}\eta_{n-1} = \eta_{n-1}$$

Thus, $\{\eta_n\}$ is a martingale. As η_n is bounded, it follows from the remark to the corollary to Theorem 5.4.1 that η_n tends almost surely to a limit. As the events A_n are exchangeable, the same conclusion can also be obtained from Theorem 5.9.3, which shows that the limit of η_n (i.e., the limit of the proportion of red balls in the urn) has uniform distribution in the interval (0, 1).

Returning to the general theory of exchangeable events, let us introduce the difference operator

$$\Delta w_k = w_k - w_{k+1} \tag{5.9.24}$$

Putting $w_0 = 0$, it follows that

$$\Delta^l w_k = \sum_{j=0}^{l} \binom{l}{j} (-1)^j w_{k+j} \tag{5.9.25}$$

Thus we get from (5.9.4) and (5.9.5),

$$\Delta^l w_k = E(\alpha^k (1-\alpha)^l) \geqq 0 \qquad \text{for } k, l = 0, 1, \ldots \tag{5.9.25'}$$

Thus the differences of all orders of the sequence $\{w_k\}$ are nonnegative: i.e., w_k is a nonincreasing sequence, it is convex, etc. Sequences $\{w_k\}$ ($k = 0, 1, \ldots$) satisfying the inequality $\Delta^l w_k \geqq 0$ for $k, l = 0, 1, \ldots$ are called *absolutely monotonic*. According to a well known theorem of Hausdorff [20], a sequence w_k ($k = 0, 1, \ldots$) such that $w_0 = 1$ is absolutely monotonic if and only if it can be represented in the form

$$w_k = \int_0^1 x^k dF(x) \qquad (k = 0, 1, \ldots) \tag{5.9.26}$$

where $F(x)$ is the distribution function of a probability distribution in the interval [0, 1]. This theorem of Hausdorff can be deduced from the results of this section as follows:

Let w_k be an absolutely monotonic sequence such that $w_0 = 1$. Let us define the joint probability distribution of the random variables $\alpha_1, \ldots, \alpha_{n_k}$, taking on the values 0 and 1 only, as follows: If $n_1, n_2, \ldots, n_k, m_1, m_2, \ldots, m_l$ are all different integers, put

$$P(\alpha_{n_1} = \alpha_{n_2} = \cdots = \alpha_{n_k} = 1, \alpha_{m_1} = \alpha_{m_2} = \cdots = \alpha_{m_l} = 0) = \Delta^l w_k \quad (5.9.27)$$

(The right-hand side of (5.9.27) is by supposition nonnegative. It follows from (5.9.27) that

$$\sum_{\varepsilon_k = 0 \text{ or } 1\ (k=1,2,\ldots,n)} P(\alpha_1 = \varepsilon_1, \ldots, \alpha_n = \varepsilon_n) = \sum_{r=0}^{n} \binom{n}{r} \Delta^{n-r} w_r \quad (5.9.28)$$

Now it is easy to see that the right-hand side of (5.9.28) is equal to 1. This can be verified most easily as follows: Let S be the shift operator, defined by $Sw_k = w_{k+1}$. Then we have $(S + \Delta)w_k = w_{k+1} + w_k - w_{k+1} = w_k$; thus $S + \Delta$ is equal to the identity operator I. As the operators S and Δ are clearly interchangeable, because

$$S\Delta w_k = \Delta S w_k = w_{k+1} - w_{k+2}$$

we get

$$1 = Iw_0 = (S + \Delta)^n w_0 = \sum_{r=0}^{n} \binom{n}{r} S^r \Delta^{n-r} w_0 = \sum_{r=0}^{n} \binom{n}{r} \Delta^{n-r} w_r \quad (5.9.29)$$

Thus, (5.9.27) is a possible definition for the joint distribution of $(\alpha_1, \ldots, \alpha_n)$. It is also easy to see that the definitions of the joint distributions of $(\alpha_1, \ldots, \alpha_n)$ and $(\alpha_1, \ldots, \alpha_{n+1})$ by (5.9.27) are compatible, because

$$\Delta^{l+1} w_k + \Delta^l w_{k+1} = \Delta^l(\Delta w_k + w_{k+1}) = \Delta^l w_k \quad (5.9.30)$$

Thus, according to Kolmogoroff's fundamental theorem, there exists a probability space S and on S an infinite sequence α_n of random variables, such that (5.9.27) is valid. As α_n takes on the values 0 and 1 only, it is the indicator of an event A_n and, by (5.9.27), for these events one has

$$P(A_{n_1} A_{n_2} \cdots A_{n_k}) = \sum_{l=0}^{n_k - k} \binom{n_k - k}{l} \Delta^l w_{n_k - l} = (\Delta + S)^{n_k - k} w_k = w_k \quad (k = 1, 2, \ldots) \quad (5.9.31)$$

Thus the events $\{A_n\}$ are in fact exchangeable, and have the de Finetti constants w_k. According to Theorem 5.9.1,

$$w_k = \int_0^1 x^k dF(x) \qquad (k = 0, 1, \ldots) \quad (5.9.32)$$

where $F(x)$ denotes the distribution function of the density α of the sequence $\{A_n\}$ considered as a stable sequence. Thus we have obtained a probabilistic proof for Hausdorff's purely analytic theorem.* Of course, conversely, one can deduce Theorem 5.9.1 from Hausdorff's theorem.

5.10 THE INVARIANCE OF LIMIT THEOREMS UNDER CHANGE OF MEASURE

We first prove the following:

THEOREM 5.10.1. *Let $\{A_n\}$ be a stable sequence of events of the probability space $S=(\Omega, \mathscr{A}, P)$ with density $\alpha=\alpha(\omega)$. Let P^* be another probability measure on the σ-algebra $\mathscr{A}$, which is absolutely continuous with respect to P. Then the sequence $\{A_n\}$ of events is stable in the probability space $S^*=(\Omega, \mathscr{A}, P^*)$ too, with the same density α.*

Proof. Let $\lambda=dP^*/dP$ denote the Radon-Nikodym derivative of P^* with respect to P. Then we have for every $B \in \mathscr{A}$ with indicator β,

$$P^*(B)=\int_B \lambda dP=E(\beta\lambda) \tag{5.10.1}$$

Thus if α_n is the indicator of the event A_n and β is that of B, we have

$$P^*(A_n B)=E(A_n \beta\lambda) \tag{5.10.2}$$

As α_n tends weakly to α, in view of Remark 2 to Theorem 5.7.2, we get immediately,

$$\lim_{n\to+\infty} P^*(A_n B)=E(\alpha_n\beta\lambda)=\int_B \alpha\, dP^* \tag{5.10.3}$$

Denoting by E^* the expectation on the probability space $S^*=(\Omega, \mathscr{A}, P^*)$, we can also write (5.10.3) in the form

$$\lim_{n\to+\infty} P^*(A_n B)=E^*(\alpha\beta) \tag{5.10.4}$$

Thus $\{A_n\}$ is stable in S^* with the same density α as in S. ∎

COROLLARY TO THEOREM 5.10.1. *If $\{A_n\}$ is a mixing sequence of events with density c $(0<c<1)$ in the probability space $S=(\Omega, \mathscr{A}, P)$ and P^* is a probability measure on $\mathscr{A}$ which is absolutely continuous with respect to P, then*

* Hausdorff's theorem and de Finetti's theorem can both be regarded as particular instances of Choquet's theorem on the representation of points in a convex compact cone as the point of gravity of distribution on the set of extremal points of the cone (see Kendall [21], Phelps [22]). It should be added that the extremal points of the set of absolutely monotonic sequences $\{W_k\}$ are the sequences $\{p^k\}$ corresponding as de Finetti constants to sequences of independent events having the same probability p.

$\{A_n\}$ is mixing in the probability space $S^* = (\Omega, \mathscr{A}, P^*)$ too, with the same density c.

Remark: If $\{A_n\}$ is a sequence of pairwise independent events in the probability space S, each having the probability c $(0 < c < 1)$, then (see Example 5.8.2) the events A_n are mixing with density c. If P^* is absolutely continuous with respect to P, then, according to the above corollary, the events A_n are mixing in $S^* = (\Omega, \mathscr{A}, P^*)$ too, but they are in general no more pairwise independent. This remark can be used for the construction of mixing sequences.

THEOREM 5.10.2. *Let $\xi_1, \xi_2, \ldots, \xi_n, \ldots$ be a sequence of independent random variables, and put*

$$\zeta_n = \xi_1 + \xi_2 + \cdots + \xi_n \qquad (n = 1, 2, \ldots)$$

Suppose that there exists a sequence of constants $B_n > 0$, $(n = 1, 2, \ldots)$ such that $\lim_{n \to +\infty} B_n = +\infty$ and the distribution of

$$\eta_n = \zeta_n / B_n$$

tends weakly to a nondegenerate limit distribution, i.e., denoting by $A_n(x)$ the event $\eta_n < x$, we have

$$\lim_{n \to +\infty} P(A_n(x)) = F(x)$$

for all points of continuity x of the probability distribution function $F(x)$. Then for each x, such that x is a point of continuity of $F(x)$ and $0 < F(x) < 1$, the sequence of events $\{A_n(x)\}$ $(n = 1, 2, \ldots)$ is mixing with density $F(x)$.

Proof. For the proof we need the following useful lemma, due to H. Cramér:

LEMMA 5.10.1. *If the distribution of the random variables η_n tends for $n \to +\infty$ weakly to a limit distribution and δ_n is a sequence of random variables which tends for $n \to +\infty$ in probability to 0, then the distribution of $\eta_n + \delta_n$ tends weakly to the same distribution as that of η_n.*

Proof. We have to show that for every bounded and continuous function $f(x)$, we have

$$\lim_{n \to +\infty} E(f(\eta_n + \delta_n)) = \lim_{n \to +\infty} E(f(\eta_n)) = \int_{-\infty}^{+\infty} f(x) dG(x) \tag{5.10.5}$$

where $G(x)$ denotes the limit distribution of the random variables η_n.

Let us suppose that $|f(x)| \leqq K$, and let us choose an arbitrary $\varepsilon > 0$. Then we can find a number A such that

$$\int_{|x| \geqq A} dG(x) < \varepsilon \tag{5.10.6}$$

The function $f(x)$ is uniformly continuous in the interval $[-A-1, A+1]$. Thus we can find a δ such that $0 < \delta < 1/2$ and $|f(x_1) - f(x_2)| < \varepsilon$ if $|x_1 - x_2| \leqq \delta$, $|x_1| \leqq A+1$, $|x_2| \leqq A+1$. Now let us choose a number n_1 such that for $n \geqq n_1$ we have

$$P(|\delta_n| > \delta) < \varepsilon \quad \text{and} \quad P(|\eta_n| > A + 1/2) < 2\varepsilon$$

This is possible because of (5.10.6). Then we have, with probability $>1 - 3\varepsilon$, for $n > n_1$,

$$|f(\eta_n + \delta_n) - f(\eta_n)| < \varepsilon$$

and therefore

$$|E(f(\eta_n + \delta_n)) - E(f(\eta_n))| < \varepsilon + 6K\varepsilon$$

As $\varepsilon > 0$ can be chosen arbitrarily small, (5.10.5) follows. ■

Now we are in the position to prove Theorem 5.10.2. According to Theorem 5.8.1, it is sufficient to show that for each fixed k,

$$\lim_{n \to +\infty} P(A_n(x)A_k(x)) = F(x)P(A_k(x)) \tag{5.10.7}$$

We may disregard those values of k (if any) for which $P(A_k(x)) = 0$, because for these (5.10.7) is trivially satisfied. (By supposition, $F(x) > 0$, thus there are at most a finite number of such values of k anyway.) If $P(A_k(x)) > 0$, we can write (5.10.7) in the form

$$\lim_{n \to +\infty} P(A_n(x) \mid A_k(x)) = F(x) \tag{5.10.8}$$

Now as by supposition the distribution of ζ_n/B_n tends to $F(x)$, and if k is fixed, the random variable ζ_k/B_n tends in probability to 0 because of $B_n \to +\infty$, it follows from Lemma 5.10.1 (with $\delta_n = \zeta_k/B_n$) that $P((\zeta_n - \zeta_k)/B_n < x) \to F(x)$. As further, $\zeta_n - \zeta_k$ is independent from ζ_k, we get

$$\lim_{n \to +\infty} P\left(\frac{\zeta_n - \zeta_k}{B_n} < x \mid A_k(x)\right) = F(x)$$

Again applying Lemma 5.10.1 (to the probability space $(\Omega, \mathscr{A}, P^*)$ where $P^*(A) = P(A|A_k(x))$ with $\delta_n = \zeta_k/B_n$, we get (5.10.8). ■

Now by combining Theorems 5.10.1 and 5.10.2 we can prove the following result:

THEOREM 5.10.3. *Let $\xi_1, \xi_2, \ldots, \xi_n, \ldots$ be a sequence of independent random variables on a probability space $S = (\Omega, \mathscr{A}, P)$ and put $\zeta_n = \xi_1 + \xi_2 + \cdots + \xi_n$. Suppose that there exists a sequence B_n of positive constants, tending to $+\infty$, such that the distribution of ζ_n/B_n tends weakly to a limit distribution. Let P^* be another probability measure which is absolutely continuous with respect to P. Then the limit distribution of ζ_n/B_n, considered as random var-*

iables on the probability space $S^* = (\Omega, \mathscr{A}, P^*)$, *exists too and is the same as that on the probability space* S.

Proof. According to Theorem 5.10.2, if $A_n(x)$ denotes the event $\zeta_n/B_n < x$, where x is a point of continuity of the limit distribution $F(x)$ and $0 < F(x) < 1$, the events $A_n(x)$ are mixing with density $F(x)$. Thus by Theorem 5.10.1 they are also mixing with the same density in S^*, and thus

$$\lim_{n \to +\infty} P^*(A_n(x)) = F(x) \quad \blacksquare$$

Remark 1: The condition that P^* should be absolutely continuous with respect to P may be replaced by a weaker condition. See Abbott and Blum [43].

Remark 2: The interest of Theorem 5.10.3 lies in the fact that the random variables ξ_n are in general not independent with respect to the measure P^*. Thus Theorem 5.10.3 makes it possible to deduce from limit theorems on sums of independent random variables (such as, e.g., the central limit theorem under Lindeberg's condition) corresponding theorems on dependent random variables (see Révész [42]). (Of course, the variables ξ_n, though not independent with respect to P^*, are nevertheless rather weakly dependent in the limit on P^*.)

It is worthwhile to formulate the following:

Corollary to Theorem 5.10.3. *Let* $\xi_1, \xi_2, \ldots, \xi_n, \ldots$ *be independent random variables on the probability space* $S = (\Omega, \mathscr{A}, P)$. *Put* $\zeta_n = \xi_1 + \xi_2 + \cdots + \xi_n$ $(n = 1, 2, \ldots)$ *and suppose that there exists a sequence* B_n *of positive numbers such that* $B_n \to +\infty$ *and*

$$\lim_{n \to +\infty} P\left(\frac{\zeta_n}{B_n} < x\right) = F(x)$$

for every point of continuity of the nondegenerate probability distribution function $F(x)$. *Let* $B \in \mathscr{A}$ *be any event such that* $P(B) > 0$. *Then one has*

$$\lim_{n \to +\infty} P\left(\frac{\zeta_n}{B_n} < x \mid B\right) = F(x)$$

for every point of continuity x *of* $F(x)$.

Proof. For the proof it is sufficient to point out that $P^*(A) = P(A \mid B)$ is absolutely continuous with respect to P. $\blacksquare$

Remark 1: Theorem 5.10.3 has been proved first for discrete random variables by Rényi [23]; this result has been generalized by Kolmogoroff [24], who removed some of the restrictions. The final result has been obtained in a much easier way by the use of the notion of mixing, by Rényi [15].

Remark 2: The content of the corollary can be formulated somewhat vaguely as follows: Normalized sums of independent random variables have, in the limit, the same distribution in every fixed subset of positive probability of the basic space. This shows that the sum of a large number of independent random variables has to fluctuate rather wildly in every "not too small" part of the basic space. The following theorem describes the same phenomenon from another point of view:

THEOREM 5.10.4. *Let $\xi_1, \xi_2, \ldots, \xi_n$ be independent random variables, put $\zeta_n = \xi_1 + \xi_2 + \cdots + \xi_n$, and let B_n be a sequence of positive constants such that $B_n \to +\infty$. Suppose that the distribution of ζ_n/B_n tends to a nondegenerate limit distribution. Then the random variables ζ_n/B_n do not converge for $n \to +\infty$ in probability (and thus neither strongly, nor almost surely) to a random variable.*

Proof. Suppose we would have

$$\zeta_n/B_n \Rightarrow \zeta \tag{5.10.9}$$

Then (see Theorem 4.2.1) the probability distribution function $F(x)$ of ζ would be identical to the limit distribution function of ζ_n/B_n. Let us choose a number x which is a continuity point of $F(x)$ and for which $0 < F(x) < 1$. (Such an x can be found, as we have supposed that the limit distribution of ζ_n/B_n is nondegenerate.) It follows from Theorem 5.10.2 that denoting by $A_n(x)$ the set on which $\zeta_n/B_n < x$, $A_n(x)$ is a mixing sequence of sets with density $F(x)$. Thus, denoting by C_ε the event $\zeta > x + \varepsilon$ where $\varepsilon > 0$ is chosen so that $F(x+\varepsilon) < 1$, we have

$$\lim_{n\to\infty} P(A_n(x)C_\varepsilon) = F(x)(1 - F(x+\varepsilon)) > 0 \tag{5.10.10}$$

On the other hand, as

$$P(A_n(x)C_\varepsilon) \leqq P\left(\left|\frac{\zeta_n}{B_n} - \zeta\right| > \varepsilon\right)$$

we get, if (5.10.9) holds,

$$\lim_{n\to\infty} P(A_n(x)C_\varepsilon) = 0 \tag{5.10.11}$$

As (5.10.10) and (5.10.11) contradict each other, it follows that (5.10.9) is impossible. ■

Remark: The condition in Theorem 5.10.4 that the limit distribution of ζ_n/B_n should be nondegenerate is necessary, because if the distribution of ζ_n/B_n tends to the degenerate distribution of a constant c, then by Theorem 4.2.1, ζ_n/B_n tends in probability to c.

EXERCISES

☐ **E.5.1.** Let the joint distribution of ξ and η be a two-dimensional normal distribution with density function

$$h(x, y) = \frac{\sqrt{AC - B^2}}{2\pi} e^{-1/2(Ax^2+2Bxy+Cy^2)}$$

where $A > 0$, $C > 0$ and $AC > B^2$. Show that

$$E(\eta \mid \xi) = -B\xi/C$$

Hint: One obtains by some calculus that the density function $f(x)$ of ξ is

$$f(x) = \frac{1}{\sqrt{2\pi}\sigma_1} e^{-x^2/2\sigma_1^2}$$

where $\sigma_1 = \sqrt{C/(AC - B^2)}$.

It follows that the density function $g(y \mid x)$ of the conditional distribution of η under condition $\xi = x$ is

$$g(y \mid x) = \frac{h(x, y)}{f(x)} = \sqrt{\frac{C}{2\pi}} e^{-C(y+Bx/C^2)/2}$$

and thus

$$E(\eta \mid \xi) = -B\xi/C$$

Remark: It is easy to see that the correlation coefficient $R(\xi, \eta)$ of ξ and η is

$$R(\xi, \eta) = B/\sqrt{AC}$$

Thus it follows from the above results that *if the joint distribution of ξ and η is normal, ξ and η are independent if and only if they are uncorrelated.* The same holds for k random variables ($k \geqq 3$) having a joint normal distribution; they are independent if and only if they are pairwise uncorrelated.

E.5.2. (a) Let $\{\xi_n\}$ be a homogeneous Markov chain with a finite number of possible values $x_1 < x_2 < \cdots < x_r$ and transition probabilities $p_{j,k} = P(\xi_{n+1} = x_k \mid \xi_n = x_j)$. Show that if $\{\xi_n\}$ is also a martingale, then one has necessarily $p_{1,1} = 1$ and $p_{r,r} = 1$, i.e., if once the value x_1 or x_r is attained, the sequence ξ_n is constant from that point onwards.

Hint: By the definition of a Markov chain

$$E(\xi_{n+1} \mid \xi_1, \ldots, \xi_n) = E(\xi_{n+1} \mid \xi_n).$$

Thus, if $\{\xi_n\}$ is also a martingale, we have $E(\xi_{n+1} \mid \xi_n) = \xi_n$. Thus the system of equations

$$x_j = \sum_{k=1}^{r} p_{j,k} x_k$$

has to be valid, which implies $p_{1,1} = p_{r,r} = 1$.

Remark: It follows from the martingale convergence theorem that $\lim_{n\to+\infty}\xi_n$ exists almost surely. This means that ξ_n has to be constant from some n onwards; the constant value may be x_1 or x_r or some other x_k for which $p_{k,k} = 1$.

(b) Suppose that a player plays the game of throwing dice. He starts with a capital of \$2 and tries to increase his capital to \$10. He stops if he attains this goal or if he runs out of money. He plays the "bold" strategy which consists in that he stakes all his money provided that, in case he wins (i.e., doubles his money), this doubled sum does not exceed \$10. Otherwise, he stakes just as much as is needed to increase his money, in case he wins, to \$10. Put $\xi_1 = 2$ and let ξ_n for $n \geqq 2$ denote the capital of the player after $n-1$ throws. Show (1) that $\{\xi_n\}$ is a Markov chain and find all the transition probabilities, (2) that $\{\xi_n\}$ is a martingale, (3) that the player reaches his aim (i.e., leaves with \$10 in his pocket) with probability 1/5, and looses all his money with probability 4/5.

Hint: Clearly ξ_{n+1} is independent from $\xi_2, \ldots, \xi_{n-1}$ if the value of ξ_n is given; thus, $\{\xi_n\}$ is a Markov chain; the possible values of ξ_n (in dollars) are 0, 2, 4, 6, 8 and 10; the possible transitions are shown by the following diagram:

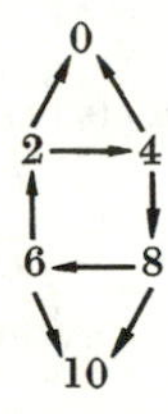

and every transition indicated by an arrow has the transition probability 1/2. Further, the transitions $0 \to 0$ and $10 \to 10$ have the probability 1. The equations $x_j = \sum_{k=1}^{r} p_{j,k}x_k$ of (a) are satisfied, thus $\{\xi_n\}$ is a martingale, and $\lim_{n\to+\infty}\xi_n = \xi$ almost surely exists. Thus, with probability 1, ξ_n has to be constant from some n onwards. The only constant values possible are 10 and 0. Denoting the corresponding probabilities by p and $1-p$, we have $E(\xi) = 10p = E(\xi_1) = 2$, i.e., $p = 1/5$.

Remark: For a general treatment of "bold" strategies see Dubbins and Savage [25].

E.5.3. Let A_n be a sequence of events in the probability space $(\Omega, \mathscr{A}, P)$ such that $\lim_{n\to+\infty} P(A_n) = 0$. Let $\mathscr{B} \subseteq \mathscr{A}$ be any sub-σ-algebra of $\mathscr{A}$. Prove that $P(A_n \mid \mathscr{B}) \Rightarrow 0$ for $n \to +\infty$.

Hint: By definition, if $B \in \mathscr{B}$, we have

$$\int_B P(A_n \mid \mathscr{B})\, dP = P(A_n B)$$

Choose for B the event $P(A_n \mid \mathscr{B}) > \varepsilon$, where $\varepsilon > 0$ is arbitrary. It follows that

$$\varepsilon P(P(A_n \mid \mathscr{B}) > \varepsilon) \leqq P(A_n) \qquad (n = 1, 2, \ldots)$$

from which the statement follows.

E.5.4. Deduce Kolmogoroff's Theorem 4.4.5 from Theorem 5.3.2.

Hint: For the sake of simplicity we sketch the proof for the special case of random variables ξ_k which are independent and identically distributed with the same distribution function $F(x)$, which is symmetric about 0, i.e., $F(-x) = 1 - F(x + 0)$ and such that $E(|\xi_k|) = 2\int_0^{\infty} |x| dF(x) = A$ is finite. Put for $k = 1, 2, \ldots,$

$$\xi^* = \begin{cases} \xi_k & \text{if} \quad |\xi_k| \leqq k \\ 0 & \text{if} \quad |\xi_k| > k \end{cases}$$

Then we have $E(\xi_k^*) = 0$, and

$$D_k^2 = D^2(\xi_k^*) = 2\int_0^k x^2\, dF(x)$$

Thus

$$\sum_{k=1}^{+\infty} \frac{D_k^2}{k^2} = 2\sum_{k=1}^{+\infty} \frac{1}{k^2}\int_0^k x^2\, dF(x) = 2\sum_{k=1}^{+\infty}\left[\int_{k-1}^{k} x^2\, dF(x)\right]\left[\sum_{j=k}^{+\infty}\frac{1}{j^2}\right]$$

As $\sum_{j=1}^{+\infty} 1/j^2 = \pi^2/6$ and $\sum_{j=k}^{+\infty} 1/j^2 \leqq \sum_{j=k}^{+\infty} 1/j(j-1) = 1/(k-1) \leqq 2/k$ for $k \geqq 2$, we get

$$\sum_{k=1}^{+\infty} \frac{D_k^2}{k^2} \leqq \frac{\pi^2}{3}\int_0^1 x^2\, dF(x) + 4\int_1^{+\infty} x\, dF(x)$$

Thus the series $\sum_{k=1}^{+\infty} D_k^2/k^2$ is convergent and therefore $1/n \sum_{k=1}^{n} \xi_k^*$ tends almost surely to 0. As the series

$$\sum_{k=1}^{+\infty} P(\xi_k \neq \xi_k^*) = \sum_{k=1}^{+\infty} 2\int_k^{+\infty} dF(x) = \sum_{k=1}^{+\infty} 2k\int_k^{k+1} dF(x) \leqq \int_{-\infty}^{+\infty} |x|\, dF(x)$$

is convergent, $\xi_k = \xi_k^*$ for all but a finite number of values of k almost surely. Thus, $1/n \sum_{k=1}^{n} \xi_k$ tends to 0 almost surely, too. The general case can be settled quite similarly.

E.5.5. Let

$$x = \sum_{n=1}^{+\infty} \frac{a_n(x)}{q_1 q_2 \cdots q_n}$$

be the Cantor series of the real number x with respect to the sequence $q_n \geqq 2$ of integers ($a_n(x)$ may take on the values $0, 1, \ldots, q_n - 1$). Let $N_n(r, x)$ denote the frequency of the digit r ($r = 0, 1, 2, \ldots$) among the numbers $a_1(x), \ldots, a_n(x)$. Prove that if $\sum_{q_k > r} 1/q_k$ is divergent, one has, for almost all x,

$$\lim_{n \to +\infty} \frac{N_n(r, x)}{\sum_{\substack{q_k > r \\ k \leqq n}} 1/q_k} = 1$$

Hint: Apply Theorem 5.3.4 to the random variables $\delta_r(k, x)$ defined as follows:

$$\delta_r(k, x) = \begin{cases} 1 & \text{if } a_k(x) = r \\ 0 & \text{otherwise} \end{cases}$$

Notice that the $\delta_r(k, x)$ $(k = 1, 2, \ldots)$ considered as random variables on the Lebesgue probability space are independent, $E(\delta_r(k, x)) = 1/q_k$ and $D^2(\delta_r(k, x)) = 1/q_k(1 - 1/q_k)$ if $q_k > r$. Thus, one has to prove only that the series

$$\sum_{q_n > r} \frac{1/q_n(1 - 1/q_n)}{(\sum_{\substack{q_k > r \\ k \leq n}} 1/q_k)^2}$$

is convergent. This follows from the inequality valid for any positive numbers $b_1, b_2, \ldots,$

$$\frac{b_n}{(b_1 + \cdots + b_n)^2} \leqq \frac{1}{b_1 + \cdots + b_{n-1}} - \frac{1}{b_1 + \cdots + b_n}$$

E.5.6. Consider the series

$$\sum_{n=2}^{+\infty} \sum_{k=1}^{n-1} c_{n,k} R_n(x) R_k(x) \tag{E.5.6.1}$$

where $R_n(x)$ denotes the nth Rademacher functions. Suppose that the series

$$\sum_{n=2}^{+\infty} \sum_{k=1}^{n-1} c_{n,k}^2$$

is convergent. Prove that the partial sums $S_n(x)$ of the series (E.5.6.1) defined by

$$S_{N(N-1)/2+r}(x) = \sum_{n=2}^{N} \sum_{k=1}^{n-1} c_{n,k} R_n(x) R_k(x) + \sum_{k \leqq r} c_{N+1,k} R_{N+1}(x) R_k(x)$$

for $N = 2, 3, \ldots; r = 1, 2, \ldots N$, converge almost everywhere in the interval $(0, 1)$.

Hint: The partial sums $S_{N(N-1)/2}(x)$, as random variables on the Lebesgue probability space, form a martingale, such that $E(S^2_{N(N-1)/2})$ is bounded; thus $\lim_{N\to+\infty} S_{N(N-1)/2}(x)$ exists almost everywhere in $(0, 1)$. From Kolmogoroff's inequality, we get for any $\varepsilon > 0$,

$$\sum_{N=2}^{+\infty} P(\max_{r \leqq N} |S_{\binom{N}{2}+r}(x) - S_{\binom{N}{2}}(x)| > \varepsilon)$$
$$= \sum_{N=2}^{+\infty} P(\max_{r \leqq N} |\sum_{k=1}^{r} c_{N+1,k} R_k(x)| > \varepsilon) \leqq \frac{1}{\varepsilon^2} \sum_{N=2}^{+\infty} \sum_{k=1}^{N} c_{N+1,k}^2 < +\infty$$

Thus by the Borel-Cantelli lemma, for almost all x, we have

$$\max_{r \leqq N} |S_{\binom{N}{2}+r}(x) - S_{\binom{N}{2}}(x)| \leqq \varepsilon$$

for all but a finite number of values of N. As this holds for every $\varepsilon > 0$, it follows that almost surely

$$\lim_{n\to+\infty} S_n(x) = \lim_{N\to+\infty} S_{\binom{N}{2}}(x)$$

Remark: The result of this exercise is due to Pál [26], who proved it by the methods of the theory of orthogonal functions in a much more complicated way.

E.5.7. Let $\{\xi_n\}$ be a homogeneous Markov chain with two possible states, 0 and 1, and a symmetric transition probability matrix, that is, $p_{0,0} = p_{1,1} = p$ and $p_{0,1} = p_{1,0} = q$ where $0 < p < 1, p + q = 1$. Put $\zeta_n = \xi_1 + \xi_2 + \cdots + \xi_n$ $(n \geq 1)$ and prove that ζ_n is normally distributed in the limit.

Hint: Put $\nu_0 = 0$, and let $1 \leq \nu_1 < \nu_2 \cdots < \nu_k < \cdots$ denote those values of n for which $\xi_n = 1$. Then it is easy to see that the variables $\delta_k = \nu_k - \nu_{k-1}$ ($k = 1, 2, \ldots$ are independent and identically distributed, having the distribution

$$P(\delta_k = n) = \begin{cases} p & \text{if} \quad n = 1 \\ q^2 p^{n-2} & \text{if} \quad n \geq 2 \end{cases}$$

Thus

$$E(\delta_k) = 2 \qquad D^2(\delta_k) = \frac{2p}{q}$$

It follows by the Moivre-Laplace theorem that

$$\lim_{k \to +\infty} P\left(\frac{\nu_k - 2k}{\sqrt{2kp/q}} < x\right) = \Phi(x)$$

Now evidently,

$$P(\zeta_n > k) = P(\nu_k < n)$$

Thus, if $k = (n/2 - x/2\sqrt{pn/q})$,

$$P\left(\zeta_n > \frac{n}{2} - \frac{x}{2}\sqrt{\frac{pn}{q}}\right) \approx P\left(\frac{\nu_k - 2k}{\sqrt{2kp/q}} < x\right)$$

and therefore

$$\lim_{k \to \infty} P\left(\frac{\zeta_n - n/2}{1/2\sqrt{pn/q}} < x\right) = 1 - \Phi(-x) = \Phi(x)$$

Remark: For $p = q = 1/2$ the result, of course, reduces to the Moivre-Laplace theorem for the symmetric case. For a general central limit theorem for Markov chains see Ibrahimov and Linnik [27].

E.5.8. Prove the strong law of large numbers for a sequence $\{A_n\}$ of exchangeable events as follows: Denoting by α_n the indicator of A_n show that the series $\sum_{n=1}^{\infty} E(((\alpha_1 + \cdots + \alpha_n)/n - \alpha)^4)$ is convergent, and use Beppo Levi's theorem.

Hint: According to (5.9.11), if k_1, k_2, k_3 and k_4 are different positive integers,

$$E((\alpha_{k_1} - \alpha)(\alpha_{k_2} - \alpha)(\alpha_{k_3} - \alpha)(\alpha_{k_4} - \alpha)) = 0$$

$$E((\alpha_{k_1} - \alpha)^2(\alpha_{k_2} - \alpha)(\alpha_{k_3} - \alpha)) = 0$$

$$E((\alpha_{k_1} - \alpha)^2(\alpha_{k_2} - \alpha)^2) = w_2 - 2w_3 + w_4$$

$$E((\alpha_{k_1} - \alpha)^3(\alpha_{k_2} - \alpha)) = 0$$

$$E((\alpha_{k_1} - \alpha)^4) = w_1 - 4w_2 + 6w_3 - 3w_4$$

It follows that

$$E\left(\left(\frac{\alpha_1 + \cdots + \alpha_n}{n} - \alpha\right)^4\right) = O\left(\frac{1}{n^2}\right)$$

E.5.9. Generalize Example 5.9.3 (Pólya's urn model) as follows: Suppose that originally there are $R+1$ red and $B+1$ black balls in the urn ($R \geqq 0$, $B \geqq 0$) and each time a ball is drawn it is replaced and one ball of the same color is added.

(a) Show that if A_n denotes the event that at the nth occasion a red ball is drawn, then the events A_n are exchangeable. Determine the de Finetti constants and the distribution function of $\alpha = \lim_{n\to\infty} (\alpha_1 + \cdots + \alpha_n)/n$ where α_n is the indicator of the event A_n.

(b) Show that if η_n denotes the proportion of red balls in the urn after the nth drawing, then η_n is a martingale.

(c) Determine the exact distribution of $S_n = \alpha_1 + \alpha_2 + \cdots + \alpha_n$.

Hint: The problem is equivalent to the following: Suppose we start with an urn containing one red and one black ball and during the first $R+B$ drawings we draw a red ball R times and a black ball B times. Determine under this supposition the conditional probabilities concerning the continuation of the process. It follows (if A_n denotes the event that a red ball is drawn at the $n+R+B$th occasion) that

$$P(A_{n_1} A_{n_2} \cdots A_{n_k}) = (R+B+1)\binom{R+B}{B} \Delta^B w_{k+R}$$

where $w_k = 1/(k+1)$ $(k = 1, 2, \ldots)$. As

$$\Delta^r w_l = \frac{r!}{(l+1)(l+2)\cdots(l+r+1)}$$

we get, denoting by $w_k(R, B)$ the de Finetti constants,

$$w_k(R, B) = \frac{(R+B+1)\binom{R+B}{B}}{(R+B+k+1)\binom{R+B+k}{B}}$$

It follows that

$$w_k(R, B) = \int_0^1 x^k f(x, R, B)\,dx$$

where

$$f(x, R, B) = \frac{(R+B+1)!}{R!B!} x^R (1-x)^B \quad (0 \leqq x \leqq 1)$$

Thus α has a beta distribution with parameters $(R+1, B+1)$. The sequence η_n remains, of course, a martingale under condition $\eta_{R+B} = (R+1)/(R+B+2)$ too. The distribution of $S_n = \alpha_1 + \cdots + \alpha_n$ is

$$P(S_n = k) = \binom{n}{k} \frac{(R+B+1)!}{R!B!} \int_0^1 x^{R+k}(1-x)^{B+n-k}\,dx = \frac{\binom{R+k}{k}\binom{B+n-k}{n-k}}{\binom{R+B+n+1}{n}}$$

for $k = 0, 1, \ldots, n$.

E.5.10. Consider the following modification* of the Pólya urn model: An urn contains one red and one black ball. We select a ball and replace it, and add one ball of the *opposite* color. Let A_n denote the event that we draw a red ball at the nth drawing, and denote by α_n the indicator of A_n. Put $S_n = \alpha_1 + \cdots + \alpha_n$ and denote by η_n the proportion of the red balls in the urn after the nth drawing.

(a) Prove that the events A_n are *not* exchangeable.

(b) Determine the behavior of S_n/n for $n \to +\infty$ using the fact that the variables S_n form a Markov chain.

Hints: (1) While by symmetry $P(A_n) = 1/2$ for $n = 1, 2, \ldots$, one has, for instance, $P(A_1 A_2) = 1/6$, but $P(A_1 A_3) = 5/24$.

(2) Clearly S_n is a Markov chain with transition probabilities

$$P(S_{n+1} = k | S_n = k) = \frac{k+1}{n+2} \qquad P(S_{n+1} = k+1 \,|\, S_n = k) = \frac{n-k+1}{n+2}$$

Evidently $E(S_n) = n/2$, and by the theorem of total expectation, putting $D_n^2 = D^2(S_n)$, we have

$$D_{n+1}^2 = D_n^2\left(1 - \frac{2}{n+2}\right) + \frac{1}{4}$$

As $D_1^2 = 1/4$, it follows by induction that $D_n^2 = (n+2)/12$. Thus $D^2(S_n/n) = (n+2)/12n^2 \to 0$, and therefore S_n/n tends strongly and thus in probability to $1/2$.

Remark: There is a striking difference between the behavior of S_n/n in the Pólya urn model and the Friedman urn model. The reason for this difference can be heuristically understood as follows: Suppose that after n steps, where n is a large number, the proportion of red balls in the urn is equal to p. During the next $n\varepsilon$ drawings, if ε is very small compared with p, the proability of drawing a red ball will be practically constant, namely equal to p. During these $n\varepsilon$ drawings, the expected number of red balls drawn will therefore be approximately equal to $np\varepsilon$. In the case of Pólya's model this will result in that, after these $n\varepsilon$ additional drawings, the proportion of red balls in the urn will be approximately equal to $(np + np\varepsilon)/(n + n\varepsilon) = p$; thus a proportion p, once established, will remain almost constant whatever the value of p is. On the other hand, in the case of Friedman's model the proportion of red balls in the urn after the $n\varepsilon$ additional drawings will be equal to

$$\frac{np + n\varepsilon(1-p)}{n + n\varepsilon} = p + \frac{\varepsilon}{1+\varepsilon}(1-2p)$$

Thus if $p < 1/2$, the proportion will increase and in case $p > 1/2$, it will decrease. In any case, it will get nearer to $1/2$ as

$$\left|p + \frac{\varepsilon}{1+\varepsilon}(1-2p) - \frac{1}{2}\right| = \left|p - \frac{1}{2}\right|\left(1 - \frac{2\varepsilon}{1+\varepsilon}\right) = \left(\frac{1-\varepsilon}{1+\varepsilon}\right)\left|p - \frac{1}{2}\right|$$

This explains the difference of the behavior of S_n/n in the two models.

* This problem is sometimes called Friedman's urn model. See Friedman [28]. (See also Rényi [29], p. 353, and Freedman [30]).

PROBLEMS

P.5.1. Let η and ζ be random variables on a probability space $S = (\Omega, \mathscr{A}, P)$ and suppose that the expectations of η, ζ and $\eta\zeta$ exist. If the random variables η and ζ are conditionally uncorrelated under every condition $B \in \mathscr{B}$ such that $P(B) > 0$, where $\mathscr{B}$ is a σ-algebra of subsets of Ω such that $B \subseteq \mathscr{A}$. then almost surely

$$E(\eta\zeta \mid \mathscr{B}) = E(\eta \mid \mathscr{B})E(\zeta \mid \mathscr{B})$$

Hint: Let us choose an $\varepsilon > 0$ and denote by C_k the set on which $k\varepsilon \leqq E(\eta | \mathscr{B}) < (k+1)\varepsilon$ and by D_l the set on which $l\varepsilon \leqq E(\zeta \mid \mathscr{B}) < (l+1)\varepsilon$ $(k, l = 0, \pm 1, \pm 2, \ldots)$. Then $C_k \in \mathscr{B}$ and $D_l \in \mathscr{B}$, and we have for any $B \in \mathscr{B}$ with $P(B) > 0$,

$$\int_B E(\zeta \mid \mathscr{B})E(\eta \mid \mathscr{B})dP = \sum_k \sum_l kl\varepsilon^2 P(BC_k D_l) + O(\varepsilon)$$

Now if $P(BC_k D_l) > 0$, we have, by supposition,

$$E(\eta\zeta \mid BC_k D_l) = E(\eta \mid BC_k D_l)E(\zeta \mid BC_k D_l)$$

By summation over k and l, we get

$$\int_B \eta\zeta \, dP = \sum_k \sum_l kl\varepsilon^2 P(BC_k D_l) + O(\varepsilon)$$

As $\varepsilon > 0$ can be chosen arbitrarily small, it follows that

$$\int_B \eta\zeta \, dP = \int_B E(\eta \mid \mathscr{B})E(\zeta \mid \mathscr{B})dP$$

which was to be proved.

P.5.2. Let τ_n $(n = 0, 1, 2, \ldots)$ be a sequence of random variables such that $\tau_0 \equiv 0$ and the random variables $\tau_k - \tau_{k-1}$ $(k \geqq 1)$ are independent and identically distributed positive random variables having an absolutely continuous distribution with density $f(x)$. Let $\zeta(a, b)$ for $0 \leqq a < b$ denote the number of values of n for which $a \leqq \tau_n < b$. We call the process $\{\tau_n\}$ (which can also be described by the family $\zeta(a, b)$ of random variables) a *renewal process*. Show that the sequence of random variables $\zeta(0, t_n)$, where $0 < t_1 < t_2 < \cdots < t_n < \cdots$, is a Markov chain for every choice of the sequence $\{t_n\}$, if and only if $\{\tau_n\}$ is a Poisson process.

Remark: The motivation behind the above definition is the following: Consider a certain apparatus containing a part which has to be renewed from time to time. For example, we may think about a portable radio or tape recorder in which the battery has to be eventually renewed. Let τ_n denote the time of the nth renewal. Then $\tau_k - \tau_{k-1}$ $(k = 1, 2, \ldots)$ are the time intervals during which the first, second, ..., etc., battery is working; it is reasonable to suppose that these time intervals are independent positive random variables having the same distribution. In this example, $\zeta(a, b)$ is the number of renewals of the battery needed in the time interval $a \leqq t < b$.

Hint: Suppose that $0 < t_1 < t_2 < t_3$. If $\zeta(0, t_n)$ is a Markov chain,

$$A(t_1, t_2, t_3) = P(\zeta(0, t_3) = 1 \mid \zeta(0, t_2) = 1, \zeta(0, t_1) = 0)$$

does not depend on t_1. Now we have

$$A(t_1, t_2, t_3) = \frac{\int_{t_1}^{t_2} f(x)\left[\int_{t_3-x}^{+\infty} f(y)dy\right]dx}{\int_{t_1}^{t_2} f(x)\left[\int_{t_2-x}^{+\infty} f(y)dy\right]dx}$$

As by supposition $\partial A(t_1, t_2, t_3)/\partial t_1 = 0$, it follows, putting for the sake of brevity $G(x) = \int_x^{+\infty} f(y)dy$, that

$$\frac{G(t_3 - t_1)}{G(t_2 - t_1)} = A(t_1, t_2, t_3)$$

Thus, $G(t_3 - t_1)/G(t_2 - t_1)$ does not depend on t_1 either, i.e.,

$$G(t_3 - t_1) = G(t_2 - t_1)G(t_3 - t_2)$$

This implies, as $G(x)$ is a decreasing function, that $G(x) = e^{-\lambda x}$ with some $\lambda > 0$, i.e., $f(x) = \lambda e^{-\lambda x}$ for $x > 0$.

P.5.3. Let ξ and η be random variables on a probability space $S = (\Omega, \mathscr{A}, P)$. Suppose that η takes on only a finite number of values $y_1, y_2, \ldots, y_r$ with the corresponding probabilities $P(\eta = y_k) = p_k > 0$ $(k = 1, 2, \ldots, r)$, however ξ may have an arbitrary distribution. The (average) conditional entropy $H(\eta \mid \xi)$ of η, given the value of ξ, is defined as follows:

$$H(\eta \mid \xi) = E(H(p_1(\xi), \ldots, p_r(\xi)))$$

where

$$H(q_1, q_2, \ldots, q_r) = \sum_{k=1}^{r} q_k \log_2 \frac{1}{q_k}$$

is the entropy of the distribution $\{q_k\}$ and

$$p_k(\xi) = P(\eta = y_k \mid \xi) \quad (k = 1, 2, \ldots, r)$$

is the conditional distribution of η for a given value of ξ. Put $H(\eta) = H(p_1, \ldots, p_r)$. Prove that

$$I(\xi, \eta) = H(\eta) - H(\eta \mid \xi) \tag{P.5.3.1}$$

Hint: Recall that, by definition (see Section 3.8),

$$I(\xi, \eta) = \sup I(f(\xi), \eta)$$

where the supremum is taken over all functions f taking on a finite number of values. Using Jensen's inequality and the convexity of the function $-\log_2 x$, we get for every $B \in A_\xi$ with $P(B) > 0$,

$$\frac{\int_B p_k(\xi) \log_2 1/p_k(\xi)\, dP}{\int_B p_k(\xi)\, dP} \leqq \log_2 \frac{P(B)}{\int_B p_k(\xi)\, dP},$$

Now let $f(x)$ be any function taking on a finite number of values $a_1, \ldots, a_N$ and denote by B_j the set on which $f(\xi) = a_j$. As, by definition,

$$\frac{1}{P(B_j)} \int_{B_j} p_k(\xi)\, dP = P(\eta = y_k \mid f(\xi) = a_j)$$

it follows that

$$H(\eta) - H(\eta \mid \xi) \geqq I(f(\xi), \eta)$$

for every f and thus

$$H(\eta) - H(\eta \mid \xi) \geqq I(\xi, \eta) \tag{P.5.3.2}$$

On the other hand, define $f_\varepsilon(x)$ so that it takes different values on the sets $B(l_1, \ldots, l_r)$ defined by $l_i \varepsilon \leqq p_i(x) < (l_i + 1)\varepsilon$ $(i = 1, 2, \ldots, r)$ where $(l_1, \ldots, l_r)$ is an arbitrary r-tuple of positive integers and $\varepsilon > 0$ is an arbitrary number. It follows that

$$\limsup_{\varepsilon \to 0} I(f_\varepsilon(\xi), \eta) \geqq H(\eta) - H(\eta \mid \xi)$$

and thus

$$I(\xi, \eta) \geqq H(\eta) - H(\eta \mid \xi) \tag{P.5.3.3}$$

It follows from (P. 5.3.2) and (P.5.3.3) that (P.5.3.1) holds.

P.5.4. Suppose that a certain type of population (e.g., of bacteria) reproduces as follows: A member of the population alive at time $t = n$ $(n = 0, 1, \ldots)$ will—independently from the past history and size of the whole population and of what happens to other members of the population—produce until time $t = n + 1$ with probability p two offspring and die; while with probability $q = 1 - p$, it dies at time $t = n + 1$ without producing new offspring. Denote by ζ_n the number of members of the population at time $t = n$ $(n = 0, 1, 2, \ldots)$. Suppose $\zeta_0 = 1$, i.e., that at time $t = 0$ the population consists of one member only. Put $m = 2p$. Prove that $\eta_n = \zeta_n / m^n$ is a martingale, and $\lim_{n \to +\infty} \eta_n$ exists almost surely if $p > 1/2$.

Hint: Clearly, $\zeta_{n+1} = \xi_1 + \xi_2 + \cdots + \xi_{\zeta_n}$, where the random variables ξ_k $(k = 1, 2, \ldots)$ are independent, each takes on the value 2 or 0 with probabilities p and q respectively, and they are independent from ζ_n. Thus

$$E\left(\frac{\zeta_{n+1}}{m^{n+1}} \,\middle|\, \zeta_1, \ldots, \zeta_n\right) = \frac{2p\zeta_n}{m^{n+1}} = \frac{\zeta_n}{m^n}$$

i.e., $\eta_n = \zeta_n / m^n$ is in fact a martingale. As

$$E(\eta_n^2) = 2(1 - p) \sum_{k=1}^{n} \frac{1}{m^k}$$

it follows that if $p > 1/2$, i.e., $m > 1$, $E(\eta_n^2)$ is bounded; thus by Theorem 5.4.1, if $m > 1$, $\lim_{n \to \infty} \eta_n$ exists with probability 1.

Remark: The process $\{\zeta_n\}$ is a special branching process. For the general theory of branching processes see Harris [31].

P.5.5. Let $\{\xi_n\}$ be a homogeneous stationary Markov chain with values $1, 2, \ldots, r$, with transition probabilities $p_{j,k} > 0$ $(j, k = 1, 2, \ldots, r)$ and with the stationary probabilities p_k. Let the random variables π_n be defined as follows: If $\xi_1 = k_1, \ldots, \xi_n = k_n$, put $\pi_n = p_{k_1} p_{k_1,k_2} p_{k_2,k_3} \cdots p_{k_{n-1},k_n}$. Thus, π_n is the probability of the n-tuple $(\xi_1, \ldots, \xi_n)$ observed. Prove that almost surely

$$\lim_{n \to +\infty} \frac{1}{n} \log_2 \frac{1}{\pi_n} = H \qquad \text{(P.5.5.1)}$$

where

$$H = \sum_{j=1}^{r} p_j \left(\sum_{k=1}^{r} p_{j,k} \log_2 \frac{1}{p_{j,k}} \right)$$

Hint: Let us realize the Markov chain $\{\xi_n\}$ on a probability space $S = (\Omega, \mathscr{A}, P)$ where Ω is the set of all sequences $\omega = (\omega_1, \omega_2, \ldots, \omega_n, \ldots)$ where all ω_i are positive integers $\leqq r$, and $\xi_n(\omega) = c_n(\omega)$ (see Example 5.5.1). Let T denote the shift transformation $T\omega = (\omega_2, \omega_3, \ldots)$. Then T leaves the probability measure P invariant. Define the function $f(\omega)$ as follows:

$$f(\omega) = \log_2 \frac{1}{p_{\omega_1,\omega_2}}$$

Then we have

$$\frac{1}{n} \log_2 \frac{1}{\pi_n} = \frac{1}{n} \log_2 \frac{1}{p_{\omega_1}} + \frac{1}{n} \sum_{k=1}^{n-1} f(T^{k-1}\omega)$$

As T is mixing and therefore ergodic, and

$$E(f(\omega)) = H$$

(P.5.5.1) follows from Birkhoff's ergodic theorem (Theorem 4.4.7).

Remark: The statement of this problem is a special case of McMillan's theorem, mentioned in Chapter 4. H is called the entropy of the Markov chain $\{\xi_n\}$.

P.5.6. Show that if $\sum_{n=1}^{\infty} c_n^2 = +\infty$, the series $\sum_{n=1}^{\infty} c_n R_n(x)$, where $R_n(x)$ is the nth Rademacher function, is almost everywhere divergent.

Hint: Using Theorem 5.3.2, we get that for $0 < \varepsilon < 1$,

$$P\left(\sup_{1 \leqq m \leqq M} \left| \sum_{n=N+1}^{N+m} c_n R_n(x) \right| > \sqrt{(1-\varepsilon) \sum_{N+1}^{N+M} c_n^2 - 1} \right) \geqq \varepsilon$$

If $\sum_{n=1}^{\infty} c_n^2 = +\infty$, we can choose a subsequence n_k of the positive integers such that $\sum_{n=n_k+1}^{n_{k+1}} c_n^2 > 4/1 - \varepsilon$. It follows that

$$P\left(\max_{n_k < n \leqq n_{k+1}} \left| \sum_{n_k+1}^{n} c_l R_l(x) \right| > 1 \right) \geqq \varepsilon > 0$$

As the sums $\sum_{n_{k+1}}^{n_{k+1}} c_l R_l(x)$ are independent, we get, by the Borel-Cantelli lemma, that for almost all x for an infinity of values of k,

$$\max_{n_k < n \leqq n_{k+1}} \left| \sum_{n_k+1}^{n} c_l R_l(x) \right| > 1$$

Thus the series $\sum_{l=1}^{+\infty} c_l R(x)$ is almost surely divergent.

We can even prove somewhat more. Let A denote the set on which

$$\liminf_{N\to+\infty} \sum_{n=1}^{N} c_n R_n(x) = -\infty$$

and let B denote the set on which $\limsup_{N\to+\infty} \sum_{n=1}^{N} c_n R_n(x) = +\infty$. It follows from the above proof that

$$\limsup_{N\to+\infty} \left| \sum_{n=1}^{N} c_n R_n(x) \right| = +\infty$$

almost everywhere. Thus $P(A+B)=1$. As, however, by symmetry, $P(A) = P(B)$ and by the zero-one law (Theorem 5.4.3), $P(A)=0$ or 1, it follows that $P(A)=P(B)=1$.

P.5.7. Let $\{\xi_n\}$ be a sequence of pairwise independent random variables and let η be an arbitrary random variable having finite variance. Show that if $D^2(\eta \mid \xi_n)$ denotes the conditional variance of η given ξ_n (i.e., if $D^2(\eta \mid \xi_n) = D^2(\eta \mid \mathscr{A}_{\xi_n})$, (see Definition 5.1.5), then we have

$$\sum_{n=1}^{\infty} D^2(\eta \mid \xi_n) \leqq D^2(\eta) \tag{P.5.7.1}$$

Hint: We have by definition

$$D^2(\eta \mid \xi_n) = E([E(\eta \mid \xi_n) - E(\eta)]^2)$$

If $g(x)$ is any Borel function of x such that $E(g(\xi_n)) = 0$ and $D(g(\xi_n)) = 1$, then the random variables $g(\xi_n)$ form an orthonormal system and thus by Bessel's inequality we have for every $\eta \in L_2(S)$,

$$\sum_{n=1}^{\infty} E^2([\eta - E(\eta)]g(\xi_n)) \leqq D^2(\eta)$$

Choosing

$$g(\xi_n) = \frac{E(\eta \mid \xi_n) - E(\eta)}{D(\eta \mid \xi_n)}$$

we get

$$\sum_{n=1}^{\infty} \frac{E^2([\eta - E(\eta)][E(\eta \mid \xi_n) - E(\eta)])}{D^2(\eta \mid \xi_n)} \leqq D^2(\eta)$$

By the definition of conditional expectation, we have

$$E([\eta - E(\eta)][E(\eta \mid \xi_n) - E(\eta)]) = D^2(\eta \mid \xi_n)$$

Thus we get (P.5.7.1)

Remark: It follows that $\lim_{n\to+\infty} D^2(\eta \mid \xi_n) = 0$, which implies that $E(\eta \mid \xi_n) \Rightarrow E(\eta)$. This means that the sequence $\{\xi_n\}$ is quasi-mixing according to Definition 5.8.4.

P.5.8. (a) Let $\{A_n\}$ be a qualitatively strongly independent sequence of events in the probability space $S = (\Omega, \mathscr{A}, P)$. Suppose that $0 < P(A_k) < 1$ ($k = 1, 2, \ldots$) and there exists a constant $K > 1$ such that for every n one has

$$\sum \frac{P^2(A_1^{\varepsilon_1} A_2^{\varepsilon_2} \cdots A_n^{\varepsilon_n})}{P(A_1^{\varepsilon_1}) \cdots P(A_n^{\varepsilon_n})} \leqq K$$

where the summation is extended over all n-tuples $(\varepsilon_1, \ldots, \varepsilon_n)$ consisting of $+1$ and -1. Suppose further that the series

$$\sum_{k=1}^{+\infty} P(A_k)(1 - P(A_k))$$

is divergent. Let α_k denote the indicator of the event A_k ($k = 1, 2, \ldots$) and put $\zeta_n = \alpha_1 + \cdots + \alpha_n$, $E_n = \sum_{k=1}^{n} P(A_k)$ and $D_n = \sqrt{\sum_{k=1}^{n} P(A_k)(1 - P(A_k))}$. Show that

$$\lim_{n \to +\infty} P\left(\frac{\zeta_n - E_n}{D_n} < x\right) = \Phi(x)$$

for every real x.

(b) Show that if we put $\gamma_n = P(A_1^{\varepsilon_1} \cdots A_n^{\varepsilon_n})/(P(A_1^{\varepsilon_1}) \cdots P(A_n^{\varepsilon_n}))$ for $\omega \in A_1^{\varepsilon_1} \cdots A_n^{\varepsilon_n}$ ($\varepsilon_i = \pm 1$, $1 \leqq i \leqq n$; $n = 1, 2, \ldots$), then $\lim_{n \to +\infty} \gamma_n = \gamma$ exists with probability 1.

Hint: Let $\mathscr{A}_\infty$ be the least σ-algebra containing all events A_n ($n \geqq 1$). Let P^* be the measure on $\mathscr{A}_\infty$, with respect to which the events A_n are independent and $P^*(A_n) = P(A_n)$. (Such a measure exists according to Theorem 3.3.2.) It is easy to see that $\{\gamma_n\}$ is a martingale on $(\Omega, \mathscr{A}_\infty, P^*)$ and, denoting by E^* the expectation with respect to the measure P^*, $E^*(\gamma_n^2) \leqq K$. Thus, $\lim_{n \to +\infty} \gamma_n = \gamma$ exists and $\gamma_n = E(\gamma \mid \mathscr{A}_n)$ where $\mathscr{A}_n$ is the algebra generated by $A_1, \ldots, A_n$. This implies $P(A) = \int_A \gamma \, dP^*$ for every $A \in \mathscr{A}_\infty$, i.e., that P is absolutely continuous with respect to P^*. Now the random variables $\alpha_n - P(A_n)$ fulfill, on $(\Omega, \mathscr{A}, P^*)$, the condition of Lindeberg's theorem. It follows that the distribution of $(\zeta_n - E_n)/D_n$ in $(\Omega, \mathscr{A}_\infty, P^*)$ tends to the standard normal distribution. According to Theorem 5.10.3, the same holds in $(\Omega, \mathscr{A}, P)$.

P.5.9. Show that there exists for every $k \geqq 1$, a sequence of events $\{A_n\}$ such that $\{A_n\}$ is k-mixing but not $(k + 1)$-mixing.

Hint: If $k = 1$, a simple example is the following: Let the sets $\{C_n\}$ be independent and each have the probability 1/2. Let us define the sequence A_n as follows:

$$A_n = \begin{cases} C_n & \text{if} \quad n \text{ is not a square} \\ \overline{C}_l & \text{if} \quad n = l^2, \ l = 1, 2, \ldots \end{cases}$$

Then $\{A_n\}$ is mixing with density 1/2, because both $\{C_n\}$ and $\{\overline{C}_n\}$ have this property, but it is not 2-mixing, because $P(BA_l A_{l^2}) = 0$ ($l = 2, 3, \ldots$) if l is not a square.

For $k = 2$ we construct the sequence $\{A_n\}$ as follows:

$$A_n = \begin{cases} C_n & \text{for all } n \text{ except for } n = l^3 \text{ where } l = 2, 3, \ldots \\ C_l C_{l^2} + \overline{C}_l \overline{C}_l & \text{if } n = l^3, \ l = 2, 3, \ldots \end{cases}$$

Then $\{A_n\}$ is 2-mixing with density 1/2 and $P(BA_{n_1}A_{n_2}A_{n_3}) \to P(B)/8$ if $n_1 \to +\infty$, $n_1 < n_2 < n_3$ and none of n_1, n_2, n_3 is a cube, but

$$\lim_{k\to+\infty} P(BA_kA_{k^2}A_{k^3}) = P(B)/4.$$

Thus, $\{A_n\}$ is not 3-mixing.

This construction can be generalized for any $k \geqq 3$. For instance, let C_l again be a sequence of independent events with probability 1/2, put $C(a, b) = C_aC_b + \bar{C}_a\bar{C}_b$ and let the sequence $\{A_n\}$ consist of the events $C(l, l+1)$ $(l = 1, 2, \ldots)$ and $C(l, l+k)$ $(l = 1, 2, \ldots)$, suitably arranged. Then $\{A_n\}$ is k-mixing but not $(k+1)$-mixing.

Remark: For further examples of this kind see [16.]

P.5.10. Let $\{\xi_n(\omega)\}$ be a sequence of independent random variables on a probability space $S = (\Omega, \mathscr{A}, P)$ such that each $\xi_n(\omega)$ has a standard normal distribution. Let $\{W_n(x)\}$ denote the system of Walsh functions and define the random variables $\eta(t)$ as follows:

$$\eta(t) = \sum_{n=1}^{+\infty} \xi_n(\omega) \int_0^t W_n(x)\,dx \qquad (0 \leqq t < 1) \tag{P.5.10.1}$$

Prove the following statements:

(a) For each s and t $(0 \leqq s < t < 1)$, $\eta(t) - \eta(s)$ has the normal distribution $N(0, \sqrt{t-s})$.

(b) If $0 \leqq s_1 < t_1 < s_2 < t_2 < \cdots < s_k < t_k < 1$, the random variables $\eta(t_j) - \eta(s_j)$ $(j = 1, 2, \ldots, k)$ are independent.

(c) One has $E(\eta(s)\eta(t)) = s$ if $0 < s < t < 1$.

(d) With probability 1, $\eta(t)$ is a continuous function of t.

Remark: The system $\{\eta(t)\}$ of random variables $(0 \leqq t < 1)$ is called a Wiener process or *Brownian movement* process. It describes, e.g., the behavior of any of the coordinates of a particle carrying out a Brownian movement, or of the dust particles described already by Lucretius (see Section 4.1).

Hint: The series defining $\eta(t)$ is almost surely convergent because, denoting by $e_t(x)$ $(0 < x < 1, 0 < t < 1)$ the indicator of the interval $(0, t)$, and taking into account that

$$\int_0^t W_n(x)\,dx = \int_0^1 e_t(x) W_n(x)\,dx$$

are the Fourier-Walsh coefficients of the function $e_t(x)$, we get from the Parseval relation (using the fact that $\{W_n(x)\}$ is a complete orthonormal system) that

$$\sum_{n=0}^{\infty} \left(\int_0^t W_n(x)\,dx\right)^2 = t$$

This also implies $E(\eta^2(t)) = t$.

The Parseval relation also yields, for $0 < s < t < 1$,

$$E(\eta(s)\eta(t)) = \sum_{n=1}^{\infty} \left(\int_0^t W_n(x)\,dx\right)\left(\int_0^s W_n(x)\,dx\right) = \int_0^1 e_s(x)e_t(x)\,dx = s$$

It follows that, for $0 < s_1 < t_1 < s_2 < t_2$,

$$E([\eta(t_1) - \eta(s_1)][\eta(t_2) - \eta(s_2)]) = 0$$

and

$$E([\eta(t) - \eta(s)]^2) = t - s \qquad \text{if } s < t$$

Now, as the sum of independent normally distributed random variables also has a normal distribution, $\eta(t) - \eta(s)$ is $N(0, \sqrt{t-s})$-distributed. Similarly, we get that the joint distribution of $(\eta(t_j) - \eta(s_j))$ $(1 \leqq j \leqq k)$ is a k-dimensional normal distribution. As the components of a k-dimensional normally distributed vector are independent if and only if they are uncorrelated (see Exercise E.5.1), it follows that the random variables $\eta(t_j) - \eta(s_j)$ $(1 \leqq j \leqq k)$ are independent.

The almost sure continuity of $\eta(t)$, as a function of t, can be proved as follows: If $2^s \leqq n < 2^{s+1}$, then $W_n(x) = R_{s+1}(x) W_{n-2^s}(x)$ where $W_{n-2^s}(x)$ is a product of some of the Rademacher functions $R_k(x)$ with $k \leqq s$, and thus is constant on every interval of the form $(r/2^s, (r+1)/2^s)$. On the other hand, the indefinite integral of $R_{s+1}(x)$ over such an interval $(r/2^s, (r+1)/2^s)$ increases linearly from 0 to $1/2^{s+1}$ and then decreases linearly to 0. It follows that

$$\sum_{n=2^s}^{2^{s+1}-1} \xi_n \int_0^t W_n(x)\, dx$$

is for fixed $\omega \in \Omega$ a continuous function of t such that on every interval $(r/2^s, (r+1)/2^s)$ it varies between $\pm\sum_{n=2^s}^{2^{s+1}-1} \xi_n \varepsilon_{nr}$, where $\varepsilon_{nr} = W_{n-2^s}((r + 1/2)/2^s)$, and thus $\varepsilon_{nr} = +1$ or $= -1$. Now the sum $\sum_{n=2^s}^{2^{s+1}-1} \xi_n \varepsilon_{nr}$ is normally distributed with variance 2^s, thus, putting $\delta_{sr} = \sum_{n=2^s}^{2^{s+1}-1} \xi_n \varepsilon_{nr}$,

$$P(\delta_{sr} > s2^{s/2}) < e^{-s^2/2}$$

It follows that

$$P(\max_{1 \leqq r \leqq 2^s} |\delta_{sr}| > s2^{s/2}) < 2^s e^{-s^2/2}$$

As the series $\sum_{s=1}^{\infty} 2^s e^{-s^2/2}$ is convergent, it follows by the Borel-Cantelli lemma that for almost all $\omega \in \Omega$, $\max_{1 \leqq r \leqq 2^s} |\delta_{sr}| < s2^{s/2}$ for all but a finite number of values of s. This implies that for almost all values of ω, one has uniformly for $0 \leqq t < 1$,

$$\sum_{s=1}^{\infty} \left| \sum_{n=2^s}^{2^{s+1}} \xi_n(\omega) \int_0^t W_n(x)\, dx \right| \leqq \sum_{s=1}^{\infty} \frac{s}{2^{s/2+1}} < +\infty$$

Thus, $\eta(t)$ is for almost all ω the sum of a uniformly convergent series of continuous functions, i.e., it is almost surely a continuous function of t.

Remark: The idea of the above sketched construction of the Wiener process is due to Wiener and Lévy. Wiener used the trigonometric system, while Lévy used Haar functions instead of Walsh functions. The continuity of $\eta(t)$ has for this case been proved by Ciesielski [51]. (See also Lamperti [32].) The construction can be easily modified to yield the Wiener process on the whole real line. As regards the uniform convergence of the series (P.5.10.1) when instead of the Walsh functions a general orthonormal system of functions is used, see Ito and Nisio [49].

REFERENCES

[1] A. N. Kolmogoroff, *Grundbegriffe der Wahrscheinlichkeitsrechnung*, Springer, Berlin, 1933.

[2] J. L. Doob, *Stochastic Processes*, Wiley, New York, 1953.

[3] K. L. Chung, *Markov Chains with Stationary Transition Probabilities*, 2nd edition, Springer, Berlin, 1967.

[4] J. Ville, *Étude Critique de la Notion de Collectif*, Gauthier-Villars, Paris, 1939.

[5] P. Lévy, *Théorie de l'Addition des Variables Aléatoires*, Gauthier-Villars, Paris, 1937.

[6] K. Krickeberg, *Probability Theory*, Addison-Wesley, Reading, 1965.

[7] J. Neveu, *Mathematical Foundations of the Calculus of Probability*, Holden-Day, Inc., San Francisco, 1965.

[8] A. Haar, "Zur Theorie der orthogonalen Funktionssysteme," *Math. Annalen*, **69**: 331–371, 1910; **71**: 33–53., 1912. (See also A. Haar, *Gesammelte Arbeiten*, Akadémiai Kiadó, Budapest, 1959.

[9] J. Hájek and A. Rényi, "Generalization of an Inequality of Kolmogorov," *Acta Math. Acad. Sci., Hung.*, **6**: 281–283, 1955.

[10] F. Riesz and B. Sz. Nagy, *Functional Analysis*, Blackie, London, 1956.

[11] Yu.V. Linnik, "An Information Theoretic Proof of the Central Limit Theorem on Lindeberg's Conditions," *Teoria Ver. Prim.*, **4**: 311–321, 1959.

[12] F. R. Gantmacher, *Matrizenrechnung*, VEB Deutscher Verlag der Wissenschaften, Berlin, 1959.

[13] A. Rényi, "On Stable Sequences of Events," *Sankhya*, **25**: 293–302, 1963.

[14] W. Schmeidler, *Lineare Operatoren im Hilbertschen Raum*, Teubner, Stuttgart, 1954.

[15] A. Rényi, "On Mixing Sequences of Sets," *Acta Math. Acad. Sci. Hung.*, **9**: 215–228, 1958.

[16] N. Friedman and A. Rényi, *On k-Stable and k-Mixing Sequences of Events* (in print).

[17] P. Révész and A. Rényi, "A Study of Sequences of Equivalent Events as Special Stable Sequences," *Publicationes Math.*, **10**: 319–325, 1963.

[18] B. de Finetti, "La Prévision: ses Lois Logiques, ses Sources Subjectives," *Annales Inst. Poincaré*, **7**: 1–68, 1937.

[19] G. Pólya, "Sur Quelques Points de la Théorie des Probabilités," *Annales Inst. Poincaré*, **1**: 117–61, 1931.

[20] F. Hausdorff, "Momentenprobleme für ein endliches Intervall," *Math. Zeitschrift*, **16**: 220–248, 1923.

[21] D. G. Kendall, "On Finite and Infinite Sequences of Exchangeable Events," *Studia Sci. Math. Hung.*, **2**: 319–327, 1967.

[22] R. R. Phelps, *Lectures on Choquet's Theorem*, Van Nostrand, Princeton, 1966.

[23] A. Rényi, "On Theory of the Limit Laws for Sums of Independent Random Variables" (in Russian), *Acta Math. Acad. Sci. Hung.*, **1**: 99–108, 1950.

[24] A. N. Kolmogoroff, "A Theorem on the Convergence of Conditional Expectations and Applications" (in Russian), *Proceedings of the First Hungarian Congress of Mathematicians*, pp. 377–386, Akadémiai Kiadó, Budapest, 1950.

[25] L. E. Dubins and L. J. Savage, *How to Gamble if You Must*, McGraw-Hill, New York, 1965.

[26] L. Pál, *Dissertation* (in print).

[27] I. A. Ibrahimov and Yu. V. Linnik, *Independent and Stationary Dependent Sequences of Random Variables*, Nauka, Moscow, 1965.

[28] B. Friedman, "A Simple Urn Model," *Comm. Pure Appl. Math*, **2**: 59–70 1949.

[29] A. Rényi, *Wahrscheinlichkeitsrechnung, mit einer Anhang über Informationstheorie*, VEB Deutscher Verlag der Wissenschaften, Berlin, 1962.

[30] D. Freedman, "L'Urne de Bernard Friedman," *Comptes Rendus Acad. Sci.* (*Paris*), **257**: 3809, 1963.

[31] T. Harris, *The Theory of Branching Processes*, Springer, Berlin, 1963.

[32] J. Lamperti, *Probability*, W. A. Benjamin, New York-Amsterdam, 1966.

[33] M. Loève, *Probability Theory*, Van Nostrand, New York, 1955.

[34] W. Feller, *An Introduction to Probability Theory and Its Applications*, Vol. 2, Wiley, New York, 1966.

[35] Y. S. Chow, "A Martingale Inequality and the Law of Large Numbers," *Proc. Am. Math. Soc.*, **11**: 107–111, 1960.

[36] D. G. Kendall, "Information Theory and the Limit Theorems for Markov Chains and Processes With a Countable Infinity of States," *Annals Inst. Statistical Math.*, **15**: 137–143, 1964.

[37] I. Csiszár, "Eine Informationstheoretische Ungleichung und ihre Anwendung auf den Beweis der Ergodizität von Markoffschen Ketten," *Magyar Tud. Akad. Mat. Kut. Int. Közl.*, **8**: 85–108, 1963.

[38] C. Ryll-Nardzewski, "Remarque sur un Théorème de A. Rényi," *Colloquium Mathematicum*, **2**: 319–320, 1951.

[39] L. Sucheston, "On Mixing and the 0-1 Law," *J. Math. Analysis Applications*, **6**: 447–456, 1964.

[40] R. M. Fischler, "Borel-Cantelli Type Theorems for Mixing Sets," *Acta Mat. Acad. Sci. Hung.*, **18**: 67–69, 1967.

[41] R. M. Fischler, "The Strong Law of Large Numbers for Indicators of Mixing Sequences," *Acta Math. Acad. Sci. Hung.*, **18**: 71–81, 1967.

[42] P. Révész, "A Limit Distribution Theorem for Sums of Dependent Random Variables," *Acta Math. Acad. Sci. Hung.*, **10**: 125–131, 1959.

[43] J. H. Abbott and J. R. Blum, "On a Theorem of Rényi Concerning Mixing Sequences of Sets," *Annals, Math. Stat.*, **32**: 257–260, 1961.

[44] E. Parzen, *Modern Probability Theory and its Applications*, Wiley, 1960.

[45] M. Rosenblatt, "A Central Limit Theorem and a Strong Mixing Condition," *Proc. Nat. Acad. Sci. USA*, **42**: 43, 1956.

[46] T. Nemetz and G. Varga, "On the Extension of Limit Theorems" (in Hungarian), *Magyar Tud. Akad. III. Oszt. Közl.*, **14**: 415–421, 1964.

[47] J. R. Blum, D. L. Hanson and L. H. Koopmans, "On the Strong Law of Large Numbers for a Class of Stochastic Processes," *Zeitschrift Wahrscheinlichkeitstheorie Gebiete*, **2**: 1–11, 1963.

[48] S. Kakutani, "On Equivalence of Infinite Product Measures," *Ann. Math.*, **49**: 214–226, 1948.

[49] K. Ito and M. Nisio, *Osaka Mathematical Journal*, **5**: 35–48, 1968.

[50] D. Waterman, "On a Problem of Steinhaus," *Studia Sci. Math. Hung.* (in print).

[51] Z. Ciesielski, "Hölder Condition for Realizations of Gaussian Processes," *Trans. Am. Math. Soc.*, **99**: 403–413, 1961.

[52] L. A. Shepp, "Radon-Nikodym Derivatives of Gaussian Measures," *Ann. Math. Statistics*, **37**: 321–354, 1966.

[53] J. Delporte, "Fonctions Aléatoires presque Sûrement Continues sur un Interval Fermé," *Ann. Inst. H. Poincaré, Sec. B*, **1**: 111–215, 1964.

APPENDIX A

ON MEASURE THEORY*

A.1 ON THE EXTENSION OF MEASURES

Several times we have used the following theorem on the extension of a measure:

EXTENSION THEOREM. *Let P_0 be a probability measure on an algebra $\mathscr{A}_0$ of subsets of a nonempty set Ω. Let $\mathscr{A}$ denote the least σ-algebra containing $\mathscr{A}_0$. Then P_0 can be extended to a measure P on $\mathscr{A}$, i.e., there exists on $\mathscr{A}$ a probability measure P which coincides with P_0 on $\mathscr{A}_0$: this measure P is uniquely determined.*

We give here a short sketch of the classical proof of this theorem—due to Carathéodory. (For the details see [1]–[3].) One first defines the outer measure $P^*(A)$, for *every* subset of Ω as follows: $P^*(A)$ is the greatest lower bound of the sum $\sum_k P_0(A_k)$ where $A_k \in \mathscr{A}_0$ $(k = 1, 2, \ldots)$ and $A \subseteq \sum_k A_k$; in other words, we consider all possible *coverings* of the set A by a finite or countably infinite sequence of (not necessarily disjoint) sets belonging to $\mathscr{A}_0$, and take the lower bound of the sum of the P_0 measure of the sets forming such a covering. The outer measure of $A \in \mathscr{A}_0$ coincides with $P_0(A)$; in general, the outer measure is not even finitely additive, but it is subadditive in the sense that one has

$$P^*\Big(\sum_k C_k\Big) \leqq \sum_k P^*(C_k) \tag{A.1.1}$$

In particular, one has for each subset C of Ω,

$$P^*(C) + P^*(\overline{C}) \geqq P^*(\Omega) = P_0(\Omega) = 1 \tag{A.1.2}$$

We call those sets C, for which there is equality in (A.1.2), i.e., for which

$$P^*(C) + P^*(\overline{C}) = 1 \tag{A.1.3}$$

* Throughout this book we have taken it for granted that the reader is familiar with the basic notions of measure theory. To help those readers who found during reading this book that they need to refresh their knowledge of measure theory, we have collected in this appendix those basic definitions and theorems which are used in this book. However, this appendix does not aim to replace a textbook of measure theory; as a rule, we do not give full proofs, we sketch only the underlying basic ideas. We give, however, quite a number of references to other books (at the end of Appendix B) which the reader should consult for a detailed treatment of measure theory, the theory of the Lebesgue integral, etc.

holds, *P^*-measurable.* Evidently, every $A \in \mathscr{A}_0$ is P^*-measurable. It can be shown that the family $\mathscr{A}^*$ of P^*-measurable sets is a σ-algebra, and thus $\mathscr{A}^* \supseteq \mathscr{A}$. It can be further shown that P^* is a measure on $\mathscr{A}^*$ and $P^*(A) = P_0(A)$ for $A \in \mathscr{A}_0$. Putting $P(A) = P^*(A)$ for $A \in \mathscr{A}$, we arrive at an extension of the measure P_0 to $\mathscr{A}$. The uniqueness of the extension is easily shown, as follows: Let P and Q be two measures on $\mathscr{A}$ which coincide on $\mathscr{A}_0$. Let $\mathscr{A}'$ denote the family of those sets $A \in \mathscr{A}$ for which $P(A) = Q(A)$. As both P and Q are by supposition measures, it follows that $\mathscr{A}'$ is a σ-algebra and as $\mathscr{A}_0 \subseteq \mathscr{A}'$, it follows that $\mathscr{A}' = \mathscr{A}$.

The extension theorem can be generalized as follows: If μ_0 is a σ-finite measure on a ring $\mathscr{R}$, it can be extended to a σ-finite measure (the value of which for any set is either a nonnegative number or $+\infty$) on the least σ-algebra $\mathscr{A}$ containing the ring $\mathscr{R}$. The proof is essentially the same; one first defines a corresponding outer measure μ^*. The definition of μ^*-measurable sets has to be modified in case $\mu^*(\Omega) = +\infty$, as follows: A subset C of Ω is called μ^*-measurable if for every subset $A \in \mathscr{A}_0$ one has

$$\mu^*(A) = \mu^*(AC) + \mu^*(A\bar{C}) \tag{A.1.4}$$

Let us consider the particular case when Ω is the real axis, $\Omega = R$, the ring $\mathscr{R}$ is the family of all subsets of R which are the union of a finite number of intervals, and μ_0 is defined by putting $\mu_0([a, b)) = F(b) - F(a)$ for $a < b$, where $F(x)$ is a nondecreasing function, continuous from the left in every point, and putting $\mu_0(\sum_{k=1}^n I_k) = \sum_{k=1}^n \mu_0(I_k)$ if the intervals I_k $(1 \leqq k \leqq n)$ are disjoint. One has to prove first that μ_0 is a measure on $\mathscr{R}$ (see, e.g., [8], pp. 40–41). After this the extension theorem can be applied; the measure μ thus obtained on the family of Borel subsets of the real line is called the Lebesgue-Stieltjes measure corresponding to the function F, and one writes

$$\mu(A) = \int_A dF(x)$$

Especially, if $F(x) = x$, we obtain the (ordinary) *Lebesgue measure*, which is the straightforward extension of the notion of length, the Lebesgue measure of an interval being equal to its length. The Lebesgue measure of subsets of the interval $[0, 1)$ can be obtained from the extension theorem for probability measures by extending the probability measure P_0 on the algebra $\mathscr{A}_0$ of subsets of the interval $[0, 1)$, consisting of those subsets of $[0, 1)$ which are the union of a finite number of disjoint subintervals of $[0, 1)$, and the measure $P_0(A)$ is defined as the sum of the lengths of the subintervals forming A. In this way we get directly the Lebesgue probability space.

Notice that the above described extension procedure through the outer measure leads to the extension of the measure P_0 to a σ-algebra $\mathscr{A}^*$ which is larger than the least σ-algebra $\mathscr{A}$ containing $\mathscr{A}_0$. For instance, the σ-algebra $\mathscr{A}^*$ has the property that it contains all subsets of a set of measure 0, because these evidently have outer measure 0 and their complement the outer measure 1; thus such sets satisfy (A.1.3), i.e., are measurable, and have measure 0. Thus the extended measure P is *complete* on $\mathscr{A}^*$. In particular, the Lebesgue measure is defined not only for all Borel sets, but for a more extensive class, called the class of Lebesgue-measurable sets, and is a complete measure on the σ-algebra of these sets.

The k-dimensional Lebesgue measure ($k = 2, 3, \ldots$) is defined similarly, by extending the k-dimensional volume: e.g., the two-dimensional Lebesgue measure is obtained by extending the area, defined first for rectangles on the (x, y) plane. The k-dimensional Euclidean space with the k-dimensional Lebesgue measure is the product of k exemplars of the real line with the Lebesgue measure, in the sense of Section 3.4.

The extension procedure preserves the invariance properties of the measure involved: If the measure P_0 is invariant with respect to the $\mathscr{A}_0$-measurable transformation T of the space Ω, then P is also invariant with respect to T (being also $\mathscr{A}$-measurable). Thus, in particular, the k-dimensional Lebesgue measure is invariant with respect to all motions of the Euclidean space R^k. The Lebesgue measure in the interval $[0, 1)$ is invariant with respect to every shift mod 1. The Lebesgue measure on the circumference of a circle (obtained by extending the arc length) is invariant with respect to the rotations of the circle.

The Lebesgue-Stieltjes measures have the following property: The measure of any measurable set A is equal to the greatest lower bound of the measures of open sets G containing A, and also equal to the least upper bound of the measures of compact sets C contained in A. If any measure P in a topological space has this property, it is called a *regular measure*.

A.2 THE LEBESGUE INTEGRAL

Let $S = (\Omega, \mathscr{A}, P)$ be a probability space. Let $f(\omega)$ be a real-valued $\mathscr{A}$-measurable function (i.e., a random variable) on S. The Lebesgue integral of $f = f(\omega)$ on Ω with respect to the measure P, denoted by $\int_\Omega f\,dP$, can be defined as follows: If I is an interval on the real line, denoting by $f^{-1}(I)$ the set of all $\omega \in \Omega$ for which $f(\omega) \in I$, we put

$$\int_\Omega f\,dP = \lim_{\varepsilon \to 0} \sum_{k=-\infty}^{+\infty} k\varepsilon P(f^{-1}([k\varepsilon, (k+1)\varepsilon)) \tag{A.2.1}$$

provided that the series on the right of (A.2.1) is absolutely convergent for every $\varepsilon > 0$ and its limit exists for $\varepsilon \to 0$. It can be shown that if the series on the right of (A.2.1) is absolutely convergent for *some* $\varepsilon > 0$, then it is convergent for *every* $\varepsilon > 0$ and its limit for $\varepsilon \to 0$ exists. In this case, f is called *integrable*. If the series on the right of (A.2.1) does not converge absolutely for any $\varepsilon > 0$, we say that f is not integrable. It can be shown that if f and $g \geqq 0$ are measurable functions on S, and $|f(\omega)| \leqq g(\omega)$ for all $\omega \in \Omega$ (or at least almost surely, i.e., except on a set of measure 0) and g is integrable, then f is also integrable and $|\int_\Omega f\,dP| \leqq \int_\Omega g\,dP$. As every constant function c is evidently integrable and $\int_\Omega c\,dP = c$, it follows that every bounded measurable function is integrable, and if $|f(\omega)| \leqq c$ almost surely, then $|\int_\Omega f\,dP| \leqq c$. It follows also from the definition that if $f(\omega)$ is a measurable function, either $f(\omega)$ and $|f(\omega)|$ are both integrable or none of them are. The main properties of the integral are the following: It is a linear functional, i.e., if f and g are integrable functions, and a and b are constants, then $af + bg$ is integrable and $\int_\Omega(af + bg)dP = a \int_\Omega fdP + b \int_\Omega gdP$. If f is almost surely nonnegative, its integral (if it exists) is nonnegative, and it is equal to 0 if and only if f itself is almost surely equal to 0. If two measurable functions are almost surely equal, then if one of them is integrable, so is the other and their integrals are equal. The integral of an integrable function f over a set $A \in \mathscr{A}$, denoted by $\int_A fdP$, is defined as follows:

$$\int_A fdP = \int_\Omega fi_A dP$$

where $i_A = i_A(\omega)$ is the indicator of the set A. It can be shown that if f is integrable, then fi_A is also integrable, i.e., $\int_A fdP$ exists. The set function

$$\mu(A) = \int_A fdP$$

is a signed measure (i.e., a real-valued σ-additive set function) on $\mathscr{A}$; if $f \geqq 0$, then $\mu(A)$ is a finite measure on $\mathscr{A}$.

The definition of the Lebesgue integral $\int_\Omega fd\mu$ for the case when μ is an arbitrary finite measure, is exactly the same. If the measure μ is not finite, only σ-finite, the same definition is still applicable. Of course, if $\mu(\Omega) = +\infty$, then the constants are no longer integrable. Another somewhat more convenient way is to reduce this case to the case of a finite measure, as follows: We start from a partition of Ω into sets $\Omega_k \in \mathscr{A}$ such that $\mu(\Omega_k) < +\infty$ $(k = 1, 2, \ldots)$ and define $\int_\Omega fd\mu$ as

$$\int_\Omega fd\mu = \sum_k \int_{\Omega_k} fd\mu$$

provided that this series is absolutely convergent.

In particular, if Ω is the real line R, $\mathscr{A}$ is the σ-algebra of Borel subsets of R, and P is the Lebesgue-Stieltjes measure corresponding to a nondecreasing function $F(x)$, then the corresponding Lebesgue integral is called *the Lebesgue-Stieltjes integral*. The Lebesgue-Stieltjes integral of a continuous function $f(x)$ is denoted by $\int_{-\infty}^{+\infty} f(x)dF(x)$. If $\xi = \xi(\omega)$ is a random variable on an arbitrary probability space $S = (\Omega, \mathscr{A}, P)$, and $F(x)$ is the distribution function of ξ, then we have

$$\int_{\Omega} \xi \, dP = \int_{-\infty}^{+\infty} x dF(x) \tag{A.2.2}$$

because both sides are the limits of the same sums (with different interpretation).

An alternative way to define the Lebesgue integral in any probability space is to define first the Lebesgue-Stieltjes integral and define $\int_{\Omega} \xi \, dP$ by taking (A.2.2) as its definition.

There are, however, still other alternative ways to define the Lebesgue integral. For instance, one may consider the product of the probability space $S = (\Omega, \mathscr{A}, P)$ and the measure space $(R, \mathscr{B}, \lambda)$ where R is the real axis, $\mathscr{B}$ is the σ-algebra of Borel subsets and λ is the Lebesgue measure. If f is any nonnegative function, one can define its integral $\int_{\Omega} f dP$ as the measure of the set of those points (ω, x) of this product space for which $0 \leqq x < f(\omega)$ whenever this set has finite measure. If f is an arbitrary measurable function, one can define $\int_{\Omega} f dP$ as follows: We put $f^+ = \max(f, 0)$ and $f^- = \max(-f, 0)$, so that $f = f^+ - f^-$ and put

$$\int_{\Omega} f dP = \int_{\Omega} f^+ dP - \int_{\Omega} f^- dP$$

(see, e.g., Kamke [4]). This approach is very intuitive, as it defines the abstract Lebesgue integral of a nonnegative function as the "area" under the "curve" of the function, i.e., analogously as the Riemann integral of a continuous function of a real variable is usually defined.

Another alternative way due to Riesz (see Riesz-Sz. Nagy [5] and Sz. Nagy [6]) of introducing the integral, in case of functions of one or several real variables, is the following: For the sake of simplicity we deal only with functions of one real variable defined on a finite interval $[a, b]$. One starts from the family of step functions, i.e., functions which take on in $[a, b]$ only a finite number of values, and each such value is taken on in an interval. If $f(x)$ is a real-valued function which is equal to a_k in the finite interval I_k $(k = 1, 2, \ldots, n)$, these intervals forming a partition of the interval $[a, b]$, then we put

$$\int_a^b f(x)dx = \sum_{k=1}^{n} a_k |I_k| \tag{A.2.3}$$

where $|I_k|$ denotes the length of the interval I_k. If $f(x)$ is a measurable function and $f_n(x)$ is a sequence of step functions, such that $f_n(x) \leqq f_{n+1}(x)$ and $\lim_{n \to +\infty} f_n(x) = f(x)$ almost everywhere, then if the integrals $\int_a^b f_n(x)dx$ are bounded, we define $\int_a^b f(x)dx$ as

$$\int_a^b f(x)dx = \lim_{n \to +\infty} \int_a^b f_n(x)dx \tag{A.2.4}$$

Of course, one has to show that the value of this limit does not depend on the choice of the sequence f_n, and that if f itself is a step function, then the limit in (A.2.4) coincides with the integral of f as a step function as defined by (A.2.3). If $f = f_1 - f_2$, where the integrals of f_1 and f_2 have already been defined, we put $\int_a^b f\,dx = \int_a^b f_1\,dx - \int_a^b f_2\,dx$. It can be shown that in this way we arrive at the class of all Lebesgue-integrable functions. This approach can also be applied for Euclidean spaces of more than one dimension. A similar way, starting from the integral of certain "elementary functions" and extending the notion by passing to the limit, can be applied in the case of an abstract space, following Daniell (see, e.g., [7]). In this approach the extension of the measure is avoided; the Lebesgue measure is introduced as a particular case of the Lebesgue integral: a set $A \in \mathscr{A}$ is called measurable if its indicator is an integrable function, and its measure is defined as the integral of its indicator. While this approach certainly has its advantages, from the point of view of probability theory the approach described first is the more natural one, because in probability theory the measure is already given when one comes to the definition of expectation. Of course, in principle, the logical order could be reversed even in this case; one could start with the notion of expectation as the basic notion of probability theory, and obtain the notion of probability as a special case (as the expectation of the indicator of the event). One may even find some historical arguments in favor of this unusual approach; the notion of expectation has historically preceded that of probability. (For instance, the gamblers were interested mainly in the expected gain.) Nevertheless, the classical way to take the notion of probability as the primary notion is more natural.

The basic limit theorems concerning the Lebesgue integral follow.

THE LEBESGUE BOUNDED CONVERGENCE THEOREM. *If $\{f_n\}$ is a sequence of measurable functions on the probability space $S = (\Omega, \mathscr{A}, P)$ which possesses a common integrable majorant, i.e., $|f_n(\omega)| \leq g(\omega)$ for almost all $\omega \in \Omega$ and $n = 1, 2, \ldots$, and $\int_\Omega g(\omega)dP$ exists, then if $\lim_{n\to+\infty} f_n(\omega) = f(\omega)$ exists almost everywhere, it follows that f is integrable and*

$$\int_\Omega f(\omega)dP = \lim_{n\to+\infty} \int_\Omega f_n(\omega)dP$$

i.e., integration and passing to the limit can be interchanged.

THE BEPPO LEVI THEOREM. *If $\{f_n\}$ is a sequence of nonnegative integrable functions, such that the series*

$$\sum_n \int_\Omega f_n dP$$

is convergent, then the series

$$\sum_n f_n(\omega)$$

is almost surely convergent, and its sum is an integrable function f such that

$$\int_\Omega f dP = \sum_{n=1}^{\infty} \int_\Omega f_n dP$$

*i.e., the series $\sum_n f_n$ can be integrated term by term.**

THE FATOU THEOREM. *If $\{f_n\}$ is a sequence of nonnegative integrable functions such that $\lim_{n\to+\infty} f_n(\omega) = f(\omega)$ exists almost everywhere and the integrals $\int_\Omega f_n(\omega)dP$ are bounded, $\int_\Omega f_n(\omega)dP \leqq C$ for $n = 1, 2, \ldots$, where C is a positive constant, then f is also integrable and*

$$\int_\Omega f(\omega)dP \leqq \liminf_{n\to+\infty} \int_\Omega f_n(\omega)dP$$

The integral of a vector-valued or complex-valued function can be reduced to the case of real-valued functions. For example, if $f = f_1 + if_2$ is a complex-valued measurable function with real and imaginary parts f_1 and f_2, then we put $\int f dP = \int f_1 dP + i\int f_2 dP$. We shall not go into a discussion of further generalizations here (e.g., if f takes on values in a group or a Banach space), as we do not need them in the present book.

An important theorem, often used in probability theory, is the following:

FUBINI'S THEOREM. *Let the probability space $S = (\Omega, \mathscr{A}, P)$ be the product of the probability spaces $S_1 = (\Omega_1, \mathscr{A}_1, P_1)$ and $S_2 = (\Omega_2, \mathscr{A}_2, P_2)$. Let $h(\omega_1, \omega_2)$ be an integrable function on S. Then the integral $g(\omega_2) = \int_{\Omega_1} h(\omega_1, \omega_2)dP_1$ (in S_1) exists for almost all ω_2 (with respect to P_2) and similarly, the integral $f(\omega_1) = \int_{\Omega_2} h(\omega_1, \omega_2)\, dP_2$ (in S_2) exists for almost all ω_1 (with respect to P_1) and one has*

$$\int_\Omega h(\omega_1, \omega_2)dP = \int_{\Omega_2} g(\omega_2)dP_2 = \int_{\Omega_1} f(\omega_1)dP_1$$

Especially, it follows from Fubini's theorem, that if $h(\omega_1, \omega_2) = h_1(\omega_1)h_2(\omega_2)$, then

$$g(\omega_2) = h_2(\omega_2) \int_{\Omega_1} h_1(\omega_1)dP_1$$

and thus

$$\int_\Omega h_1(\omega_1)h_2(\omega_2)dP = \left(\int_{\Omega_1} h_1(\omega_1)dP_1\right)\left(\int_{\Omega_2} h_2(\omega_2)dP_2\right)$$

* Notice that the Beppo Levi theorem contains as a special case the first part of the Borel-Cantelli lemma. If $\sum P(A_n) < +\infty$, then $P(\limsup A_n) = 0$. As a matter of fact, if α_n denotes the indicator of the event A_n, then $P(A_n) = \int_\Omega \alpha_n dP$ and thus the convergence of $\sum P(A_n)$ implies that the series $\sum \alpha_n$ is almost surely convergent, i.e., with probability 1 only a finite number of the events A_n occur simultaneously.

A.3 THE RADON-NIKODYM THEOREM

Let ν be a finite signed measure and μ a finite measure on the σ-algebra $\mathscr{A}$ of subsets of a set Ω. The signed measure ν is called *absolutely continuous* with respect to μ (denoted by $\nu \ll \mu$) if $\mu(A) = 0$, $A \in \mathscr{A}$ implies $\nu(A) = 0$. A necessary and sufficient condition for $\nu \ll \mu$ is that to every $\varepsilon > 0$ there should exist a $\delta > 0$ such that if $\mu(A) < \delta$, then $|\nu(A)| < \varepsilon$.

THE RADON-NIKODYM THEOREM. *If $\nu \ll \mu$, then there exists an $\mathscr{A}$-measurable, μ-integrable function $f = f(\omega)$ $(\omega \in \Omega)$ such that*

$$\nu(A) = \int_A f\, d\mu \quad \textit{for all } A \in \mathscr{A} \tag{A.3.1}$$

The function f is uniquely defined up to a set of measure 0, i.e., if g is another function such that $\nu(A) = \int_A g\, d\mu$ for all $A \in \mathscr{A}$, then f and g are (μ-) almost everywhere equal. The function f is called the Radon-Nikodym derivative of ν with respect to μ and is denoted by $d\nu/d\mu$.

Conversely, if f is a μ-integrable function and the set function ν is defined by (A.3.1), then ν is a signed measure and $\nu \ll \mu$.

The theorem remains valid also in the more general case when μ and ν are not finite, only σ-finite. In this case, f is not necessarily integrable on Ω, only on every set $A \in \mathscr{A}$ for which $\nu(A)$ is finite.

In particular, if Ω is the real line R, $\mathscr{A}$ is the family of Borel subsets of R, μ is the Lebesgue measure and ν is the Lebesgue-Stieltjes measure corresponding to the distribution function $F(x)$, then the Radon-Nikodym derivative $d\nu/d\mu$ is equal to $F'(x)$, provided that $F(x)$ is absolutely continuous with respect to the Lebesgue measure, i.e., for every $\varepsilon > 0$, there exists a $\delta > 0$ such that if the intervals (a_k, b_k) are disjoint $(1 \leqq k \leqq n)$ and $\sum_{k=1}^{n} (b_k - a_k) < \delta$, then $\sum_{k=1}^{n} F(b_k) - F(a_k) < \varepsilon$. In this case, $F(x)$ is equal to the indefinite integral of $F'(x)$.

The Radon-Nikodym derivative possesses essentially the same properties as the ordinary derivative; e.g.,

$$\frac{d(\nu_1 + \nu_2)}{d\mu} = \frac{d\nu_1}{d\mu} + \frac{d\nu_2}{d\mu} \qquad \frac{d(c\nu)}{d\mu} = c\frac{d\nu}{d\mu}$$

if c is a constant; further, if $\nu \ll \mu$ and $\mu \ll \lambda$ where μ and λ are σ-finite measures and ν is a σ-finite signed measure, we have almost everywhere (with respect to λ), $d\nu/d\lambda = (d\nu/d\mu)(d\mu/d\lambda)$.

Thus the formula for the transformation of integrals is also valid: If λ and μ are finite measures, $\mu \ll \lambda$ and the function f is μ-integrable, then $f\, d\mu/d\lambda$ is λ-integrable and

$$\int_\Omega f\, d\mu = \int_\Omega f\,(d\mu/d\lambda) d\lambda$$

If ν and μ are two arbitrary, finite measures on the σ-algebra $\mathscr{A}$ and ν is *not* absolutely continuous with respect to μ, then ν can be decomposed into the sum $\nu = \nu_1 + \nu_2$ where ν_1 and ν_2 are finite measures on $\mathscr{A}$, $\nu_1 \ll \mu$ and ν_2 is orthogonal to μ, i.e., there exists a set $A \in \mathscr{A}$ such that $\mu(A) = 0$ and $\nu_2(\bar{A}) = 0$.

In particular, each distribution function $F(x)$ on the real line can be decomposed in the form

$$F(x) = pF_1(x) + (1 - p)F_2(x)$$

where $0 \leqq p \leqq 1$, $F_1(x)$ is an absolutely continuous distribution function and $F_2(x)$ is a singular distribution function, i.e., such that $F_2'(x) = 0$ almost everywhere. The function $F_2(x)$ can be decomposed further as $F_2(x) = cF_3(x) + (1 - c)F_4(x)$ $(0 \leqq c \leqq 1)$ where $F_3(x)$ is the distribution function of a discrete probability distribution and $F_4(x)$ is a continuous singular distribution function.

Two measures ν and μ on the σ-algebra $\mathscr{A}$ are called *equivalent* (denoted by $\nu \sim \mu$) if simultaneously $\nu \ll \mu$ and $\mu \ll \nu$. In this case $(d\mu/d\nu)(d\nu/d\mu) = 1$ almost surely, i.e., if $f = d\nu/d\mu$, then f is almost everywhere positive and

$$\frac{d\mu}{d\nu} = \frac{1}{f}$$

If μ and ν are equivalent, then every set having μ measure 0 has ν measure 0, and conversely, i.e., what is true μ-almost everywhere is true ν-almost everywhere too.

For the standard proof of the Radon-Nikodym theorem see Halmos [1]. In Appendix B we shall sketch a different proof (due to von Neumann) using notions of Hilbert space. For another variant of this proof see Krickeberg [9].

APPENDIX B

ON FUNCTIONAL ANALYSIS*

B.1 METRIC SPACES

A nonempty set X is called a *metric space* if there is given on X a function $d(x, y)$ $(x \in X, y \in X)$ having the following properties:

M.1. $d(x, y) \geqq 0$ and $d(x, y) = 0$ if and only if $x = y$.

M.2. $d(x, y) = d(y, x)$.

M.3. $d(x, y) \leqq d(x, z) + d(z, x)$ for all x, y, z in X.

Such a function $d(x, y)$ is called a *metric*. For given x and y the positive number $d(x, y)$ is called the *distance* of x and y.

The set $\{x : d(x_1 x_0) < r\}$ is called a *sphere* with center x_0 and radius $r > 0$. Sets which are the union of spheres are called *open*; their complementary sets are called *closed*. The least σ-algebra of subsets of a metric space containing all open sets is called the family of *Borel sets* of the space x. A sequence $\{x_n\}$ of elements of X is called *convergent* to the limit $x \in X$ if $\lim_{n \to +\infty} d(x_n, x) = 0$. A sequence $\{x_n\}$ $n = 1, 2, \ldots$ for which $\lim_{\substack{n \to +\infty \\ m \to +\infty}} d(x_n, x_m) = 0$ is called *fundamental*. The metric space X is called *complete* if every fundamental sequence has a limit in X (which is necessarily unique). A metric space X is called *separable* if X has a countable subset Y such that for any $x \in X$ there exists a sequence of elements of Y which converges to X. The set of points $x = (x_1, \ldots, x_n)$ of the Euclidean space R^n is a complete and separable metric space with the metric $d(x, y) = \sqrt{\sum_{k=1}^{n} (x_k - y_k)^2}$.

* The aim of this appendix is—similarly as that of Appendix A—to state certain definitions and theorems from functional analysis, the knowledge of which is necessary for a full understanding of the present book. As a rule, full proofs are not given, but some proofs are sketched. However, references to some textbooks are given which the reader should inspect if he feels that he needs to refresh his knowledge of functional analysis.

B.2 LINEAR SPACES

A set X is called a *linear space* (over the field of real numbers) if an operation $x + y$ is defined in X with respect to which X is an Abelian group,* further, the multiplication of elements $x \in X$ and of real numbers $a \in R$ (denoted by ax) is defined and has the following properties:

L.1. $a(x + y) = ax + ay$.

L.2. $(a + b)x = ax + bx$.

L.3. $(ab) \cdot x = a(bx)$.

L.4. $1 \cdot x = x$.

It follows that $-1 \cdot x = -x$ is the solution of the equation $x + z = 0$ and $a \cdot 0 = 0$.

A linear space X is called *normed* if on X a function $\|x\|$ (called the *norm* of x) is defined, having the following properties:

N.1. $\|x\| \geqq 0$ and $\|x\| = 0$ if and only if $x = 0$.

N.2. If a is a real number $\|ax\| = |a| \cdot \|x\|$.

N.3. $\|x + y\| \leqq \|x\| + \|y\|$.

A normed linear space X is a metric space with respect to the metric $d(x, y) = \|x - y\|$. If the metric space thus obtained is complete, we call the normed linear space X a *Banach space*. Thus a Banach space is a complete normed linear space.

The family of all random variables ξ on a probability space $S = (\Omega, \mathscr{A}, P)$ having finite expectation is a Banach space $\mathscr{B}$ with respect to the norm $\|\xi\| = E(|\xi|)$, provided that two random variables which are almost surely equal are considered as identical.†

The family of continuous functions $f(x)$ in the closed interval $[0, 1]$ is a Banach space with respect to the norm $\|f(x)\| = \max_{0 \leqq x \leqq 1} |f(x)|$.

A real-valued function $A(x)$ on a Banach space X is called a *linear functional* if for $a \in R$, $b \in R$, $x \in X$, $y \in X$ we have $A(ax + by) = aA(x) + bA(y)$. A linear functional is called *bounded* if there exists a constant c such that $|A(x)| \leqq c\, \|x\|$. On the Banach space of continuous functions on the interval $[0, 1]$ every bounded linear functional $A(f)$ can be represented in the form

$$A(f) = \int_0^1 f(x)dF(x)$$

* That is, the operation $x + y$ is commutative and associative, there is an element 0 such that $x + 0 = x$ for all x, and for every $x \in X$ and $y \in X$ the equation $x + z = y$ has a solution $z \in X$.

† This means that each element of $\mathscr{B}$ is a class of equivalent random variables.

where $F(x)$ is a function of bounded variation in $[0, 1]$. The bounded linear functionals of a Banach space form another Banach space if addition and multiplication with real numbers of functionals is defined in the obvious way and the norm of the functional $A = A(x)$ is defined as

$$\|A\| = \sup_{\|x\| \leqq 1} |A(x)|$$

The Banach space X of bounded linear functionals of the Banach space X is called the *dual space* (or conjugate space) of X. A functional with norm $\leqq 1$ is called a contraction. A Banach space can be interpreted as a space of (generalized) *vectors*, the norm $\|x\|$ corresponding to the *length* of the vector x.

B.3 HILBERT SPACES

A Banach space is called a (real) Hilbert space if the following identity is valid:

$$\|x + y\|^2 + \|x - y\|^2 = 2(\|x\|^2 + \|y\|^2)$$

(i.e., the sum of squares of the diagonals of a parallelogram is equal to the sum of squares of the sides). In this case, putting

$$(x,y) = 1/4(\|x + y\|^2 - \|x - y\|^2)$$

the function (x, y) (called the *scalar product* or *inner product* of x and y) has the following properties:

S.1. $(x,y) = (y, x)$.

S.2. $(x,x) = \|x\|^2$ and thus $(x,x) \geqq 0$ with equality if and only if $x = 0$.

S.3. For fixed y, $A(x) = (x, y)$ is a linear functional.

A Hilbert space can also be defined by starting from a scalar product. If in the linear space H a real-valued function (x, y) of two variables is given, having the properties S.1–S.3, then, defining the norm $\|x\|$ by $\sqrt{(x,x)} = \|x\|$, H is a normed linear space. If H is complete, then H is a Hilbert space.

In a Hilbert space the inequality of Schwarz, $|(x, y)| \leqq \|x\| \cdot \|y\|$, is valid. Two elements x and y are called *orthogonal* if $(x,y) = 0$. More generally, if x and y are any elements of H such that $x \neq 0$, $y \neq 0$, one may define the angle ϑ between x and y (interpreted as vectors) by

$$\cos \vartheta = \frac{(x,y)}{\|x\| \, \|y\|}$$

(ϑ is defined only mod π). Thus x and y are orthogonal if and only if they form an angle $\vartheta = \pi/2$. A subset H_1 of a Hilbert space H is called a subspace if it is a linear space under the same operations. A subspace is called closed if it is a complete linear space, in which case H_1 is also a Hilbert space. The notion of Hilbert space is a straightforward generalization of the notion of Euclidean spaces: it may be interpreted as a Euclidean space of infinite dimension. As a matter of fact, if we take, in a Hilbert space, n linearly independent elements $x_1, \ldots, x_n$, i.e., such elements that for any real numbers $c_1, \ldots, c_n$ one has $c_1x_1 + c_2x_2 + \cdots + c_nx_n = 0$ if and only if $c_1 = c_2 = \cdots = c_n = 0$, then the *subspace* H_1 of H spanned by the elements $x_1, \ldots, x_n$, i.e., the set of all elements of the form $\sum_{k=1}^{n} c_k x_n$ with real c_k $(1 \leqq k \leqq n)$, is isomorphic to R^n.

In this book we have repeatedly used the following theorem: If H_1 is a closed subspace of a Hilbert space H, then each element x of H can be represented uniquely in the form: $x = x_1 + x_2$ where $x_1 \in H_1$ and x_2 is orthogonal to every element of H_1. The operation which maps x to x_1 is called the *projection* of H on H_1 and is denoted by $P_{H_1}x = x_1$. The projection of x on H_1 is the element of H_1 which has the least distance from x. The proof of the existence of the decomposition $x = x_1 + x_2$ is proved just by this property of the projection: if $x \notin H_1$ then $\inf_{y \in H_1} \|x - y\| = d$ is positive. If one takes a sequence $y_n \in H_1$ for which $\lim_{n\to\infty} \|x - y_n\| = d$, it is easy to show that $\lim y_n = y$ exists and $\|x - y\| = d$; further, as H_1 is closed, $y \in H_1$, and $x - y$ is orthogonal to any $z \in H_1$. Thus we may put $x_1 = P_{H_1}x = y$ and $x_2 = x - x_1$.

Another theorem, which is often needed, is the *theorem of Riesz*, according to which all bounded linear functions $A(x)$ in a Hilbert space H can be represented in the form $A(x) = (x, z)$, where z is a suitably chosen and uniquely determined element of H. This theorem can be proved by the projection theorem: If $A(x)$ is a (not identically zero) bounded linear functional, let H_1 denote the set of all elements y of H for which $A(y) = 0$. Evidently, H_1 is a closed linear subspace of H. Let z_1 be an element of H which does not belong to H_1. Then every x can be represented in the form $x = x_1 + cz_1$, where $x_1 \in H_1$ and $c = A(x)/A(z_1)$. As a matter of fact, one has $A(x - A(x)z_1/A(z_1)) = A(x) - A(x) = 0$, i.e., $x - A(x)z_1/A(z_1) \in H_1$. (In other words, the subspace H_2 of elements of H which are orthogonal to H_1 consists of all scalar multiples of a single element z_1.) From the representation $x = x_1 + cz_1$ $(x_1 \in H_1)$ it follows that, putting $z = z_1A(z_1)/\|z_1\|^2$, one has

$$(x,z) = cA(z_1)\frac{\|z_1\|^2}{\|z_1\|^2} = A(x)$$

which was to be proved.

If $S = (\Omega, \mathscr{A}, P)$ is a probability space, taking the family of all random variables ξ on S for which $E(\xi^2)$ is finite, then identifying random variables which are almost surely equal and putting $(\xi,\eta) = E(\xi\eta)$, we get a Hilbert space $L_2(S)$.

The fact that the space $L_2(S)$ thus obtained is complete, i.e., every fundamental sequence has a limit, is the famous *Riesz-Fischer theorem*. It can be deduced from the limit theorems concerning the Lebesgue integral (the Beppo Levi theorem and Fatou's theorem).

As an application of the Riesz representation theorem we sketch the proof —due to von Neumann (see, e.g., Yosida [10])—of the Radon-Nikodym theorem, for the case when P and Q are probability measures on the experiment $(\Omega, \mathscr{A})$ and $Q \ll P$. (The general case can be reduced to this special case.)

Let us put $P^*(A) = 1/2(P(A) + Q(A))$; then P^* is also a probability measure on $\mathscr{A}$. Let us consider the probability space $S^* = (\Omega, \mathscr{A}, P^*)$, and let us consider in the Hilbert space $L_2(S^*)$ the functional $A(\xi) = \int_\Omega \xi \, dQ$. The functional $A(\xi)$ is evidently linear and it is also bounded, because by the Schwarz inequality and the definition of P^*, we have

$$|A(\xi)| \leqq \left(\int \xi^2 \, dQ\right)^{1/2} \leqq \left(2 \int \xi^2 \, dP^*\right)^{1/2}$$

Thus there exists an element η of $L_2(S^*)$ such that $A(\xi) = (\xi, \eta) = \int_\Omega \xi\eta \, dP^*$. This implies that

$$\int_\Omega \xi(2-\eta) dQ = \int_\Omega \xi\eta \, dP \tag{B.3.1}$$

for all $\xi \in L_2(S^*)$. It follows easily that with probability 1 (with respect to P) $0 \leqq \eta \leqq 2$, further, that denoting by C the set on which $\eta = 2$, one has $P(C) = Q(C) = 0$. Now let B_n denote the set on which $\eta < 2 - 1/n$ ($n = 1, 2, \ldots$); it follows that $\sum_{n=1}^\infty B_n = \Omega - C$. If $A \subseteq B_n$ for some n, denoting by α the indicator of the event A, putting $\xi = \alpha/(2-\eta)$ and taking into account that ξ is bounded, and thus $\xi \in L_2(S^*)$, we get from (B.3.1), putting $= \eta/(2-\eta)$,

$$Q(A) = \int_A f \, dP \tag{B.3.2}$$

Thus if $\mathscr{R}$ denotes the ring of all sets $A \in \mathscr{A}$ such that there exists an n such that $A \subseteq B_n$, then the two measures $Q(A)$ and $Q^*(A) = \int_\Omega f \, dP$ coincide on $\mathscr{R}$, and thus, by the extension theorem, they coincide on the σ-algebra of all sets $A \in \mathscr{A}$ such that $A \subseteq \Omega - C$. As $P(C) = Q(C) = 0$, it follows that the two measures $Q(A)$ and $Q^*(A)$ coincide on A, i.e., (B.3.2) holds for all $A \in \mathscr{A}$, which was to be proved.

As regards a thorough treatment of the concepts mentioned in this section, see [3], [5], [6], [10]–[13].

REFERENCES TO APPENDICES A AND B

[1] P. Halmos, *Measure Theory*, Van Nostrand, Princeton, 1950.

[2] J. Neveu, *Mathematical Foundations of the Calculus of Probability*, Holden-Day, Inc., San Francisco, 1965.

[3] H. L. Royden, *Real Analysis*, MacMillan, New York, 1963.
[4] E. Kamke, *Das Lebesgue-Stieltjes Integral*, Teubner, Leipzig, 1956.
[5] F. Riesz and B. Sz. Nagy, *Functional Analysis*, Ungar Publishing Co., New York, 1955.
[6] B. Sz. Nagy, *Introduction to Real Functions and Orthogonal Expansion*, Oxford University Press, New York, 1964.
[7] G. E. Silov and B. L. Hurewicz, *Integral, Measure and Derivative*, Prentice Hall, New York, 1966.
[8] A. Rényi, *Wahrscheinlichkeitsrechnung, mit einer Anhang über Informationstheorie*, VEB Deutscher Verlag der Wissenschaften, Berlin, 1968.
[9] K. Krickeberg, *Wahrscheinlichkeitstheorie*, Teubner, Stuttgart, 1963.
[10] K. Yosida, *Functional Analysis*, Springer, Berlin, 1965.
[11] G. Choquet, *Topology*, Academic Press, New York, 1966.
[12] N. Dunford and J. T. Schwartz, *Linear Operators*, Vols. 1 and 2, Interscience, New York, 1958, 1964.
[13] R. E. Edwards, *Functional Analysis*, Holt, Rinehart and Winston, New York, 1965.

AUTHOR INDEX

SUBJECT INDEX

EXPLANATION OF NOTATIONS AND ABBREVIATIONS